INTRODUCTION TO ENGINEERING DESIGN

BOOK 9, FOURTH EDITION

ENGINEERING SKILLS AND HOVERCRAFT MISSIONS

Vincent M. Brannigan
Kevin Calabro
James W. Dally
William L. Fourney
Bruce Jacob
Wesley G. Lawson
Gary A. Pertmer
Guangming Zhang

University of Maryland, College Park

College House Enterprises, LLC
Knoxville, Tennessee

This textbook is intended to provide accurate and authoritative information regarding the various topics covered. It is distributed and sold with the understanding that the publisher is not engaged in providing legal, accounting, engineering or other professional services. If legal advice or other expertise advice is required, the services of a recognized professional should be retained.

The manuscript was prepared using MS Word with 11 point Times New Roman font. Ingram and Spark prints this book "on demand."

College House Enterprises, LLC.
5713 Glen Cove Drive
Knoxville, TN 37919, U. S. A.
Phone and/or FAX (865) 947 6174
Email jwd@collegehousebooks.com
http://www.collegehousebooks.com

10 Digit ISBN 0-9792581-9-7
13 Digit ISBN-978-0-9792581-9-0

PREFACE

This book is the fourth edition of the ninth textbook in this series dealing with Introduction to Engineering Design. Jim Dally, working with College House Enterprises and faculty members in the Clark School of Engineering, has prepared eight previous books in this series—a new one almost every academic year—for the first-year engineering students of the University of Maryland at College Park. Several other Colleges of Engineering have adopted one or more of the books in this series to introduce design and engineering skills for their first or second year students. The design, build and test of a hovercraft model, described in Book 9, is such an interesting and challenging project that this is the fifth year we have used it (eight previous semesters) with approximately 5,000 students. The project is also assigned for the first year Introduction to Engineering Design at the University of Nevada @ Reno where it is in its third year.

PROCESS AND CONTENT

The procedure followed in offering a design experience to first or second year engineering students is to:

- Teach the class in moderate size sections.
- Divide the class into product development teams with five or six members per team.
- Assign a project entailing the development of a prototype that will require the entire semester.
- All student teams develop the same product—this year it is a model of a hovercraft.
- In the product realization process, the students:

  - Design
  - Manufacture or procure parts for the prototype
  - Assemble the prototype
  - Test and evaluate the prototype

- In developing the prototype, the students have the opportunity to learn:

  - Communication skills
  - Team building skills
  - Engineering graphics
  - Software applications including CAD
  - Design methods and procedures

The textbook is used to support the students during a semester-long project. Some of the material may be covered in lecture, recitation or in a computer laboratory or a model shop. Additional material is covered with reading assignments. In other instances, the students use the text as a reference document for independent study. Exercises, provided at the end of each chapter, may be used for assignments when the demands of the project on the students' time are not excessive.

The book is organized into five parts to present many topics that first year engineering students should understand as they proceed through a significant portion of the product realization process. Product and system development processes is introduced in Part I. An Introduction to the course and to the textbook is provided in Chapter 0 to alert the students to the demands of the course and to introduce them to the problems frequently encountered by other students in designing, building and testing a model of a hovercraft. Information on team skills and the importance of the product development process is covered in Chapter 1 and 2. Several hovercraft missions are also presented in Chapter 2 together with a description of the design concepts involved in hovercraft development. By assigning a demanding project, a holistic approach is employed in the student's first engineering experience that motivates them. Design of a hovercraft enables the instructor an opportunity to integrate a spectrum of knowledge about many topics. The student's hands-on participation in a design, building and testing a hovercraft significantly enhances their learning process.

The theoretical background needed to conduct elementary design analyses for the hovercraft is presented in Chapter 3. Chapter 4 describes basic electric circuits, sensors and batteries to provide technical background helpful for the design and control of the hovercraft. Concepts of statics and dynamics are introduced in Chapter 5 to enable the students to understand the forces and moments and their effect on controlling the motion of their hovercraft. Finally, an introduction to programming in RobotC is included in Chapter 6.

Part II presents two chapters on engineering graphics using computer programs. Pro/Engineer, Wildfire-Version 5.0-M030, a computer aided design (CAD) program is described in detail in Chapter 7. The use of tables and graphs in communicating engineering information using Microsoft Excel is presented in Chapter 8.

Part III treats the very important topic of communications. Chapter 9, on technical reports, describes many aspects of technical writing and library research. The most important lesson here is that a technical report is different than a term paper for the History or English Departments. An effective professional report is written for a predefined audience with specific objectives. The technical writing process is described, and many suggestions to facilitate composing, revision, editing and proofreading are given. Chapter 10, on design briefings, shows the distinction among speeches, presentations and group discussions. Emphasis is placed on the technical presentation and the importance of preparing excellent visual aids. PowerPoint slides are usually employed as visual aids in a design briefing.

Part IV introduces the students to the engineering profession. Chapter 11 describes engineering disciplines, on-the-job activities, salary statistics and registration information for your PE license. A useful student survival guide is also included in Chapter 12.

Part V contains three chapters dealing with engineering and society. A historical perspective on the role engineering played in developing civilization and on improving the lives of the masses is presented in Chapter 13. In this chapter, we move from the past into the present and indicate the current relationship between business, consumers and society. The twenty greatest engineering achievements of the 20th century are briefly described. Chapter 14 discusses the balance between safety and performance. Methods to evaluate and recognize risky environments are discussed. The chapter includes a listing of hazards, which is important in identifying the many different ways users of a product can be injured. Chapter 15 on ethics, character and engineering includes a large number of topics so the instructor can select from among them. A description of the Challenger and Columbia accidents is also given because both of these fatal crashes provide excellent case histories covering safety related conflicts between management and engineers.

ABET AND EDUCATIONAL OUTCOMES

ABET's Criteria 2000 was considered in writing this series of books on Introduction to Engineering Design. It is believed that a first course in engineering design could include the educational outcomes listed below:

Communication Skills:

1. Engineering Graphics
    - Understand the role of graphics in engineering design.
    - Understand orthographic projection in producing multi-view drawings.
    - Understand three-dimensional representation with pictorial drawings.
    - Understand dimensioning and section views.
    - Demonstrate capability of preparing drawings using computer methods.
    - Demonstrate understanding of engineering graphics by incorporating appropriate high-quality drawings in the design documentation.
2. Design Briefings:
    - Within a team format, present a design review for the class using appropriate visual aids.
    - Each team member demonstrates briefing skills.
3. Design Reports:
    - The team's design is documented in a professional style report incorporating time schedules, costs, parts list, drawings and an analysis.

Team Experience:

- Develop an awareness of the challenges arising in teamwork.
- Demonstrate teamwork in the product realization process through a systematic design concept selection process involving participation of all team members.
- Demonstrate planning from conceptualization to the evaluation of the prototype.
- Understand and demonstrate sharing responsibility among team members.
- Demonstrate teamwork in preparing design reports and presenting design briefings.

Software Applications:

- Demonstrate entry-level skills in using spreadsheets for calculations and data analysis.
- Show a capability to prepare graphs and charts with a spreadsheet.
- Show a capability to prepare professional quality visual aids.
- Understand entry-level skills in a feature-based solid modeling program.
- Demonstrate these computer skills in preparing appropriate materials for design briefings and design reports.

Design Project:

- The design project is the overarching theme of the course.
- Utilize all the skills listed above to assist in the product development process.
- Demonstrate competence in defining design objectives.
- Generate design concepts that meet the design objectives.
- Understand the basis for design for manufacturing assembly and maintenance.
- Manage the team and the project effectively.

ACKNOWLEDGEMENTS

Acknowledgments are always necessary in preparing a textbook because so many people are involved in many helpful ways. First, it is important to recognize the contributions of many dedicated faculty members and teaching fellows at the University of Maryland. To date nearly 100 different faculty members have taught this introductory engineering course on at least one occasion over the past twelve years. Many of them have made suggestions for improving the material in this book. In particular, thanks are due to Dr. Gary Pertmer and Dr. William L. Fourney for their commitment in providing the leadership necessary to operate a multi-section offering of this course and in providing support for the many different instructors and teaching fellows involved each semester. We appreciate the information posted on the ENES 100 "Blackboard" provided by Peter Sunderland, Bruce Jacob, Kevin Calabro, Wes Lawson, R. D. Gomez, Ken Kiger, Chris Cadou and Evandro Valente. The slides on the "Blackboard" are available for review by the students 24-7 and support the lectures and this textbook. We also appreciate the support of several Deans of Engineering of the Clark School of Engineering who have provided, for 15 years, the significant resources necessary for a hands-on course like this one to be successful.

The development of this textbook has become a team effort with instructors preparing new chapters or revising chapters from previous editions. Dr. Guangming Zhang authored Chapter 7 on Pro/ENGINEER in its entirety. Kevin Calabro edited and revised Chapters 1 and 2 and combined two previous versions of chapters dealing with Tables and Graphs and Excel to produce a new updated version in Chapter 8. Wes Lawson made extensive revisions to Chapter 4 on Basic Electrical Circuits and Sensors. Bruce Jacob wrote Chapter 6 Building a Hovercraft Control System in RobotC, in which he introduced concepts for programming in C. Vince Brannigan made major revisions to Chapters 14 and 15. Vince is a faculty member in our Fire Protection and Safety Department and brings an experienced safety background to the table. His changes made a major improvement in the content of these two chapters. I also enlisted the help of two undergraduate students, McKenzie C. Primerano and Emi P. DiStefano to edit Chapter 12—A Student Survival Guide—and to check its content from an undergraduate viewpoint.

Thanks are also in order for Anne S. McClain from the University of Alabama at Birmingham who shared experiences and class notes with us. A hovercraft project involving lift capabilities and speed was used in a first year engineering course at the University of Alabama in the spring 2003 semester.

Vincent M. Brannigan, Kevin Calabro, James W. Dally, William L. Fourney, Bruce Jacob,
Wesley G. Lawson, Gary A. Pertmer, Guangming Zhang and many others
University of Maryland at College Park
June 2010

CONTENTS

CHAPTER 3 FLUID MECHANICS AND DESIGN ANALYSIS

CHAPTER 4 BASIC ELECTRIC CIRCUITS AND SENSORS

CHAPTER 5 HOVERCRAFT MOTION AND CONTROL

CHAPTER 6 BUILDING A HOVERCRAFT CONTROL SYSTEM IN ROBOTC

PART II ENGINEERING GRAPHICS

CHAPTER 7 COMPUTER AIDED DESIGN (CAD)

CHAPTER 8 MICROSOFT EXCEL

PART III COMMUNICATION

CHAPTER 9 TECHNICAL REPORTS

CHAPTER 10 DESIGN BRIEFINGS

PART IV   ENGINEERING AND SUCCESS SKILLS

CHAPTER 11 THE ENGINEERING PROFESSION

CHAPTER 12 A STUDENT SURVIVAL GUIDE

PART V ENGINEERING AND SOCIETY

CHAPTER 13 ENGINEERING AND SOCIETY

CHAPTER 14 SAFETY AND PERFORMANCE

CHAPTER 15 ENGINEERING ETHICS AND DESIGN

APPENDIX A: FORMS FOR DESIGN TEAMS

PART I

PRODUCT AND SYSTEM DEVELOPMENT PROCESSES

CHAPTER 0

INTRODUCTION

0.1 GETTING THE MOST OUT OF A FIRST YEAR DESIGN COURSE

An "Introduction to Engineering" is commonly found in the engineering curriculum for most accredited colleges in the U.S. However, "Engineering Design" is rarely offered to first year students. Many faculty and administrators believe students cannot design until they have mastered two or three years of coursework in mathematics, science and engineering. We concur that meaningful design analysis during the first year of study is challenging, but we believe that you are capable of conceiving design concepts, producing sketches of your ideas, performing the necessary analysis to design a system, and, with exposure to introductory CAD content, producing engineering drawings of your design.

Design analysis is not neglected in this course. The basic equations you will need to design an autonomous hovercraft capable of meeting a challenging set of product specifications are presented. Many of these equations will be derived and discussed in great detail during courses you will take later in your program. In this course, you will gain a very basic exposure to concepts in fluid mechanics, electronics, sensor and control systems, and dynamics. Some examples are presented in this textbook to illustrate the method of engineering design analysis. For example, the processes for determining levitation pressure and flow rate requirements, propulsion and frictional forces, vehicle motion characteristics, and electrical power requirements are presented. Conducting these analyses will enable you to explore the design solution space to properly size the components that are required to build, power and control your hovercraft.

Requiring students to work on teams to design, build and test a hovercraft during their first year is a significant challenge to you, the faculty and the administration. It is a challenge to you because it is a very difficult project. Your team's hovercraft is required to autonomously navigate over a predefined track and to negotiate a set of obstacles. You are required to design it together with several other students that are organized into a design team with several smaller sub-teams. It is an early experience in teamwork and project management that in itself is challenging for many students. Your design team will compete with other design teams in your section and with design teams[1] in the entire course.

A first year engineering design course is a challenge for the instructors and the teaching fellows that provide guidance for you. They must help you understand the equations needed for your analysis. They serve as monitors for the design teams and offer advice as your team develops design

---

[1] The administration expects that student design teams, with 8 to 10 students per team, will design, build and test about 80 hovercrafts in the 2010-11 academic year.

concepts. They make timely suggestions and offer counsel. Teaching a design course is much more difficult than presenting lectures in a more traditional course.

Engineering design courses also present a challenge to the administration. Design courses are expensive to conduct. They require immense amounts of faculty time, extensive floor space for laboratories and expensive software, tools and equipment. Fortunately, the administration recognizes the importance of a solid engineering design experience for our first year students.

0.2 DESIGN ISSUES

The purpose of this section is to describe some common problems that teams encounter as they design their hovercraft. The majority of your team's design analysis will be conducted during the first half of the semester. Your initial design analysis will conclude with a preliminary design review ("PDR"). The PDR process requires each team to provide a formal presentation and to generate a design report. Following the PDR, your team will construct a prototype hovercraft. Once constructed, system and subsystem testing will be performed to confirm your design analysis. Any subsequent design modifications required can be made during the testing phase of the project. The final stage of your design experience requires a final design review ("FDR") in which the final details of your product development are documented and discussed in detail.

Since this is likely your first engineering course, you will be exposed to the technical information necessary to successfully design a hovercraft "just in time" for when this content is needed. This will require your team to continually reiterate on your design until all of the necessary technical information is known and final design decisions can be confidently made. You are encouraged to automate this process wherever possible. This can be done by creating design calculators using spreadsheet software such as Microsoft Excel.

The hovercraft is typically thought to have six major subsystems. In order to design a product that meets all of the product specifications set forth, not only does each subsystem need to be properly designed, but the interfaces between subsystems must also be properly designed. In fact, one of the most common design failures is in failing to consider the interrelatedness between each subsystem. For example, teams in the past have designed perfectly good propulsion and power systems, but have failed to consider that the propulsion fans need to be powered by the battery selected by the power sub-team. If the sub-teams designing these two subsystems are not in constant communication regarding design choices and requirements, a fatal design error can result. While these errors can be fixed when construction begins, the correction is usually costly and will typically result in a delay to the project schedule. The six major subsystems are listed below with a description of the major design tasks associated with each subsystem:

- Structural system
  - Estimate vehicle weight; specify construction materials and methods; specify vehicle configuration (size and shape); specify placement of components to insure a balanced craft
- Levitation system
  - Specify desired hover height; calculate required levitation fan pressure and flow rate; procure a fan that meets these requirements
- Propulsion system
  - Specify vehicle's desired linear and angular acceleration rates; estimate frictional force and mass moment of inertia; specify type, number and placement of fans; select and procure fans

- Power system
    - Determine electrical power requirements (voltage, current, and capacity); select and procure an auxiliary power unit (battery); select power modulation system (voltage regulators, relays, transistors)
- Sensor system
    - Specify type, number and placement of sensors aboard craft; procure and calibrate sensors
- Controls and navigation system
    - Generate pseudocode logic that integrates all system limitations; code, test and debug software package generated

One of the reasons why this project is so challenging is because an error in the design of any one of the subsystems listed above can easily result in a craft that is unable to satisfy all product specifications. In order to ensure your team is making progress, an external milestone process has been initiated. Each team must follow the milestone process which includes project reporting milestones (presentations and reports), construction milestones, and testing milestones. These milestones serve as a very good indication of the progress your team must make in order to be successful. Your team should work very hard to meet each milestone by the deadline assigned.

0.3 CONSTRUCTION ISSUES

The purpose of this section is to describe some common problems that design teams encounter as they develop their hovercraft. We trust that you can avoid or minimize these common construction problems if you are aware of them at the beginning of the course.

You will not be using a kit developed by others to construct your hovercraft. Instead you will procure all of the materials and components from outside suppliers such as Lowe's, Home Depot, Radio Shack, etc. and online vendors such as Digi-Key, McMaster-Carr, etc. The only exception to this is the microprocessor used to control the hovercraft which is rented to your team. The current microprocessor is the NXT which can be obtained for a $15 rental fee + $35 refundable deposit. You may use your own NXT, but the Bill of Materials (parts list) must reflect a $50 charge for the use of the NXT. The total cost of the hovercraft **must be less than $400.00**. The cost must be tabulated in your Bill of Materials, in which the fair market value of each component must be listed along with the part number, vendor and quantity. Donated and/or used components may be incorporated, but the fair market value of a new equivalent component must be listed in the Bill of Material. The costs incurred in procuring the components and materials must be shared equally among the team members.

We expect your construction efforts to be neat and to reflect good design practice. We also insist that the hovercraft be safe and that you follow all safety rules during the construction phase. Some examples of hovercraft constructed by student design teams in the past are presented below:

The hovercraft presented in Fig. 0.1 utilizes two pieces of foam to form the body and hard skirt of the hovercraft. Four fans with fixed orientation are employed in an effort to decouple propulsion and turning. Note the wire screens on the fans to prevent injury to fingers. A duct is used to carry the air flow from the lift fan to the plenum chamber. This choice reduces the effect of the spin instability caused by the start-up torque and rotational airflow associated with a deck mounted axial lift fan. Auxiliary power is provide by two battery packs. Four light sensors are mounted below the deck to measure the light reflected from either the black tape or the white surface of the track in a very controlled environment. This model is an example of careful attention to detail.

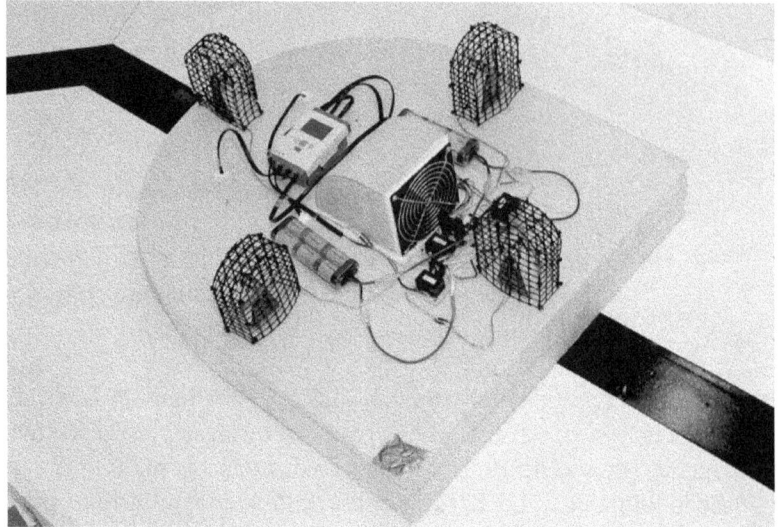

Fig. 0.1 Hovercraft model on the test track.

     The hovercraft, illustrated in Fig. 0.2, is designed with a deck fabricated from a sheet of foam which was cut using a hotwire cutter. The deck is adhesively bonded to a soft skirt sewn from fabric. Three light sensors extend from the front end of the craft to measure the light reflected from the tape or the track. Two fixed axial fans provide the propulsion force and one axial fan is ducted to provide the lift force. The power to the two propulsion fans is modulated using a series of relays. Two battery packs are located aft. The crude wiring and the excessive use of duct tape shows a lack of attention to detail.

Fig. 0.2 A hovercraft negotiating a bend in the track.

The hovercraft, illustrated in Fig. 0.3, is designed with a deck fabricated from a sheet of foam. The deck is duct taped to a soft skirt made out of rubber. One light sensor extends from the bow of the craft and two ultrasonic distance measuring sensors extend from the starboard side of the craft. Two propellers driven with small electric motors provide the propulsion/steering force and one centrifugal fan is ducted to provide the lift force. One battery pack powers the craft and the power to each fan is modulated using voltage regulators and transistors. The use of duct tape in attaching the sensor arms shows a lack of attention to detail in construction. On a bright note, the electronic components are soldered to a perforated board and each component is protected using heat sinks and shrink wrap tubing. A bit more attention to detail should have been displayed to clean up the loose wires atop the craft, but the construction methods utilized are safe, reliable and robust.

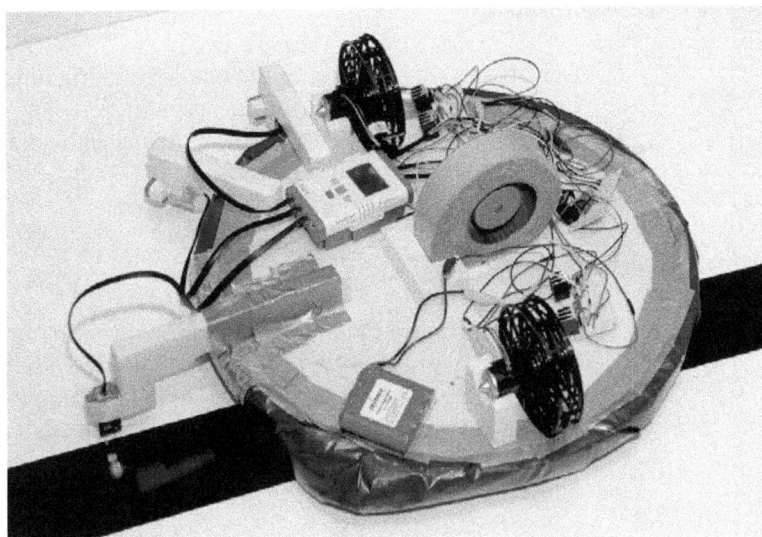

Fig. 0.3 A hovercraft with propellers to develop propulsion forces and ultrasonic sensors.

0.4 OPERATING ISSUES

The test course, presented in Fig. 0.4, is located indoors on a flat surface. The track is fabricated with a smooth white surface. A single 100 mm wide solid black tape line is placed along the track's centerline to aid navigation. The track is fitted with two wooden sidewalls that are spaced 1.219 m apart which can also be used to aid navigation. The course includes a 2.0 m rough patch ("the desert"). This section of track consists of fine grain sand with variable diameters ranging up to ¼ mm and represents a friction obstacle. The black tape centerline does not extend over this section of the track.

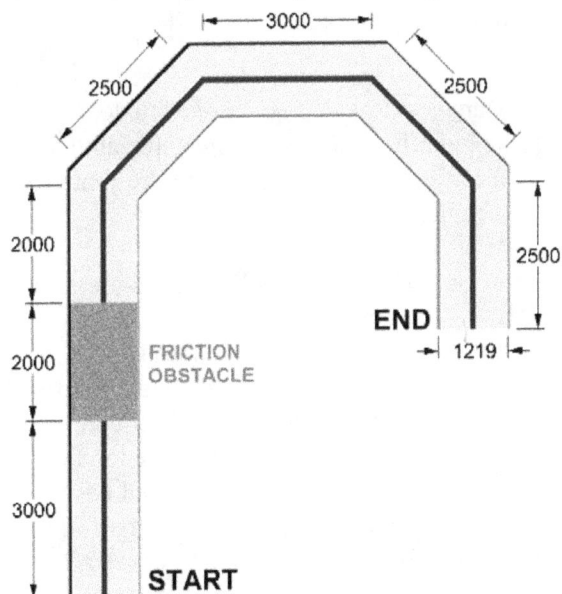

Fig. 0.4 The test course for the Fall 2010 semester at the University of Maryland at College Park. Dimensions are in mm.

The hovercraft project has been used by the University of Maryland at College Park and the University of Nevada at Reno for several years[2]. The instructors and the teaching fellows have observed the successes and failures of several hundred student design teams. Typical accomplishments and shortcomings of these design teams are listed below:

---

[2] The track configuration and the obstacles change from semester to semester, but the basic concept of design, build, test and compete remain the same.

Accomplishments:

1. Design and construction of the hovercraft structure.
2. Levitating for the required time period.
3. Controlling dynamic instabilities (balance and spin)[3]
4. Moving the hovercraft with propulsion forces.
5. Writing line and wall following programs.

Shortcomings:

1. Failure to control the hovercraft following skirt contact with the track's surface/tape.
2. Failure to avoid contact with the track's side walls after fishtailing.
3. Failure to account for the time delay in adjusting the propulsion forces.
4. Failure to maintain the battery power supplies in a ready-to-test condition.
5. Failure to design a robust system that can overcome irregularities in track surfaces.
6. Failure to sense and control speed.

We will provide guidance in Section 0.5 on methods to improve the control of the hovercraft. Hopefully your design team can learn from these past shortcomings to avoid the problems encountered by the majority of teams over the past few years. If your design team does fail, and many do for one reason or another, learn from the failure. It is a painful (to one's ego) way to learn, but it is an effective way.

0.5 CONTROL ISSUES

Many student design teams fail to reach the end of the track because they lose control of the hovercraft. The requirement for autonomous control has proven to be the most difficult product specification to meet. There are many reasons why a hovercraft loses control. Common causes include:

- The vehicle either moves too fast or too slow to overcome imperfections in the track and to remain controllable[4].
- The hovercraft is not stable (i.e., balance and/or spin problem not resolved).
- The response time for a fan to generate the desired thrust force is much slower than the speed at which a nearly frictionless hovercraft is traveling out of control.
- When the craft is moving too fast, it is difficult for the controller to recover following a large deviation from the desired position (all light sensors crossing the tape line completely or the craft getting too close or far from a wall).
- When the skirt touches the track, an unexpected friction force may develop which causes the vehicle to spin out of control.

Control of the path followed by the hovercraft is maintained by the sensors selected, the steering mechanism and the algorithm programmed into the NXT. Most design teams employ two or three light sensors to follow the black tape as shown in Figs. 0.1 and 0.2. The time required for the

---

[3] This was an early shortcoming, but the increased use of well designed soft skirt systems has dramatically improved the stability of hovercraft prototypes.

[4] No large surface is truly level and vehicles experience undesirable behavior when traveling even slightly up or downhill.

algorithm and the NXT to issue commands to adjust the steering to correct for the undesirable spin is less than 5 ms. Unfortunately, many fans require significantly more time to reach their operating speed (and thrust) or to turn off. The hovercraft can deviate significantly from the desired path even after the appropriate steering commands have been issued by the NXT.

Most design teams use two fans to propel and to steer their hovercraft. They change the speed of one or both fans to turn to port or starboard. In theory this technique should be effective if the time to adjust the speed of the fans is not an issue. However, the time required for the fans to change speed is an issue. The time to change the RPM of the fans ranges from 1 to 3 seconds depending on the size of the impeller, characteristics of the motor driving the impeller and the inertia of the rotating mass.

An alternative solution that many teams have successfully experimented with is to change the direction of the propulsion force vector and maintain its magnitude constant. This approach holds the power to the fans constant, but provides a mechanism to vector the thrust force from the fan. This can be done by adjusting the axis of the fan or fans, as shown in Fig. 0.4, or by using a rudder system. Adjusting the orientation of the fan's axis can be achieved by mounting the fans on a servo motor that can be controllably rotated through ± 90°.

Fig. 0.4 Rotate the propulsion fans to adjust steering to compensate for the friction force and to correct for the unanticipated rotation.

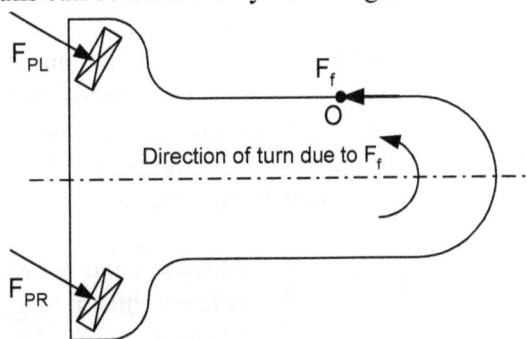

Another alternative that teams have begun to experiment with is using multiple fans to accomplish different tasks. Teams have included side thruster fans solely used for making course corrections in addition to one or more main propulsion fans. By decoupling forward motion from both lateral and rotational motions, teams have had much better success in controlling their craft's speed and making appropriate and swift course corrections.

Having explored some of the most common control issues that design teams have faced in the past, you are further cautioned that the product specifications change each year. While the solutions noted in the above two paragraphs have been quite successful, they have been solutions to a different problem. The current product specification includes a high-friction patch which must be overcome. In considering this new challenge, design teams are encouraged to give greater thought than in the past to sensing speed and generating controllers that can intelligently provide the appropriate levels of thrust at the appropriate times to complete the course. Further complicating things, the 2-m friction patch does not include a black centerline to aid navigation. Novel solutions and critical thought are necessary to meet the challenges of the new set of product specifications.

0.6 TEAMWORK ISSUES

Teams are effective because they meet together to focus their wide range of disciplinary skills and natural talents toward the solution of a set of well-formulated problems. The team meeting is the format for synergistic efforts of the team. Unfortunately, not all teams are effective because some members destroy the cohesiveness of the team. Bonding of team members is vital if the team is to successfully solve the multitude of problems that arise in the product development process.

Teams **fail** because of three main reasons. First, they deviate from the goals and objectives of the development, and cannot meet the milestones on the development schedule. Second, the team members become alienated and the bonding, trust and understanding critical to the success of the team never develops. Third, when things begin to go wrong and adversity occurs, "finger pointing" begins and an attempt to fix blame on someone replaces creative actions.

Clearly, many problems will arise to impair the progress of a development team. It is recommended that each team establish a set of simple rules it considers necessary to govern the team. Once established, each team member should sign and date this contract[5]. Suggestions for your team to consider are provided below:

- Members will acknowledge problems and deal with them
- Members will be supportive rather than judgmental
- Members will respect differences
- Members will provide responses directly and openly in a timely fashion
- Members will provide information that is specific and focused on the task and not on personalities
- Members will not discount the ideas of others
- Each member is responsible for the success of the team experience
- Every member of the team will contribute to the team's success
- Members recognize the importance of team meetings to the success of the group, and accordingly agree to:

  1. Always attend the scheduled team meetings when possible, but when attendance is impossible, notify the team leader and as many other members as possible in advance
  2. Use meeting time wisely – it should be a time for providing status updates and making decisions, not performing the necessary work for the project
  3. **Start meetings on time**
  4. Utilize an agenda and keep focused on the meeting's goals
  5. Avoid side issues, personality conflicts and hidden agenda
  6. Avoid making phone calls or engaging in side conversations that interrupt the team
  7. **End meetings on time**
  8. Communicate action items with deadlines immediately following the meeting to all team members

0.7 TIME ISSUES

This course places many requirements on the student design teams and cuts into their time to a significant degree. Time management and effective team meetings are mandatory. The course is designed with one (one hour) lecture per week for the entire class and two recitation/lab (two hour) periods per week. The first half of the course is devoted to covering analysis, programming and CAD methods described in the first eight chapters of this textbook. Much of the second half of the course is spent in the laboratories designing, building and testing your team's hovercraft.

---

[5] An example of a development contract is presented in the Appendix. It contains the team structure, development constraints, milestone dates, scoring rules and an agreement to four simple teaming rules.

The time involved to successfully complete ENES 100 depends largely on the ability of the team to work together and the effectiveness of the team leader. Much of Chapter 1 deals with forming and leading effective design teams. Our experience with student design teams indicates that most of you have not been adequately prepared to work on an engineering team. High schools emphasize individual academic achievement. Athletic programs do emphasize team work, but many students have not participated in structured team sports.

We suggest you reflect on the information provided in Sections 1.6 to 1.11, and make every effort to insure that your design team is efficient and effective.

Time management is also important. How do you spend your time? Do you schedule your day and your week? Have you allocated study time? Do you even consider time management or do you roll with the flow? If you want to learn important life long skill in time management you are invited to study Sections 12.2 to 12.4 and begin to implement estimating time requirements and scheduling to accommodate them.

0.8 SUMMARY

This chapter was inserted into the fourth edition to provide you with a preview of the instructor's expectations and the problems these expectations will pose for you and your design team. Construction issues usually do not present a problem for student design teams. Procurement of structural materials and hardware must be achieved in a timely manner and within the budget limitations provided. Most teams do not struggle to satisfy this requirement. However, your team can quickly exceed its budget if appropriate procurement decisions are not made in the design stage. A thorough and complete design analysis is necessary before you begin construction of your craft.

The most significant problem for design teams over the past four years has been a loss of control caused by the inherent instabilities associated with the autonomous control of a nearly frictionless vehicle. This problem is compounded by the requirement to do so in as fast a time as possible. It is important that your team recognize this as a serious problem and develop a strategy for coping with it. Otherwise your team will join a long list of teams that either neglected to adequately recognize the problem or that failed to develop an effective method for coping with it.

Effective and efficient team work is critical to the successful development of a hovercraft and the successful completion of its mission. Unfortunately, experience with hundreds of student design teams indicates that your team meetings are not very effective. It is important that you quickly understand the expectations for team behavior and take the steps necessary to hold highly effective meetings. Much of Chapter 1 in this text is devoted to describing how effective teams operate and how team members contribute.

Time issues and teamwork issues are closely related. It is understood that ENES 100 is a demanding and time consuming course. It is designed to prepare you for the engineering curriculum that is also demanding of time and talent. ENES 100 also has an objective to introduce you to time management, which is skill important for your success in college and in your professional career. You will be required to manage your own time within this course as well as your project development time using tools such as a Gantt chart.

There is a new challenge in the hovercraft mission this semester—the sand trap. Give it careful consideration and develop an effective design strategy for crossing it, while maintaining control of your team's hovercraft. We are fully aware of how challenging this project is, but are confident that you will rise to meet this new challenge. Good luck this semester and have fun!

EXERCISES

1. Prepare a list of the skills that you expect to acquire in this course during the semester. Place these skills in priority order.
2. Prepare a list of challenges you will encounter in successfully completing ENES 100.
3. What will be the most difficult challenge for you to master? Why?
4. What materials do you believe would be the best to use in constructing the structure for the hovercraft?
5. What sensors do you believe would be the most effective for line following and for crossing the sand trap? Do you know a method for using two different sensors to guide the hovercraft?
6. Prepare a design brief for your team detailing a strategy for regaining control of the hovercraft after skirt contact with the track's surface causes a deviation from the desired path?
7. Measure the time required to rotate a LEGO motor through 90°. Your instructor may have to provide the instruments needed for this measurement. Why is it necessary to understand the time response for the motor?
8. Write an assessment of your team work skills. Provide a description of previous experiences working on a team. Identify the objective of the team. Was it a pleasant experience?
9. In your previous experiences with teaming, were you the leader or a follower? Did you have trouble following? Do you always want to be the team leader? What is you reaction if someone else is selected as the team leader?
10. Develop a project charter for your team describing the scope and objectives of the project, as well as each team member's roles and responsibilities.
11. Describe the characteristics of a good team member and a destructive team member.
12. Have you developed effective time management skills? Do you believe they are important?
13. Do you have an agenda that identifies the tasks and the time required for each task that you plan to accomplish today?
14. At the close of each day, do you evaluate how effective you have been in completing the task identified on your agenda?
15. Have you developed a schedule for each day of the week?

CHAPTER 1

DEVELOPMENT TEAMS

1.1 INTRODUCTION

The development of quality, leading edge, high-performance products requires the coordinated efforts of many skillful individuals from different disciplines over an extended period of time. This group of individuals is organized within a company to act in an integrated manner to successfully complete the development process. The organizational structure employed varies from one corporation to another; it also depends on the size of the company engaged in the development. For relatively new firms in the entrepreneurial stage, team structure is usually not an issue. These firms are small with a single product and only ten to twenty individuals are involved in the development process. Everyone on the payroll is deeply committed to this product, and communication is usually accomplished around the lunch table. On the other hand for very large corporations, where the number of employees may exceed 100,000, the company is often organized into divisions or operating groups. These large companies have many products or services that are offered by each division, and in some instances one division may actually be a direct competitor of another division. For example, General Motors with its five automotive divisions produces many different models of each of its lines that compete for the same market segment.

In a large corporation, the total number of new products that are introduced each year may exceed 100. Communication is often very difficult because hundreds or even thousands of miles sometimes separate personnel involved in the development of a given product. Sometimes different divisions are in different countries, and business is conducted in different languages. Clearly, in a large firm, organization of the product development team and the physical location of its team members becomes extremely important.

In the past decade, there have been many examples [1] demonstrating that a cross disciplinary team provides a very effective organizational structure for product development. The idea of forming a cross disciplinary team to design a new product appears simple until one examines the existing organizational structure of all but the smaller of the product oriented corporations. Larger corporations are usually organized along functional lines. The organization chart presented in Fig. 1.1 shows many of the functional departments that provide personnel needed on a typical product development team. Each department has a functional manager. Also, several departments are often grouped to form a division, and of course, each division has a director. Titles for the division leaders vary from firm to firm—but Director or Vice President are commonly employed.

The functional organization with its leadership structure and its ongoing operations presents a management problem when forming product development teams. As a team member, do you report to the team leader (product manager) or to the functional manager? When important decisions must be made, are they made by the product manager, by one or more of the functional managers or by the division directors? Clearly, the establishment of a product development team within a company that is organized along functional lines creates several operational problems for management. Functional organizations may also create problems for working engineers attempting to serve several different managers. In the next section, these problems are addressed, and a popular team structure employed in many corporations that are organized along functional lines is described.

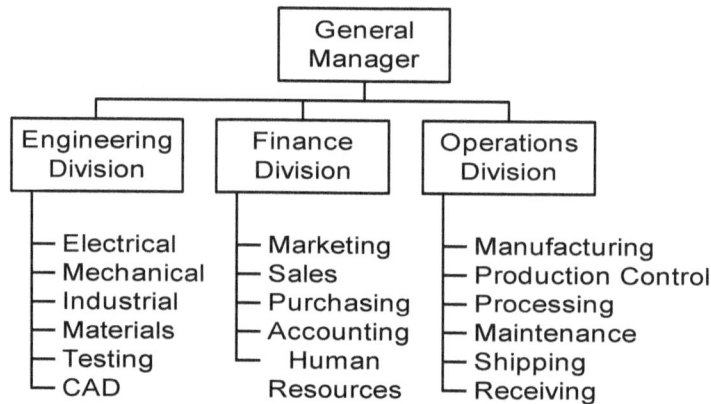

```
                        ┌─────────────┐
                        │   General   │
                        │   Manager   │
                        └─────────────┘
         ┌─────────────────────┼─────────────────────┐
  ┌─────────────┐       ┌─────────────┐       ┌─────────────┐
  │ Engineering │       │   Finance   │       │ Operations  │
  │  Division   │       │  Division   │       │  Division   │
  └─────────────┘       └─────────────┘       └─────────────┘
    ├─ Electrical         ├─ Marketing          ├─ Manufacturing
    ├─ Mechanical         ├─ Sales              ├─ Production Control
    ├─ Industrial         ├─ Purchasing         ├─ Processing
    ├─ Materials          ├─ Accounting         ├─ Maintenance
    ├─ Testing            └─ Human              ├─ Shipping
    └─ CAD                   Resources          └─ Receiving
```

Fig. 1.1 Typical organization chart showing the disciplinary functions. The organization includes three divisions and three levels of management.

1.2 A BALANCED TEAM STRUCTURE IN A FUNCTIONAL ORGANIZATION

There are several organizational arrangements frequently used in forming teams within companies organized along functional lines. They include:

- Functional teaming
- Modified functional teaming
- Balanced team structure
- Independent team organization

Each of these techniques for forming product development teams in a functional organization has advantages and disadvantages. A complete discussion of these four different team structures is given in reference [2]. To introduce you to the concept of forming teams within a functional organization, the balanced team structure will be described.

The balanced team structure is represented in the organization chart presented in Fig. 1.2. The team is formed with members drawn from the functional departments; however, in most situations, the team members are located in close proximity in a separate room or a building wing reserved for the team. The team members are usually dedicated to a single project, and their responsibilities are coordinated and focused. The team members retain their reporting relationship to the functional manager, but the contact is less frequent than in functional teaming arrangements, and the relation is less intense. Members of a balanced team often consider the product manager to be more important than their functional manager. The product manager is frequently a senior technical administrator with more experience, status and rank than the typical functional manager. The status of the product manager is dependent on the product development costs and the size of the development team. The product manager reports to the general manager, and usually is equal in status to the division directors in the organization. The senior program manager, with significant status and clout, leads the team and insures that resources will be available as required to meet schedules. The team members recognize the freedom from functional constraints that the balanced team structure implies. They assume ownership of the product specifications and commit fully to the success of the development activities. Communication is accomplished through the use of division and department coordinators and daily meetings. The team makes most of the routine decisions and only the higher-level decisions require the approval of the appropriate division directors.

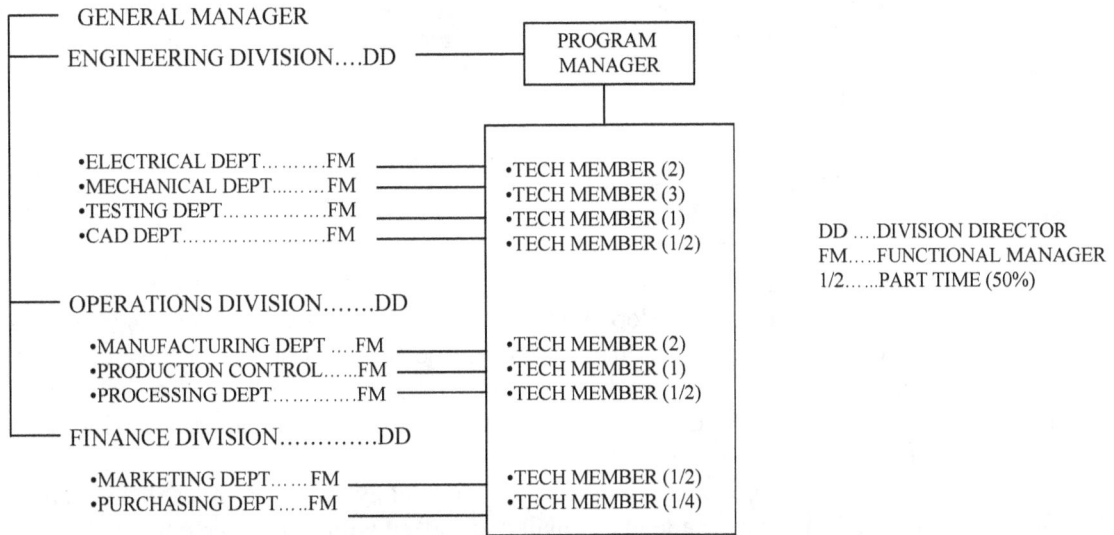

```
┌──── GENERAL MANAGER                          ┌──────────────┐
├──── ENGINEERING DIVISION....DD ──────────────│   PROGRAM    │
│                                              │   MANAGER    │
│                                              └──────────────┘
│                                         ┌─────────────────────────┐
│     •ELECTRICAL DEPT.........FM ────────│ •TECH MEMBER (2)         │
│     •MECHANICAL DEPT........FM ─────────│ •TECH MEMBER (3)         │      DD ....DIVISION DIRECTOR
│     •TESTING DEPT..............FM ──────│ •TECH MEMBER (1)         │      FM.....FUNCTIONAL MANAGER
│     •CAD DEPT....................FM ────│ •TECH MEMBER (1/2)       │      1/2......PART TIME (50%)
│                                         │                         │
├──── OPERATIONS DIVISION.......DD         │                         │
│                                         │                         │
│     •MANUFACTURING DEPT ....FM ─────────│ •TECH MEMBER (2)         │
│     •PRODUCTION CONTROL......FM ────────│ •TECH MEMBER (1)         │
│     •PROCESSING DEPT.............FM ────│ •TECH MEMBER (1/2)       │
│                                         │                         │
└──── FINANCE DIVISION.............DD      │                         │
      •MARKETING DEPT......FM ────────────│ •TECH MEMBER (1/2)       │
      •PURCHASING DEPT.....FM ────────────│ •TECH MEMBER (1/4)       │
                                          └─────────────────────────┘
```

Fig. 1.2 Organization with a balanced team structure.

The primary disadvantage of the balanced team structure is the dedication of the team members to a single project. Many talented team members are needed by the functional managers to support design activities across several product lines. With some of the functional resources dedicated to a single product development, other products may not receive adequate attention. The balanced team structure enhances the development capability of a given product at the cost of reducing the technical capability that can be applied across the company's entire product line. A second disadvantage is that some team members from a given department or division may not have sufficient expertise to perform adequately. With loose and distant functional supervision and review, this fact may not be apparent. The new design may not be at the cutting edge of technology.

1.3   DEVELOPMENT PLANNING

As the development team becomes more independent and moves away from the functional departments, planning becomes more important. Team planning documents are often prepared by a small group of senior team members with talents in business, marketing, finance, engineering design and manufacturing. These planning documents incorporate detailed descriptions of the product, market analysis, business strategy for capturing market share, product performance specifications, development schedules and budgets, team staffing requirements, material selections, design strategies, technology prerequisites, manufacturing processes, tooling requirements, inspection procedures, test specifications, facility availability and distribution capabilities. An element of the plan covers techniques and personnel that are to be employed to insure quality and product reliability. Finally, the plan includes a section specifying the deliverables and other sections on methods to measure performance of the product and the productivity of the team.

The development planning documents are extremely important since they outline strategies for achieving the company's goals and objectives in developing a new product. Management may use some of these documents during periodic design reviews to monitor progress and to judge the team's performance. The plan is a well-defined contract between the team and the divisional managers involved in the product development. The team members are expected to sign-on; to commit to the level of effort required to meet the schedule, achieve the goals and to produce the deliverables. Management is expected to fund the program, to provide adequate staffing and to provide the necessary facilities, computers, software, tooling and equipment in a timely manner.

1.4 STAFFING

Cross-disciplinary teams are necessary in staffing a product development team regardless of the structure employed by the organization. Usually, at least five disciplines or functions should be represented on the team including: marketing, finance, design engineering, operations/manufacturing, and quality control. The level of participation of each team member depends on the product being developed and the stage of the development. In some cases, team members work on more than one product and have responsibilities on two or more different projects. In other instances, several engineers from the same discipline are needed to complete long and complex tasks to meet the schedule for a single product. These members are usually dedicated to a single project.

Staffing on a product development team usually changes over the duration of the development period. Initially the team is small and staffed with senior personnel with demonstrated skills and talent. This small cadre may stay with the project for its duration. As the development proceeds, additional design engineers are required as technical strategies are converted into design concepts, then to design proposals and finally into detailed engineering drawings. Manufacturing engineers are needed in larger numbers when the early prototypes are being produced. After release of the design by engineering, the number of designers is reduced, and the emphasis is shifted to production where staffing from operations (production control, quality control, purchasing and plant maintenance) increases sharply. Clearly, the staffing of the development team is dynamic. Usually only a small fraction of the total team is committed to the project on a full time basis from its initiation to its completion. In some companies, individuals are committed to the development team for the duration of the project, but at varying levels of participation. This approach enhances ownership of the project by the individual team members.

1.5 TEAM LEADERS

The experience and leadership skills required of the product manager are strongly dependent on the team structure employed as well as the size and cost of the product development. Some products can be developed with a relatively small commitment of staff (say less than 10 members). In these circumstances, the team established is often structured along functional lines, and the product manager is often one of the functional managers. The functional managers are experts in their respective disciplines, skilled in managing personnel and knowledgeable (but not expert) in the other disciplines. They are also familiar with the company's entire product line.

Other product developments require a larger commitment of personnel (say a team numbering 10 to 50), and the balanced team structure shown in Fig. 1.2 is probably the most suitable organization.

The product manager is a seasoned staff member well-known for his or her expertise and respected for performing at a level that exceeds expectations. The appointment as a product manager (equal in rank and status to the functional department managers) often represents a promotion for the individual involved and a first experience in managing a technical group. The functional managers and the divisional directors provide close support through periodic reviews that monitor progress and reveal problems early in the development process.

As the size of development team becomes even larger (100 or more members), team leadership assumes major importance. The product manager in this situation is a senior member of the company's executive management with a well-established record of accomplishments. He or she is senior to the managers of the functional departments, and is at least equal in status to the division directors. The authority and responsibility given the product manager is significant, and his or her career depends strongly on the success of the product in the market and the productivity of the development team.

1.6 TEAM MEMBER RESPONSIBILITIES

Members of a product development team have two different sets of responsibilities—one to his or her disciplinary function and the other to the team. A team member, from say the electrical engineering department, provides the technical expertise in his or her specialty and ensures that the product is correctly designed with leading edge, state-of-the-art technology. The team member acts to bring all of the important functional issues that affect product performance to the attention of the team. The team member also represents the functional department to ensure that the product development team maintains the principles, goals and objectives of the functional discipline.

The most important team activity for the individual member is to share responsibility. Highly effective teams that routinely succeed have team members that hold themselves mutually accountable for the overall success of the product. Next, the team member must recognize and understand all of the product features, and fully participate in the techniques employed to meet the design objectives. Individual team members must assess team progress and participate in improving team performance. A team member must cooperate in establishing all of the reporting relationships required to maintain the communication among the team and functional departments. In many of the team structures, an individual may report to three different managers—the Division Director, the Department Manager and the Program Manager. The matrix organization structure (Fig. 1.2), inherent in forming development teams, creates complex reporting relations. Flexibility and cooperation on the part of the individual team members are required for the team structure to be effective.

Finally, the team members must be able to communicate in a clear concise manner in all three modes: **writing, speaking and graphics**.

1.7 TEAM MEMBER TRAITS

When a group of individuals work together on a team to achieve a common goal, they can be extremely effective [3]. The team interaction promotes productivity for several different reasons. First, meeting together is synergistic in that one's ideas freely expressed stimulate additional ideas by other team members. The net result is many more original ideas than would have been possible by the same group of individuals working independently. Another advantage is in the breadth of knowledge available in the team. The team is cross-disciplinary and the very wide range of skills necessary for the product development process is included within the team. Each member of the team is different with some combination of strengths and weaknesses. Acting together the team can build on the strengths of each member and compensate for any weaknesses. The grouping of individuals provides social benefits that are also very important to the individuals and to the corporation. There is a bonding that occurs over time and a support system develops that team members appreciate. The team develops a sense of autonomy valued by each of its members. The team develops solutions, implements them and then monitors the results with little or no direction from management. Empowered teams assume ownership of the project. They commit to the product development process and consistently perform beyond expectations.

While cross disciplinary teams have been employed in many different companies for at least two decades, little has been done by the educational system to develop team skills. On the contrary, both the secondary school and many college systems tend to encourage competitive attitudes and independence. Students compete for the "A" grades that might be given in a course and are discouraged from cooperating on assignments. This system prepares students to work independently, but team-building skills are not addressed.

The industrial workplace is much different. Team members compete, but not within the team. They compete with a similar development team from a rival corporation. Team members bond, cooperate and consistently help each other by sharing assignments. Competition between team members is discouraged. Recognition and rewards go to the team as a whole much more often than to select individuals.

There is a set of characteristics that describe a good team member, and another set that depicts an individual who can destroy the efforts of a team.

The characteristics of a good team member are:

1. Treat every team member with respect, trust their judgment and value their friendship.
2. Maintain an inquiring attitude free of predetermined bias so that team members will all participate and share knowledge and opinions in an open and creative manner. Listen carefully to the other team members.
3. Pose questions to those hesitant team members to encourage them to share their knowledge and experience more fully. Share your experiences and opinions in a casual, easy-going manner. Help other members to relax and enjoy the interactive process involved in team cooperation. Participate but do not dominate.
4. Observe the body language of the other team members because it may indicate lack of interest, defensive-attitudes, hostility, etc. Act with the team leader to defuse hostility and to stimulate interest. Disagree if it is important, but with good reason and in good taste.
5. Emotional responses will occur when issues are elevated to a personal level. It is important to accept an emotional response even when you are opposed. It is also important to control your emotions and to think and speak objectively. You should be self confident in your discussions, but not dogmatic.
6. Allocate time for self-assessment so the team can determine if it is performing up to expectations. Make suggestions during these assessment periods to build team skills.
7. Be comfortable with your disciplinary skills but continue to study to improve your capabilities. Communicate effectively by speaking clearly, writing clearly and concisely, and illustrating effectively with modern graphics.

The characteristics of a destructive team member are:

1. No member of the team should ever participate in a conversation that is derogatory about a person on the team or in the corporation. If you cannot make complimentary remarks about a person, keep the negative thoughts to yourself. Respect, trust and friendship are vital elements in the founding of a successful product development team. Derogatory remarks destroy the foundation necessary for respect, trust and friendship.
2. Arguments among team members should be avoided. You are encouraged to introduce a different opinion and participate in a discussion with a different viewpoint, but the discussion should never degenerate into an argument. Every member on the team is responsible for quickly resolving arguments that arise among the team members.
3. All members of the team are responsible for on-time attendance at their meeting. If a member is absent, the leader should know the reason beforehand and explain it to the team. You must be certain that all members appreciate and respect the reason for everyone's role on the team and the importance of every member attending every meeting.
4. Team progress is hindered when one member dominates the meeting. These people are often overly critical, intimidating and stimulate confrontations. If you have these characteristics, work hard to suppress them.
5. The team leader must be extremely careful about intervention. If the team is moving and working effectively, the leader should be quiet because intervention in this instance is counter productive. When the team is having difficulty, intervention may be necessary depending on the problem. In the event a team member becomes hostile, intervention should be quick, and the disagreement producing the hostility should be dealt with immediately. When a team is seeking consensus, the process may require time and the leader should wait 5 to 10 minutes before intervening.
6. No member of the team can be opinionated including the leader. He or she is not a judge determining the correct solution. The leader seeks to facilitate so that the team can reach a consensus on the correct solution. It is only when the team accepts the solution that implementation can begin.

Compare your traits with those listed above. You will probably exhibit both good and bad characteristics. To be effective team members, you must work to enhance the favorable traits and to suppress those that are destructive to the efforts of the team.

1.8 EVOLUTION OF A DEVELOPMENT TEAM

If a number of workers or students are assigned to a new team, it is possible to observe several behavioral phases as the team bonds, melds and matures [4]. In the very beginning (Phase I), the team members usually exhibit both excitement and concern. They are excited about working on a high-profile project with a new group of talented people. It should be an opportunity to learn, make new friends, advance in position and/or stature and have fun in the process. At the same time, the team members exhibit signs of concern. They are worried about meeting and understanding other team members. The tasks assigned to the team are probably extremely vague at this stage of the development, and vagueness leads to uncertainty. The skill levels and areas of expertise of fellow team members have yet to be determined. The personality of the team leader is unclear. There is worry over the new reporting relations that new team membership implies. In this orientation phase, team members are searching for their role in the team and evaluating their possibilities for success or failure. It is a tense time.

Dissatisfaction is the next phase in the evolution of the team (Phase II). You have met your fellow team members, and the news is not good. You recognize the differences in personalities and in work habits. A few of the team members do not understand how to tell time and never arrive on time when attending team meetings. Member schedules make it very difficult for the entire team to meet. Some team members lack social graces. Some members are inexperienced and unable to cope with their assignments. You have met the team leader and he or she is very difficult (stressed, demanding, exacting, impatient and abrupt). The schedule calls for the team to leap over mountains on a daily basis. The budget for the project is totally inadequate. Surely you are in a lose-lose situation. Your thoughts are dominated by schemes to transfer to a better team. During this phase, the progress of the team toward scheduled milestones is exceedingly slow.

With time progress is made, and the team members start learning how to work together. This is the resolution phase (Phase III). Everyone agrees to attend meetings and arrive on time. Personality conflicts are smoothed over. Team members may never be true friends, but everyone manages to get along and show mutual respect. Team members have come to terms with the schedule, and have committed to the extra effort that it requires. The manager has mellowed, or team members have learned how to handle his or her impatient demands. Solutions for reducing development costs have been found. There appears to be a good possibility of meeting the product development plan. The team is beginning to perform well, and more members are committed.

The team has melded into an efficient unit. Conflicts rarely arise. Lasting friendships begin to be formed. The strengths of each team member are fully utilized. The manager recognizes the talents of the team. Tasks are completed ahead of schedule and under budget. Upper level managers recognize the accomplishments of the team. You now find yourself in a win-win situation. This is the production phase of the development team (Phase IV).

The project has been completed and the ribbons have been cut. The team prepares to disband. In this termination phase (Phase V), you reflect on your experiences, good and bad, over the duration of the project. You examine your individual performance and evaluate various factors that improved the effectiveness of the team. You meet with the manager for a performance evaluation. The team members share a sense of accomplishment in achieving the goals and objectives of the development.

1.9   A TEAM CONTRACT

Teams are effective because they meet together to focus their wide range of disciplinary skills and natural talents toward the solution of a set of well-formulated problems. The team meeting is the format for synergistic efforts of the team. Unfortunately, not all teams are effective because some members are so disruptive they destroy the cohesiveness of the team. Bonding of team members is vital if the team is to successfully solve the multitude of problems that arise in the product development process.

Teams **fail** because of three main reasons. First, they deviate from the goals and objectives of the development, and cannot meet the milestones on the development schedule. Second, the team members become alienated and the bonding, trust and understanding critical to the success of the team never develops. Third, when things begin to go wrong and adversity occurs, "finger pointing" begins and an attempt to fix blame on someone replaces creative actions.

Clearly, many problems will arise to impair the progress of a development team. It is recommended that each team establish a set of simple rules it considers necessary to govern the team.

Once established, each team member should sign and date this contract[1]. Suggestions for your team to consider are provided below:

- The information discussed by team members will remain confidential
- Members will acknowledge problems and deal with them
- Members will be supportive rather than judgmental
- Members will respect differences
- Members will provide responses directly and openly in a timely fashion
- Members will provide information that is specific and focused on the task and not on personalities
- Members will be open, but will respect the right of privacy
- Members will not discount the ideas of others
- Members are each responsible for the success of the team experience
- The team will have the resources needed to solve the problems that arise
- Every member of the team will contribute to its success
- Members recognize the importance of team meetings to the success of the group, and accordingly agree to:

    1. Always attend the scheduled team meetings
    2. When attendance is impossible, notify the team leader and as many other members as possible in advance
    3. Use meeting time wisely
    4. Start on time
    5. Limit breaks and return from them on time
    6. Keep focused on the goals
    7. Avoid side issues, personality conflicts and hidden agenda
    8. Share responsibility for briefing members when they miss a meeting
    9. Avoid making phone calls or engaging in side conversations that interrupt the team

1.10 EFFECTIVE TEAM MEETINGS

Leadership is important to ensure that the team remains focused on the overall project goals/objectives and on the agenda at each meeting. Creating the correct team environment or atmosphere is essential so that the team members cooperate, share ideas and support each other to achieve solutions [5]. It is important for the team leader to assess the performance of the entire team at each meeting. When the team fails to make progress, it is critical for the team leader or possibly an individual member to make changes as necessary to enhance the team's effectiveness. Some typical difficulties encountered during team meetings are identified in the following paragraphs, along with suggested corrective actions team members can take to enhance the team's productivity.

One common problem impairing team productivity occurs when the goals and objectives of the meeting are not clear or they appear to be changing. While in many cases team meetings are planned well in advance and are supposed to follow an agenda, new topics are introduced and the meeting drifts from one item to another. This team behavior indicates that either some of the team

---

[1] An example of a development contract is presented in Appendix A-1. It contains the team structure, development constraints, milestone dates, scoring rules and an agreement to four simple teaming rules. This development contract was for the Spring 2010 semester and the new contract for later semesters may differ to some degree.

members do not understand the goals, or they do not accept them. Instead, they are marching to the beat of their own drum and trying to take the team with them. Unless corrective actions are taken to focus the team on the original goals, this team meeting will fail and time and effort will be lost. A focused team stays on the agenda, and its team members accept the meeting objectives. Team discussions should generate a wide set of solutions to the problem being considered, and should continue until team members reach a consensus solution. A meeting is successful when an issue has been resolved and all of the team members accept the solution. This allows the team to move forward by organizing its talents to address the next problem.

Leadership is an essential element for team effectiveness. The most effective teams are usually democratic with a large amount of shared leadership. While there is a recognized leader, with a considerable degree of responsibility and authority, the team leader is supported by each team member. The team member with the appropriate expertise (relative to the problem being considered) will usually lead the discussion and effectively act as leader of the team during this period. However, in some instances, the team leader maintains strict control of the meeting and never shares the leadership role with any team member. Some team members resent this style of leadership and may not participate as fully as possible. The result is that complete utilization of the resources of the team does not occur.

Attitude of the team members is an important element in the success of the team. Are the team members committed? This commitment is evidenced in their attitude. If they come to the meeting table exhibiting interest and willingness to participate, they are committed and will make a positive contribution. However, if they are bored and indifferent, they will not become engaged in a meaningful way. In fact, if they are sufficiently disinterested, they may initiate side conversations or arguments and destroy the atmosphere of trust necessary for an effective meeting. It is important to evaluate each team member, and to secure their commitment to the goals and objectives of the project.

As the meeting progresses, it is important to assess the discussions that are occurring. For a meeting to be successful, the discussion should include everyone on the team. A discussion should stay focused on an agenda item with few if any deviations to unrelated topics. The team members must listen carefully to one another and give all of the ideas presented a serious hearing. Teams tend to accomplish less when the discussion is dominated by only a few of the participants. These members tend to intimidate others in their efforts to control and dominate the team. The result is disastrous because they essentially eliminate the contributions of other team members.

Teams do not operate with total agreement on every issue. There must be accommodation for disagreements and for criticism. All of the ideas or suggestions made by every team member at any meeting will not be outstanding. Indeed, some of them may be ridiculous; criticism of these ideas must occur. However, the criticism should be frank and without hostility. Personal attacks must not be a part of the critique of an idea. If the criticism is to improve an idea or to eliminate a false concept, then it is of benefit to the progress of the team. The criticism should be phrased so that the team member advancing the flawed idea is not embarrassed. When disagreements occur, they should not be suppressed. Suppressed disagreements breed hostility and distrust. It is much better for the team to deal with disagreements when they arise. The root causes of a disagreement should be ascertained, and the team should take the actions necessary to resolve them. On some occasions voting is a mechanism used to resolve conflict. This procedure must be used with care particularly if the vote indicates the team is split almost equally in their opinion. A better practice is to discuss the issues involved and attempt to reach a consensus. Reaching a team consensus may require more time than a simple vote, but the results are worth the effort. Consensus implies that all members of the team are in general agreement and willing to accept the decision of the team. When the team votes, a simple majority is sufficient to resolve an issue. However, the minority members may become resentful if they are always on the short end of the vote. In a short time, they will not accept the outcome, and they will not commit to the actions necessary to implement the team's decision.

The team meetings must be open. The agenda should be available in advance of the meetings and subject to change with added topics introduced as "new business". A sample agenda is shown in Table 1.1. Sample forms for the agenda are provided in Appendix A.

Team members should believe that they have the authority to bring new topics before the team. They should feel free to discuss procedures used in the team's operation. Secret meetings of an inside group should be carefully avoided. When the fact leaks that secret meetings are being held and that issues are prejudged by a select few, the effectiveness of the team is destroyed.

Table 1.1:
Weekly meeting of the Mighty Terrapin Team
'September 14, 2010 7:59 AM

Weekly status report	Team Leader
Review of outstanding action items	
Action item #8	Member responsible
Action item #9	Member responsible
Action item #N	Member responsible
Report on progress	
Subsystem – Propulsion	Member responsible
Subsystem – Electrical	Member responsible
Subsystem – Mechanical	Member responsible
Subsystem – Control	Member responsible
Market study results	Marketing Representative
Identify new problems	All member participate
Assignment of action items	Team Leader and volunteers
New business	All member participate
Summary	Team Leader
Adjourn at 8:59 AM or sooner	

At least once during a meeting it is useful to evaluate the progress of the team. Is the team following the agenda? Is someone dominating the meeting? Is the discussion to the point and free of hostility? Is everyone properly prepared to address his or her agenda items? Are the team members attentive, or are they bored and indifferent? Is the team leader leading or is he or she pushing? If a problem in the operation of the team is identified during the pause for self-appraisal, it should be resolved immediately through open discussion.

Finally, the team must act on the issues that are resolved and the problems that are identified. To discuss an issue and to reach a consensus is part of the process, but not closure. Implementation is required for closure, and implementation requires action. When decisions are made, team members are assigned **action items**. These action items are tasks to be performed by the responsible individual; only when the tasks are completed is an issue considered closed. It is important that the action items are clearly defined, and the role of each team member in completing their respective tasks is understood. A realistic date should be set for the completion of each action item. The individual responsible should be clearly identified; he or she must accept the assignment without objection or qualification. A checking system must be employed to follow up on each action item to insure timely completion. If there is a delay, the schedule for the entire product development cycle may be at risk. It is important to deal with delays immediately and for the team to participate in the development of plans to eliminate the cause of the delay. A form has been provided in Appendix A-3 to facilitate the assignment of action items and the follow-up on each item.

1.11 PREPARING FOR MEETINGS

Effective team meetings do not just happen. It is necessary to prepare for the meeting and to execute post-meeting activities to insure success. The preparation usually involves selecting, arranging and equipping the meeting room, scheduling the meeting so that the necessary personnel are in attendance, preparing and distributing an agenda in advance of the meeting and conducting the meeting following rules established by the team. Equally important are actions taken by the team members following the meeting to implement the decisions that have been reached.

The meeting room and its equipment also affect the team's progress. The room should be sized to accommodate the team and any visitors that have been invited. Rooms that are too large permit the members to scatter and the distance between some members becomes too long for effective and easy communication. The seating arrangement should be around a table so that everyone can observe each other's face and body language. It is very difficult to engage anyone in a meaningful conversation if they have their back to you. Water, coffee, tea or soft drinks should be available if the meeting duration exceeds an hour. Smoking is strictly prohibited. The equipment that will be needed for presentations should be available, and its operation should be checked prior to the start of the meeting. A computer with projection capability is of growing importance in a well-equipped meeting room. The availability of a computer during the meeting permits one to draw from a large database and to modify the presentation in real time. Additionally, it allows results of analyses or experiments to be clearly displayed in graphical form for the entire team to review.

Prior to the scheduled meeting, it is important to make careful preparations. Minutes from the previous meeting should be distributed with sufficient time for review before the next scheduled meeting. A detailed agenda, similar to the one shown in Table 1.1, is distributed to inform team members of the topics and/or problems that will be addressed in the next meeting. Individual team members responsible for specific agenda items are identified. The details are clear to all and responsibility shared by individual members is defined. In some instances, information will be needed from corporate employees, suppliers or visitors that are not members of the team. In these cases, arrangements must be made to invite these people to the meeting so that they can provide the necessary expertise and respond to questions from all of the disciplines represented on the team.

In conducting the meeting, it is important to start on time. It is very annoying to the majority of the members to wait five or ten minutes for a straggler or two. The team leader must make certain that everyone involved understands that 8:00 am means 8:00 am and not 8:08 or 8:12 am. The objectives of the meeting should be clearly understood and the time scheduled for each agenda item should be estimated. A team member should be assigned as the timekeeper, and another should act as a secretary to record notes and to prepare the minutes of the meeting. If team meetings are frequent and held over a long time (several months), the duties of the timekeeper and the secretary should be shared with others on the team. The meeting should follow a set of rules that govern the behavior of individual members.

As the meeting draws to a conclusion, the team leader should take a few minutes to summarize the outcome of the discussions. This summary gives an ideal opportunity to insure that assignments are understood, responsibility accepted and completion dates established. The meeting must be completed on time and everyone's schedule should be respected. If the meeting is not periodic (e.g. every Tuesday at 8:00 am), then it is important that the time and place for the next meeting be scheduled.

After the meeting, it is productive for the team leader to make certain that the complete minutes have been prepared and distributed. If any member was not able to attend the meeting, the team leader should brief that person on the team's progress. Finally, the team leader should follow up on each action item to ensure that progress is being made, and that new or unanticipated problems have not developed. It is clear that effective meetings do not happen by accident. Many members of the team work diligently before, during and after the meeting to make certain that the goals and objectives are clearly defined, that the issues and problems are thoroughly discussed and that the solutions developed result in assignments that are executed with dispatch.

1.12 SUMMARY

Arguments for the importance of development teams in industry have been presented. Experience has shown that effective development teams are vital if a corporation is to develop highly successful products and introduce them in time to win a major share of the market.

Functional organizations that exist within most corporations have been described. Reasons why a functional organizational structure often interferes with rapid low-cost product development have been given. The balanced team structure commonly employed in establishing development teams within large functionally organized corporations has been described.

The importance of location of team members in either enhancing or inhibiting communication has not been considered. However, you should be aware of the important role distance plays in effective and timely communication. The rule is simple. To enhance communication, minimize the distance between those needing to communicate with each other. The ideal situation is to place the entire development team in the same room.

Development planning and staffing of the development team has been covered. The success of the team depends strongly on generating a comprehensive plan. The plan is used initially to justify the development budget and its schedule. During the course of the development, the plan provides guidance for monitoring the progress and for judging both the team and product performances. Team staffing is cross-disciplinary with adequate representation from both the engineering and finance divisions. Engineers interact with the business personnel from marketing, sales and purchasing. Responsibility is shared between disciplines and functions.

The team's organizational structure and the team leader's management status have been discussed. This relationship is important because of executive authority that corresponds with a higher level of management. Higher-level managers carry more authority, many decisions can be made more rapidly, communication is more effective and team progress is often enhanced if the leader has status and clout associated with executive management.

Next, team member responsibilities and team member traits were discussed in considerable detail. Team members work in a matrix organization and often report to two or more managers. This arrangement is sometimes difficult particularly when you get mixed signals from different managers. It is important to be flexible and remain cooperative with the program manager and the functional managers. Team member traits are very important to your career. Evaluate your traits in an honest self-assessment. This is a critical first step in building team skills.

Most design teams evolve over the duration of a development project. Five phases usually experienced by the team in this evolution have been described. Not all of the phases provide pleasant experiences. Early in the evolution, a team encounters the dissatisfaction phase with very slow progress, low productivity and member gloom. To minimize the time in this phase, it is suggested that the team prepare and execute a contract that provides guidelines for member behavior.

Finally, guides for conducting successful team meetings have been provided. You will spend more time than you can imagine in team meetings. From the outset, learn the techniques for a successful meeting. Reasons for the failure of some meetings and the success of others have been presented. The role of the team leader and the behavior of the team members are described. The meeting room is more important than most engineers imagine. Features and furniture arrangements in a room affect the ability of a team to function effectively. Recommendations are made for meeting room size and for arrangements to enhance team productivity during meetings. The progress of the project is dependent on the team's ability to control its meetings.

REFERENCES

1. Smith, P. G., D. G. Reinertsen, Developing Products in Half the Time, Van Nostrand Reinhold, New York, NY, 1991.
2. Schmidt, L. C. and Zhang, G., Dieter, G., Cunniff, P. F., and Herrmann, J. W., Product Engineering and Manufacturing, 2nd Edition, College House Enterprises, Knoxville, TN, 2001.
3. Barczak, G. and Wilemon, D., "Leadership Differences in New Product Development Teams," Journal of Product Innovation Management, Vol. 6, 1989, pp. 259-267.
4. Lacoursiere, R. B. The Life Cycle of Groups: Group Development State Theory, Human Service Press, New York, NY, 1980.
5. Barra, R. Putting Quality Circles to Work, McGraw Hill, New York, NY, 1983.
6. Foxworth, V., "How to Work Effectively in Teams," Class Notes ENME 371, University of Maryland, College Park, September 5, 1997.
7. Clark, K. and Fujimoto, T., Product Development Performance, Harvard Business School Press, Boston, MA, 1991.
8. Wheelwright, S. C. and Clark, K. B., Revolutionizing Product Development, Free Press, New York, NY, 1992.

EXERCISES

1.1 Write an engineering brief describing why team structure in a large corporation is so important in the product development process. Add a second paragraph indicating why team structure is much less important in a very small company.

1.2 Prepare an organization chart, like the one shown in Fig. 1.1, for the University, School or College that you are attending. Describe the logic, as you see it, for this organizational structure.

1.3 List the advantages and disadvantages of the balanced team structure.

1.4 Outline a development plan that you would prepare if you were a senior team member representing an engineering function (discipline) on a newly formed development team. The product to be developed is a new portable digital reader to compete with Apple's iPad tablet.

1.5 Write an engineering brief describing staffing required for the development team for the digital reader. Give the reasons for changing the staff as the digital music player evolves during the development process.

1.6 Describe the experience and status of the product manager for the balanced team structure.

1.7 Describe a matrix organization structure. Explain why this organization impacts an engineer assigned to a product development team with a balanced structure. If you were this engineer, how would you respond to inquires from your functional manager?

1.8 You (a male/female figure) are attending weekly team meetings and are always seated next to Sally/Bill who is young and very attractive. Sally/Bill is apparently interested in you because she/he frequently involves you in side conversations that are not related to the ongoing team discussions. Write a plan describing all of the actions you will take to handle this situation.

1.9 Brad, the team leader, is a great guy who believes in leading his team in a very democratic manner. He encourages open discussion of the issue under consideration for a defined time period, usually 5 or 10 minutes. At the end of the time period, he intervenes and calls for a team vote to resolve the issue. Write a critique of this style of leadership.

1.10 You together with Rob and Jean are senior members on a development team with six other members. The three of you are really very knowledgeable and experienced. You get along famously, and have developed a habit of gathering together at the local watering hole the evening before the scheduled team meeting. During the evening you discuss the issues on the agenda and arrive at some design decisions. Write a brief describing the consequences of continuing this behavior.

1.11 Sue has a wonderful personality, and as a team leader is skillful in promoting free and open discussion by all of the team members. After the team reaches a consensus, she calls for volunteers to implement the required action. When two or more people volunteer, she assigns them as a group to handle the action item. When no one volunteers, she assigns the least active person on the team to the action item and moves promptly to the next item on the agenda. Write a critique of this leadership style.

1.12 You are the leader of an ten-member team meeting for the first time to develop a new hair dryer. During this initial meeting the following events occur.

- The temperature in the meeting room is 88 °F.
- The room is set up for a seminar speaker.
- The meeting is scheduled to begin at 9:30 am and at 9:35 am only five of the ten members have arrived.
- During the meeting Horrible Harry begins to verbally abuse Shy Sue.

- At 9:50 am Talkative Tom begins to describe his detailed positions on the new health care plan and is still going at 10:10 am.
- The team breaks for coffee at 10:15 am and has not returned at 10:30 am.
- Sleepy Steve begins to snore.
- Procrastinating Peter refuses to accept an action item after leading the discussion on the issue for 15 minutes.

Describe the approach that you, acting as the team leader, would follow in handling each of these situations.

1.13   Prepare a list of ground rules that should be followed by all the members of a team in conducting an effective meeting.

1.14   You are a member of a team with an ineffective leader. What can you do to improve the productivity of the team? In responding to this question consider the irresponsible actions on the part of certain team members that are listed in Exercise 1.12. Remember in your response that you are not the leader and the most that you can do is to share the leadership role from time to time.

1.15   Meet with your team and discuss a list of rules that will be used in guiding the team's conduct during the semester. Incorporate the team's list of rules in the form of a contract. As a final step, each member of the team is to sign the contract indicating his or her commitment to the rules of conduct.

CHAPTER 2

PRODUCT DEVELOPMENT AND HOVERCRAFT MISSIONS

2.1 THE IMPORTANCE OF PRODUCT DEVELOPMENT

Hundreds of products are used to perform many functions every day. Products surround us. This morning I prepared breakfast by toasting bread, frying eggs and making tea. How many products did I use in performing this simple task? The products used included a toaster (Phillips), a frying pan (Calphalon), a microwave (General Electric) to heat the tea and a gas cook top (Kitchen Aid) to heat the frying pan. These are all products that are designed, manufactured and sold to customers both in America and abroad. Some products are relatively simple, such as the frying pan with only a few parts, yet some are much more complex, such the microwave with over a hundred parts.

Corporations worldwide continuously develop their product lines with minor improvements introduced every year or two and with more major improvements every four or five years. Product development involves many engineering disciplines and is a major responsibility of engineers entering the work place. An examination of Table 2.1 shows that designing, manufacturing, and product sales represent more than 80% of the responsibilities of mechanical engineers in their initial position in industry. This distribution of activities is also typical of many of the other engineering disciplines.

Table 2.1
Responsibilities of Mechanical Engineers in their first position [1]

ASSIGNMENT	% of TIME	
Design Engineering	40	
Product Design		24
Systems Design		9
Equipment Design		7
Plant Engineering / Operations / Maintenance	13	
Quality Control / Reliability / Standards	12	
Production Engineering	12	
Sales Engineering	5	
Management	4	
Engineering		3
Corporate		1
Computer Applications / Systems Analysis	4	
Basic Research and Development	3	
Other Activities	7	

The financial success of many corporations depends on the introduction of a steady stream of winning products to the marketplace. A successful product must meet the customer needs and provide robust and reliable service at a competitive price.

2.1.1 Development Teams

Interdisciplinary teams develop products with members drawn from engineering, marketing, manufacturing, production, purchasing, etc. A typical organization chart for a team developing a relatively simple electro-mechanical product is shown in Fig. 2.1. The team leader coordinates and directs the activities of the designers—mechanical, electronic, industrial, etc. A marketing specialist supports the team to ensure that the evolving product meets the needs of the customer. Financial, sales and legal assistance is usually provided on an as-needed basis by corporate staff or outside contractors.

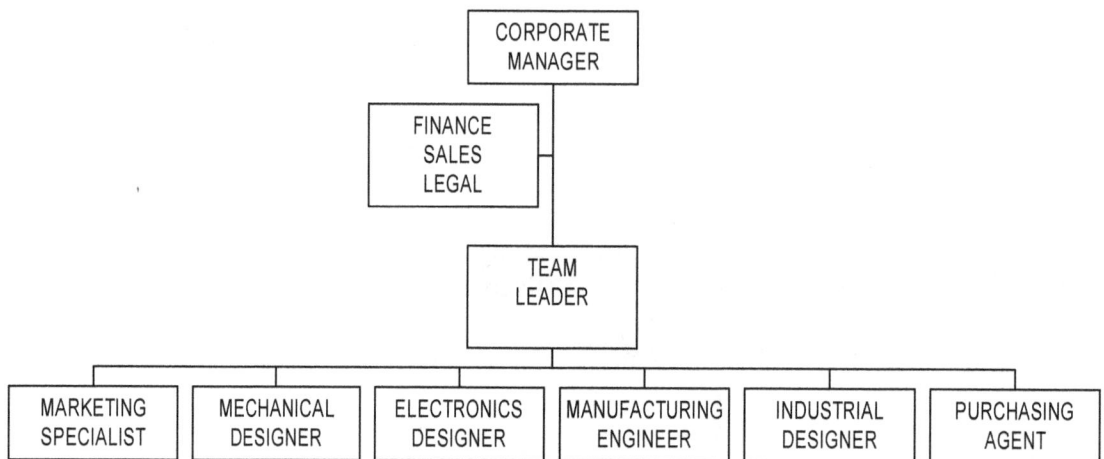

Fig. 2.1 Organization of a product development team for a relatively simple product.

The team leader is also supported by a purchasing agent who establishes close relations with the suppliers who provide materials and components used in the final assembly of the product. In the past decade, the relationship between suppliers and end-product producers has changed significantly. Suppliers become part of the development team; they often provide a significant level of engineering support that is external to the internal (core) development team. In some developments, these external (supplier funded) development teams—when totaled—are larger than the internal development team. For example, the internal development team for the Boeing 777 airliner involved 6,800 employees and the external teams totaled about 10,000 persons from many suppliers [2].

2.2 DEVELOPING WINNING PRODUCTS

There are two primary aspects involved in developing successful products that win in a competitive marketplace. The first is a combination of quality, performance and price of the product. The second involves cost and time. The time and the cost of a product development, and the manufacturing costs incurred during the production cycle are vitally important.

2.2.1 Quality, Performance and Price

Let's discuss the product first. Is it attractive and easy to use? Is it durable and reliable? Is it effective and efficient? Does it meet the needs of the customer? Is it better than the products now available in the marketplace? If the answer to all of these questions is an unequivocal **yes**, then the customer **may** want to buy the product if the price is right. Next, you need to understand what is implied by product cost and its relation to the price actually paid for the product. Cost and price are distinctly different quantities. Product cost clearly includes the cost of materials, components, manufacturing and assembly. The accountants also include less obvious costs such as the prorated costs of capital equipment (the plant and its machinery), the price of the tooling, the development cost, and even the expense of maintaining the inventory in establishing the **total cost** of producing the product.

   Price is the amount of money that a customer pays to buy the product. The **difference** between the **price** and the **total product cost** is **profit**, which is usually expressed on a per unit basis.

$$\text{Profit} = \text{Product Price} - \text{Total Product Cost} \qquad (2.1)$$

This equation is the most important relation in engineering and in business. If a corporation cannot make a profit, it soon is forced into bankruptcy, its employees lose their positions and the stockholders lose their investment. It is this profit that everyone employed by a corporation seeks to maximize while maintaining the strength and vitality of the product lines. The same statement can be made for a business that provides services instead of products. If a business is to make a profit and prosper, the price paid by the customer for a specified service must be more than the cost to provide that service.

2.2.2 The Role of Time in the Development Process

Let's now discuss the role of the development process in producing a line of winning products. Developing a product involves many people with technical expertise in different disciplines. Also, a product development takes time and requires significant sums of money. Let's first consider development time. Development time, as it is used in this context, is time to market—the time from the product development kickoff to the introduction of the product to the market. This is a very important target for a development team because of the many significant benefits that follow from being first to market with a new product. A corporation realizes many competitive advantages with a rapid development capability. First, the product's market life is extended. For each month cut from the development schedule, a month is added to the life of the product in the marketplace with an additional month of sales revenue and profit. The benefits of being first to market on sales revenue are shown in Fig. 2.2. The shaded area between the two curves to the left side of the graph is the enhanced revenue due to the longer sales life.

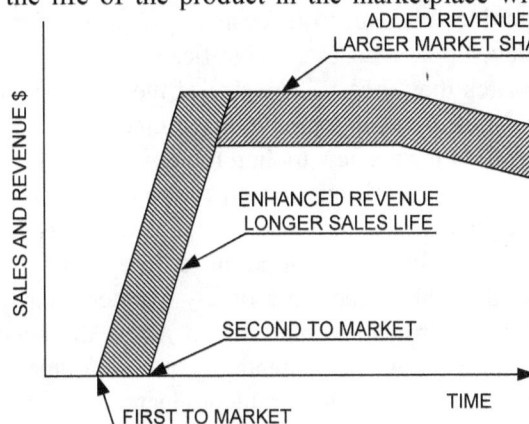

Fig. 2.2 Increased sales revenue due to extended market life and larger market share.

A second benefit of early product release is increased market share. The first product to market has 100% of market share in the absence of a competing product. For products with periodic development of "new models," it is generally recognized that the earlier a product is introduced to compete with older models—without sacrificing quality and reliability—the better chance it has for acquiring and retaining a large share of the market. The effect of gaining a larger market share on sales revenue is also illustrated in Fig. 2.2. The shaded area between the two curves at the top of the graph shows the enhanced sales revenue due to increased market share.

A third advantage of a short development cycle is higher profit margins. If a new product is introduced prior to availability of competitive products, the manufacturer is able to command a higher price for the product, which enhances the profit. With time, competitive products will be introduced forcing price reductions. However, in many instances, relatively large profit margins can still be maintained because the company that is first to market has additional time to reduce manufacturing costs. Their employees learn better methods for producing components and reduce the time needed to assemble the product. The advantage of being first to market, with a product where a learning curve for manufacturing exists, is shown graphically in Fig. 2.3. The learning curve reflects the reduced cost of manufacturing and assembling a product with time and experience in production. These cost reductions are due to many innovations introduced by management and workers after beginning production. With time and manufacturing experience, it is possible to drive down production costs.

Fig. 2.3 The development team bringing a product to market first enjoys a higher initial price and cost advantages from manufacturing efficiencies.

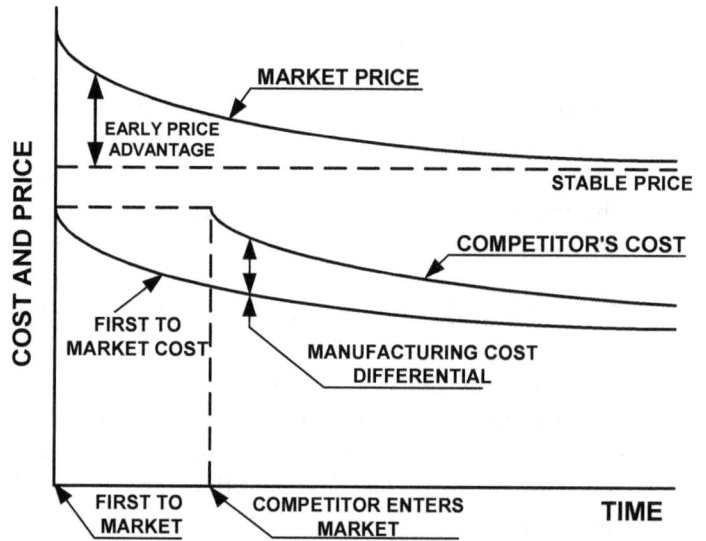

Let's next consider development costs that represent a very important investment for the companies involved. Development costs include the salaries of the members of the development team, money paid to subcontractors, costs of pre-production tooling, expense of supplies and materials, etc. These development costs are significant, and most companies must limit the number of developments in which they invest. The size of the investment may be appreciated by considering the development cost of a new automobile is estimated at $1 billion, with an additional expenditure of $500 to $700 million for the new tooling required for high-volume production [2].

This discussion on time and cost of product development has been included to help you begin to appreciate some of the business aspects of developing winning products. Companies involved in the sale of products depend completely on their ability to continuously introduce winners to the marketplace in a timely manner. To win market share, the development team must bring a quality product to market that meets the needs of the customer. The development costs must be minimized while maintaining a development schedule that permits an early (preferably first-to-market) introduction of the product.

2.3 LEARNING ABOUT PRODUCT DEVELOPMENT

The best way to learn about product design is to work on a development team and build a prototype of some product. For this reason, your instructor is encouraged to form student development teams that will engage in the design, construction and testing of a prototype during the semester. A prototype is the first working model of a product. In some instances, companies develop several prototypes before finalizing a particular product design. However, time available during this semester will limit each team to a single prototype of the product selected by the class instructor.

The development of a prototype of a new product is an intimidating task, particularly when the assignment is given so early in the engineering curriculum. However, experience with several thousand students beginning their studies of engineering indicates that students gain significantly from a multidisciplinary product development experience early in their education. In this class you will learn to work effectively within a team structure, to make difficult design decisions, and to design and build successful prototypes. Extensive surveys from students completing this course clearly indicate that they spent many hours each week on the project, had fun in the process, learned a lot about engineering and appreciated the lessons learned in working as a member of a student development team.

2.4 HOVERCRAFT

The product described in this textbook is a hovercraft, which is a vehicle that moves on a cushion of air. As a hovercraft is without wheels, it is propelled by thrust usually generated by auxiliary fans. A photograph of a large commercial hovercraft is presented in Fig. 2.4.

Fig. 2.4 A large commercial hovercraft used to transport people at high speed over water.

Let's examine this hovercraft to identify some of its design features. First note the black rubber tube (skirt) that encircles the hovercraft, shown in Fig. 2.4. This tube serves two purposes—it provides buoyancy to support the hovercraft on water and it provides the walls of a plenum chamber that contains a cushion of air which supports the hovercraft while it is in motion. The two large axial fans located aft provide propulsion and steerage. If both fans are operating at the same speed, the thrust developed propels the hovercraft forward. Operating only one fan causes the hovercraft to turn either to port or starboard. The deck and cabin form the top of the plenum chamber. The attachment of the skirt to the deck and cabin must have an airtight seal because the air cushion in the plenum chamber is pressurized. The centrifugal blower that supplies pressurized air to the plenum chamber is hidden in Fig. 2.4.

The design of a hovercraft that operates on water differs from one that operates only on land. For operation on water, the skirt is made from a hollow tube (usually fabricated from reinforced rubber). The volume of this tube is sized to provide a buoyancy force sufficiently large to keep the hovercraft from sinking when the air pressure in the plenum chamber goes to zero, as shown by the top illustration in Fig. 2.5. However, when the hovercraft is moving under normal operations, the plenum pressure is increased to the point where the skirt is lifted until the hovercraft is skimming over the surface of the water, as shown in the bottom illustration of Fig. 2.5. The advantage of operating with plenum pressure is that the drag force is significantly reduced, thus allowing higher speeds to be reached.

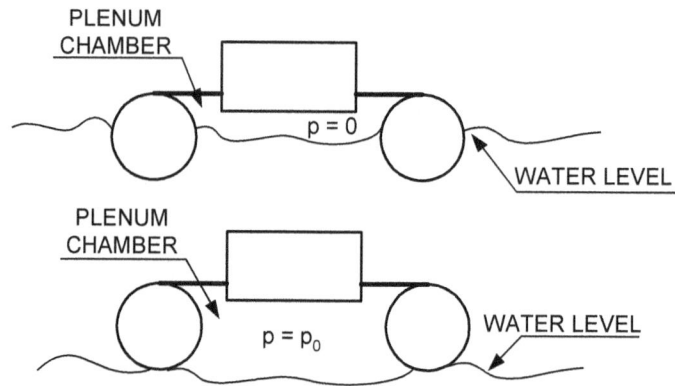

Fig. 2.5 Effect of plenum pressure on a hovercraft operating in water.

When the plenum chamber is pressurized, the rubber tube does not provide buoyancy. The lift required to keep the hovercraft from sinking is provided by the upward force due to the plenum chamber pressure. Some of the pressurized air may escape around the bottom of the skirt; however, a centrifugal blower supplies a sufficient quantity of air to maintain adequate plenum chamber pressure.

Hovercraft Operation on Land

The primary difference between hovercraft operating on land and water is in the design of the skirt. On water relatively large tubes are required to provide buoyancy when the plenum pressure goes to zero. For a hovercraft operating on a solid surface, this design requirement vanishes. A skirt is required as part of the plenum chamber, but it does not necessarily have to be tubular. As an example of a skirt suitable for land based hovercraft, examine the small radio controlled model shown in Fig. 2.6, which was designed and fabricated by students at California Institute of Technology. The plenum chamber and skirt for this model was fabricated from a Frisbee. Note the small axial fans used for lift, propulsion and steering. Also observe the screens used to prevent accidental injury to the fingers of the operators.

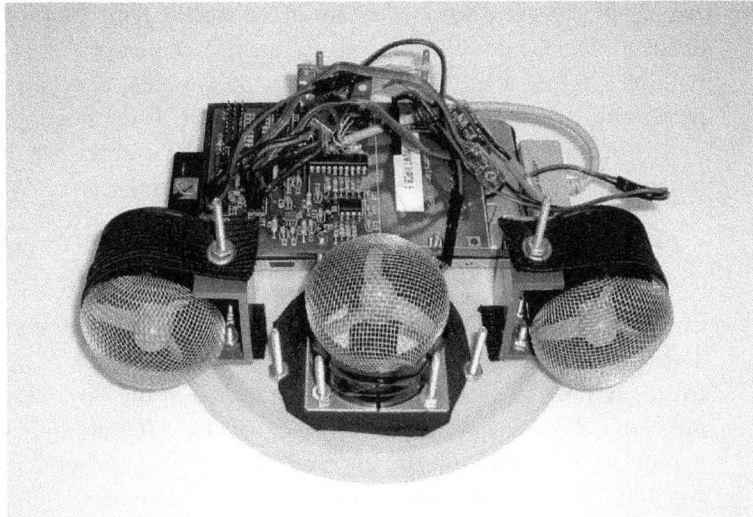

Fig. 2.6 A small model
hovercraft that incorporates a
Frisbee as a plenum chamber.

2.5 HOVERCRAFT MISSIONS

The design of a hovercraft will depend on its mission. Clearly, the design will depend upon whether it is to operate on land, on water, or on both surfaces. However, the design will also depend upon its intended application. Is it a military hovercraft with missions on both land and water? Is it a commercial model used to carry passengers rapidly over water for relatively short distances? Is it a toy to be used by middle school children? Is it a small hovercraft intended for high speed racing by sports enthusiasts? The designs for each of these applications will differ significantly.

We recognize that in a first year engineering design class, time, resources and facilities do not exist for building a full scale hovercraft for any of these applications. However, several missions are described that involve the design and construction of relatively small models. All of these missions require autonomous operation of the hovercraft. This requirement implies that you press a start button to begin the mission, but from that instant onward the operation is totally automatic.

2.5.1 Land Race Mission

The land race mission is basically a time trial where a team's hovercraft traverses the track defined in Fig. 2.7. The track is laid out on a smooth surface such as a parking lot. The boundaries of the track are formed using a light colored tape. The exact geometry of the track is dependent upon local conditions and your instructor may change the dimensions and the configuration shown in Fig. 2.7.

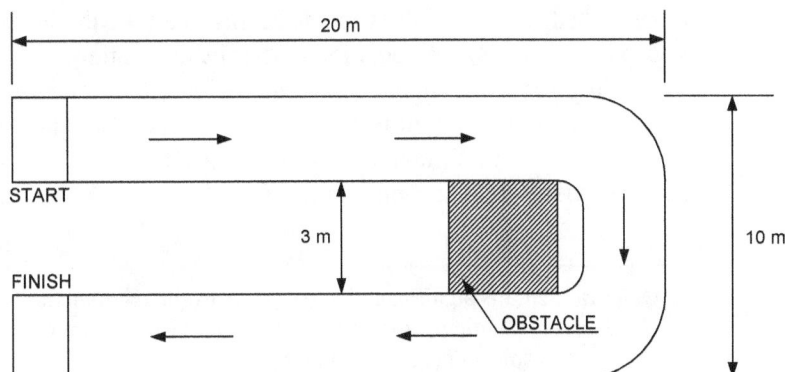

Fig. 2.7 Suggested track for
the hovercraft land race.

An obstacle is shown near the turn in the track. This is a rigid obstacle and if the hovercraft strikes it at speed damage could occur. Consequently, a penalty of 15 seconds will be imposed on the team if its hovercraft strikes the obstacle. At the conclusion of the race, no penalty will be imposed if the hovercraft overshoots the finish line; however, if the hovercraft stops with 0.2 m of the line a bonus of 15 seconds will be given to the team. A 10 second penalty will be imposed each time the entire hovercraft moves outside the boundaries of the track.

2.5.2 Water Race Mission

The water race[1] is also a time trial, but it is conducted in a swimming pool or a water tight box with a shallow covering of water. In this instance, the track is not defined. Instead the hovercraft starts from a defined position circles around a float at the far end of the pool and then returns to the finish position. The navigation and steering of the hovercraft is autonomous, although navigational aids deployed along the route are allowed. A suggested arrangement of the track is illustrated in Fig. 2.8. No dimensions have been provided in this illustration because they depend on the size of the pool available for the time trials; however, an Olympic size pool is recommended. An Olympic swimming pool is 25 m wide with a minimum depth of 2.0 m. It is 50 m in length between touch panels. The 25 m width allows for 8 lanes, each 2.5m wide, with 2 spaces each 2.5m wide outside lanes 1 and 8.

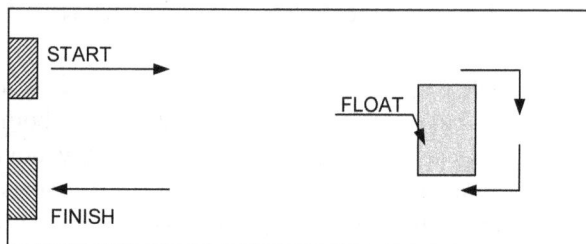

Fig. 2.8 Suggested track for the water race mission.

A float is placed near the far end of the pool to serve as a turning point. This is a relatively rigid obstacle and if the hovercraft strikes it at speed damage might occur. Consequently, a penalty of 10 seconds will be imposed on the team if its hovercraft impacts the float. Also a 10 second penalty will be imposed each time the hovercraft strikes the far wall of the pool behind the float. At the conclusion of the race no penalty will be imposed if the hovercraft misses the finish mark; however, if the hovercraft stops with 0.4 m of the mark a bonus of 5 seconds will be given to the team.

2.5.3 The Navigation Mission

The navigation mission entails steering the hovercraft around a number of square obstacles that are arranged in a 3 by 3 array, as shown in Fig. 2.9. The obstacles are large (3 to 4 m square) and the passages between them are relatively wide (about 2 m). The surface is level and relatively smooth. The prescribed course involves ten 90° turns that take the hovercraft around three of the obstacles as it proceeds from the stating position to the finish location.

This mission can be conducted with either one or two vehicles. If one hovercraft is employed, the time required to traverse the course from start to finish is the basic criterion to judge the design team's performance. If two vehicles are employed, it is suggested that one runs the standard course while the other runs the same course but starting from the opposite side. The

---

[1] The water race entails significant danger to electronic equipment mounted on the hovercraft. If the hovercraft sinks, the electronic equipment may be destroyed. Also it is important that the electrical equipment be shielded from water due to splashing or wave action.

hovercraft with the better performance will complete the entire track and return to where it started the fastest.

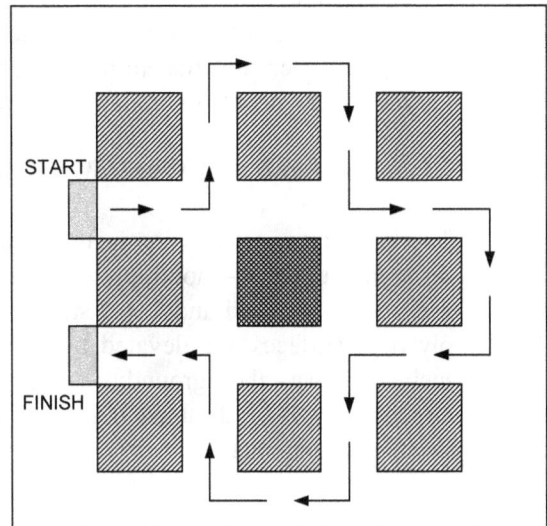

Fig. 2.9 Suggested track for the navigation mission.

The center obstacle, shown with double hatching, must be avoided. If a hovercraft strikes this obstacle a penalty of 15 seconds will be imposed on the team. At the conclusion of the race, no penalty will be imposed if the hovercraft overshoots the finish line; however, if the hovercraft stops within 0.2 m of the finish line a bonus of 12 seconds will be given to the team. A 10 second penalty will be imposed each time the entire hovercraft moves outside the boundaries encircling the track.

2.5.4 The Cargo Mission

The cargo mission involves a hovercraft designed to carry freight from a starting position to an unloading dock located about 50 to 60 m from the starting line as shown in Fig. 2.10. The track for the cargo mission is laid out on a parking lot with a smooth surface. The boundaries of the track are formed using a light colored tape. The exact geometry of the track is dependent upon local conditions and your instructor may change the dimensions and the configuration shown in Fig. 2.10.

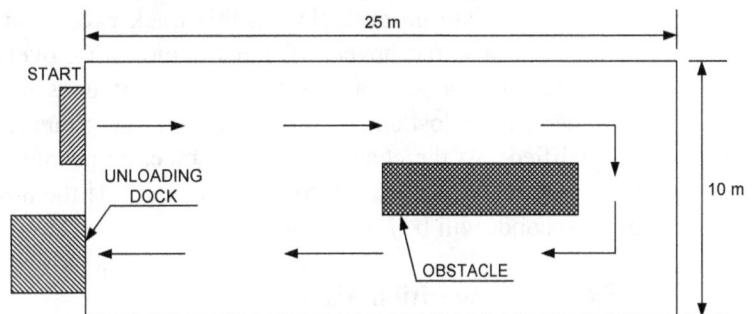

Fig. 2.10 Suggested track for the cargo mission.

The unloading dock is fixed in place and is 3 m wide, 1 m deep and 100 mm high. Your mission is to deliver cargo to this unloading dock. You may place the cargo on the deck of the hovercraft at the starting position, but it must be unloaded automatically at the loading dock. The size and weight of the cargo is decided by the instructor. The time taken to complete the delivery is the main judging parameter, with the handling of the cargo as a secondary parameter.

An obstacle is shown near the turn in the track. This is a rigid obstacle and if the hovercraft strikes it at speed damage to the cargo could occur. Consequently, a damage penalty of 15 seconds will be imposed each time the hovercraft strikes the obstacle. At the conclusion of the race, a 10 second penalty will be imposed if the cargo does not remain on the unloading dock and a 30 second penalty will be imposed if the cargo is damaged during the mission. Also a time penalty of 15 seconds will be imposed each time the entire hovercraft moves outside the boundaries of the track. Note that your instructor may decide to modify the criteria for judging the performance of the cargo hovercraft.

2.5.5 The Land and Water Mission

The land and water mission is also a time trial, but in this mission the hovercraft must traverse two different surfaces—smooth plywood and water. The track, shown in Fig 2.11 is fabricated using 4 × 8 ft sheets plywood and 2 × 4 studs. The plywood surfaces are elevated by about 4 inches above the ground; hence, the hovercraft will be disabled if it runs over the edge of the track.

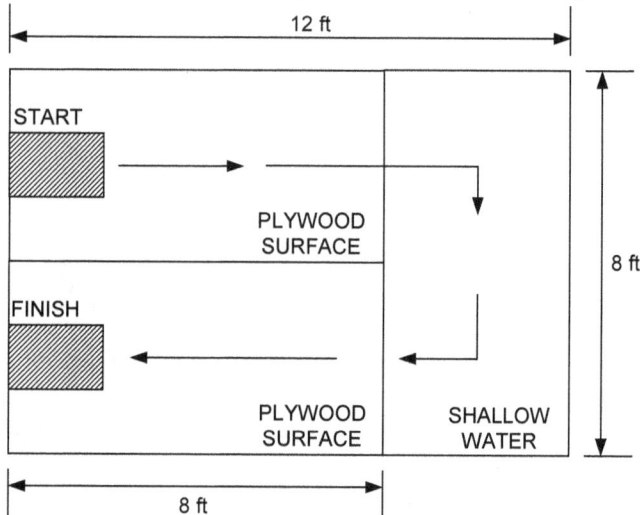

Fig. 2.11 Suggested test arrangement for the land and water mission.

The water portion of the track is formed with a 4 × 8 sheet of plywood with 2 × 4 studs as sides. The joints are sealed with caulking compound to provide a water tight tray. When filled the surface of the water in this tray is flush with the plywood surface of the solid part of the track. The water in the tray is about 3.5 inches deep.

There are no obstacles on this track except for the boundaries which are elevated about 4 inches. Clearly, the hovercraft must avoid going over the edge. We do not believe that this will occur because as soon as the lip of the skirt goes over the edge, the air pressure in the plenum chamber will be lost and the hovercraft will be grounded. If a hovercraft does go over the edge it is disqualified. At the conclusion of the race, no penalty will be imposed if the hovercraft overshoots the finish line if it does not go over the edge. If the hovercraft stops within the finish area, a bonus of 15 seconds will be given to the team.

2.5.6 The Demolition Mission

As the name implies, the demolition mission is about survival of your team's hovercraft. These vehicles (three are suggested) are placed in a walled area about 3 × 4 m in size, as indicated in Fig. 2.12. The hovercrafts are to engage one another until a failure occurs. A hovercraft is declared a failure when it cannot move. During the demolition event, each hovercraft is required to be in constant motion. They can move in a circle, back and forth or side to side, but their motion cannot stop. Loss of plenum pressure even for a short period of time is also a cause of failure.

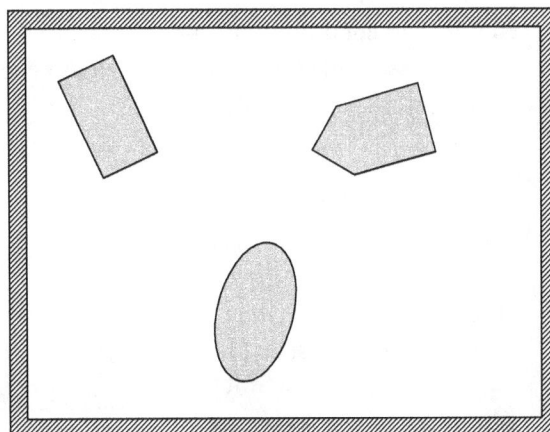

Fig. 2.12 Walled enclosure for the demolition
mission.

The area of engagement is constructed from 2 × 4 studs nailed together to form a suitable wall around a rectangular area like that shown in Fig. 2.12. Placing the enclosure on a parking lot provides a relatively smooth and level surface. There is no penalty for striking the wall as the damage that may be imparted to the hovercraft is a sufficient detriment. It is anticipated that the design of the fighting hovercraft will involve both offensive and defensive devices, as well as programming for engagement or retreat. The winner of the demolition mission is the sole surviving hovercraft.

2.5.7 Summary

The six missions described above are only brief sketches of the potential design challenges involved in developing a prototype hovercraft. It is expected that the instructor will provide additional material describing the details of the track and the restrictions (such as weight or size) placed on the development of the hovercraft for a specific mission. The instructor is responsible for designing the track and arranging for its construction. The instructor can make it larger or smaller, add obstacles, add or remove turns, make the surface rough, etc. Also, the parameters used to judge the adequacy of the design of the hovercraft will depend upon the instructor. The instructor is free to change the parameters, consider others and to modify the penalties and bonuses. Regardless, final product specifications should be provided very early in the design stage to ensure sufficient time to complete the selected mission.

2.6 HOVERCRAFT DESIGN CONCEPTS

Lift

The operation of a hovercraft over land or water depends on a cushion of air generated by a fan to lift and support the vehicle. As indicated in Fig. 2.13, this cushion of air is supplied by a fan that produces a high volume of air at a relatively low pressure. The pressure acts on the lower deck of the hovercraft and provides an upward lift force that is sufficiently large to overcome the weight of the structure, fans and other equipment. Under ideal conditions, the hovercraft lifts up and a gap forms between the skirt and the surface, as illustrated in Fig. 2.13. Air flow through this gap is considered leakage. At an equilibrium state, the airflow into the plenum chamber equals the air flow due to leakage out of the plenum chamber and the lift force equals the hovercraft weight. This is the equilibrium condition for both forces and flow rate. When a gap forms around the entire perimeter of the hovercraft, the structure is floating on air and the hovercraft moves without the frictional constraints that occur with normal sliding.

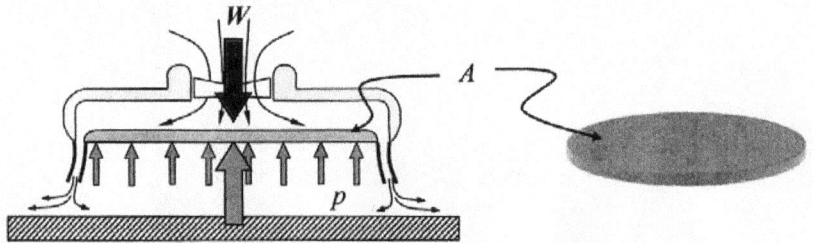

Fig. 2.13 Representation of an equilibrium state for operation of a hovercraft on a solid surface.

The height of the gap around the periphery of the plenum chamber is an important parameter. Under ideal conditions, the gap height is a constant about the periphery of the plenum chamber. However, if the hovercraft is not in perfect balance, it will tip and the skirt may contact the surface at some point on its periphery. When contact with surface is made, a friction force develops and the hovercraft will begin to drag and rotate about this point. To pilot the hovercraft in a straight line, it is essential to either avoid contacting the skirt with the surface of the course over which the hovercraft is moving or to control the propulsion forces and correct for the adverse affects of the friction force on the hovercraft's intended path. Increasing the gap height reduces the probability of the skirt contacting the surface and the loss of control of the hovercraft.

Fan Characteristics

The fan that delivers the air upon which the hovercraft floats is an essential component of the vehicle. Thus, it is important to understand the relationship between the flow rate and the pressure of the air delivered by the fan. A typical curve that depicts this relationship is presented in Fig. 2.14. Examination of this curve shows that the pressure is a maximum when the flow rate is zero (the outlet of the fan is blocked), while the flow rate is a maximum when the pressure is zero and the fan is exhausting unobstructedly into the atmosphere.

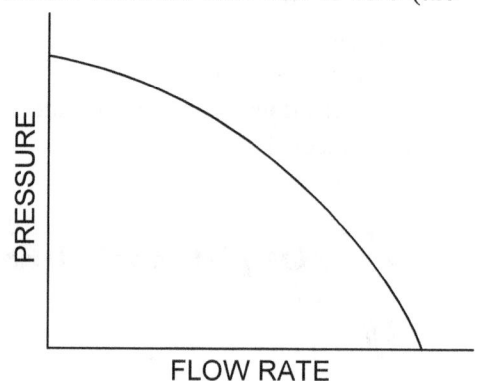

Fig. 2.14 The general shape of the fan characteristic curve showing flow rate as a function of output pressure. The exact shape of this curve is either determined experimentally or found on the web as manufacturer's data.

It is clear from Fig. 2.14 that there are many different pressure/flow rate combinations at which the fan can operate. The operating point is established by examining the rate of air flow (leakage) through the gap formed between the skirt and the surface. As the pressure in the plenum chamber increases, the flow rate through the gap increases, as illustrated in Fig. 2.15. The operating point is established at the intersection of the fan characteristic curve and the plenum chamber's leakage curve. At the operating point, the plenum pressure is p_0 and the flow rate is Q_0.

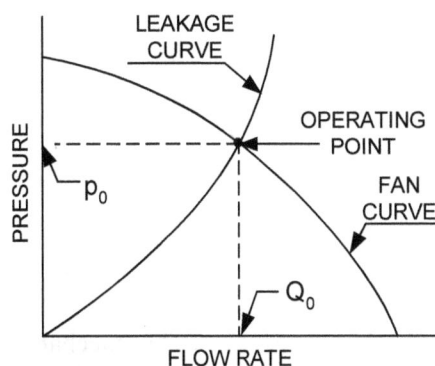

Fig. 2.15 The intersection of the fan and leakage curves gives the operating point for the fan. The leakage curve is dependent on the flow channel and the size of the gap.

Gap Size

When the plenum pressure p is sufficient to develop a lift force equal to the total weight of the hovercraft, the hovercraft lifts off of the ground and a gap with a height h_{gap} is formed between the skirt and the ground. The height h_{gap} of the gap is dependent on the total weight of the hovercraft and the fan's characteristic curve. The leakage curves, shown in Fig. 2.16, illustrate the effect of gap height. When the gap is small, the flow rate through the reduced area between the skirt and the ground is relatively low and the fan is able to maintain a higher plenum pressure. However, when the gap is large, the area between the skirt and the ground increases, the flow rate increases and the plenum pressure maintained by the fan decreases. The controlling factor in establishing the gap height is the weight of the hovercraft. The plenum pressure developed by the fan will produce a lift force equal to the hovercraft's weight if the fan is capable of developing this pressure.

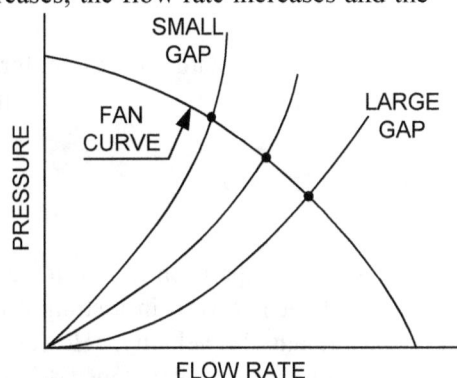

Fig. 2.16 The influence of gap size on the operating point of the fan.

Steering and Braking

Under ideal conditions the hovercraft floats on a cushion of air. It does not have wheels or brakes. How do we steer it? Let's consider a boat as we explore this question of steering. A boat floats on water and does not have wheels or brakes. A boat is steered with a rudder or with its propulsion unit. Let's consider each of these methods. The action of a boat's rudder is illustrated in Fig. 2.17. The rudder is located beneath the boat and is usually oriented along its axis. When turning the boat to the right (starboard) the rudder is rotated to the position shown in Fig. 2.17. The flow of water is interrupted by the rudder and a component of force F develops that acts on the rudder at a position some distance d from the axis of the boat. This force creates a moment M = (F)(d) that turns the boat to the right (starboard).

The rudder on an airplane acts in a similar manner except that the fluid is air and the velocity of the airplane is significantly higher than that of a boat. Will a rudder be effective on a model of a hovercraft? Note that the rudder would be located above the deck of the hovercraft and would interfere with the flow of air not water. Also be aware that the velocities of the hovercraft your team will build for these missions will be significantly lower than that of an airplane.

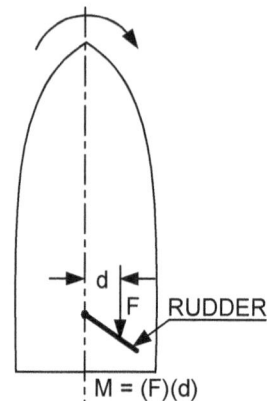

Fig. 2.17 The mechanics of steering a boat with a rudder.

Boats are also steered with thrust directional control or thrust vectoring. The propeller on an outboard motor is rotated in the water to provide a thrust force F that makes an angle with the axis of the boat. This force creates a moment to turn the boat to the right (starboard), as illustrated in Fig. 2.18.

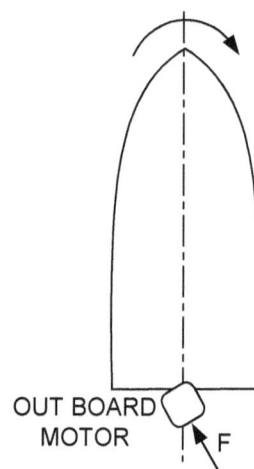

Fig. 2.18 Rotating the thrust force relative to the axis of the boat causes it to turn.

A second approach to steering a boat is possible if it is equipped with twin propellers. In this case, the boat is driven in a straight line when both propellers are turning at the same angular velocity. However, if one propeller is slower than the other, the thrust forces developed differ as shown in Fig. 2.19. If $F_2 > F_1$, as shown, the boat will turn to starboard.

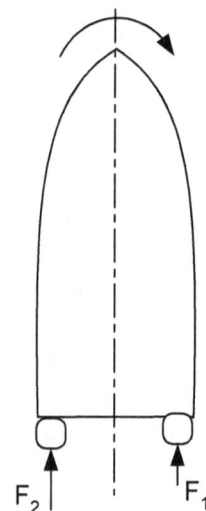

Fig. 2.19 Using different angular velocities on twin propellers to turn a boat.

One problem with changing the velocity of the propellers in Fig. 2.19 is the time required to change their speed. If a sudden change in force is required from a propulsion fan it might take a second or two for the fan to respond and control of the hovercraft could be lost due to this delay. A better approach is to rotate the propeller, as shown in Fig. 2.18, because a few degrees of rotation can be achieved in much less time than changing the propeller RPM by a significant amount.

It is also possible to steer by causing the skirt on one side of the hovercraft to touch the surface momentarily. When contact with the surface is made, a drag force is created that will cause the hovercraft to turn. This procedure is difficult to control as balance of the hovercraft must be reestablished as soon as the turn is complete. Otherwise, the hovercraft will continue to rotate until the gap between the skirt and the surface is reestablished. The choice of a method of steering is yours. The commercial designs introduced earlier in this chapter employ twin fans at the stern of the hovercraft to steer the boat using the same principle indicated in Fig. 2.19. However, these are large and heavy vehicles that respond much more slowly to changes encountered as they move over either water or a solid track.

If the hovercraft is operating on a solid surface, braking is easy. Turn off the plenum fan and the plenum pressure quickly decreases. The hovercraft will drop to the surface and skid to a stop. On water it is much more difficult to brake. It is possible to turn the fan off, but hovercraft designed for water applications have skirts that provide buoyancy for the vehicle. When the plenum pressure decays, the hovercraft settles into the water and the drag forces increase; however, the vehicle will continue to move. If you are using fans for propulsion, your can turn the hovercraft through 180° and use the fans to counteract the vehicles momentum.

Balance

Balance of the hovercraft is essential particularly if it is to operate on a solid surface. If the hovercraft is not in balance, it will tilt to one side, causing the skirt to drag on the surface and steering to become very difficult if not impossible[2]. To insure balance, the center of pressure and the center of gravity must coincide. The center of pressure is the centroid of the area over which the pressure acts. For regular shapes, the center of pressure is obvious as indicated by the black dots presented in Fig. 2.20. For areas with two axes of symmetry, the center of pressure occurs at the intersection of these two axes. For an area with only one axis of symmetry, such as the triangle shown in Fig. 2.20, it is necessary to compute[3] the location of the center of pressure.

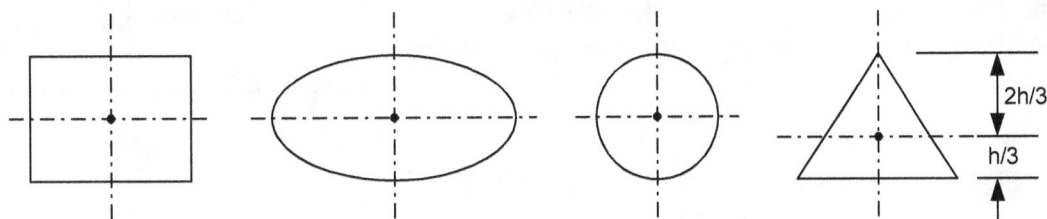

Fig. 2.20 The dot locates the center of pressure for common geometrical shapes.

The center of gravity pertains to the distribution of weights over the deck area of the hovercraft. Suppose the hovercraft is designed with the layout of components illustrated in Fig. 2.21. The components include two axial fans, a centrifugal fan, the NXT microcontroller, sensors, relays and

---

[2] Loss of balance is the most perplexing problem the design teams encounter in testing their hovercraft models. The skirt touches the surface and the vehicle begins to spin. Changing the speed of the propulsion fans is usually too slow to stop the spin and bring the hovercraft back on track.

[3] Locating centroids and centers of pressure will be covered in a course on Statics (ENES 102) at the University of Maryland.

an auxiliary battery pack. The figure also includes the centerlines for the rectangular deck with the center of pressure located with a black dot. The axial fans used for propulsion and steering are placed at the aft end of the hovercraft, but they are symmetrical with respect to the longitudinal axis. The sensors are mounted in the bow and are also symmetrical with respect to the longitudinal axis. The NXT and the relay cluster are located in the aft portion of the craft on the longitudinal centerline. The outlet nozzle of the centrifugal fan was placed at the intersection of the centerlines. However centrifugal fans usually are not symmetrical and as a consequence its center of gravity may be shifted relative to the longitudinal axis.

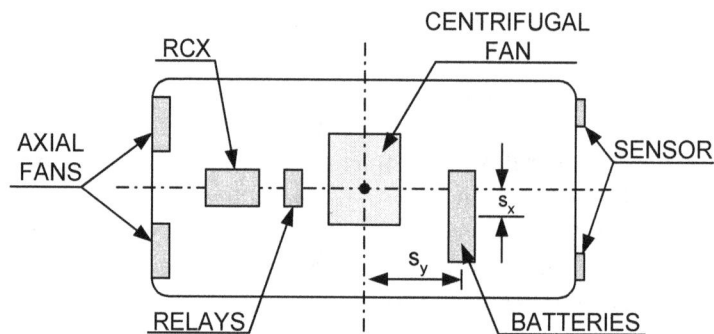

Fig. 2.21 Layout of components on
model hovercraft.

As shown above, the hovercraft is not in balance. The components to the left of the transverse centerline will cause the aft end to tilt downward. Also the weight of the centrifugal fan is not symmetrically distributed relative to the longitudinal centerline. To balance the hovercraft model, the battery pack is positioned with s_x and s_y adjusted so that the moments about the longitudinal and transverse centerlines are in balance.

In practice, balance can be achieved by suspending the model with a piece of string or wire attached at its center of pressure, as shown in Fig. 2.22. All of the components are mounted on the model except the battery pack. The battery pack is then placed on the deck as shown in Fig. 2.22 and shifted in both the x and y directions until the deck is level. The distances s_x and s_y are measured to establish the exact location of the battery pack that provides the necessary balance to prevent tipping when operating the hovercraft.

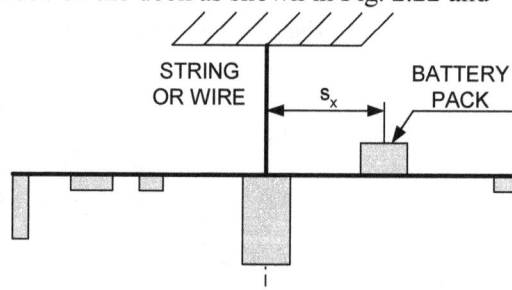

Fig. 2.22 Placement of the battery pack to achieve
balance of the hovercraft.

Unbalance Due to Dynamic Forces

The procedure illustrated in Fig. 2.22 is effective for establishing a static balance of the vertical forces acting on the hovercraft. However, when the hovercraft is in motion dynamic forces in the horizontal direction act to propel and steer the vehicle. These forces produce an unbalanced condition and if they are sufficiently large, the hovercraft will tilt and its skirt will contact the track's surface at one or more points. A friction force F_f develops at each contact point in the direction that opposes the motion. Let's consider contact at one point with the development of a friction force F_f, as illustrated in Fig. 2.23. This force generates a torque that causes the hovercraft to turn left in an uncontrolled manner.

To restore control, a sensor must be employed to detect the deviation from the specified track along which the hovercraft is to follow. The sensor's output is a feedback signal that is used to adjust the thrust developed by the two propulsion fans.

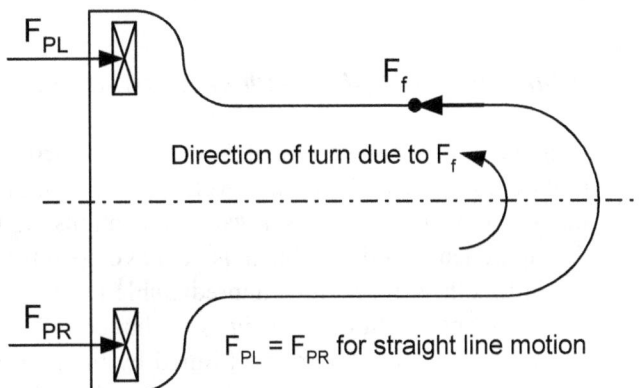

F_{PL}

F_f

Direction of turn due to F_f

Fig. 2.23 Friction force applied at contact point at the skirt surface interface causes the hovercraft to execute an unscheduled turn and lose control.

F_{PR}

$F_{PL} = F_{PR}$ for straight line motion

The adjusted propulsion forces F_{PL} and F_{PR}, shown in Fig. 2.24, create a torque that offsets the torque produced by the friction force F_f. and tends to turn the hovercraft back to the right. This sounds like a good plan to maintain control of your vehicle, but it does not take into account the delay time in changing the angular velocity of the two propulsion fans.

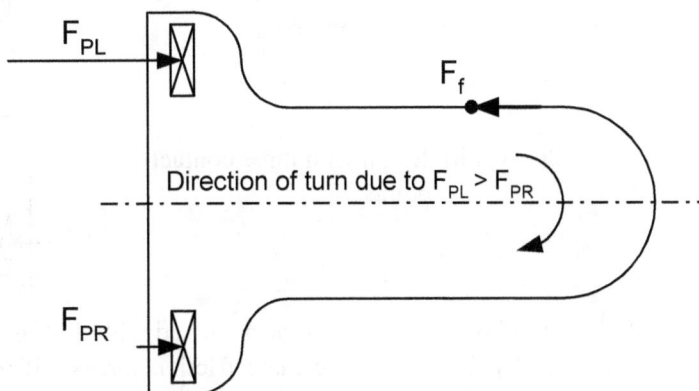

F_{PL}

F_f

Direction of turn due to $F_{PL} > F_{PR}$

Fig. 2.24 Changing the speed of the propulsion fans to control the hovercraft's natural instabilities.

F_{PR}

A better approach is shown in Fig 2.25, where the fans are maintained at constant angular velocity, but are rotated about their mounts to provide the forces necessary to correct for the frictional drag force.

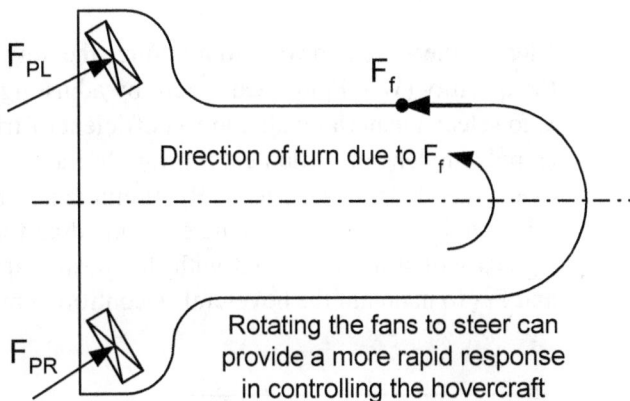

F_{PL}

F_f

Direction of turn due to F_f

Fig. 2.25 Changing the orientation of the propulsion fans to control the hovercraft's natural instabilities.

F_{PR}

Rotating the fans to steer can provide a more rapid response in controlling the hovercraft

Again sensors are necessary to detect when the hovercraft has realigned with the target track. At this instant, the output from the sensors is employed as feedback signals to adjust the propulsion forces so that $F_{PL} = F_{PR}$, with their orientation (rotational axis) along the centerline, which is necessary for straight line tracking when the hovercraft regains balanced condition.

A Strategy for Dealing with an Unbalanced Hovercraft

Experience has shown that it is extremely difficult to maintain perfect balance as the hovercraft moves along a specified track even on a very smooth surface. Moreover, unexpected friction forces that occur at contact points at various locations along the skirt disrupt motion and steering strategies. An approach to this problem is to develop either a control strategy or a hovercraft design that minimizes the effects of the unpredictable friction forces.

Consider first the design of a hovercraft with a skirt that is configured so that three contact points extend below the plane formed by the base of the skirt, as illustrated in Fig. 2.26. If the lift force F_L becomes slightly less than the weight W of the hovercraft, the skirt will contact the surface of the track at these three specified points. Friction forces will develop at these contacts, but the magnitude of these friction forces can be minimized and the turning torques they generate will tend to cancel out.

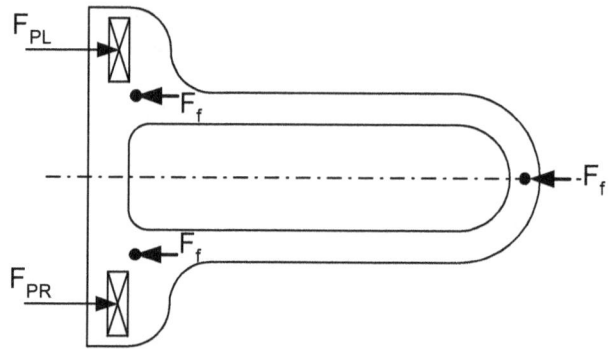

Fig. 2.26 A skirt design with three contact points.

Better stability in steering can be achieved with this design; however, the penalty is the drag force F_D produced by the sum of the three friction forces. If we assume the friction forces are equal, it is possible to write the equation for the drag force F_D as:

$$F_D = 3F_f = \mu(W - F_L) \tag{2.2}$$

where μ is the coefficient of friction between the contact points and the surface of the track.

Clearly, there are two ways to minimize the drag force. The first way is to adjust the lift force so that the quantity $(W - F_L)$ is nearly zero by adjusting the air flow rate and the gap size. The second way is to select a material with a low coefficient of friction μ for the contacts on the skirt. Materials such as polyethylene or Teflon exhibit low friction coefficients in the range of 0.10 to 0.15 when sliding over smooth hard surfaces. By minimizing the quantity $(W - F_L)$ and using contact supports fabricated from low friction material, the drag force can be minimized[4]. Sensors providing feedback signals will still be required with this design approach in order to adjust the propulsion forces F_{PL} and F_{PR} to maintain the hovercraft's controlled motion along the track.

---

[4] Your instructor may object to this approach because it produces a model that is not truly a hovercraft, but a levitated sled.

A second strategy for dealing with the unbalanced condition that causes deviation from straight line tracking involves controlling the speed of the hovercraft. If the speed of the hovercraft is relatively high when the unbalance condition occurs, the vehicle turns abruptly and the deviation from the track is significant. When the deviation from the track is excessive, the sensors used to maintain control are no longer in a position to measure the tracking target and systematic control is impossible. To avoid these excessive deviations, the speed of the hovercraft should be limited so that the turns, which occur when the vehicle becomes unbalanced, are less abrupt. More time is available for corrective actions (sensing, feedback and adjusting the power to the propulsion fans or orienting the fans to compensate for the frictional forces) when the speed is limited. With additional time, adjustments to the propulsion forces can be made bringing the hovercraft back on to the target track before the sensors are rotated out of their tracking positions.

Autonomous Control

Autonomous control means that the entire mission is automatic. You interact with the hovercraft only once—to start the process. Autonomous control may be achieved in a number of different ways, but they all involve a microprocessor (or microcontroller). In this textbook, the LEGO NXT unit has been selected as the controller to be employed in achieving autonomous control. A photograph of the NXT is presented in Fig. 2.27 and a thorough description of its features is presented in Chapter 6.

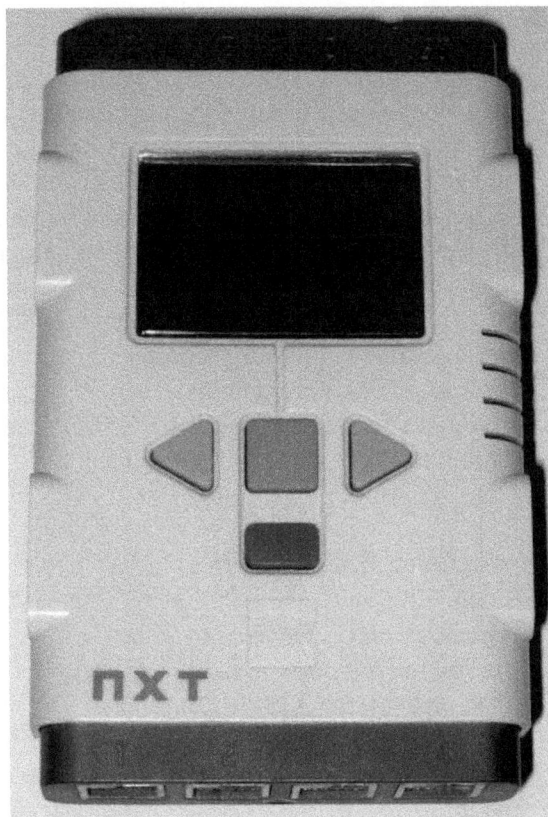

Fig. 2.27 The NXT controller.

The NXT controller uses a 32-bit, 48MHz ARM7 processor with 256KB flash memory and 64KB of RAM. The controller is housed inside a battery-powered brick that is equipped with ports for sensors and motors. The NXT case contains a tray for six AA batteries or a rechargeable lithium battery. It has four input ports to accommodate sensors, three output ports to power motors and lights and a standard USB 2.0 port. A monochrome LCD monitor with a graphical user interface controlled by four buttons lets the user select and run stored programs.

If the NXT controller is procured with a Mindstorms kit, three servo-motors are provided to power your inventions. These motors can also be used to sense rotation to within one degree. The kit also includes four sensors: ultrasonic for range-finding and motion detection; sound for measuring noise level; touch, which responds to contact by activating a momentary switch; and light, which measure the intensity of light reflected from a surface. The motors and sensors are connected to the NXT controller's ports by wires with RJ12 connectors that resemble (but don't match) Ethernet plugs.

Programming the NXT may be accomplished using several different codes. The most popular methods for programming are based on LabVIEW software from National Instruments. Recently National Instruments has released a toolkit to allow standard LabVIEW users to create and download NXT programs. LabVIEW software employs an icon oriented programming environment where icons are dragged onto a working diagram to create a sequence of commands. These commands include loops, forks and jump and lands. Programming is also performed with a structured language such as C or modified versions of C. This textbook describes the details of programming in ROBOT C in Chapter 6. Firmware to control the sensors and the motors are downloaded from a computer to the NXT controller by USB cable or wirelessly through Blue tooth communications.

Sensors

Autonomous operation of the hovercraft requires either timing each maneuver necessary to complete the mission or controlling actuators based on feedback signals from sensors. Timing a maneuver is the easiest method to program because the NXT controller contains a clock that is accurate to 0.1 seconds. With time of operation used to control forward motion or turns, sensors are not required. However, the different times, which are programmed to control the mission, must be known precisely for each operation and each operation must be reproducible.

Sensors that indicate angle of rotation, contact, temperature, light intensity, proximity, etc. are available to provide feedback signals. These feedback signals are interpreted by the NXT and, depending on its programming, actuators are employed to adjust the course of the hovercraft so as to complete the assigned mission. Sensor feedback control provides a reliable means of control when each operation is not easily reproducible.

\

Auxiliary Power Supplies

The NXT unit is powered by either six AA alkaline or NiMH batteries that provide adequate power for the microprocessor and for operation of small motors and actuators found in the MindStorm kit marketed by the LEGO Educational Division. The six alkaline batteries provide a 9 V supply voltage and the six NiMH batteries provide a 7.2 V supply voltage. However, the missions described previously require fans that will draw more power than the NXT can safely supply. The output ports on the NXT can deliver about 500 mA at about 9 volts into a low resistance circuit before entering a thermal shutdown mode to prevent damage to internal components in the NXT. Because the current draw by the levitation fan is several amps, it is clear that an auxiliary power supply will be required. The auxiliary power supply (i.e., battery) is sized to provide the necessary current at the voltage required by the fans (probably 12 Vdc). The 500 mA output from the NXT controller is more than sufficient to activate a transistor or relay (switch) that connects the battery supply to one or more fans.

2.7 HOVERCRAFT SKIRTS

Skirts are added to the plenum chamber to provide a deeper cushion of air upon which the hovercraft rides, as indicated in Fig. 2.28. The purpose of the skirt is to enable the hovercraft to pass over rough surfaces while maintaining its normal speed. The skirt is often fabricated from a flexible material so that it can deform easily (low-force contact) when an obstacle is encountered. The skirt is often semi-circular in shape so that retains its form as the plenum pressure is increased. The semi-circular shape is conducive to inward deformation when obstacles are encountered yet resistive to the deformation due to the plenum pressure.

If the surface over which the hovercraft travels is very smooth, skirts fabricated from rigid materials have the advantage of more precise geometry and simplicity of construction. The edge of the skirt can form a near perfect plane that enables the gap size to be uniform about its periphery. The disadvantage is the large contact forces generated when he skirt strikes an obstacle causing loss of control.

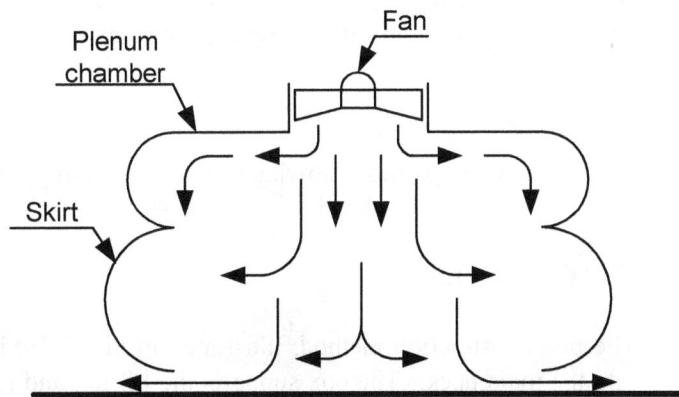

Fig. 2.28 Adding a skirt to the plenum chamber to enable the hovercraft to move over obstacles.

2.8 HOVERCRAFT STRUCTURES

The hovercraft structure must be light and strong. Weight is important because the plenum pressure—which is limited by the fan your team selects—must be sufficiently high to lift the weight of the structure, the hovercraft's components, and any cargo it is to carry. Aircraft modeling methods and materials are often employed to fabricate very light structures. Materials used for airplane models include balsa wood, foam, and fabrics that are light and strong. However, these airplane-like structures are not robust and may not withstand impacts that are inevitable during the hovercraft's mission.

More suitable materials include thin gage plywood, foam board, polypropylene and styrene sheets. Styrene sheets can be cold formed or vacuum formed when heated. Polypropylene sheets can easily be cut with a hobby knife and cold formed. Pieces of both of these plastics can be joined with suitable adhesives. Epoxy sheets reinforced with either glass or carbon fibers are also available. The reinforced epoxies exhibit a very high strength to weight ratio, but they are expensive when compared to other plastic sheet materials. Closed cell foams are often employed because this material is light, reasonably strong and easy to cut.

Rubber tubes and waterproof fabric are often used to fabricate soft skirts because they are flexible and will deflect when the hovercraft encounters an obstacle. Attachments are made with staples, cord lacing and/or adhesives.

2-22 — Chapter 2
Product Development and Hovercraft Missions

2.8.1 Types of Hovercraft Structures

Raft Construction

There are several methods used to construct the hovercraft structure. The raft construction method, illustrated in Fig. 2.29, incorporates a flat raft-like piece of foam or foam board under the main deck. The raft is attached rigidly to the main deck with stringers located at intervals about its periphery. It is also used to support the edges of the flexible skirt. The attachment to the flexible skirt can be made with lacings or with a continuous piece of fabric or thin plastic sheet. If fabric or a thin plastic sheet is used, holes must be provided in the sheet/fabric to permit the air flow from the lift fan to enter the plenum chamber under the raft. The raft construction method is often used when the hovercraft is expected to traverse a body of water. The volume of the foam used in constructing the raft is sufficient to provide the buoyancy needed to prevent the hovercraft from sinking in the event of a loss of power to the lift fan. The raft also contributes to the stiffness of the main deck if the stringers are placed on sufficiently close centers.

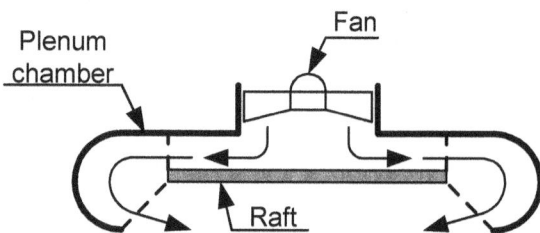

Fig. 2.29 Cross section showing raft construction details.

Box Construction

The box construction method, illustrated in Fig. 2.30, incorporates a box-like structure that is inset into the main deck. The box supports the lift fan and extends both forward and aft to provide space for transporting cargo. It is tied into the main deck about its periphery and contributes to the stiffness of the deck. Holes are provided in the sides of the box to enable the air flow from the lift fan to inflate the bag skirt. It is also used to support the edges of the flexible (bag) skirt. The attachment to the skirt can be made with lacings or with a continuous piece of fabric or thin plastic sheet, as described above.

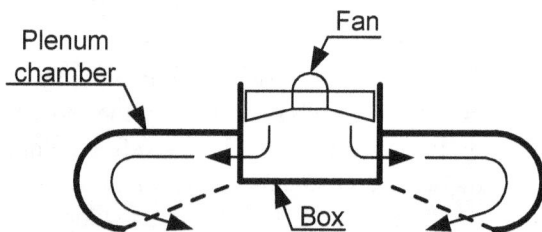

Fig. 2.30 Cross section showing box construction details.

Shell Construction

The box construction method, illustrated in Fig. 2.31, incorporates a rigid shell that serves as the main deck and the closure for the plenum chamber. A flexible skirt can be added if the course for the hovercraft mission is rough or if it includes obstacles. However, if the surface for the course is smooth, a skirt is not required. From a construction viewpoint, the shell design is the simplest because it does not involve building an internal structure or a flexible skirt. However, fabricating the structure involves molding the shell, which requires considerable skill. An alternative to molding is to fabricate the rigid skirt from a plastic pipe or foam pipe insulation cut longitudinally to form a C-section. The C-section can be formed into a circular ring and attached to the deck with adhesive to produce the shell like structure.

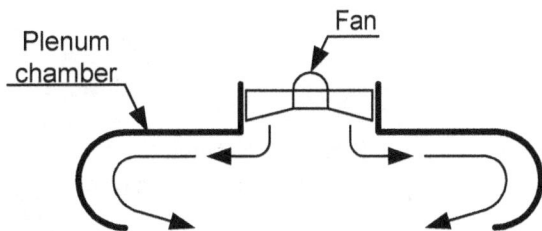

Fig. 2.31 Cross section showing shell
construction details.

Double Deck Construction

The double deck construction method, illustrated in Fig. 2.32, is similar to the raft construction method because it incorporates a flat deck-like board under the main deck. The two decks are spaced apart by a predetermined dimension and fastened together with screws. One edge of a bag-like skirt is attached to the upper deck and the other edge is attached to the lower deck. Some of the air from the lift fan is deflected by the lower deck and inflates the bag skirt. A hole cut in the lower deck permits the remaining air to flow into the plenum chamber to lift the hovercraft. The spacing between the two decks and the size of the hole in the lower deck control the pressure in the skirt and the plenum pressure. The double deck hovercraft is relatively easy to build and the dimensions of the hole or the spacing between the decks are easy to modify if the skirt and plenum pressures require adjustment.

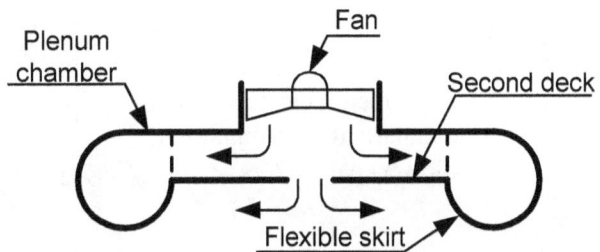

Fig. 2.32 Cross section showing double
deck construction details.

2.9 HOVERCRAFT COMPONENTS

The layout of the components, presented in Fig. 2.21, identifies many of the primary components that will be used in the design and fabrication of your team's prototype hovercraft. These components include fans, the NXT, relays, batteries and sensors. The selection of these components is a significant part of the design process. For this reason, a brief review of each of the primary components is presented in the paragraphs below:

Fans

Centrifugal Fans or Blowers

There are two basic types of fans—axial flow and centrifugal flow. The centrifugal fan, often called a blower, is presented in Fig. 2.33. It has an inlet port, a small motor that drives a rotor, and an output port. The outlet is tangent to the rotor and perpendicular to the axis of the motor. The centrifugal fan generates a higher pressure than the axial fan and is often used for developing the lift force. It has the disadvantage that it is relatively large, heavy and requires significant power to generate high flow rates.

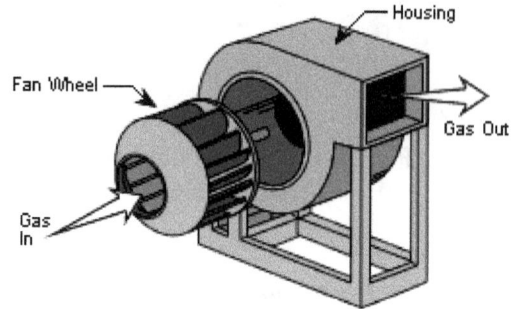

Fig. 2.33 Schematic illustration of a
centrifugal fan (blower).

The characteristic curves showing pressure, efficiency and power requirements as a function of flow rate for a typical centrifugal fan are presented in Fig. 2.34.

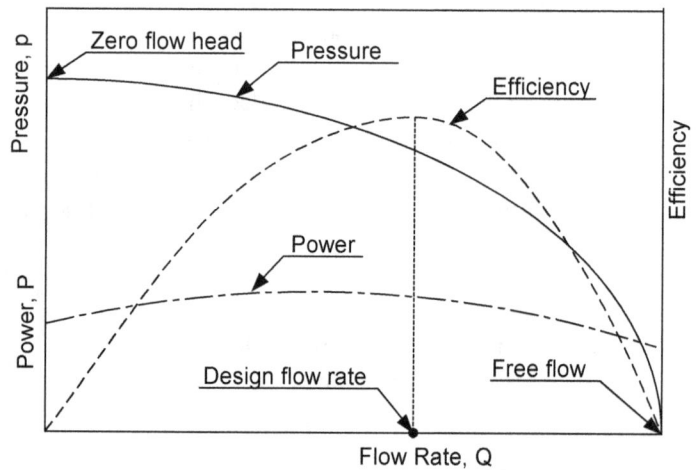

Fig. 2.34 Characteristic curves that describe important centrifugal fan (blower) parameters.

A large number of fan characteristic curves can be found by performing an internet search for fan vendors. The authors are particularly partial to EBM Papst web page at http://ebmpapst.us/. When you select a fan, the manufacturer should be able to provide you with the characteristic curves that define the performance that you can expect. The curves provide the data needed to estimate battery requirements, pressure for levitation, and flow rate.

Axial Fans

An axial flow fan, depicted in Fig. 2.35, does not use centrifugal forces to develop air pressure. Instead it has a motor embedded in its center hub that drives fan blades that propel air down the axis of the fan. These fans are lighter than the centrifugal fans for the same flow rates; however, the pressures developed are significantly lower.

The characteristic curve for a 12 Vdc axial fan 120 × 120 × 38 mm in size is presented in Fig. 2.36. Note at the nominal 12 V, the maximum flow rate is about 105 CFM; however, when the flow rate is zero the static pressure is only 0.275 in. of H_2O. The weight of this fan is only 0.62 pounds and the rated current is 0.58 A. Axial flow fans are used for a variety of applications that require large volumes of low pressure air, such as for cooling of electronic equipment.

Fig. 2.35 A small axial fan with direct motor drive.

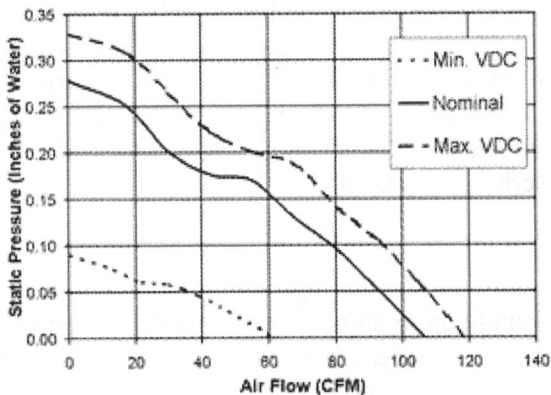

Fig. 2.36 The pressure as a function of flow rate for an axial fan.

When your team selects a fan, be certain to study the characteristic curves as they will markedly affect the performance of the hovercraft. The analysis of hovercraft performance that uses flow rate and pressure from the fan's characteristic curve will be described in Chapter 3.

Relays

Relays are electromechanical switches that employ electromagnets to close one or more sets of contacts. In general, the purpose of a relay is to use a small amount of power to energize an electromagnet which moves an armature that is capable of switching much larger power levels into a circuit. For example, you might want activate the electromagnet using 6 V and 40 mA (240 mW), while the armature can support 12 Vdc at 12 amps (144 W).

As indicated in Fig. 2.37, relays are relatively simple devices that consist of only four parts:

- Electromagnet
- Armature
- Spring
- Set of electrical contacts

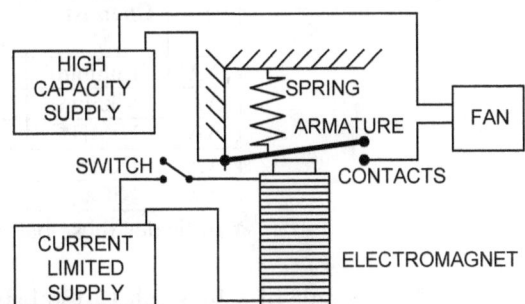

Fig. 2.37 Schematic diagram illustrating the use of a relay.

In this figure, it is evident that a relay consists of two separate and completely independent **circuits**. The first circuit, shown at the bottom of Fig. 2.37, drives the electromagnet. In this circuit, a switch controls current flow to the electromagnet. When the switch is closed, the electromagnet is activated. The armature acts as another switch in the second circuit. When the electromagnet is activated, the armature closes on the contacts completing the second circuit that turns on the fan. When the current to the electromagnet is interrupted, the spring pulls the armature up, separating the contacts and opening the fan circuit.

When procuring[5] relays, it is necessary to check on several design features:

- The voltage and current needed to activate the electromagnet.
- The maximum current capacity of the contacts.
- The number of armatures (generally one or two)
- The number of contacts (poles).
- If the contact or contacts are normally open (NO) or normally closed (NC).

Another switching method employs transistors that can be activated with small currents at low voltages. Transistors are capable of switching large current flows on and off at an incredibly fast pace. They have the advantage of reduced weight and more efficient use of power when switching. Transistors for this and other applications are described in Chapter 4.

Batteries

While the NXT has three output ports that can supply up to 500 mA, this amount of current is not sufficient to drive the fans required to lift and propel your team's hovercraft. An auxiliary power supply consisting of one or more batteries will be required. There are a large number of different types of batteries available and each type comes in many different sizes. Which type is the most suitable for the mission? What size battery will minimize weight while providing energy for the duration of the mission? Let's explore the type of batteries currently available and some of their characteristics.

Table 2.2 lists several different types of batteries. It is possible to compare the voltages, capacities and relative energy output normalized by the weight of a cell, because the data listed in Table 2.2 is for AA size batteries. The lithium battery exhibits the highest performance with a capacity relative to its weight of 300 W-h/kg. These batteries are relatively expensive; they are usually employed with digital camera or watches where long life with low current drain is the primary design requirement.

Table 2.2
Comparison of the characteristics of several types of batteries

Size	Chemistry	Cell Voltage	Capacity A-h	Capacity W-h	W-h/kg	Cost $
AA	Lithium	1.5	2.9	4.35	300	3.70
AA	Alkaline	1.5	2.85	4.28	185	0.60
AA	NiMH*	1.2	2.0	2.4	83	2.48
AA	CZn	0.8	1.1	0.88	62	0.45
AA	NiCad*	1.2	1.0	1.2	52	2.21

* Indicates these batteries are rechargeable.

An example of a 14.8 volt Li-Ion battery pack is presented in Fig. 2.38. This pack consists of four lithium-ion cells with a total capacity of 2200 mAh with a weight of 6.0 oz. The maximum

[5] An excellent source of many electronic components including batteries, electromagnetic relays and transistors is Digi-Key (www.digikey.com).

discharge current is 5A. Following proper procedures the battery pack can be recharged. The pack contains a built-in IC chip to prevent the batteries from over charging and over discharging.

Fig. 2.38 A Li-Ion battery pack.

The Nickel-Metal-Hydride (NiMH) and the Nickel-Cadmium (NiCad) batteries, also relatively high in cost, are both rechargeable. They are used in applications where the current and power demands are relatively high and the battery life is extended by recharging. These batteries can be recharged several hundred times before they fail. Thus their relatively high cost is mitigated.

The lowest cost batteries are the common carbon-zinc and the alkaline types; however, these batteries cannot be recharged. They are commonly used in flashlights, smoke detectors, door openers, toothbrushes, etc. The superiority of the alkaline battery is evident as its capacity approaches that of the lithium battery at a much more modest cost.

Another important type of battery is the sealed lead acid (SLA) that is used as a power source for automobiles. The battery is available in a box like package, with six individual cells connected in series to produce an output of 12 Vdc. The cell size is varied to produce batteries with many different capacities. The smallest SLA commonly available has a capacity of 1.3 A-h, while a much larger SLA battery on your automobile can deliver 25 A for more than two hours. High current drain is possible with this type of battery, with cold cranking amperes quoted in excess of 500 A. It is also rechargeable with a relatively long lifespan (4 to 6 years). With a capacity relative to its weight of 52 W-h/kg, it is comparable to a Ni-Cad battery.

When your team selects a battery, consider voltage, capacity, current drain, weight, cost and the advantages of recharging. Making this selection will provide you with an opportunity to perform a design trade-off study.

Sensors

Sensors are required in most autonomous missions to provide feedback signals that enable course corrections to be made. Sensors represent significant business opportunities in today's market. They are commercially available for measurements of force, torque, distance, strain, temperature, light intensity, air quality, water quality, proximity, etc. Several different types of sensors are described in detail in Chapter 4. Also included in Chapter 4 is a discussion of the methods of interfacing sensors with the NXT. The NXT contains a microprocessor, four sensor input ports, and three power output ports that allow you to create an autonomous hovercraft, which can be programmed to perform specified tasks. The NXT can be programmed with a wide variety of software programs such as LabVIEW, NXC (Not Exactly C) and le JOS NXJ (Java), but in this textbook, we will describe ROBOT C as the programming approach. Two way communications with the NXT is also possible with Bluetooth signals from transmitters and receivers located in the NXT and your computer.

Several sensors have been developed by the LEGO Educational Division to interface with the input ports on the NXT. These sensors include touch (contact switch), light (LED and photodiode), temperature (thermistor), distance (sonar), sound (microphone) and rotation. Several other sensors are available from third party suppliers. Unless time is used to control the action of the propulsion fans, one or more sensors will be required to provide the feedback signals necessary for maneuvering the hovercraft over the track associated with the assigned mission.

Mechanical Components

LEGO, over the last twenty years, has evolved from a hands-on children's toy to a surprisingly complete prototyping tool for relatively complex engineering systems. In addition to structural elements, such as plates, beams, bars and columns, its accessories include motors, sensors, gears, wheels, belts, toggles, controllers and infrared transmitters and receivers. Compared to other prototyping alternatives, LEGO components are relatively inexpensive. The availability of components and the familiarity that most students have with the LEGO concept allows design teams to rapidly construct models without significant laboratory support or other college infrastructure. Also rapid construction permits the student teams to try several different design approaches enabling significant improvements to be implemented in the limited time available during a semester.

You should be aware that we do not consider the LEGO components as toys. They provide a design space sufficient to teach both mechanical and electrical concepts of power, motion, communication and control. Have fun, but learn about gears, power transmission, sensors, motors, microprocessors, IR communication, programming, feedback loops, electrical circuits and control in the process.

2.10 HOVERCRAFT DEVELOPMENT

When designing the hovercraft, we usually divide the development process into four major phases, which include:

1. The design of a system consisting of the many subsystems needed to meet the mission requirements.
2. The preparation of an assembly kit and the construction of the hovercraft prototype.
3. The generation of a control algorithm capable of autonomously controlling the hovercraft's motion.
4. The verification of the hovercraft prototype's performance through testing.

We will facilitate this process by providing you with descriptions of the hovercraft to be designed, the mission that the system is to perform and the time constraints on both the development of the system and its ability to complete the mission. The instructor will provide the test environment; however, it will be necessary for you to schedule testing times in advance because many design teams share the same facilities. Your instructor will schedule the time of the final trials of your hovercraft model. It will be the responsibility of each design team to have their hovercraft operational and ready for testing at the designated time.

We suggest that you make full use of the "blackboard", library and the Internet to search for literature and patent files. There is absolutely nothing to be gained by reinventing the wheel. Copying a design is common practice—it is called reverse engineering. Reverse engineering involves a complete dissection of a competitor's product to gain a detailed understanding of all aspects of each component. Each component is then redesigned to improve its performance, the appearance of the new product and to reduce its cost. If you are able to find a description of a previous hovercraft in the literature and decide to reverse engineer it, be certain to understand the function of each component in the system. In your design, make a serious attempt to improve performance by making design changes to optimize your vehicle's performance for the particular product specifications you have been given.

You can understand the design requirements for the hovercraft by reviewing the product specification provided at the beginning of the course[6]. The product specification, written relatively early in the development process, is usually prepared in tabular format. It tersely states design requirements and design limits or constraints. Remember, we always design with constraints and/or limits. The specification also includes the performance criteria and design targets for weight, size, power, cost, etc.

We will place explicit design constraints that limit your flexibility in designing the various subsystems needed in the assembly of the hovercraft. First, there is a maximum limit[7] of $25.00 per team member for the cost of materials, supplies and components that may be purchased for use on the hovercraft. The idea is to keep the out-of-pocket expenses for team members at an affordable level, while giving the team adequate funds to build high-quality, high-performance systems for each mission. We envision that three sub-teams each with two to four members will work to develop the three subsystems that are integrated to develop a complete hovercraft. With 10 members, the maximum amount that can be expended on a hovercraft model is $250.00. Other design constraints on size, weight and power requirements may be imposed by your instructor.

2.10.1 Design Strategy

After each member of the development team understands the design specification and design limitations, you are ready to generate a design strategy. There are two interrelated aspects to the design strategy. The first entails the mechanical features such as lift fans, propulsion fans, relays, auxiliary power supplies, as well as structural details. The second involves sensors, control logic and the programming to execute that logic. Design strategies are simply ideas for performing each function involved in the operation of a system that performs a specified task within the constraints imposed upon the system. These design strategies enable your team to meet the product specifications. To illustrate design strategies, let's examine the functions (systems and processes) involved in designing a hovercraft.

1. The structural system supports all of the components, such as fans, the NXT, batteries, sensors, etc. It also contains the cushion of air upon which the hovercraft floats.
2. The lift system includes the levitation fan and the controls for turning this fan on and off.
3. The propulsion system includes a means of providing thrust, the control of the magnitude and direction of the thrust vector and the time required to change the direction and magnitude of the force vector.
4. The power system includes auxiliary power supplies and control of these supplies.

---

[6] An example of the product specifications for the hovercraft design in the spring semester of 2010 is presented in the Appendix.
[7] Your instructor may choose to increase or decrease this arbitrary limit on student expenditures depending on the system design requirements.

5. The sensing system includes the sensors required to generate feedback signals enabling control of the hovercraft's speed and direction.

6. The master control system enables autonomous control to successfully complete the mission.

We anticipate that Team A will take responsibility for functions 1 and 2, Team B will be responsible for functions 3 and 4, and Team C will be responsible for functions 5 and 6. It is essential that the three subsystem teams communicate continuously during the design phase, because all the subsystems developed must be successfully integrated to produce an efficient hovercraft that can complete the mission without mishaps.

2.10.2 Preparing the Assembly Kit

When your design is complete, prepare a parts list that indicates the name of each part, size and quantity that will be used in constructing the hovercraft. To demonstrate the technique used in preparing a parts list, consider the example presented in Fig. 2.39.[8]

After the parts list is complete, the parts can be procured from the appropriate suppliers[9] and arranged in an assembly kit. When the procurement activity is concluded and the assembly kit is complete, the team can begin to assemble the hovercraft. Assembly will be rapid because the design is finalized and all the parts have been identified and placed in a convenient container.

PARTS LIST FOR A HOVERCRAFT

NUMBER	NAME OF ITEM	DESCRIPTION OR DRAWING NO.	QUANTITY	PRICE
1	LIFT FAN	SUPPLIER NAME AND MODEL NO	1	$ 38.00
2	PROPULSION FAN	SUPPLIER NAME AND MODEL NO	2	$ 8.75
3	DECK	1/4 X 4 X 4 FOAM BOARD --DWG. NO. 100-01	1	$ 1.15
4	SKIRT	4 in. DIA. DRAINAGE PIPE	8 FT	$ 6.25
5	RELAYS	SUPPLIER NAME AND MODEL NO	3	$ 2.85
6	MAIN BATTERY	SUPPLIER NAME AND MODEL NO	1	$ 12.30
7	SECOND BATTERY	SUPPLIER NAME AND MODEL NO	1	$ 2.50
8	PC BOARD	SUPPLIER NAME AND MODEL NO	1	$ 7.45
9	TERMINALS	RADIO SHACK --- ASSORTED SIZES	1	$ 1.75
10	WIRE	RADIO SHACK --- STRANDED #16	25 FT	$ 1.75
11	WIRE	RADIO SHACK --- STRANDED #22	25 FT	$ 1.10
12	EYE SCREWS	HOME DEPOT ---- SMALL WITH 1/2 in. EYE	12	$ 0.60
13	SENSOR	SUPPLIER NAME AND MODEL NO	1	$ 32.00
14	SEALANT	HOME DEPOT ---- LARGE TUBE	1	$ 3.85
15	RCX BRICK	LEGO EDUCATIONAL DIVISION	1	$125.00

Fig. 2.39 Parts list for a model hovercraft.

---

[8] This parts list is an example and is not complete. Your team's parts list is to contain all of the parts required to build a complete model as well as the source and model number of each part.

[9] Information describing the characteristics of many of the components required to design the hovercraft can be found on the web. These components can be ordered on-line or purchased at local retail outlets such as Home Depot, Lowe's, or Radio Shack.

2.10.3 Programming for System Control

The NXT controls the actions of the hovercraft that your team designs. To exercise this control, the NXT is programmed (given instructions to stop or start output fan motor ports in response to sensor signals, etc.) The program, written on a computer, is transmitted by cable or Bluetooth signals to the microcontroller. The microcontroller is essentially the brains of the NXT. Programming may be accomplished in several languages such a C, not exactly C (NEC), ROBOT C or with a graphical language. Three different graphical languages are commonly employed including: the MindStorms' code NXT-G, ROBOLAB 2.9 and a LabVIEW Tool Kit for the NXT. We have selected ROBOT C as the programming environment for this course because of the interest of several engineering departments to introduce a form of structured programming language, such as C, into the first year curriculum.

Programming in ROBOT C will be described in detail in Chapter 6.

2.10.4 Verifying System Performance

When your hovercraft is complete and you have written the program to control its actions, it is time to verify its performance. Ideally, verification experiments should be conducted several times during the design process. Subsystems can be operated to check their performance prior to assembly of the entire system. Preliminary checks at several stages in the design process permit the team to detect design flaws early and to remedy them. Finding errors early in the process is important. If you ship a product with serious flaws and a company must issue a recall, it is a very expensive proposition. Even more serious is an accident due to a faulty product. Litigation is expensive even if the company wins the case. If the company loses, punitive damages are often many millions of dollars[10]. Testing products released to the public to insure their safety is an essential part of doing business.

Verifying system performance in the restrictive design environment created in this course is much easier. You will know in advance the design requirements and the performance targets that the system is to achieve. When the system design and assembly is complete, operate the system in a simulated environment. Measure the time required for the hovercraft to meet the specified mission requirements. Observe the response of the system. Does it operate at the correct speed? Is the program efficient? Are the sensors providing the necessary feedback signals? Is the deck sufficiently rigid? Is the skirt contacting the surface uniformly? Is the flow rate from the lift fan sufficient? Is the pressure developed in the plenum chamber adequate? Do the propulsion fans respond quickly enough? Is the hovercraft in balance, and other questions of similar nature? The time involved to verify the performance of a prototype is usually relatively low, thus allowing sufficient time to redesign and introduce improvements.

2.11 DESIGN CRITERIA

Your design experience this semester is constrained to some degree by the number of components that you can locate, study, analyze, discuss and procure in a relatively short period of time. Also, you cannot ask a company to develop a product specifically for your application. For this reason, you will not have to deal with several design goals encountered when designing in a much larger design space. An unconstrained design space permits you to design with any standard component or

---

[10] More information on recalls, and fines imposed on companies for delaying a recall may be found by reading about the brake problems encounter by several Toyota models in 2008 to 2010.

with specialized components that can be manufactured according to your specifications. How many standard components are there in the marketplace? The answer probably depends on the definition of a standard part, but it is reasonable to consider that about 200,000 mechanical parts are available from a well-stocked supply house and probably another 200,000 components are available from an electronic supply outlet. If your company has manufacturing facilities, more specialized components can be produced to meet your design requirements. Clearly, a more challenging design space exists in the real world.

To prepare you for this world, you should be aware of design criteria with which engineers are concerned with as they develop products. These include:

1. We design products that meet the needs of a specific set of customers.
2. We always seek a quality product. Quality means different things to different people. In this instance, we define a quality product as one that is reliable, meets the specified performance requirements, and is durable. Your team may wish to add to this definition.
3. We often attempt to achieve high-performance in a product with additional power, installing superior controls, reducing weight or by using new higher strength materials.
4. We strive to design a product that can be produced at low cost. However, balance between cost, performance and quality must be achieved. Low cost that is achieved by sacrificing reliability and/or performance will result in eventual failure of the product in the marketplace.
5. Our products should be reliable. They should be capable of meeting the performance specifications without the need for repair. For higher priced products, a limited amount of periodic maintenance may be justified if it prolongs the product's life significantly.
6. Products should be designed for ease of manufacturing and assembly. Manufacturing and assembly are separate operations and could be listed separately. They are both very important as manufacturing costs and assembly time markedly affect the total cost of the product.
7. We attempt to design products for easy maintenance. In many situations, spare parts and skilled technicians are in short supply or non-existent. Easy maintenance is an important element in customer satisfaction.
8. Product weight is important in most designs. Usually lighter is better because using less material results in lower costs for raw materials and shipping.
9. Product size is also important with small typically being better than large. This is particularly important in electronic products where performance increases while cost decreases as feature sizes are reduced.
10. Last, but certainly not least, the product must be safe to operate. Hazards associated with products are covered in detail later in this text. Before beginning the design of any of your systems, we **strongly recommend** that you read Section 14.7 and understand the **many different hazards** associated with the design and operation of products.

Your team should develop many different ideas to support the strategy upon which the design of your hovercraft is based. Each idea or concept should be discussed in considerable detail prior to incorporating it in the model's final design. The more completely you understand the design concepts, the better prepared you will be to conduct design trade-off analyses. In a design trade-off analysis, you compare several design ideas, select the best one and discard the others.

2.12 DESIGN TRADE-OFF ANALYSIS

To perform a design trade-off analysis, we consider each design concept individually and list its strengths and weaknesses. Factors usually considered in this analysis include the design criteria listed above such as: size, weight, cost, ease of manufacturing, performance, appearance, etc. After the strengths and weaknesses of each design idea have been listed, we can compare and evaluate the design concepts and select the most suitable concept for our prototype development.

Design concepts are generated for each function of the system. You will have several functions to consider in the design of the overall hovercraft system with many design decisions involved with each function. For example, the hovercraft involves the six functions listed below. There are many design features that must be considered in designing the components (parts) necessary to meet each functional requirement. We will list some of the key design decisions associated with each of the six major functions.

1. **Structural System** – Supports all of the components, such as fans, NXT, batteries, sensors, etc. It also contains the cushion of air upon which the hovercraft floats.

   - Type of structure
   - Deck material
   - Size and shape of the deck
   - Size and location of openings in the deck
   - Attachment of components to the deck
   - Fabrication of the deck
   - Skirt material
   - Dimensions of the skirt
   - Attachment of the skirt to the deck
   - Sealing to minimize air leakage
   - Locating the main battery supply to balance the deck

2. **Levitation System** – Includes the fan and controls for that fan.

   - Development of a weight estimate
   - Analysis of required plenum pressure
   - Analysis of leakage flow
   - Selection of a lift fan or impeller to provide adequate pressure and flow rate.
   - Estimate of power requirement for the lift fan or impeller
   - Determine control method for the lift fan or impeller
   - Procure lift fan or impeller

3. **Propulsion System** – Includes a means of providing thrust and the control of the magnitude and direction of the thrust vector.

   - Develop strategy for steering
   - Analysis of forces required for propulsion
   - Selection of fans or other propulsion devices
   - Design mounting methods for fans or propellers
   - Determine power requirements
   - Specify control method for the propulsion fans or propellers
   - Procure propulsion fans or propellers

4. **Power System** – Includes external auxiliary power supplies and the control of these supplies.

- Determine mission duration
- Calculate power requirements (voltage, current and wattage)
- Determine maximum current drain for each battery
- Select method of controlling auxiliary power supplies
- Perform design trade-off study to select battery chemistry and size
- Procure the batteries required
- Design mounting system for batteries and relays or transistors.

5. **Sensing System** – Provides feedback signals enabling the control of the hovercrafts speed and direction.

- Analyze the mission requirements to establish a control strategy
- Develop a data file that provides specification for various sensors
- Conduct a design trade-off analysis to establish the most suitable sensing system
- Procure the sensor or sensors used to provide feedback signals to the NXT
- Calibration of sensors for use with NXT
- Specify any auxiliary power requirements if necessary
- Design mounting system for the NXT and sensors

6. **Master Control System** – Enables autonomous control to successfully complete the mission.

- Procure a NXT unit to use as the microcontroller.
- Procure a copy of the ROBOT C software to program the NXT
- Plan the strategy for executing the mission
- Review sensor data to insure that the sensors will provide the necessary feedback signals
- Write the program that will lift the hovercraft, propel it along the mission's course, steer it through the required turns, and stop it at its final destination.

The procedure in design is to generate ideas for performing each function, expanding these ideas with more detail until they are understood by all of the team members and finally to perform a design trade-off analysis where the merits and faults of each concept are evaluated. When the design proposals are selected for each of the different functions involved in the selection process, the winning concepts must then be integrated into a seamless system that meets the mission requirements.

We often refer to the part of a product that performs some function as a subsystem. The collection of all of the subsystems constitutes a complete system, which when manufactured and assembled, becomes a prototype. The integration of the various subsystems is often difficult, because one subsystem influences the design of the others. The manner in which the subsystems interact is defined as the interface between subsystems. In a mechanical application, the fit of a shaft in a bearing and/or the fit of a seal against a shaft are examples of interfaces. In an electronic application, the timing of the communication signals between subsystems is an example of an interface. Frequently, interfaces between the subsystems are troublesome and difficult to manage. It is vitally important that your team control the many interfaces created by the six different functions to enable seamless integration of all of the subsystems without loss of effectiveness and efficiency of the hovercraft system.

2.13 PREPARING DESIGN DOCUMENTATION

The final step in designing a hovercraft involves preparing the design documentation package. The package is an extended engineering document and contains:

- Engineering sketches or photographs (preferably digital) illustrating the design concepts.
- Engineering assembly drawings describing the arrangement of various components that fit together to form the complete system.
- Engineering drawings of all of the component parts in sufficient detail to permit the component to be manufactured by anyone capable of reading an engineering drawing or to be procured by a purchasing agent.

You are to define the names of all the unique parts to enable your team members to properly identify them. Assembly drawings are important because we must be able to convey the construction sequence for both subassemblies and the complete model. However, with the advent of digital photography it is possible to replace the assembly drawings with assembly photographs. Sketching will always be important as a quick means of conveying your ideas. We encourage you to use sketches in discussions with team members to convey your ideas more clearly in a graphical format.

We recognize that many of you may need instruction in graphics, and have included three chapters in this text to help you learn the techniques required to prepare three-view and pictorial drawings and to construct tables and graphs.

The design documentation package also contains a parts list that identifies every unique component employed in the assembly of the model. A typical example of a parts list for the hovercraft was presented previously in Fig. 2.39. This figure shows the quantity of each part required for the assembly of a single prototype.

An engineering report is also included as part of the design package. This report supports the design by describing the strategy and the key features incorporated in the hovercraft's development. Your report should treat each function involved in the design and describe the concepts that your team considered. The rationale for each design concept that was adopted, based on systematic design trade-off studies, is an essential element of the report. The report should also present a flow diagram of the program written for the controller. If the program is complex, a description of the programming strategy should also be included. Additionally, design analyses that predict the performance of your hovercraft are included. These analyses are completed before actually testing the model and the results from the analyses and the tests are useful in assessing the merits of your design. More information on the preparation of an engineering report is included in Chapter 9.

2.14 FINAL DESIGN PRESENTATION

The design phase of the product development process is concluded with a final design review. This is a formal review of the design of the hovercraft system and is presented to the class (peers), to the instructor and to any others in the audience. It is the team's responsibility to describe all of the unique features of the design and to predict the performance of the product. It is the responsibility of the class (peer group) and the instructor to question the feasibility of the design and the accuracy of the predicted performance. If you note a shortcoming or a flaw in the design during the design briefing, identify the problem to the team presenting their work. As a peer reviewer of another team's design, be tactful in your critique. Criticism, when provided for a colleague, is always a difficult and delicate task. Offer constructive suggestions in good faith and in good taste. If you are on the receiving end of criticism, do not become defensive. The individual offering the critique is not attacking your capabilities; he or she is actually trying to give you a suggestion that might help you achieve a better design.

The purpose of the design review is to locate design deficiencies and errors. It is much better to correct errors in the paper stage of the design process, not later in the hardware phase when it is considerably more difficult and costly to fix problems due to a poor design.

2.15 HOVERCRAFT ASSEMBLY

When the assembly kit is complete and checked against the parts list, you begin the final design phase—model evaluation. The first part of the evaluation is to assemble or build the model. This step is often called "first article build," because it is the first time that we have attempted to fit together all of the components used in the assembly. The "first article build" is successful if:

- All of the parts are available
- The parts all fit together properly
- The tolerances on each part and each feature are correct
- The surface finish on all of the parts is acceptable

Product design is often judged against the four F's—form, fit, finish and function. The "first article build" permits us to assess how well the team performed with regard to the first three of the Fs—form, fit and finish. Form, fit and finish of the various parts that your team manufacturers are dependent on your team's tool room skills and patience. Most of the components required for the hovercraft may be procured directly from suppliers. Only a few parts will require your skill with hand tools. Work smart and carefully to develop a model that looks professional. Function depends on the merits of your design. If your team develops functional subsystems and if they are properly integrated, the hovercraft will perform well and the mission will be conducted successfully.

We anticipate that each team will make a few errors in the preparation of the design drawings and changes will be required. The natural tendency is to remove and replace the offending parts to correct (modify) the design and move on with the "first article build." This behavior is acceptable only if the team revises the drawing of this part to reflect the modification made to correct the design deficiency. In the real world, the integrity of the drawing package is much more important than the prototype. The second and all subsequent assemblies will be fabricated from the details shown in your engineering drawings and not by examining the prototype. The prototype is often scrapped after it has been built and tested.

2.16 MODEL EVALUATION

After the assembly of the hovercraft is complete, it should be carefully inspected to insure that it is safe. Even simple products can be **dangerous.** For example, a large button that can be pulled from a doll is a potential danger if a child swallows it. In this course, the dangers are relatively small because we are dealing with low voltages and small, light-weight models. However, be certain that you have avoided introducing pinch points. Pinch points are openings that close due to motion of one or more parts when a product is operated. Fingers, hair and/or clothing caught in pinch points clearly represent a hazard). **Also, carefully shield any fan openings to avoid the possibility of a person inserting his or her finger into a rotating fan blade.**

The final step in this development process is to test your hovercraft to determine if it is functional and efficient. Typically the instructor prepares a test facility and schedules a time interval for evaluating each team's model. Your hovercraft will be judged based on the criteria announced by your instructor. He or she usually will select criteria from the following listing:

1. Performance: Ability to achieve the specified mission.
2. Control: Ability of system to perform quickly and effectively.
3. Cost: Minimize part count and avoid using expensive components.
4. Design innovation: Novelty and creativeness.
5. Quality of assembly: craftsmanship.
6. Appearance: Style

2.17 TEAMWORK

In this course, you are required to participate on a large development team. There are three reasons for this requirement. First, the projects are too ambitious for an individual to complete in the time available. You will need the collective efforts of the entire team to complete the design, construction and testing of a hovercraft model during the semester. Your instructor will press each development team to complete each phase of the design on schedule.

Second, it is a course objective that you to begin to learn teamwork skills. Experience shows that most students entering the engineering program have not fully developed these skills yes. From k-12, the educational process has focused on teaching you to work as an individual, often in a setting where you competed against other students in your class. We will insist that you begin functioning as a team member where cooperation, following, and listening are at least as important as individual effort. Leadership is important in a team setting, but cooperation and following the lead of others are also critical elements for successful team performance.

The final reason is to better prepare you for the real world that you will enter upon graduation with a B. S. degree in engineering. You will probably be assigned to a development team very early in your career if you take a position in industry. A recent study [4] by the American Society Of Mechanical Engineers (ASME), the results of which are shown in Table 2.3, ranked teamwork as the most important skill to develop in an engineering program. Teamwork was also the first skill, in a list of 20, considered important by managers in industry. We believe that this course will be instrumental in introducing you to the team working skills so necessary for a successful career in engineering.

Table 2.3
Important Skills for New Mechanical Engineers with Bachelor Degrees
Priority Ranking by Industrial Representatives

1. Teams and Teamwork	11. Sketching and Drawing
2. Communication	12. Design for Cost
3. Design for Manufacture	13. Application of Statistics
4. CAD Systems	14. Reliability
5. Professional Ethics	15. Geometric Tolerancing
6. Creative Thinking	16. Value Engineering
7. Design for Performance	17. Design Reviews
8. Design for Reliability	18. Manufacturing Processes
9. Design for Safety	19. Systems Perspective
10. Concurrent Engineering	20. Design for Assembly

As you work as a team member in the development of a hovercraft system, you will be introduced to several of the 20 topics listed in Table 2.3. As you begin to understand the design process, we trust that you will develop an appreciation for these skills.

2.18 OTHER COURSE OBJECTIVES

While your experiences in designing a model hovercraft with autonomous control are the primary objectives of this course, there are several other important educational objectives. You will quickly recognize a critical need for graphics as you attempt to describe your design concepts to fellow team members. We have included two chapters on engineering graphics to help you learn the basic skills required at this entry level. These chapters include materials on the spreadsheet Excel for preparing tables and graphs and a computer aided design program known as Pro/ENGINEER.

As you proceed with the design of your hovercraft, we will require you to communicate both orally and in writing. Design reviews before the class, provide you the opportunity to learn presentation skills such as style, timing and the preparation of visual aids. Preparing the design report gives you experience in developing complete, high-quality engineering drawings and in writing technical reports that contain text, figures, tables and graphs.

Another objective of the course is to introduce you to structured programming. A chapter has been included to describe the usage of ROBOT C to control your hovercraft model.

The final objective is to begin the development of your design analysis skills. We understand that you are beginning your studies. For this reason, we will present key equations that mathematically model the electrical and mechanical components used in many of the subsystems that you employ in the design of your hovercraft. We do not expect you to completely understand all of the theoretical aspects of the relatively complicated subsystems involved. Most of you will take several courses later in the curriculum dealing with these subjects in much greater detail. However, at this stage of your career, we want you to begin to appreciate the relationship between analysis and design. To help you with the analyses, we have included chapters on fluid mechanics, electrical circuits and sensors, and motion and control. While the coverage is very brief and introductory, it will provide you with a better understanding of both mechanical and electrical systems.

REFERENCES

1. Valenti, M. "Teaching Tomorrow's Engineers," Special Report, Mechanical Engineering, Vol. 118, No. 7, July 1996.
2. Ulrich, K. T., S. D. Eppinger, 2nd Edition, Product Design and Development, McGraw-Hill, New York, NY, 2000.
3. Cross, N., Engineering Design Methods, 3rd Edition, Wiley, New York, NY, 1994.
4. Anon, "Integrating the Product Realization Process into the Undergraduate Curriculum," ASME report to the National Science Foundation, 1995.
5. Macaulay, D., The New Way Things Work, Houghton Mifflin Co., Boston, 1998.
6. Horowitz, P. and W. Hill, The Art of Electronics, 2nd Edition, Cambridge University Press, New York, 1989.
7. Dally, J. W., Riley, W. F. and K. G. McConnell, Instrumentation for Engineering Measurements, 2nd Edition, Wiley, New York, 1993.
8. Martin, Fred G., Robotics Explorations: A Hands-On Introduction to Engineering, Prentice-Hall, Upper Saddle River, NJ, 2001.

9. Capozzoli, P. and C. B. Rogers, "LEGOs and Aeronautics in Kindergarten through College," AIAA, Vol. 96, 1996, p. 2245.

10. Cyr, M., V. Miragila, T. Nocera and C. Rogers, "A Low Cost Innovative Methodology for Teaching Engineering Through Experimentation," Journal of Engineering Education, Vol. 86, No. 2, 1997, pp. 167-171.

11. Portsmore, M., M. Cyr and C. Rogers, "Integrating the Internet, LabVIEW, and Lego Bricks into Modular Data Acquisition and Analysis Software for K-College," Proceedings of the ASEE 2000 Annual Meeting, Session 2320, St. Louis, June 19-23, 2000.

LABORATORY EXERCISES

2.1   Measure the static head developed by a lift fan or by a suitably confined impeller.

2.2   Measure the thrust developed by a small axial fan.

2.3   Calibrate a proximity (sonar) sensor over operational range of the sensor.

EXERCISES

2.1   List ten products that you have used today and the companies that manufacture and market them.

2.2   Suppose that you do not want to work on product development, manufacturing or sales. What opportunities remain for you in engineering?

2.3   Write an engineering brief describing the characteristics of a winning product.

2.4   Write the most important equation in engineering and business.

2.5   Why is the Ford Motor Company reluctant to develop a completely new model of an automobile?

2.6   What is a design concept? How many concepts will your team generate in developing a propulsion system for the hovercraft model?

2.7   What are the main subsystems in the development of a hovercraft model?

2.8   Why is it important to prepare a parts list during a development program?

2.9   What is an assembly kit and why do we prepare one in the development of the first model?

2.10 Why do engineering executives invest scarce company funds in assembling a prototype?

2.11 What are the safety considerations with which you must be concerned in testing your hovercraft models?

2.12 Why do you believe industry representatives ranked teams and teamwork as the most important skill for new engineers?

2.13 Prepare a milestone schedule showing the date when each major design phase will be completed to ensure overall project success.

2.14 Prepare a strategy to guide the programming of the hovercraft to win the land race mission.

2.15 Prepare a strategy to guide the programming of the hovercraft that will successfully negotiate the specified path on the navigation mission.

2.16 Prepare a strategy to guide the programming of the hovercraft to quickly traverse the course on the land and water mission.

2.17 Prepare a strategy to guide programming the hovercraft to win the demolition mission.

2.18 Prepare a strategy to guide programming the hovercraft to win the water race mission.

2.19 Prepare a strategy to guide the programming the hovercraft to win the cargo mission.

CHAPTER 3

FLUID MECHANICS AND DESIGN ANALYSIS

3.1 INTRODUCTION

Developing new products requires a close relationship between design and analysis. When designing, we select components, layout the structure, place components, test subsystems, etc. At the same time engineers perform analyses to insure that all aspects of the design will function correctly and that the system will meet specifications. The hovercraft development provides an opportunity for your team to design a unique vehicle and to perform several analyses to confirm the validity of the design. For example, before you can select a lift fan, an analysis must be performed to establish the pressure required to develop sufficient force to lift the hovercraft. You must also determine if the flow rate of the fan is adequate to sustain the leakage rate. The selection of the propulsion fans also requires prior analysis to insure that they will develop sufficient thrust to propel the hovercraft. If your hovercraft is to operate on water, you must insure that it has sufficient buoyancy to float in the event that the levitation fan malfunctions.

In design you often find that several different solutions are possible. Again analysis aids you in selecting the solution that is the most suitable, while simultaneously considering the constraints on your design. As we introduce analysis methods, we will describe solution space. Solution space covers a very wide range of analytical possibilities. At its center are the parameters that insure a successful design. At its extremes are the parameters that will lead to a design that may possibly fail to meet its objective.

In this chapter we will introduce the theories necessary to perform several different design analyses associated with the development of a hovercraft. We will then perform the analyses to insure that the functions (systems and processes) involved in designing a hovercraft operate satisfactorily. These functions are listed below:

1. The structural system supports all of the components, such as fans, impellers, propellers, the NXT, batteries, sensors, etc. The structural system also contains the cushion of air upon which the hovercraft floats.
2. The lift system includes the lift fan or impeller and the controls for that fan.
3. The propulsion system includes a means of providing thrust and the control of the magnitude and direction of the thrust vector.
4. The power system includes external auxiliary power supplies and control of these supplies.
5. The sensing system provides feedback signals enabling control of the hovercraft's speed and direction.

6. The master control system enables autonomous control to successfully complete the mission.

3.2 PRESSURE-MASS DENSITY-HEIGHT RELATIONSHIPS

Let's consider a body of water that is at rest as shown in Fig. 3.1. From the surface, let's place a reference axis y that points downward into the depth of the water. Next examine a small element that is Δy thick with a cross sectional area of ΔA, located on the left side of this figure. A pressure p acts downward on its top surface and a pressure $p + \Delta p$ acts upward on its bottom surface. Note the pressure always act perpendicular to the surface upon which it acts and is increasing with the depth of the water. In addition to these pressures, a downward force ΔW acts on the element, this is due to the weight of the water contained within the element.

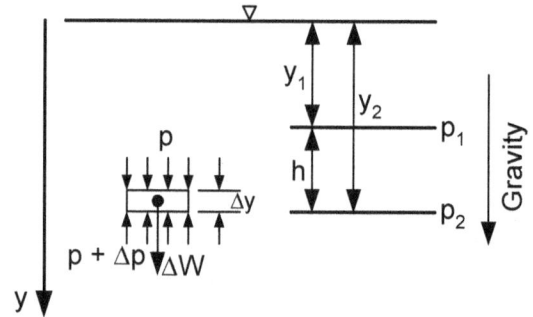

Fig. 3.1 Pressures created at depth in a static pool of water.

Let's write the equilibrium equation for the elemental volume of water depicted in Fig. 3.1. The equilibrium equation implies that the sum of the forces upward equal the sum of the forces downward. Thus, we write:

$$(p + \Delta p)\Delta A \text{ (upward)} = p\,\Delta A + \Delta W \text{ (downward)}$$

This equation reduces to:

$$\Delta p \Delta A = \Delta W \tag{3.1}$$

The weight of the water in this elemental volume is given by:

$$\Delta W = \gamma\,\Delta A\,\Delta y \tag{a}$$

where $\gamma = \rho g$ is the specific weight, g is the gravitational constant and $\rho = \gamma/g$ is the mass density (mass/volume) of water.

Substituting Eq. (a) into Eq. (3.1) yields:

$$\Delta p = \gamma\,\Delta y = \rho g \Delta y \tag{3.2}$$

Summing both sides of Eq. (3.2) gives:

$$\sum_{p_1}^{p_2} \Delta p = \gamma \sum_{y_1}^{y_2} \Delta y \tag{b}$$

Equation (b) reduces to:

$$p_2 - p_1 = \gamma\,(y_2 - y_1) = \gamma\,h = \rho g h \tag{3.3}$$

Equation (3.3) indicates that the pressure at some point in a liquid with a specific weight γ is dependent only on the height of the liquid above that point. This fact allows the vertical height or "head" of a specified liquid to be used as a measurement of pressure.

3.3 MANOMETERS

There are many transducers that can be employed to measure pressure. However, if the pressure is relatively low and constant, a manometer is a simple and effective instrument. To show this simplicity, consider the U-tube manometer illustrated in Fig. 3.2. The U-tube, a transparent glass or plastic tube, is filled with a colored liquid and then inserted into a chamber with an unknown pressure p_x.

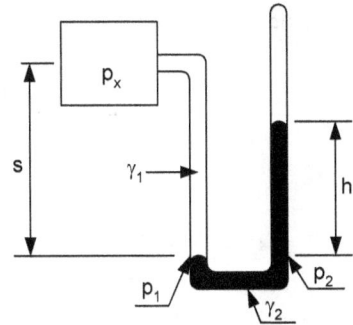

Fig. 3.2 A U-tube manometer measuring the unknown pressure p_x.

The chamber pressure forces the liquid down on one leg of the U tube and up on the other leg. It is evident that $p_1 = p_2$ because they are at the same datum plane (level). From Eq. (3.3) we can write:

$$p_1 = p_x + \gamma_1 s \qquad \text{and } p_2 = 0 + \gamma_2 h \qquad (a)$$

Equating p_1 and p_2 leads to:

$$p_x = \gamma_2 h - \gamma_1 s \qquad (3.4)$$

If the liquid in the manometer is colored water and the fluid in the pressure chamber is air, then it can be assumed that the specific weight $\gamma_1 \ll \gamma_2$ and Eq. (3.4) reduces to:

$$p_x = \gamma_2 h = \rho_2\, gh \qquad (3.5)$$

It is clear from Eq. (3.5) that a simple U-tube manometer can be used to measure the air pressure in the plenum chamber under steady state conditions if the dimension h can be measured.

3.4 LIFT FORCES

Let's next consider the lift forces that are generated when we introduce pressure into the plenum of the hovercraft. The hovercraft will be in equilibrium in the vertical direction when the force developed by the pressure equals the total weight W of the hovercraft, as indicated in Fig 3.3.

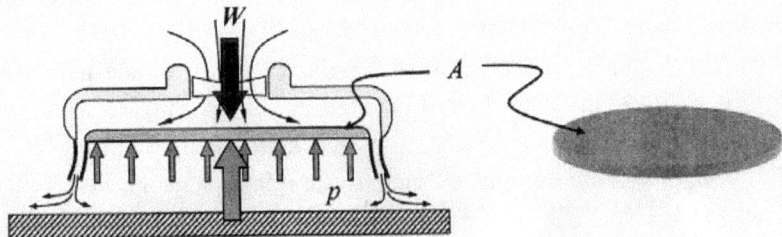

Fig. 3.3 Equilibrium diagram for the hovercraft (vertical direction).

For the hovercraft to be in equilibrium and floating on a cushion of air, the lift force due to the pressure must equal the downward force due to the weight W of the hovercraft structure and all of the components mounted on its deck. The lift force F_L is given by:

$$F_L = pA_D \qquad (3.6)$$

where A_D is the area of the deck of the hovercraft exposed to the pressure p.

The floating condition for the hovercraft based on equilibrium is:

$$F_L \geq W \qquad (3.7)$$

Let's explore the range of lift forces that can be generated by various combinations of the plenum pressure p and the deck area A_D. First suppose from our information on fans that pressures from nearly zero to 2.5 in. of water are possible depending on the choice of the fan and the flow rate that it generates. To begin this analysis, let's convert the pressure measurement given in inches of water to pounds per square inch (psi).

From Eq. (3.5), we write:

$$p = \gamma h = \left[\frac{62.4\,\text{lb}/\text{ft}^3}{1728\,\text{in.}^3/\text{ft}^3} \right] h = \left(0.03611\ \text{lb}/\text{in.}^3 \right) h \qquad (3.8)$$

The results from Eq. (3.8) are shown in the second column of the spreadsheet presented in Table 3.1 for p varying in steps of 0.1 from zero to 2.5 inches of water. Using Eq. (3.6), the lift force F_L was calculated for hovercraft deck areas ranging from 10 to 1000 in.^{2} The results show that relatively high lift forces can be generated (in excess of 90 lb), but these larger lift forces require large deck areas and high plenum pressures.

Before it is possible to proceed with either the design or the analysis of the design, it is necessary to estimate the weight of the hovercraft structure and all of its components. A preliminary weight estimate is shown in Table 3.2. It includes a selection of the components that were described previously in Chapter 2. The results listed show an estimated weight of 6.70 lbs for this hovercraft. This first compilation of the weight may be excessive. More time devoted to finding lighter weight components may be required to improve the design. It may prove necessary to modify this listing and add or change components if the design analysis of shows that the flow rate is not sufficient to provide a large enough gap between the skirt of the hovercraft and the surface over which the hovercraft's mission is to be executed.

An examination of the results of this estimate indicates that most the weight is due to the three fans, the main battery and the NXT (65.2%). The estimates of these heavy components are accurate because the weight of each component was available from the suppliers. The weight of several other components, such as the relays, sensors, fasteners, tie downs, etc., was estimated using conservative numbers. The question is what should be the design target for the total weight. Let's select 7.5 lbs, which provides a safety factor of 1.12 in achieving lift-off.

Table 3.1
Lift force as a function of plenum pressure and deck area.

DETERMINATION OF LIFT FORCE AS A FUNCTION OF PRESSURE AND AREA								
p	p	A=10 in^2 Lift Force	A=20 in^2 Lift Force	A=50 in^2 Lift Force	A=100 in^2 Lift Force	A=200 in^2 Lift Force	A=500 in^2 Lift Force	A=1000 in^2 Lift Force
in.of H2O	psi	lb	lb	lb	lb	lb	lb	lb
0.0	0	0	0	0	0	0	0	0
0.1	0.0036	0.04	0.07	0.18	0.36	0.72	1.81	3.61
0.2	0.0072	0.07	0.14	0.36	0.72	1.44	3.61	7.22
0.3	0.0108	0.11	0.22	0.54	1.08	2.17	5.42	10.83
0.4	0.0144	0.14	0.29	0.72	1.44	2.89	7.22	14.44
0.5	0.0181	0.18	0.36	0.90	1.81	3.61	9.03	18.06
0.6	0.0217	0.22	0.43	1.08	2.17	4.33	10.83	21.67
0.7	0.0253	0.25	0.51	1.26	2.53	5.06	12.64	25.28
0.8	0.0289	0.29	0.58	1.44	2.89	5.78	14.44	28.89
0.9	0.0325	0.32	0.65	1.62	3.25	6.50	16.25	32.50
1.0	0.0361	0.36	0.72	1.81	3.61	7.22	18.06	36.11
1.2	0.0433	0.43	0.87	2.17	4.33	8.67	21.67	43.33
1.3	0.0469	0.47	0.94	2.35	4.69	9.39	23.47	46.94
1.4	0.0506	0.51	1.01	2.53	5.06	10.11	25.28	50.56
1.5	0.0542	0.54	1.08	2.71	5.42	10.83	27.08	54.17
1.6	0.0578	0.58	1.16	2.89	5.78	11.56	28.89	57.78
1.7	0.0614	0.61	1.23	3.07	6.14	12.28	30.69	61.39
1.8	0.0650	0.65	1.30	3.25	6.50	13.00	32.50	65.00
1.9	0.0686	0.69	1.37	3.43	6.86	13.72	34.31	68.61
2.0	0.0722	0.72	1.44	3.61	7.22	14.44	36.11	72.22
2.1	0.0758	0.76	1.52	3.79	7.58	15.17	37.92	75.83
2.2	0.0794	0.79	1.59	3.97	7.94	15.89	39.72	79.44
2.3	0.0831	0.83	1.66	4.15	8.31	16.61	41.53	83.06
2.4	0.0867	0.87	1.73	4.33	8.67	17.33	43.33	86.67
2.5	0.0903	0.90	1.81	4.51	9.03	18.06	45.14	90.28

Table 3.2
Estimated weight of hovercraft structure and components

WEIGHT TABLE			
Component	Quantity	Weight lb	Total Weight lb
Lift Fan (Impeller)	1	1.5	1.5
Propulsion Fan	2	0.6	1.2
Relays	3	0.1	0.3
NXT	1	0.7	0.7
Sensors	2	0.1	0.2
Main Battery (SLA)	1	1.0	1.0
Auxiliary Battery	1	0.1	0.1
Deck	1	0.75	0.75
Skirt	1	0.5	0.5
Fasteners	20	0.005	0.1
Wiring	4 ft	0.025	0.1
Tie Downs	10	0.01	0.1
Mounting Hardware	3	0.05	0.15
Total Weight (lb)			6.70

3.5 SIZING THE STRUCTURE

At this point in the analysis, we have nearly enough information to size the deck of the hovercraft. The information needed to complete this phase of the analysis is related to the characteristic curve of the lift fan. Let's assume that a fan has been selected with its characteristic operating curve, as shown in Fig. 3.4. Examination of this curve shows that the pressure from the fan drops steadily as the flow rate increases. To avoid the lower pressures at the higher flow rates, let's assume[1] the fan operates at a minimum pressure of 1.0 in. of H_2O and a maximum flow rate Q of 72 cubic feet per minute (CFM).

Examination of Table 3.1 shows a number of different pressure area combinations that yield lift forces that exceed 7.5 lbs. If we study the results for a plenum pressure of 1.0 in. of H_2O or 0.0361 psi, it is evident that the deck area of the hovercraft will have to be slightly larger than 200 in^2. From Eq. (3.6), it is possible to determine the required deck area A_D of the hovercraft for a pressure of 0.0361 psi (or 1.0 in. of H_2O) as:

$$A_D = \frac{F_L}{p} = \frac{7.5}{0.0361} = 208 \text{ in}^2 \qquad (a)$$

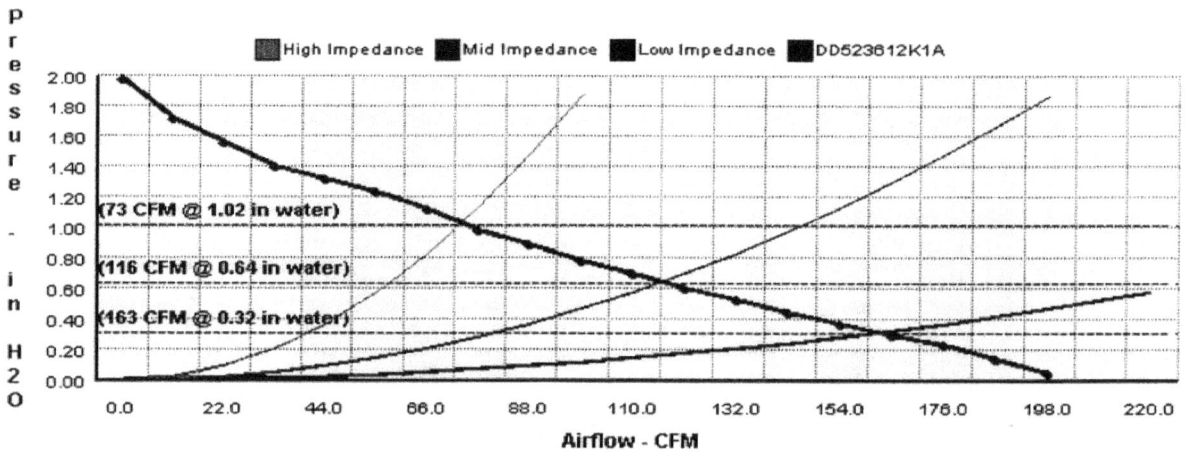

Fig. 3.4 Characteristic operating curve for a fan.
Comair Rotron Model DD523612K1A Part No. 039832.

It is clear from the data listed in Table 3.1 and Table 3.2 that the fan selected[2] is sufficient to lift the hovercraft if the deck area is 208 in^2 or larger providing a plenum pressure of 0.0361 psi (1.0 inches of H_2O) is developed. The next question to consider in this design analysis is if the flow rate is sufficient to maintain a reasonable size gap between the skirt on the hovercraft and the surface over which it must travel.

---

[1] The operating point for the impeller or fan depends on the impedance of the flow channel. This impedance is largely controlled by the size of the gap between the hovercraft's skirt and the surface. As this gap goes toward zero, the pressure generated by the fan or impeller will increase toward its static limit.
[2] Many other fans are available with higher and lower flow rates. Design trade-off studies can be conducted to balance the benefits of increased flow rates with the increased weight and power requirements of the larger capacity impellers.

3.6 DETERMINING THE LEAKAGE RATE

Continuity Equation

To determine the leakage rate through the air gap around the perimeter of the hovercraft, it is necessary to introduce two very important equations used in fluid mechanics—the conservation of mass relation and Bernoulli's equation. Let's consider the flow of a fluid through a stream tube as illustrated in Fig. 3.5. Under steady state conditions, the mass[3] of fluid entering the stream tube is the same as that leaving it.

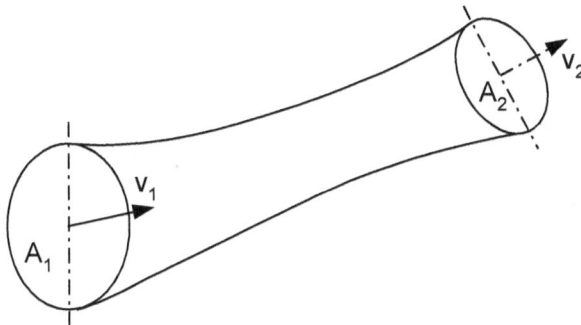

Fig. 3.5 Flow into and out of a stream tube.

Because no fluid is lost or stored as it passes through the steam tube, we can write the equation for the conservation of mass as:

$$A_1 v_1 \rho_1 = A_2 v_2 \rho_2 \qquad (3.9)$$

where v is the velocity of the flow.

If we assume the flow is incompressible,[4] $\rho_1 = \rho_2$ and Eq. (3.9) reduces to:

$$A_1 v_1 = A_2 v_2 = Q \qquad (3.10)$$

where Q is the flow rate.

It is evident from Eq. (3.10) that the **product** of the velocity of the air flow and the cross sectional area of the stream tube is a constant everywhere along its length.

Bernoulli's Equation

Bernoulli's equation is based on the principle of conservation of energy. To illustrate this concept, consider frictionless flow though the piping system shown in Fig. 3.6. At station 1, the fluid flowing with a velocity v_1 is at a pressure p_1 and a specific weight γ_1. As the fluid travels from station 1 to station 2, the pipe changes elevation and the cross sectional area of the pipe decreases. The fluid at station 2 is flowing with a velocity v_2 and it exhibits a pressure p_2 with a specific weight γ_2. If energy is not added or lost as the fluid moves from station 1 to station 2, then the principle of conservation of energy leads to Bernoulli's equation that is written as:

---

[3] This statement is based on the principle of conservation of mass.
[4] In this analysis, we assume that air is incompressible. Of course air is compressible, but we can make this assumption because the changes in pressure encountered when pressurizing the plenum chamber are very small. In other words, the low pressure air generated by the fan can be treated as an incompressible fluid without introducing serious errors in the analysis.

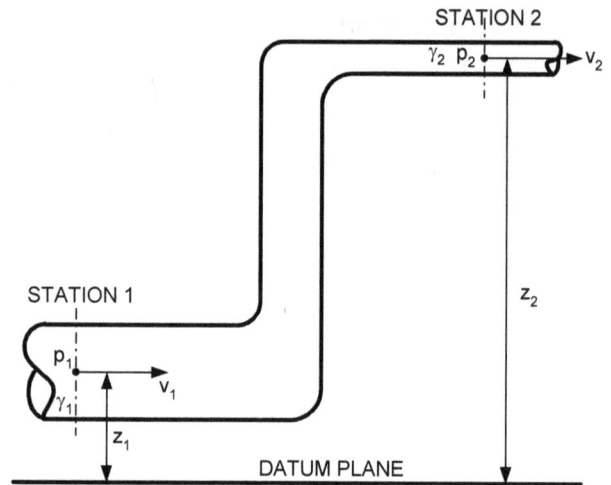

Fig. 3.6 Flow conditions in a piping system without energy added or lost.

$$\frac{p_1}{\gamma_1} + \frac{v_1^2}{2g} + z_1 = \frac{p_2}{\gamma_2} + \frac{v_2^2}{2g} + z_2 \qquad (3.11)$$

The first term in Bernoulli's equation $\frac{p}{\gamma}$ may be considered as the work that can be performed by a fluid with a pressure p. The second term $\frac{v^2}{2g}$ is related to the fluid's kinetic energy. Finally, the third term z is the fluid's potential energy relative to a defined datum plane. Each of these terms can also be considered as heads—the pressure head, the velocity head and the potential head. It is evident from Eq. (3.11) that the **sum** of the three terms involving pressure, velocity and height above a datum plane are constant at every point along a stream tube.

Leakage Flow Rate

Let's use Bernoulli's equation to derive a simple equation to employ in estimating the leakage flow rate. First, let's draw a model of the hovercraft's plenum chamber, as illustrated in Fig. 3.7. Assume the air pressure is sufficient to lift the hovercraft forming a gap between the surface and the edge of the skirt. The flow exits this gap at a velocity v_2. As the flow vents to the atmosphere, the pressure at the exit is given by $p_2 = 0$. Inside the plenum chamber, we assume that the lift fan is maintaining a constant pressure p_1. We also assume that the plenum chamber is large relative to the size of the fan's duct so that the velocity of the air within the plenum is $v_1 = 0$. The change in the elevation between station 1 and 2 is negligible.

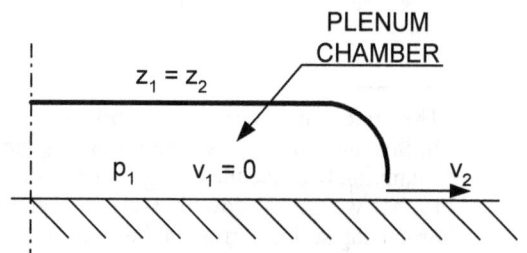

Fig. 3.7 Leakage flow from the plenum chamber through the gap between the skirt and the surface.

Substituting the conditions on the pressure, velocity and elevation at stations 1 and 2 into Bernoulli's equation and simplifying the resulting expression gives:

$$v_2 = \sqrt{2g\frac{p_1}{\gamma_1}} = \sqrt{\frac{2p_1}{\rho_1}} \tag{3.12}$$

where γ_1 is the specific weight of air at atmospheric pressure (14.7 psi), which is 0.0763 lb/ft^3 or 4.416×10^{-5} lb/in^3, $\rho = 1.144 \times 10^{-7}$ s^2–lb/in^4, and g is the gravitational constant 32.2 ft/s^2 or 386 in./s^2.

Substituting the results for γ_1 and g into Eq. (3.12) yields:

$$v_2 = \sqrt{2g\frac{p_1}{\gamma_1}} = \sqrt{\frac{(2)(386)10^5}{4.416}p_1} = 4.181 \times 10^3 \sqrt{p_1} \tag{3.13}$$

where the units for v_2 and p_1 are in/s and psi, respectively.

The results in Eq. (3.13) indicate that the velocity of the air leaking from the gap between the skirt of the hovercraft and the surface increases as the square root of the plenum pressure. Recall from Section 3.5 that we assumed the fan generated a minimum pressure of 0.0361 psi (1.0 in. of H_2O) and a flow rate of 72 CFM. Substituting this value into Eq. (3.13) yields:

$$v_2 = 794.4 \text{ in/s} = 3,972 \text{ ft/min} \tag{a}$$

Next recall Eq. (3.10) and note that $Q_{Fan} = 72$ CFM at an operating pressure of 0.0361 psi (1.0 in. of H_2O). Then from Eq. (3.10) we obtain to area of the gap A_g as:

$$A_g = \frac{Q_{Fan}}{v_2} \tag{3.14}$$

Substituting numerical values for v_2 and Q_{Fan} into Eq. (3.14) yields:

$$A_g = \frac{Q_{Fan}}{v_2} = \frac{72}{3,972} = 0.0181 \text{ ft}^2 = 2.610 \text{ in.}^2 \tag{b}$$

The result in Eq. (b) indicates that the fan operating at 72 CFM can sustain a total gap area of 2.610 in^2. Next, let's establish the height of the gap between the skirt and the surface. This dimension is important, because the gap must be sufficiently large to permit the hovercraft to travel over slightly rough surfaces or to permit the hovercraft to tilt as it adjusts to compensate for dynamic forces.

In Section 3.5 the deck area A_D was calculated as 208 in^2, if a lift force of 7.5 lb was required. However, we did not establish the shape of the deck. Is it to be square, rectangular or circular? To minimize leakage, it is important to minimize the perimeter of the deck. It is well known that the optimum shape to maximize area while minimizing perimeter is a circle. Hence, let's design the hovercraft structure with a circular plenum chamber. The diameter D of a circle with an area A_D is given by:

$$D = \sqrt{\frac{4A_D}{\pi}} = \sqrt{\frac{4(208)}{\pi}} = 16.27 \text{ in.} \qquad \text{(c)}$$

The circumference C of this circle is given by:

$$C = \pi D = \pi\,(16.27) = 51.13 \text{ in.} \qquad \text{(d)}$$

Finally, the dimension g of the gap is given by:

$$g = \frac{A_g}{C} \qquad (3.15)$$

Substituting the results from Eq. (b) and Eq. (d) into Eq. (3.15) gives:

$$g = \frac{A_g}{C} = \frac{2.610}{51.13} = 0.0511 \text{ in.} \qquad \text{(e)}$$

For a lift fan operating at Q_{Fan} = 72 CFM at a pressure of 0.0361 psi (1.0 in. of H_2O), a gap of 0.0511 in. can be maintained in a steady state. If the surface over which the hovercraft travel is rough, there is a strong possibility that the skirt will snag and possibly cause the hovercraft to turn or to stop. With a gap this small, tilt of the hovercraft, as it adjusts to dynamic forces, will likely cause contact of the skirt and loss of control of vehicle's the motion. Is it possible to decrease the weight of the hovercraft by selecting lighter weight components and by designing a lighter weight structure? Is it possible to increase the gap by changing the operating point so that the fan generates higher flow at lower pressure? Is it possible to add a second fan and to double the flow rate thereby increasing the clearance between the hovercraft skirt and the surface of the course? Another option is to select a larger fan with a higher flow rate. The larger fans weigh more, require more deck area for mounting and draw higher currents.

The fan selected for the analysis presented above is one of many commercially available. It may not be the most suitable choice for the hovercraft your team is designing. However, the analysis method demonstrated above enables your team to explore solution space and to select a more suitable lift fan or impeller.

3.7 DETERMINING THE PROPULSION FORCE

Let's determine the propulsion force or forces necessary to drive and steer the hovercraft. Recall that the hovercraft is floating on a cushion of air so it can be moved with relatively small forces. Drag forces that are usually a significant factor in propulsion are not a consideration in the design of a hovercraft that is to travel at relatively low velocities over short distances while floating over a relatively smooth surface. However, friction forces develop when the skirt contacts the track's surface, which must be taken into account.

For a hovercraft, propulsion forces are generated by momentum transferred due to air flow. This air flow can come from a number of sources—the most common of which are aft mounted fans as indicated in Fig. 2.4. Opening an orifice and bleeding air from the plenum chamber would also provide a thrust force due to the momentum transferred by that air flow. Finally, tilting the hovercraft forward so that the gap about the hovercraft's circumference becomes larger in its aft segment generates non-symmetrical air flow that in turn produces an axial force on the hovercraft driving it forward.

Let's consider an axial fan similar to the one shown in Fig. 2.30 to provide a propulsion force F_P. According to the impulse- momentum principle, we can write:

$$\mathbf{F} \, \Delta t = \Delta(\mathbf{M}) = \Delta \mathbf{mv} \tag{3.16}$$

where $\mathbf{F} \, \Delta t$ is the impulse due to the propulsion force F_P.

$\Delta \mathbf{M} = \Delta \mathbf{mv}$ is the change in momentum of the air as it passes through the axial fan.

Using Eq. (3.10) it is possible to express the change in momentum as:

$$\Delta mv = (A_0 \rho_0 v_0 \,) \Delta t \, v_0 - (A_i \rho_i v_i \,) \Delta t \, v_i \tag{3.17}$$

where the subscripts i and o refer to the inlet and outlet conditions of the flow at the fan.

Note that the incompressibility assumption indicates that mass density $\rho_0 = \rho_i = \rho$. We also assume that the velocity v_i of the hovercraft is low compared to the air's exit velocity v_0 of the fan; hence, Eq. (3.17) reduces to:

$$\Delta mv = (A_0 \, \rho \, v_0 \,) \Delta t \, v_0 = Q \, \rho \, v_0 \, \Delta t \tag{3.18}$$

Substituting Eq. (3.18) into Eq. (3.16) yields[5]:

$$F_P = Q \rho \, v_0 \tag{3.19}$$

Let's employ Eq. (3.19) to determine the propulsion force from a single axial fan. To begin refer to the characteristic curve of an axial fan, which is illustrated in Fig. 3.8.

Reference to Fig. 3.8 indicates that flow rates from the fan of about 80 to 100 CFM at low pressures are possible. The fan will be exhausting into the atmosphere so small pressures will be required to achieve these flow rates. Let's use 80 CFM as a conservative estimate of the flow rate from this axial fan, which has a flow area $A_0 = 12$ in^2 or 0.0833 ft^2. The output velocity v_0 corresponding to this flow rate is given by Eq. (3.10) as:

$$v_0 = \frac{Q}{A_0} = \frac{80}{0.0833} = 960 \text{ ft / min} \tag{a}$$

The mass density for air at standard temperature and pressure is given by:

$$\rho = \frac{\gamma}{g} = \frac{0.0765}{32.2} = 2.376 \times 10^{-3} \frac{lb - s^2}{ft^4} = 0.66 \times 10^{-6} \frac{lb - min^2}{ft^4} \tag{b}$$

Finally, we substitute these results into Eq. (3.19) to obtain the propulsion force F_P as:

---

[5] In this relation A and ρ are scalar quantities and F_P and v_0 are vectors. Because the flow is uniaxial, the directions of both F_P and v_0 are collinear and coincide with the axis of the fan. Because of this fact we can treat them as scalar quantities.

$$F_P = Q\rho\, v_0 = (80)(0.66 \times 10^{-6})(960) = 0.0507 \text{ lbs} \tag{c}$$

The force produced by this fan is relatively small—slightly less than one ounce. Will this be sufficient to drive the hovercraft, and if so what velocities can be achieved?

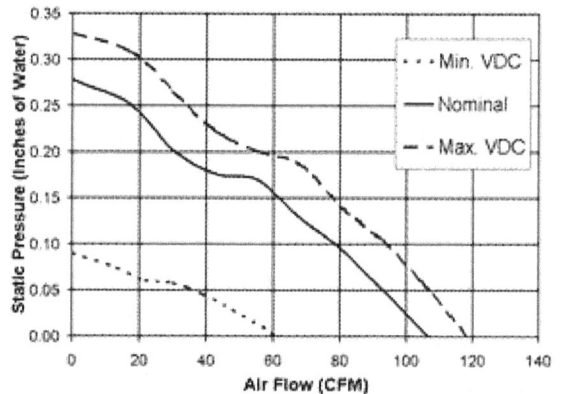

Fig. 3.8 Pressure as a function of air flow for a
120 × 120 × 38 mm axial fan.

3.8 SIZING THE AUXILIARY POWER SUPPLY

It is clear that the power required to drive the lift and propulsion fans must be supplied from an auxiliary power source. What is the required capacity of this source? If we select a source that has excess capacity, we incur a weight penalty. If we select one that is too small, we may exhaust the battery before the mission is complete. The current drawn by the fan motor varies with the flow rate. At a flow rate of 72 CFM, a 12 Vdc motor driving a medium size fan draws nearly 3 A. The smaller propulsion fans are rated at 6 W at 12 Vdc, which implies a maximum current draw of 0.5 A for each motor. Consequently, we will need an auxiliary power supply that can deliver about 4 A for sufficient time to complete a mission[6].

Let's assume that your team has a strategy for completing a mission in 6 minutes (1/10 h). The capacity C_B of the battery is given as the product of the amperes multiplied by the time that this level of current is delivered.

$$C_B = I\, t \tag{3.20}$$

where C_B is the capacity measured in (Ah) and I is the current draw in A and t is the time in h.

For a 6 minute mission, the capacity of the auxiliary supply should be $C_B = 4 \times 1/10 = 0.4$ Ah = 400 mAh. This value is small compared to the capacity of the battery in your automobile, but you certainly do not want to deal with the weight of an auto battery.

There are four types of batteries that provide this capacity and current drain—Li-Ion, Nickel Metal Hydride, Nickel Cadmium and Sealed Lead Acid (SLA). All four of these types are rechargeable, which is a requirement to avoid battery replacement after each mission. Go online and explore the battery sizes available, weight of the battery packs and the cost of the installation before

---

[6] We assume that you will not rely on the battery supply in the NXT to power any of the fans.

making your selection. We found a SLA battery that delivers 0.8 Ah at 12 V that weighs 1.0 lb at a cost of about $10.00. Can your team find a better battery for this mission? A Li-Ion battery like the one shown in Fig. 2.38, which is rechargeable, will delivery more power and weigh much less, but it will cost more. The selection of the battery for the power supply is an important design decision for your team.

3.9 DESIGNING FOR BOUYANCY

The requirements for the design of the hovercraft structure changes if the vehicle is to traverse a body of water. While the hovercraft should be able to float across the water on its cushion of air, loss of power to the lift fan would be disastrous. The hovercraft model would sink and the electronics mounted on the deck would be destroyed. To avoid this potential accident, the structure of the hovercraft is designed with sufficient buoyancy to prevent the deck from submerging.

The buoyancy force F_B developed by a body submerged in water or some other liquid is given by:

$$F_B = \gamma V = \rho g V \tag{3.21}$$

where γ is the specific weight of water 62.4 lb/ft^3 or 0.03611 lb/in^3 , V is the volume of the submerged body and ρ is the mass density of water 0.1617 lb-s^2/in^4.

Consider a ring structure for the hovercraft body that is presented in Fig. 3.9. The skirt is fabricated from a closed cell plastic foam that has inside and inside radii r_i and an outside radius of r_o, respectively. The weight on the deck of the hovercraft forces the structure into the water by a distance h developing buoyancy forces that total F_B. The structure is in equilibrium where:

$$F_B = W \tag{3.22}$$

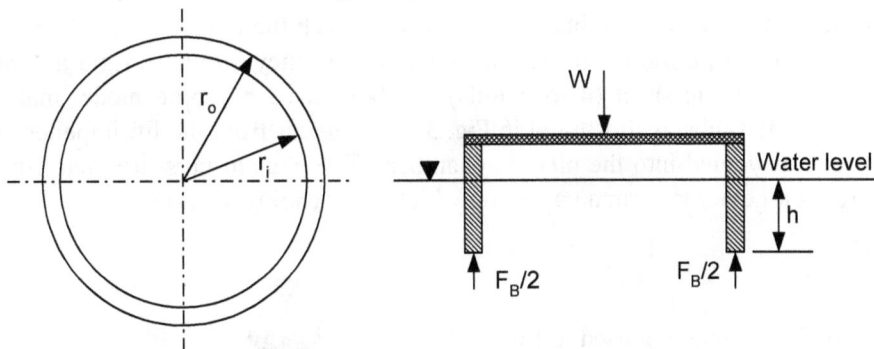

Fig. 3.9 Cross section and top views of a partially submerged hovercraft structure.

The volume of the submerged portion of the foam ring is given by:

$$V = \pi(r_o^2 - r_i^2)h \tag{3.23}$$

Substituting Eq. (3.23) into Eq. (3.21) and then substitute the result into Eq. (3.22) to obtain the submerged depth h as:

$$h = \frac{W}{\gamma\pi(r_o^2 - r_i^2)} \tag{3.24}$$

Let's consider a structure with r_o and r_i equal 24 and 22 in., respectively with a weight of 7.5 lbs. The depth of submergence is given by Eq. (3.24) as:

$$h = \frac{W}{\gamma\pi(r_o^2 - r_i^2)} = \frac{7.5}{(0.03611)\pi\left[(24)^2 - (22)^2\right]} = 0.719 \text{ in.} \tag{a}$$

The results of this example show that a relatively small depth of submergence was sufficient to prevent the deck of the hovercraft from submerging even with a fairly `high weight W = 7.5 lb. However, it is important to account for wave actions and to maintain sufficient free board to prevent waves from sweeping over the deck of the hovercraft and damaging the electronic equipment.

3.10 DESIGN CONSIDERATIONS FOR THE DECK AND SKIRT

The two main structural elements of the hovercraft are the deck and the skirt. The deck should be sufficiently strong and rigid to carry the weight of the fan, the batteries and all of the other lighter components. Reference to the preliminary weight listings in Table 3.2 indicates that the weight of the deck was estimated at 0.75 lb and as such it is only about 10% of the total weight target of 7.5 lb. It is probably advisable to use thin plywood or foam board as deck material and attempt to achieve a deck weight of less than 0.75 lb.

The skirt design is dependent on the mission. If the mission includes traveling over water, the skirt must provide a buoyancy force that is sufficient to float the hovercraft. In this case, the skirt is usually made from an inflatable tube or closed cell foam plastic. The foam plastic has the advantage of easier attachment to the deck.

If the mission is over a solid surface, there are more options available for skirt materials. An inflatable tube is still an option, but it is not easy to attach the tube to the deck in an air tight manner. Rings of closed cell plastic foam may be employed but they are more rigid and prone to snagging. Relatively thick plastic sheet (4 to 6 mils) has been used by some model makers[7] to produce a relatively large flat tube as illustrated in Fig. 3.10. The air from the lift impeller inflates this plastic tube and then is vented into the plenum chamber. The plenum pressure then lifts the cushions and the hovercraft together to form a gap along which the escaping air flows.

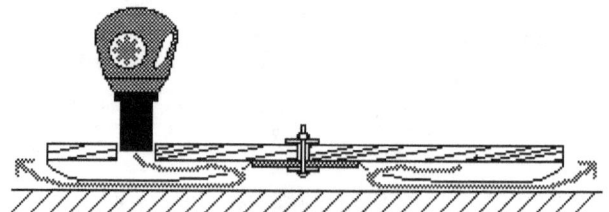

Fig. 3.10 Plastic sheeting used to form a cushion like skirt. Design and drawing by William J. Beaty.

---

[7] This design is described by William J. Beaty a research engineer at the University of Washington. He spends his free time running SCIENCE HOBBYIST, a large website for amateur science and science education. You can locate an interesting article on hovercraft construction by Mr. Beaty on www.amasci.com/me.html.

The arrangement for the skirt shown in Fig. 3.11, utilizes a ¾ segment of corrugated tubing for the skirt. This tubing, readily available from building supply outlets, is made from a relatively thin plastic material that is easy to cut and to staple. The tube is stapled to the top of the deck and then the reentrant corner is filled with a sealant to form an air tight plenum chamber. Slicing the tube to form the ¾ segment provides a structure that is semi-rigid, yet sufficiently flexible to adjust for slight surface irregularities that might be encountered.

Fig. 3.11 A segment of corrugated tubing fastened and sealed to the deck serves as a skirt.

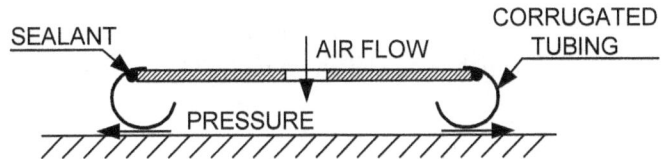

Another approach is to use either a container[8] or the lid to a container found in a kitchenware department of a retail outlet. These containers are unusually made from relatively thick polyethylene and they come in a large number of different shapes and sizes. The walls of say a dishpan can be employed to form the skirt. The bottom of the dishpan may not be sufficiently rigid for the deck, but it can be locally reinforced with thin pieces of plywood. Attachment of the plywood sheet to the plastic pan is easily accomplished with staples, tacks or small screws.

It is clear that there are a number of different approaches to designing and constructing a hovercraft structure that has a strong and rigid deck, an air tight plenum chamber and skirt with sufficient flexibility to accommodate modest surface roughness. The merits of the design of the hovercraft structure other than serving it function are its weight and the ability of its skirt to accommodate an occasional contact with the surface of the track.

3.11 SENSOR SELECTION

Sensor selection depends on the mission requirements. If the mission calls for the hovercraft to move from point a to point b, a line following strategy is often employed. In this case, a line is made delineating the course between points a and b. This line is usually made with black tape on a white background to provide the maximum contrast for detecting reflected light. The light reflected from the tape is measured with a light sensor to provide a feedback signal. Other missions may require either contacting obstacles or avoiding them. In these cases, either touch sensors or proximity sensors are used to provide feedback signals. Regardless of the mission, the sensors used must interface with the NXT controller. Information pertaining to the design of sensors that interface with the NXT is provided in Chapter 4.

Another approach is to procure sensors from the LEGO Educational Division because they are designed specifically to interface with the NXT. Sensors available from the LEGO Educational Division include:

- Touch sensors—are normally-open limit switches housed in a peg-mounted package.

---

[8] Such as dishpans or storage containers of all sizes.

- Temperature sensors are temperature-sensitive resistors (thermistors). They are housed within a small probe extending out from the short side of the peg-mounting block. They measure temperatures from –20° C to 50° C. The NXT converts the resistance measurement to a temperature, which is displayed on its LCD.

- Light sensors—are active devices (requiring power) that include a light emitting diode (LED) and a phototransistor built into a peg mounted package. The sensor can be used either to detect incident light or to detect reflected light from the LED.

- Rotation sensors, incorporated in the LEGO servo motors, include an optically coupled encoder with a resolution of 1.0 degrees.

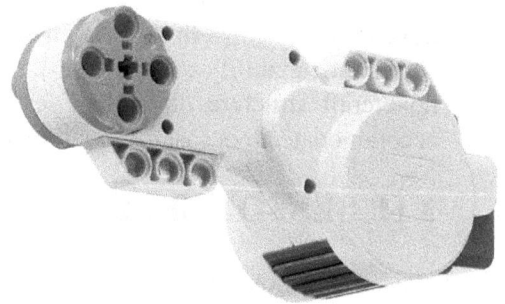

- Ultrasonic sensors are distance measuring instruments. They send out an ultrasonic signal and measure the time of transit of the signal to and from a specific target. From the speed of sound and this time measurement, the distance to the target is determined.

- Sound sensors, incorporated in a peg-mounted block, includes a microphone and circuitry to measure sound levels in dB and dBA . It is capable of recognizing sound patterns and tones.

- Compass sensors measure the Earth's magnetic field to provide accurate information for navigation. In the read mode it provides your current heading. In the calibration mode it compensates for externally generated magnetic field anomalies (like those found near motors or batteries).

This listing of sensors for the NXT microprocessor, available from the Lego Educational Division, is not complete. A complete listing of their sensors is found at http://www.legoeducation.us/store/.

3.12 MASTER CONTROL SYSTEM

The NXT serves as the master controller. Its output ports A, B and C can be used to control the relays or transistors that provide 12 Vdc with sufficient current to operate the lift and propulsion fans. Sensors are connected to the input ports 1, 2, 3 or 4 on the NXT. The type of sensors used will depend on the mission and the strategy for accomplishing it. The signals that they provide will serve as the basis for controlling the power to the fan motors.

The program for controlling the operation of the hovercraft's motors to successfully complete the assigned mission is to be written in ROBOTC for Mindstorms 1.10. A detailed description of programming in ROBOTC is presented in Chapter 6.

3.13 SUMMARY

In this chapter we introduced the theory necessary to perform several design analyses associated with developing a hovercraft. We performed several analyses to insure that the functions (systems and processes) involved in designing a hovercraft would operate satisfactorily.

Because the pressures developed by fans are often expressed in terms of inches of water, we derived the pressure height relation based on the equilibrium equation. Then we showed how a manometer measures pressure with the height of a column of water. Next we determined the lift force that could be created with a plenum chamber at a pressure p and a deck area of A_D. The results were tabulated for a large number of pressures and areas to provide a "solution space" for lift requirements. An example of the first estimate of the weight of the hovercraft was presented. The operating characteristics of a fan were used to select an operating flow rate and plenum pressure. These values were then used to determine the deck area of the hovercraft.

The continuity relation was derived based on the principle of conservation of mass and Bernoulli's equation was derived based on conservation of energy. These two equations were then used to determine the leakage flow rate. By equating the leakage flow rate with the flow rate generated by the fan, the gap area was determined. A gap dimension of 0.0511 in. was determined for a plenum chamber with a circular base for the fan selected and the assumptions made in the analysis.

The propulsion force developed by a small axial flow fan was calculated based on the momentum-impulse principle. Using this result, we then determined the velocity and distance traveled by a hovercraft as a function of time of operation.

If the hovercraft is to operate over water, its structure must be designed with sufficient buoyancy to float the vehicle in the event of failure of the levitation fan. The theory for buoyancy was developed and an example presented to demonstrate the method for determining the depth of submergence of the hovercraft.

Guidance was provided for sizing the auxiliary power supply and the selection of the battery chemistry. Several design concepts for the hovercraft's skirt were introduced and materials and methods of fabrication were briefly described. Next, a section describing sensor selection was included. The LEGO sensors which interface easily with the NXT were described. Also we discussed a proximity sensor, which issues a signal when objects about 240 mm away are encountered. This sensor, which has been adapted to the NXT, may be useful for several of the missions.

Finally, the NXT was identified as the master controller. It will be necessary to develop a strategy for completing the specified mission and to program the NXT in ROBOTC to execute this strategy.

LABORATORY EXERCISES

3.1   Construct a manometer capable of measuring a head of 120 mm of water. Could you use this manometer to measure the outlet pressure from a tank of propane used as a fuel for a barbeque grill?

3.2   Construct a plenum chamber and pressurize it with your team's lift fan. Measure the plenum pressure with a manometer as you vary the gap between the plenum chamber and a smooth flat surface. Vary the gap from zero to 6.0 mm in steps of 0.5 mm.

3.3   Using the plenum chamber from Laboratory Exercise 3.2, measure the flow rate of your lift fan as a function plenum pressure. This can be accomplished by cutting a series of circular holes in the side of the plenum chamber to serve as orifices. The flow rate can be determined for each orifice diameter, if the plenum pressure is measured.

EXERCISES

3.1   Determine the pressure in units of pounds per square inch (psi) of the water in a swimming pool at a depth of 6 feet.

3.2   What is the pressure of air in Pascal (Pa) if a mercury barometer is reading 76.8 mm of Hg?

3.3   Reproduce Table 3.1 using SI units instead of U. S. Customary units.

3.4   Prepare a drawing of the hovercraft including its skirt. Discuss the materials selected and describe the manufacturing methods your team intends to employ in fabricating the hovercraft.

3.5   Based on the assigned mission, prepare a strategy for control of the hovercraft to complete the mission in a minimum time.

3.6   Select the sensor or sensors necessary to implement the strategy prepared in Exercise 3.5.

3.7   A piston weighing 1.40 lb slides freely within a cylinder as illustrated in Fig. Ex 3.7.

   a.   What is the pressure in the cylinder in psi and Pa if the piston has a diameter of $D = 4.0$ in? [Draw a Free Body Diagram (FBD) to show the relevant forces acting piston]

   b.   What is the pressure in the cylinder in psi and Pa if the piston is square with a 4.0 in. side?

   Fig. Ex3.7

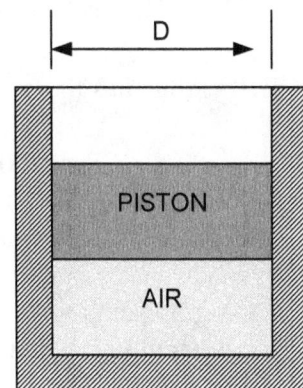

3.8   Air at 300 °K flows through a cylindrical pipe with an area change as illustrated in Fig. Ex3.8. The velocity at station 1 is 12 m/s and the pressure is 2 atm.

   (a) What is the velocity at station 2 if the pressure there is 1 atm?

   (b) What is the volume flow rate at station 1?

(c) What is the volume flow rate at station 2?
(d) What is the mass flow rate at stations 1 and 2?

Fig. Ex3.8

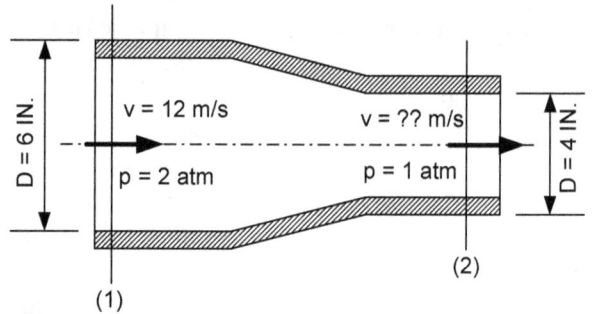

3.9 Consider the rectangular hovercraft plenum illustrated in Fig. Ex3.9.

    a. Estimate the plenum's weight assuming that it is constructed using 1/16 in. thick plywood, which has a density of 500 kg/m³.

    b. What is the required pressure in the plenum? (You are advised to draw a FBD).

    c. Derive an expression for the hover height in inches as a function of air flow rate in CFM (ft³/min).

    d. Use Excel and your result from (c) to plot the hover height as a function of air flow rate. What is the effect of increasing the plenum mass by 10%?

    e. If the mass and total area of the plenum must remain fixed, how could you change the design of the plenum to maximize hover height for a particular air flow rate?

Fig. Ex3.9

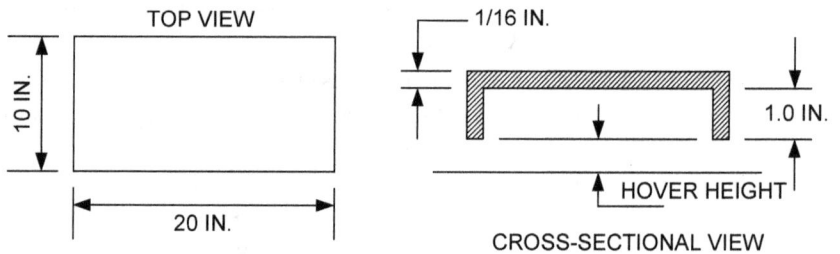

3.10 A fan has the pressure and flow rate characteristics described by:

$$P = 50,000(0.004 - Q^2)$$

where p is given in Pa and Q is given in m³/s

A graphical representation of this equation is given in Fig. Ex3.10.

Fig. Ex3.10

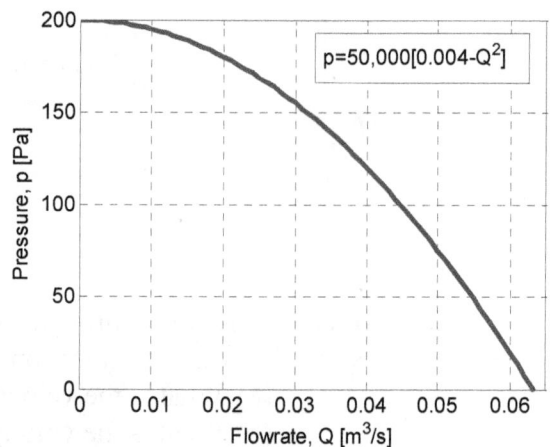

For a circular planform area of 0.218 m^2, a vehicle mass of 3.5 kg and the density of air at 1.17 kg/m^3, what is the hover height of the vehicle? You may either use the equation shown in the graph or estimate the proper values from reading the graph directly. Indicate your operating point on the graph.

3.11 a) For a hovercraft mass of 2.4 kg and a circular planform with a diameter of 0.45 m, specify an axial fan which enables the hovercraft to levitate with a target hover height of 3.5 mm. Assume an air density of 1.2 kg/m^3. To narrow the possibilities, select a fan from Emb Papst at the following location: http://www.ebmpapst.us/

    b) Calculate the required plenum pressure and volumetric flow rate. Select a fan with a plenum pressure that is 20% higher than the calculated value to allow for frictional losses.

    c) For the selected fan show its characteristics, and indicate your hovercraft operation point on the diagram.

    d) Report the fan model number, description, weight, and power consumption.

    e) Find a supplier for the fan (give web link) and price.

3.12 Consider a circular hovercraft that is 0.50 m in diameter. Its mass is set at 2.8 kg and the minimum gap required for levitation is 4 mm.

    a) Calculate the pressure differential required at the fan and its flow rate

    b) If the weight is doubled, how are the previous results affected?

3.13 To design your hovercraft, you will need to consider many competing parameters that relate to optimal levitation performance, cost, and maneuvering your vehicle through the course. From a levitation perspective, the parameters of the total mass, overall size, hover height, and power consumed by the levitation fan are the main drivers for the design. The following question is intended to help you consider some of these factors. Consider a circular hovercraft planform operating in an atmosphere with a density of 1.2 kg/m^3 and a constant hover height of h_{gap} = 0.003 m. For each parameter, examine the parametric dependence of the value in question by creating a table for the possible combinations of mass and diameter, given discrete values of m = 1, 2 and 3 kg, and D = 0.2, 0.4 and 0.6 m (There are $(3)^2$ = 9 values to calculate in each table, as shown on the next page). Use of Excel is recommended. Show the equations used, the details of one sample calculation, and the table in your solution.

    a. Calculate the required plenum pressure, Δp, in Pa.

    b. Calculate the volume flow rate, Q, in m^3/s. State how the flow rate varies with the size (diameter) of your vehicle. Explain any trends you notice in the data.

    c. Calculate the power required to drive the air through the hovercraft, P=ΔpQ, in W =N-m/s. Make a recommendation for your hovercraft design in terms of its target size and mass based on the previous calculations. What other considerations from the specifications play a limiting role in selecting these quantities?

    d. Considering the hull to be a disk of constant thickness, the mass of the hovercraft hull should scale as D^2. Using this relation, make an estimate of how the power consumption should vary with vehicle size.

mass (kg)	diameter (m)		
	0.2	0.4	0.6
1			
2			
3			

3.14 A lift fan for the hovercraft shown in Fig. Ex3.14a has the pressure and flow rate characteristics given by:

$$p = 50,000(0.005 - Q^2)$$

where p is given in Pa and Q is in m^3/s.

Fig. Ex3.14a

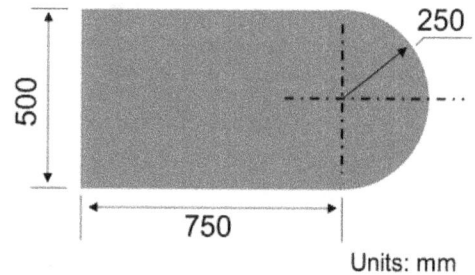

Units: mm

Top view of hovercraft planform

The flow rate as a function of pressure for this fan is presented in Fig Ex3.14b.

Fig. Ex3.14b

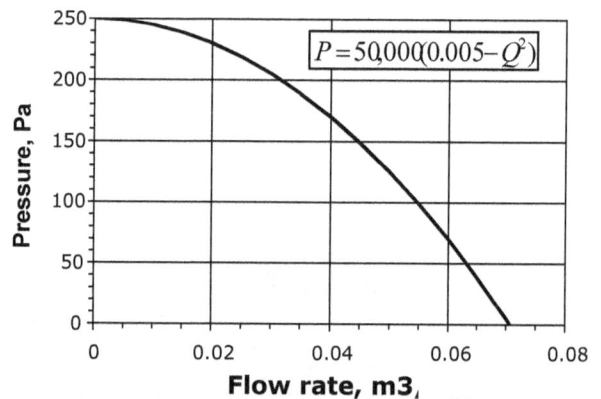

What is the maximum permissible hovercraft mass, in kg, if the minimum acceptable hover height is specified as 2.0 mm? Assume the density of air is 1.2 kg/m^3. You may either use the equation shown in the graph or estimate the proper values from reading the graph directly. Indicate your operating point on the graph.

CHAPTER 4

BASIC ELECTRIC CIRCUITS AND SENSORS

Wesley G. Lawson

4.1 INTRODUCTION

This chapter introduces you to the basic quantities, concepts, and physical devices used in electric circuits. The discussion begins with definitions of charge, current, voltage, and other basic quantities used to describe circuit characteristics. Then, key passive components are introduced, including resistors, capacitors, inductors and switches. Next, the equations needed to solve circuit problems are provided. A basic description of important electronic (semiconductor) devices, such as diodes, transistors, voltage regulators and operational amplifiers, follows. The chapter concludes with a number of circuit examples that have application to the hovercraft project.

4.2 BASIC ELECTRICAL QUANTITIES AND CONCEPTS

Electricity involves charge carriers—either electrons, which are negative charge carriers, or holes which are positive charge carriers. Since we cannot see, smell or touch an electron, electricity has always been somewhat mysterious to most of us. Holes, which can be thought of as atoms with an electron missing from their outer ring, are even more obscure. Nevertheless, we will try to introduce many of the basic electrical concepts in a simple, straightforward manner to uncover the mystery.

Electricity can be understood most simply by drawing an analogy between charge flowing through a circuit and water flowing through a pipe. The analogy is not perfect, but it provides a great deal of insight into the basic operation of electricity.

4.2.1 Charge

Charge, denoted $q(t)$, is the fundamental quantity of electricity. Charges are the substances of which current is made; just as water molecules (H_2O) are the substances of which water current is made. Charge is measured in units of coulombs (C). One electron has a charge of -1.602×10^{-19} C. We can think of charge as a substance (a large number of "free" electrons) that is contained within an object such as a wire. This is analogous to a hose full of water. Charge does not really **do** anything until it moves. Excess charge can create an electrostatic force, but nothing happens until the charge moves.

4.2.2 Current

Current, denoted $I(t)$, is charge in motion. The amount of current flow is measured in amperes (A), which is the same as coulombs per second. Just as we speak of the rate at which a volume of water moves through a pipe, we can think of current as a flow rate. The higher the current, the higher the rate at which charge is moving past a point in a conductor. This concept is expressed in equation form as:

$$I(t) = dq/dt \qquad (4.1)$$

In other words, the current in a specific wire is the (instantaneous) rate of movement of charge passing through that wire per unit time.

Despite the fact that negatively charged electrons are the carriers of current in wires, it has become customary to think of the current in a circuit as the flow of positive charges. All that one needs to remember is that the effect of electrons moving in one direction nets the same result as positive charges of identical magnitude flowing in the opposite direction.

4.2.3 Voltage

Voltage, denoted V(t), is another fundamental electrical quantity. The unit of voltage is volts (V). Voltage is always measured between two points and represents how much energy a unit charge will gain or lose as it travels between those two points. As such, it is the electrical equivalent of pressure. Just as a pump creates pressure to force water through a pipe, a voltage source creates pressure to push charge through a circuit.

An ideal voltage source is a circuit element that delivers a specified voltage (electrical pressure) regardless of the current flowing. The most basic *real* voltage source is a battery. A flashlight battery pushes charge through a tungsten wire that comprises the incandescent light bulb filament, creating friction within the filament, and causing the bulb to heat up and glow. A 6-V battery pushes four times as hard as a 1.5-V battery. The symbol for a constant voltage source (called a **direct current**, or dc, source) such as a battery is shown in Fig. 4.1a. A more general voltage source symbol is shown in Fig. 4.1b. A positive charge that enters the negative terminal of the battery (in Fig. 4.1a) leaves the positive terminal after it gains energy proportional to V_1. A positive charge traveling in the other direction gives up V_1 volts.

Sometimes we refer to voltage as the potential difference between two points. Just as the water pressure is reduced on the either side of a kink in the hose that restricts flow, the voltage drops across any part of the circuit that presents a resistance to current. This is a key concept that will be explored more in Section 4.5.

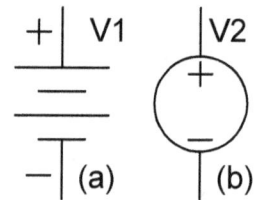

Fig. 4.1 (a) dc voltage source symbol
(b) general voltage source symbol.

4.2.4 Electric Energy and Power

Electric energy, denoted w(t), describes the potential of a circuit or component to do work. The units are the same as for all types of energy: Joules (J). Some components, like batteries, convert other forms of energy to electric energy, thereby "generating" electric energy. Other components, like capacitors and inductors, can take in and store electric energy, to be released at some later time, but they can not generate energy on their own. Still other components, like resistors or fans, can only dissipate electric energy, which is to say that they convert electric energy to other forms of energy, such as heat or kinetic energy. Any component that does not generate electric power is called a "passive" component, in contrast to the sources that are represented in Fig. 4.1.

Electric Power, denoted P(t), is the instantaneous rate of change of energy in a given object:

$$P(t) = dw/dt \qquad (4.2)$$

As such, electric power describes the rate that electric energy flows in a circuit, i.e. the rate that energy enters or leaves circuit components. The units for power are joules/second or watts (W). The instantaneous power in any given component is equal to the current entering that component times the voltage drop that current experiences before it exits the component. In other words,

$$P(t) = [V(t)] [I(t)] \tag{4.3}$$

Because of this relationship, electrical engineers almost always concentrate on finding voltages and currents in a circuit before they look to evaluate other physical quantities.

4.2.5   Assembling the Basic Circuit Equations

There are two types of equations that are used to solve for voltages and currents in a given circuit. The first type, discussed in Section 4.5, have to do only with the way that electrical components are connected to each other in a particular circuit (i.e. the topology of the circuit) – we call these equations Kirchhoff's Laws, after their "discoverer." These equations don't "care" which particular component is placed in which particular location in the circuit. The other type of equations does care about specific components, since those equations describe the intrinsic interrelation between voltage and current in a particular component. These equations are called terminal relationships and will be discussed in Section 4.4, after we describe the materials used to make electric and electronic components.

4.3 MATERIALS FOR ELECTRIC AND ELECTRONIC DEVICES

4.3.1 Conductors

Conductors are materials that have a lot of "free" electrons and therefore allow a lot of current to flow in them when a voltage difference is impressed across to sides of the conductor. To quantify what we mean by "a lot", we define the conductivity of any material to be how many amps will flow in the material per applied voltage per meter. The "per meter" part is a little tricky, but you will see how that works in the next section when we talk about resistors. We use the Greek letter σ to signify conductivity, and the units of conductivity are Amps/volt/meter. Copper is a standard material used to make wire, and the conductivity of copper is fifty-eight MILLION Amps/volt/meter… so that is what we mean by a lot of current. Remember that all of this current in conductors is realized by the movement of electrons – negative charges.

Conductors for the most part are metals, and typical conductors used in electric circuits include copper and aluminum, silver and gold, lead and tin, etc. Most wire is made of copper, because it is a very good, relatively cheap conductor. Aluminum is not as good, but it is cheaper and lighter weight. It is harder to make good permanent connections between aluminum wires, and that is why, when you walk into an electronics store, you will probably only find copper wire. Silver is the best conductor, but more expensive than copper. Gold, though very expensive, is good for making contacts. Tin and lead have low melting points and are often used to make permanent connections between wires and components.

The key to why conductors work as they do is wrapped up in the meaning of "free" electrons. Conductors, like all materials, are made up of individual atoms that are normally charge-neutral, which means that they have the same number of electrons as protons. The electrons at room temperature are normally "bound" to the atom by electric forces and are not free to move arbitrarily far away from the nucleus. When the temperature of an atom is raised, the electrons become more energetic, and at some temperature, the electron which is most weakly bound to the atom gains enough energy to leave the atom, making it a "free" electron. In even the smallest chunk of some solid material, there are so many atoms that the bonds between an electron and its atom nucleus are complicated by neighboring atoms. In conductors, normal room temperature gives enough energy to the electrons that many of them are free to move anywhere inside the chunk of conductor. Thus, when small voltages are applied to conductors, very large currents can flow. You should remember that the conductors are still overall charge-neutral, so for every charge that exits a component in an electric circuit, another equal charge enters that component somewhere else (usually from the other side).

4.3.2 Insulators

Insulators are materials that contain very few free electrons at room temperature. Examples of insulators used in electric circuits are glasses, plastics, ceramics, and rubber. The conductivity of glass, for example, is about twenty orders of magnitude smaller than the conductivity of copper. Insulators are used in electric circuits to protect people and circuit connections by covering wires and other components. They are also used inside some components, particularly in capacitors, as will be discussed in the next section.

4.3.3 Semiconductors

Semiconductors, as the name suggests, have conductivities that fall between those of conductors and insulators. Typical semiconductor materials include the elements silicon and germanium, as well as compounds like Gallium-Arsenide, Cadmium-Sulfide, etc. The intrinsic room-temperature conductivity of pure silicon is about ten milliamps/volt/meter, or in other words 9-10 orders of magnitude smaller than copper. For some applications pure silicon is adequate, but by far, most electronic components are made from doped semiconductors—semiconductors in which impurities are deliberately introduced to adjust the properties of the material. The impurities generally come in two classes. If the impurity has more electrons in its outer valence band than the semiconductor material (a standard example is phosphorous implanted in silicon), the resultant material has an excess of free electrons and is called an n-type semiconductor. Hence, electrons carry the bulk of the current in n-type materials. On the other hand, if the impurity has fewer electrons in its outer valence band than the semiconductor material (e.g. boron implanted in silicon), the resultant material has shortage of electrons, or more accurately, an excess of free "holes" and is called a p-type semiconductor. So, in contrast, positively-charged "holes" carry the bulk of the current in p-type materials. Whether an n-type or p-type material, the conductivity increases with increasing number of impurities, allowing more and more current to flow for a given applied voltage. The impurity concentrations in a semiconductor device depend on the application and the required properties (i.e. voltage, current, power, etc.) Devices based on n-type, p-type, and combinations of these two materials are described in Section 4.6 below.

4.4 BASIC ELECTRIC COMPONENTS

4.4.1 Terminals and Reference Directions

Every electric component has two or more places where it can be connected to other components in a circuit. These connection points are called "terminals" and are almost always made of metal. The terminals of each component are connected to the terminals of other components to construct the circuit. Permanent connections are often made via solder, which are metals that melt at low temperatures; temporary connections are often made with "wire nuts" or "circuit breadboards." In modern surface-mount printed circuit boards (PCBs) these terminals are small metal pads, but in most other circuits the terminals are metal wires that extend from the ends of the component. For this reason, we draw a generic two-terminal "passive" component as in Fig. 4.2. The terminal wires are represented by the two lines that extend from the rectangular box, which in turn represents the passive component. Note that the terminal connections for voltage sources were represented by the lines (wires) in Fig. 4.1.

Neither current nor voltage is a simple number. Current refers to a direction of charge flow, so in addition to the current value, we need to indicate the direction of flow, as we have done in Fig. 4.2 with the arrow. Voltage represents a potential energy gain between two different points in a circuit, and so we indicate the point of lower energy with a minus sign and the point of higher voltage with a plus sign (as shown in Fig. 4.2.). These voltage symbols are the same as those used for voltage sources. Together, the current direction and voltage polarity are known as reference directions. There are two important concepts to remember about reference directions. First, you usually need to assign reference directions to components in a circuit before you actually can solve for voltages and currents. You normally do this by guessing in which direction a current will flow, etc. If you guess wrong, don't worry. All that will happen is that when you solve for that current, for example, is that the value you calculate will be negative. Any negative value for a current or voltage is just fine – it simply means that the current really flows in the opposite direction or that the point of lower voltage corresponds to the plus terminal. Second, when you calculate power, a positive power means that energy is flowing into a device *only if* the current and voltage references are coordinated as shown in Fig. 4.2. In other words, the current is flowing into the positive voltage terminal of the device. This coordination is called the passive sign convention, because it is the standard orientation for passive components. Whether or not you chose to adopt the passive sign convention is not important. What is important is that you understand the reference direction scheme you choose, so that you understand when energy is entering components and when it's going out of components.

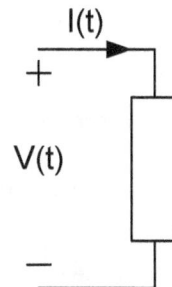

Fig. 4.2 A generic two terminal device showing reference directions.

4.4.2 Resistors

Resistors are passive components that convert electrical power into heat. Resistors are characterized by their resistance, which measures the difficulty of pushing current from one place to another. The unit of resistance is ohms (Ω), which is equal to volts/amps (V/A). The higher the resistance of a device, the more difficult it is to push current through it. Pumping water through a small-diameter pipe requires more pressure to achieve the same flow rate as pumping through a larger-diameter pipe. A high-resistance electrical device is analogous to a small-diameter pipe, resisting more strongly the flow of charge. Likewise, a low-resistance device is analogous to a large-diameter pipe that offers little resistance to flow.

An ideal resistor is a circuit element that has a fixed resistance regardless of the input voltage. The symbol for a resistor is shown in Fig. 4.3, where the resistor is connected to a battery to make a simple circuit. The value of the resistance completes the definition of a resistor and must always be shown on the side of the resistor symbol. The connecting lines between the voltage source terminals and the resistor terminals represent wires, which can be thought of as resistors with nearly zero resistance. In our water analogy, wires are extremely large pipes that offer little resistance to water flow.

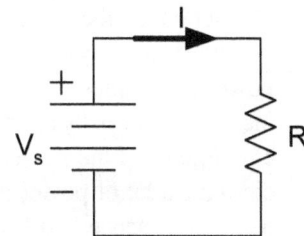

Fig. 4.3 A simple circuit loop containing a dc voltage source and a resistor.

The relation between the applied voltage V_s, the resistance R, and the current flow I is given by:

$$V_s = I\,R \tag{4.4a}$$

$$I = V_s\,/R \tag{4.4b}$$

Equations (4.4a) and (4.4b) are both expressions of Ohm's law. When V_s is expressed in volts and I in amperes, R is given in ohms (Ω).

The resistance R of a conductor is given by:

$$R = \rho L/A \tag{4.5}$$

where ρ is the specific resistivity of the metal conductor (Ω-cm).
L is the length of the conductor (cm).
A is the cross-sectional area of the conductor (cm^2).

Note that the resistivity is the reciprocal of the conductivity, σ, defined in the previous section. We can fabricate resistors by winding very-small-diameter, high-resistance wire on coils, by mixing carbon powder with insulating "filler" material, or by evaporating thin films of metal on ceramic substrates. The relation in Eq. (4.5) allows us to select the metal or carbon percentage, and to size the conductor to provide a specified resistance. Potentiometers are mechanically-tunable variable resistors.

For a resistor connected to a battery (with a constant output voltage) the electric power dissipated as heat can be re-written from Eq. 4.3 as:

$$P = V_d\,I \tag{4.6}$$

where P is the power dissipated and V_d is the voltage drop across the resistor.

Substituting Eqs. (4.4a) and (4.4b) into Eq. (4.6), we obtain useful expressions for power loss in a resistor:

$$P = I^2\,R = V_d^2\,/\,R \tag{4.7}$$

We have used the subscript "s" with the voltage in Eqs. (4.4a) and (4.4b) and the subscript "d" with the voltage in Eq. (4.7). For the simple circuit shown in Fig 4.3, the supply voltage V_s is equal to the voltage drop across the resistor V_d. In this case, the subscripts can be interchanged. Clearly, it is easy for us to determine the power lost if we know the current flowing through a resistor or the voltage drop across the resistor.

Many applications, perhaps including your hovercrafts, use resistors that are cylindrical in shape. You'll note that the size of the resistor has nothing to do with the resistance of the component—the same package size can be used from single digit resistances up to millions of ohms. The package size, instead, is a function of the maximum power that the resistor can dissipate without overheating. The smallest packages can only handle ¼ W or less, so it is important to know both the exact resistance and minimum power requirements before purchasing components. One last important item to note is that the resistivity increases (normally) with increasing temperature, so the resistance might change with time in a real circuit, even if the power level is below the maximum allowed. Sometimes resistors that are designed to dissipate a lot of power require physical contact with a heat sink – some (usually metal) block with high heat conductance and heat capacity to keep the temperature to an acceptable level. Sometimes the heat sink needs to be cooled with a fan. (A good example of this is the cooling system for the microprocessor in your desktop computer.)

Note that resistors are not fabricated perfectly, nor are they fabricated with any value that you want. Carbon composition resistors typically have tolerances of ±5% or ±10%. Metal film resistors can have tolerances of ±1% (or less). Resistance values are spaced so that given the tolerances, theoretically all values of resistors could be made. In reality, some resistance values are "less standard" than others and therefore harder to find, but for most applications, obtaining the required resistors is not a difficult problem.

4.4.3 Capacitors

Capacitors are passive components that can store electric energy. They are characterized by their capacitance, which is a bit more difficult to understand than the previous quantities. However, it also has a water-related analogy. It is used to characterize devices that have a degree of "springiness" – devices in which charge can be pumped in but which tend to push charge back when the external pressure is removed. The ease with which charge can be pushed into the device measures its capacity. The unit of capacitance is the farad (F).

An ideal capacitor is a passive electrical component that has a fixed capacitance, regardless of the amount of voltage across its terminals. The circuit symbol for a capacitor is shown in Fig. 4.4. The capacitor ordinarily consists of two flat, parallel-plate electrodes separated by a dielectric, which also serves as an electrical insulator. The approximate capacitance C of a flat-plate capacitor is given in units of picofarad and is determined by:

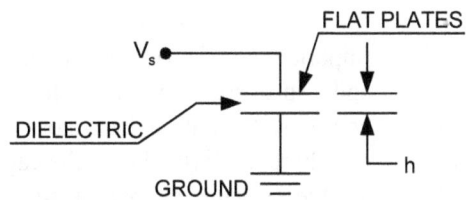

Fig. 4.4 A flat parallel plate capacitor.

$$C = k \, \varepsilon \, A/h \qquad (4.8)$$

where ε is the dielectric constant, A is the area of the plates, h is the distance between the two plates as shown in Fig. 4.4, and k is a proportionality constant: 0.00885 for dimensions in millimeters.

When a voltage V is applied across the capacitor, it stores a charge q that is given by:

$$q(t) = C \, V(t) \qquad (4.9)$$

where C is the capacitance in farads and q is the charge on the capacitor given in coulombs.

When a voltage is first applied across the terminals of a capacitor, current flows and it becomes charged. However, when the capacitor is fully charged and maintained at a constant voltage, no current flows through it. The capacitor conducts current only when the voltage across it changes with respect to time. The idea is expressed mathematically as the terminal relationship for a capacitor:

$$I(t) = C \, dV/dt \qquad (4.10)$$

The water analogy for a capacitor is a balloon or membrane stretched across the cross-section of a pipe to block the flow of water. As pressure is applied to attempt to pump water through the pipe, the balloon begins to stretch, which allows a small amount of water to flow. As the balloon stretches, it resists more and more until the backpressure of the balloon resisting the flow equals the forward pressure of the pump. At that point, the water ceases to move.

The electrical energy stored in a capacitor [w(t)] is determined from Eqs. (4.2), (4.3) and (4.10):

$$w(t) = C \, [V(t)]^2 \, /2 \qquad\qquad (4.11)$$

where w(t) is the stored energy in joules when C is expressed in Farads.

Similar to resistors, the capacitance is not the only parameter that you need to know before you purchase a capacitor. For capacitors, it is the maximum voltage that you also need to know. The spacing between capacitor plates is often very small, and that leads to large electric forces. If the forces get too high, they can damage or destroy the capacitor.

The shapes, sizes, and insulators also vary quite a bit. The size increases both with increasing capacitance and with increasing "working' voltage, i.e. the maximum allowable voltage. Some very high capacitances (tens or even hundreds of microfarads) can be found in small packages. Be careful – these capacitors are probably polarized, which means that one terminal must always be kept at a lower voltage than the other terminal –they **will** be destroyed if even a small voltage is applied with the wrong polarity. The positive terminal is usually marked with a plus sign and sometimes the negative terminal is marked with a minus sign. Capacitor tolerances are generally ±10% or ±20%, but sometimes they are even worse.

In addition to storing energy, capacitors have another property that gives them an additional important use. Capacitors tend to "absorb" short "spikes" of electrical energy. Without a capacitor, a small spike in energy can cause short-lived, but large, voltage spikes in a circuit that can destroy electronic components. With a capacitor, the energy is "diverted" to the capacitor and the voltage spikes can be greatly reduced (the larger the capacitance, the greater the reduction, of course). For this reason, capacitors are the principle element in the surge protector that you connect your computers to prevent them from being damaged by voltage spikes.

4.4.4 Inductors

Inductors are passive components that can store magnetic energy. They are characterized by their inductance, which relates the flux of magnetic field they generate to the current passing through the inductor. The symbol for the inductor is shown in Fig. 4.5. The inductance of an inductor is hard to calculate in general, but is well-defined when you buy commercial units. One way to produce an inductor is to wind a coil of small-diameter, insulated wire about a magnetic (e.g. iron or ferrite) core. The inductance L is given by:

$$L = 4\pi N^2 \, A\mu/\ell \times 10^{-9} \qquad\qquad (4.12)$$

where   L is the inductance in Henrys (H)
        N is the number of turns of wire in the coil
        μ is the permeability of the magnetic core material
        A is the cross sectional area of the coil
        ℓ is the length of the coil

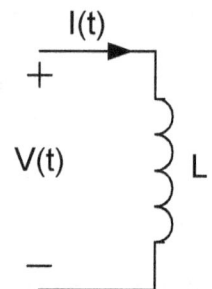

Fig. 4.5 The symbol for an inductor.

When a step voltage is first applied across an inductor, it acts like an open circuit and does not permit current to flow through it. Gradually, current flow through the inductor begins to increase until the inductor offers no resistance at all. In a sense, the inductor has an internal inertia with respect to current. The inertia must be overcome for current to flow. When current begins to flow, the inertia of the inductor tends to keep the current flowing even when the external voltage source is removed. The inductance L is a measure of this electrical inertia. This concept is summarized mathematically as the terminal relationship for an inductor:

$$V(t) = L \, di/dt \qquad\qquad (4.13)$$

The water analogy for an inductor is a water wheel in a channel through which water flows. At first, the inertia of the stationary wheel prevents the water from moving through the channel and past the wheel. As the water pushes on the wheel, the wheel begins to turn and speed up, allowing more and more water to pass. Eventually, the wheel is turning at the rate the water would flow without the presence of the wheel. If the water pressure on the upstream of the wheel is turned off, the rotational inertia of the wheel keeps it turning. It pulls water from the upstream side and pushes it downstream. With time, the wheel begins to slow since there is no external pressure, until the wheel and the water flow both cease.

The magnetic energy stored in an inductor (w(t)) is determined from Eqs. (4.2), (4.3) and (4.13):

$$w(t) = L \, [I(t)]^2 \, /2 \qquad\qquad (4.14)$$

where w(t) is the stored energy in joules when L is expressed in Henrys.

Similar to resistors, the inductance is not the only parameter that you need to know before you purchase an inductor. For inductors, it is the maximum current that you also need to know. The magnetic core of an inductor is a very nonlinear material, which means that the permeability, μ, is not necessarily a constant, and hence L is not necessarily a constant. The permeability is usually nearly constant for small magnetic fields, and then decreases as the field increases. For an inductor, this means that the inductance is nearly constant for small currents, but then L decreases as the current increases. Companies define differently the maximum allowable current through an inductor, so make sure that you read the product specification carefully. Large inductances usually require lots of wire to make many coil turns, and sometimes, to keep the inductor small, this means that very small wire is used. This wire may have a significant resistance (see the following section), and so this resistance is a third parameter that you need to consider when selecting an inductor. In general, the larger the inductor, the greater the weight, the smaller the resistance and the larger the maximum allowed current. Inductor tolerances generally range from ±10% to ±20%, but higher and lower values exist.

4.4.5 Wire

Wire comes in many sizes, shapes and materials. Most wires come surrounded by an insulating sleeve that can withstand the application of a few hundred volts without being destroyed. They are not usually heat resistant and can be accidentally damaged during soldering.

The most common wire is round (circular cross-section) copper wire. This wire comes in many sizes and two configurations: solid and stranded. Stranded wire is more flexible but there are some connections that are easier to make with solid wire. Either style is probably fine for the hovercraft application. The sizes are dictated by the American Wire Gauge number, or AWG #. The larger the number, the smaller the diameter of the wire and the larger its resistance per unit length. Each size has a maximum suggested current, above which heating and related problems may occur if the current is applied continuously. Sample wire sizes that may be of use for the hovercraft problem are given in Table 4.1.

Table 4.1.
A selection of standard wire sizes and properties. 1000 mils = 1 inch

AWG #	Wire diameter (mils)	Weight (oz/foot)	Resistance (mΩ/foot)	Max continuous current (A)
14	64.1	0.20	2.5	15
18	40.3	0.08	6.4	5.8
22	25.4	0.03	16.1	2.3
26	15.9	0.012	40.8	0.92
30	10.0	0.0055	103.2	0.36

4.4.6 Switches

A switch is a device that is used to connect various parts of an assembled circuit together. Simple switches are two terminal passive components which have two states: open and closed. The symbol for an open switch is shown in Fig. 4.6 (a). No current flows between the two terminals. The symbol for a closed switch is shown in Fig. 4.6 (b). Ideally, there is no voltage across the two terminals. There are many different styles of switches. Here we will talk briefly about manual switches, which are useful for connecting batteries to the circuits that they are designed to power. Electronic switches, such as relays and transistors, are discussed in later sections.

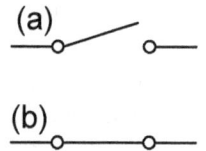

Fig. 4.6 The symbol for a switch: (a) open and (b) closed.

Switches are characterized in terms of how much current they can carry, how many volts they can withstand (when open), how many poles and positions they have, and whether the switch action is permanent or momentary. In a momentary switch, only one of the states is "stable". For example, in a normally-open (NO) switch, the switch is open. When you push the switch button, the switch closes. When you release the button, the switch returns to the open state. There are also normally-closed (NC) switches. Power switches normally are not momentary and the switch action toggles between the two states. Some devices are connected to more than one switch, so that one mechanical action (pushing the button) simultaneously changes the state of several switches. This may be useful, for example, if one needs to connect two different batteries to a circuit at the same time. The number of switches connected in a device is referred to as the number of poles. Sometimes switches have more than two terminals. The most common example of this is the "double throw" switch, which has three terminals. There is a central terminal, and in one state, the central terminal and one other terminal form a closed switch while the central terminal with the third terminal form an open switch. When the switch state is toggled, the roles of the two pairs of switches are reversed. Switches can normally withstand hundreds of volts when in the open state. Maximum currents can be only a few amps or less, so this parameter may be important to consider for the hovercraft application.

4.4.7 Batteries

Batteries are portable devices that convert some form of energy (typically chemical) into electrical energy. Ideal batteries produce a constant potential difference between their two terminals irrespective of the amount of current that passes through the battery and irrespective of the past history and use of the battery. **Real batteries do not - the potential difference between the output terminals depends on many factors!!!** Characteristics of different types of real batteries were described in Chapter 2. Here, we list and briefly discuss some important electrical properties common to all types of real batteries. *For real batteries:*

(a) *The output voltage decreases with increasing battery current.* A good approximate model for a real battery is shown in Fig. 4. 7. The real battery model has an ideal battery connected to a resistor. This resistor will cause the real voltage to decrease as the current increases. Recall that resistors generate heat – this is why batteries get hot if you try to drain too much current.

Fig. 4.7 Real battery model

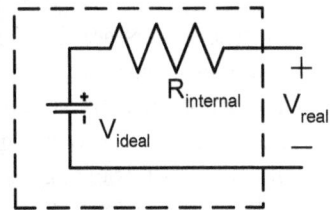

(b) *The output voltage decreases with time during continuous use.* The model for a real battery in Fig. 4.7 is quite simplified. In reality, the ideal (zero current) voltage and the internal resistance of the real battery depend on the current state of the battery. As the battery energy is drained, the output voltage slowly decreases due to the nature of the physical processes occurring in the battery. As the battery becomes nearly completely drained, the output voltage begins to drop off rapidly.

(c) *There is a maximum discharge current that should not be exceeded.* Drawing more current from the battery than it is designed to allow can cause permanent damage related to excess heat. In extreme cases, fire or other potentially dangerous conditions can occur. Some batteries are protected from excessive current, but all batteries should be selected to have performance capabilities that exceed the circuit specifications. The batteries should also always be handled with care, and accidental shorting of the battery terminals should always be avoided.

(d) *The amount of usable stored energy decreases with increased output current.* In reality, the internal losses in a battery are not proportional to the current, so the total amount of energy that can be extracted from a battery decreases with increasing current, so that the battery lifetime decreases more rapidly than corresponding increases in current. This effect is strong in some batteries and weaker in others, so check your particular battery's specifications.

(e) *The amount of usable stored energy decreases with age.* Batteries lose energy just sitting around, without any connections to an electric circuit. Primary batteries, or non-rechargeable batteries, usually have expiration dates labeled on the package as a consequence of this fact. Secondary batteries or rechargeable batteries have a finite lifetime in terms of the number of charging cycles they can handle. The re-charging process does not completely reverse the discharging process, due to damage and other real considerations. This means that advertised energy capacities given for batteries are the maximum that you can expect, and the max stored energy will decrease with time. The best way to minimize this effect is to follow all directions and restrictions with respect to the charge and discharge cycles.

(f) *The recharging cycle depends on the battery chemistry.* Never put a secondary battery in a charger that was not designed to handle it. Never put a primary battery in a charger.

(g) *Problems may occur when the battery is nearly discharged.* Some batteries will experience damage if discharged completely; others work better if completely discharged before a recharge cycle is initiated. If multiple batteries are used in the same circuit, care must be taken that all batteries discharge at about the same rate, or at least that no battery is allowed to remain in the circuit when discharged.

(h) *Electrical combinations of different battery types might lead to problems.* If different types of batteries are used in the same circuit, the danger of having discharged batteries in the circuit increases. If unavoidable, it is important to make sure that none of the specifications are exceeded for any of the batteries and that batteries are replaced or recharged well before they lose all of their stored energy.

4.4.8 Motors

Electric motors are devices that convert electric energy into kinetic energy (and heat, since they are not 100% efficient) via magnetic forces. Normally, the torque required to spin the motor comes from the interaction of a fixed permanent magnet and a rotating electromagnet (i.e. an inductor). The spinning electromagnet has to change the direction of its current periodically so that the torque is always in the same direction. If this change is done mechanically, the motor can be run backwards by reversing the applied voltage; if it is done electronically, the motor might not be reversible (make sure that your fan is reversible, if you plan to reverse the direction of air flow).

If it sounds like the electrical properties of a fan are complicated, they are, but there are only a few facts that you need to understand in order to successfully use your fan. First, when running at a constant speed, the fans can be modeled as a resistor, with the resistance given by the applied voltage divided by the average current needed. Usually, there is a minimum voltage needed to start a fan and a maximum voltage that will work before the fan burns out. You can work anywhere between those two limits and adjust the fan speed via the voltage. Also, fans usually require more current to start up than they do to run at their normal operating speed. Finally, the current-switching in fans usually results in electrical "noise" which manifests itself as a high-frequency voltage variation on the fan terminals. These variations can be very large, particularly at moments when power is being connected or disconnected to the motor. Since one main method of controlling fan speed is to periodically "pulse" the voltage applied to the fan, electrical fan noise may be something that you have to reduce.

Figure 4.8 shows oscilloscope (time) traces related to a pulsed-control of the motor speed. The quantitative details of these traces aren't important, and so only a qualitative description is given here. The independent axis represents time. The lower curve shows the voltage applied to the motor. The voltage is normally zero volts, but periodically there are brief 7V pulses. Note that the speed is controlled by changing the width of this pulse — hence the name of this technique: Pulse-width modulation, or PWM.

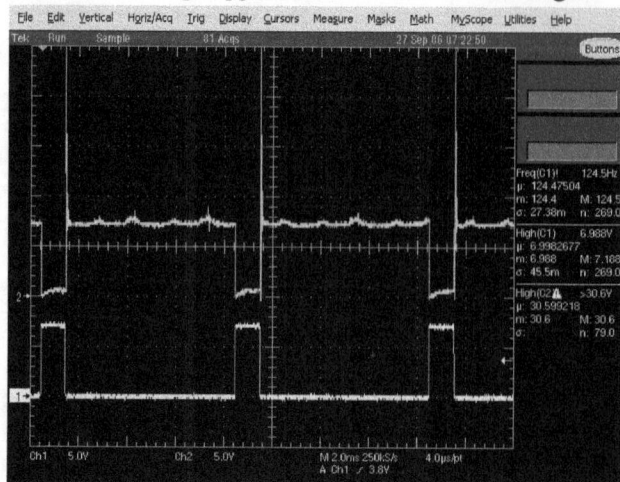

Fig. 4.8. An oscilloscope trace demonstrating important noise spikes from a PWM-controlled motor.

The upper curve in Fig. 4.8 shows the voltage of a switch that controls the fan. The important thing to note is that when the voltage is removed from the fan, there is always a very large, short voltage spike. For this example, the spike is at least seven times larger in voltage than the normal operating voltage, and this spike can damage circuit components if not suppressed. We will explore how to suppress this voltage spike later in this chapter.

4.4.9 Relays

Relays were described in Chapter 2; here we only summarize briefly the main electrical properties. Relays typically vary from single-pole, single throw (SPST) switches to triple-pole (or even quadruple-pole) double throw (3PDT or 4PDT)) switches. While the electromagnet that controls the switch location(s) is certainly an inductor, the applied voltage is usually a constant (dc – direct current), so it is the resistance of the coil winding that is most important and therefore the quantity normally provided in the specifications. These electromagnets can be energized by a range of voltages; with 5V and 12V being two standard nominal relay voltages. This is not too convenient for the hovercraft application, since the NXT output voltage is in between those two values. Still, relays have a range of acceptable voltages. There is a minimum voltage, above which the relay is guaranteed to work. There is a maximum voltage above which the relay will overheat and burn out. Usually both values are given in the specifications, so you can judge before buying if the relay will meet your needs. Relays take a finite amount of time, after the electromagnet is energized, before the switch contacts have come to rest in their new locations. This time should also be specified, but normally it is on the order of a few milliseconds or a few tens of milliseconds. While this seems pretty fast to many people, it normally means that relays are good for turning fans on and off, but not very useful for PWM speed control.

4.5 KIRCHHOFF'S LAWS

4.5.1 Topology: Nodes and Loops

The terminal relations given above (Eqs. 4.4, 4.10 and 4.13) provide one-half of the equations needed to solve for all voltages and currents everywhere in a circuit. The other 50% of the equations come from Kirchhoff's Laws – rules derived from the connections made between components in a circuit. These rules come from very simple physical principles, yet it is useful to state a few definitions first.

A node is a connection in which two or more components are electrically joined together, either by solder or some non-permanent means. A trivial node has only two components connected. Most people think of nodes as points, but actually nodes in a circuit drawing often look like wires, since they are a means to connect two points together without any voltage drop. Examples of nodes are given in the next section. A loop is any closed path in a circuit that passes through two or more components. If the closed path does not physically enclose any other components (that are not part of the loop), then the loop is also called a mesh. A trivial mesh contains exactly two components. Examples of meshes are given in the next section.

4.5.2 Kirchhoff's Voltage Law

We introduced Ohm's law with a simple circuit loop containing a voltage source and a resistor as shown in Fig. 4.2. This is a good beginning; however, what happens when you begin to insert additional components in the circuit? You will need to develop a few more tools useful in analyzing slightly more complex circuits. Let's begin by adding one additional resistor to the simple circuit loop as shown in Fig. 4.9a. We then determine the relation for the current flowing in this slightly more complex circuit. To approach this problem, we need to use **Kirchhoff's voltage law (KVL),** which is derived from the concept of a conservative electric field. This law states:

KVL: The algebraic sum of the voltage drops around any circuit loop is zero.

To understand how to apply this rule, we need to understand the key terms "algebraic" and voltage "drop." Before we can do anything, we must make sure that the reference directions for all components are marked on the loop and that the passive sign convention is used for all passive components. KVL implies that to find the relationship between voltages in a loop, we must "travel" around the loop, summing up the voltages that we encounter along the way. The direction of travel doesn't matter, but to make it simpler to compare answers, let's always agree to travel clockwise around a loop. While traveling around the loop, if we first encounter the positive reference terminal, pass through the component, and exit the negative terminal, we have experienced a voltage drop. If the opposite is true – we enter the negative terminal and leave the positive terminal, we have experienced a voltage rise. The "algebraic sum" requires that a plus sign is placed in front of the voltage drops and a minus sign is placed in front of the voltage rises.

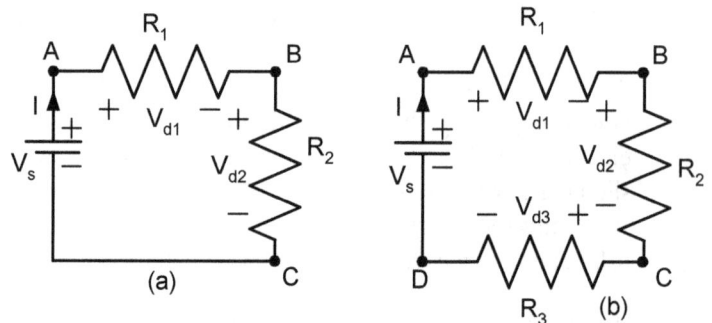

Fig. 4.9 Circuits containing two and
three resistors.

Let's use the system above to apply KVL to the mesh in Fig. 4.9a. Let's start at the lower-left corner of the mesh (the only corner not identified by a letter). We go clockwise, so we first pass through the voltage source, entering the negative terminal and leaving the positive one. This means that we have experienced a voltage rise, and we put a minus sign in front of V_s. We next pass through the first resistor, R_1, experiencing a voltage drop along the way. We also see a voltage drop passing through R_2, so both voltages V_{d1} and V_{d2} have plus signs in front. Finally, the wire on the bottom has no voltage drop across it, so we return to the starting point and set the voltages summed up algebraically along the way to zero, to get the following equation:

$$V_s + V_{d1} + V_{d2} = 0 \qquad (4.15)$$

Note that this circuit has only one mesh, and so there is only one KVL equation that we can write. It turns out, that no matter how complicated the circuit, you only need to write one KVL equation for every mesh in the circuit... writing a KVL equation for any loop that is not a mesh would not give you any new information.

Note also that this circuit has three nodes – three places where parts are connected together. We have labeled the nodes A, B, and C. Finally, note that node C extends all the way across the bottom of the circuit to connect R_2 to the negative terminal of the voltage source.

As a final note, we have only labeled one current in the circuit, even though there are three components and technically we need three reference currents. Since all of these nodes are trivial, the same current, I, passes through all of the components. Components connected by trivial nodes are said to be in *series*. Using Ohm's Law for the two resistors, we can replace the resistor voltage drops with the current, I, times the resistance, and then solve for the current through the resistors:

$$V_s - I\,R_1 - IR_2 = 0 \qquad\qquad (4.16)$$

$$I = V_s\,/(R_1 + R_2) \qquad\qquad (4.17)$$

If we apply the same analysis techniques to the circuit with the three resistors shown in Fig. 4.9b, it is easy to add another term to the relation shown above and write:

$$V_s - I\,R_1 - IR_2 - IR_3 = 0 \qquad\qquad (4.18)$$

$$I = V_s\,/(\,R_1 + R_2 + R_3) \qquad\qquad (4.19)$$

Examine Equations (4.17) and (4.19), and observe that the resistors in a series arrangement in the circuit add together to act like a single resistor with an equivalent resistance R_e given by:

$$R_e = R_1 + R_2 + R_3 = \sum R \qquad\qquad (4.20)$$

Equation (4.20) leads to a well-known resistor rule for resistors connected in a series arrangement. The effective resistance of several resistances in series is the sum of the individual resistances. The equivalent resistance is used to simplify the circuit diagram by replacing the three-resistor circuit with a single equivalent resistor as shown in Fig. 4.10.

4.5.3 Kirchhoff's Current Law

What is the effect if the resistors are not all connected in a series arrangement? Another simple configuration occurs when we have a trivial mesh. The two components in a trivial mesh are said to be in *parallel*. What if we have a parallel arrangement of the resistors such as shown

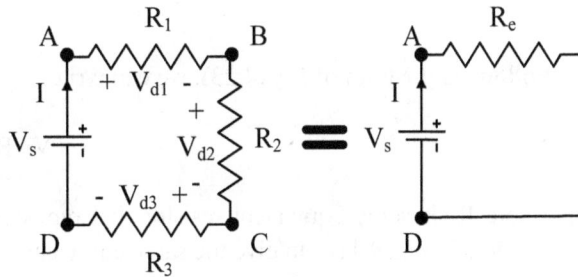

Fig. 4.10 An equivalent circuit with an equivalent resistor replacing a more complex circuit.

in Figs. 4.11a or 4.11b? To analyze this type of a circuit, we need to introduce **Kirchhoff's current law (KCL)**. This simple law, which comes from the physical concept of charge conservation, states:

KCL: The sum of all currents flowing into a node must be equal to the sum of all currents flowing out of that node.

Let's try KCL out for the circuit in Fig. 4.11a. There are two nodes in the circuit: node A and node B. Node A goes along the entire top of the circuit and the voltage source and both resistors are connected to the node. All three components are also connected to the bottom node. The current through the first resistor, I_1, for example, flows out of node A and into node B. It is easy to see for this example that there will be only one unique KCL equation that we can write for this circuit. Even for more complex circuits, one can show that you need to find one KCL equation less than there are nodes, if you want to find all voltages and currents in the circuit. The current I is flowing into node A, and I_1 and I_2 are flowing out, so an application of KCL for this node yields:

$$I = I_1 + I_2 \qquad (4.21)$$

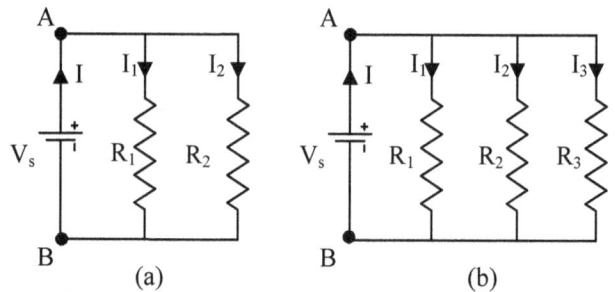

Fig. 4.11 Circuits with parallel arrangements of resistors.

Likewise, KCL for node B in the circuit in Fig. 4.11b yields:

$$I_1 + I_2 + I_3 = I \qquad (4.22)$$

KVL, applied to a trivial node, reveals that two components in parallel have the same potential difference across their terminals. Since all resistors in Fig. 4.11 are in parallel with the voltage source, they all have the potential difference of V_s across their terminals. Applying Ohm's Law (Eq. 4.4b) to the resistors in the circuits of Fig. 4.11 yields:

$$I_1 = V_s/R_1 , \quad I_2 = V_s/R_2 \ \text{and} \ \ I_3 = V_s/R_3 \qquad (4.23)$$

Following the form of Eq. (4.23), we can write:

$$I = V_s/R_e \qquad (4.24)$$

where R_e is some equivalent resistor that replaces the two parallel resistors in Fig. 4.11a (or the three resistors in Fig. 4.11b) to give the same equivalent circuit as shown in Fig. 4.12.

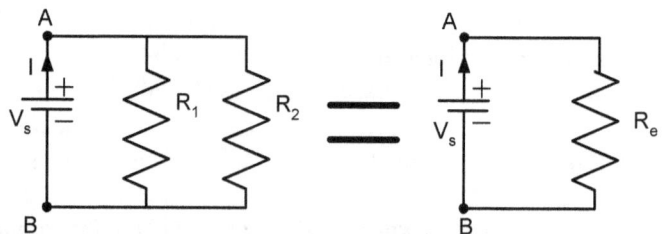

Fig. 4.12 An equivalent circuit contains an equivalent resistor that replaces the two parallel resistors.

Let's substitute Eqs. (4.23) and (4.24) into Eq. (4.21) to obtain:

$$V_s/R_e = V_s [(1/R_1) + (1/R_2)] \qquad (4.25)$$

Eliminating V_s from Eq. (4.25) gives:

$$1/R_e = (1/R_1) + (1/R_2) \qquad (4.26a)$$

or

$$R_e = (R_1 \, R_2) \, / \, (R_1 + R_2) \qquad\qquad (4.26b)$$

If we find three resistors in parallel as in Fig. 4.11b, we follow the same procedure in the analysis of the circuit and show that the equivalent resistance for the three parallel resistors is:

$$1/R_e = (1/R_1) + (1/R_2) + (1/R_3) \qquad\qquad (4.27)$$

While resistors connected in series were summed to give the equivalent resistance, resistors in parallel follow a different rule. For parallel resistors, the reciprocals are summed and set equal to the reciprocal of the equivalent resistance. **This fact implies that the equivalent resistance for a parallel arrangement of resistors is less than that of the smallest individual resistance.** The conductance, G, is defined to be the reciprocal of the resistance: $G = 1/R$. Given this definition, for parallel resistors, the conductance of the equivalent resistor is equal to the sum of the conductances of the parallel resistors.

4.6 BASIC ELECTRONIC COMPONENTS

4.6.1 Diodes

The diode is another important circuit element. It is constructed by combining a p-type semiconductor material with an n-type semiconductor. While no particular electrical quantity is associated with diodes, they act in some respects like an automatic current switch or a check valve. A diode allows current to flow freely in one direction (ideally, just like a wire) but does not permit current to flow (ideally) in the opposite direction (we call this reverse-biased). Current can flow from the p-region to the n-region but not vice-versa. In a sense, a diode is the electrical equivalent of a check valve in a pipe carrying water. The symbol for a diode is shown in Fig. 4.13. The direction of the arrow indicates the direction in which the diode allows current to flow. We will discuss a common use of diodes later in the chapter.

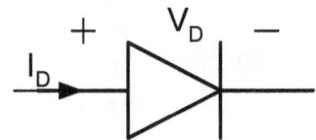

Fig. 4.13 Circuit symbol for a diode.

Real diodes have a voltage drop (V_D) across their terminals when current is passing through the diode. This voltage depends on the particular diode and the exact current (the current increases exponentially with increasing voltage), but it is rarely less than ½ volt and rarely more than 4 V. There is also a limit in terms of how much current can flow through the diode (I_D), as the voltage drop results in the diode dissipating power in the form of heat. For example, if the maximum diode power is 1 W, and the voltage drop is $V_D = 0.5$ V, then the current must be kept below $I_D = 2$ A.

Real diodes also have a maximum negative voltage that can be applied to it. If V_D becomes more negative than this maximum (which is usually tens of volts if not hundreds of volts), current will begin to flow in the opposite direction quickly destroying the diode. This condition is called voltage breakdown. There are some special diodes that are constructed to survive breakdown. The most common diode of this type is the Zener diode, which exploits this property to regulate voltages in a circuit. Real diodes do have a small reverse current when V_D is negative, but below the breakdown threshold. Normally this current is no more than a few microamps, and has little impact on most circuits.

Diodes are our first example of a non-linear device. Resistors, capacitors, inductors, and voltage sources are all linear devices. One important consequence of a linear circuit is that if you change the sign of every voltage source (from plus to minus and vice-versa), you can get the solution for all voltages and currents in the "new" circuit just by flipping the signs of those values in the original solution. Another consequence is that if all sources are constant, or vary at the same frequency, ALL voltages and currents in the circuit have the same time dependence as the sources. A diode is clearly not linear: you cannot simply change the direction of the current through the diode. As non-linear components, diodes are very useful for converting constant dc circuits to time-varying circuits and vice-versa.

4.6.2 LEDs

LED stands for Light-emitting diode, and the name describes well the function of the device. They are diodes, so current flows only in one direction, but not all of the electric power loss goes into heat – some of it is converted into light. The diodes are usually packaged in transparent or translucent plastic packages to allow the light to illuminate the surrounding area. Their forward diode voltages are typically $V_D > 1.3$ V. The required current is on the order of tens of milliamps. LED performance varies quite a bit, not only with respect to color and brightness, but also with respect to the angular coverage of the light. Viewing angles range from under $10°$ to over $180°$. Thus, some thought should go into exactly what you need the LED to do, before you go searching through the thousands of different models that can be bought online to find the one that's right for your project. Once you have LEDs on hand, they should be tested under realistic operating conditions to make sure that they will perform their intended tasks.

Because the current I_D increases exponentially with V_D, some care must be taken when designing the circuit to insure that the diodes don't burn out from too much current. A very simple way to achieve this is shown in Fig. 4.14. The series resistor's value must be selected to get the correct LED current, I_D. Using KVL for the only loop in the circuit, we can find an expression for this resistance:

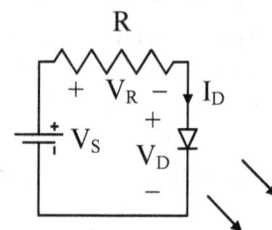

Figure 4.14 A simple LED circuit.

$$R = (V_S - V_D) / I_D$$
(4.28)

For a 9.6 V power supply and a diode voltage of $V_D = 2.2$V, for example, if the desired current is 20 mA, the required resistance is:

$$R = (9.6 - 2.2) /0.02 = 370 \ \Omega$$
(4.29)

A 370 Ω resistor is not standard, but either the closest carbon-composition resistance, 360 Ω, or the nearest metal-film resistance, 374 Ω, would certainly be adequate.

4.6.3 Photodiodes and Photoresistors

A photodiode is a semiconductor device that does the opposite of an LED – it converts light into electric energy. A photoresistor is another semiconductor device – often Cadmium-Sulfide or CdS – which acts like a variable resistance resistor. The variation is achieved by changing the amount of light that shines on the device. In Section 4.3, we mentioned that the number of free electrons in a material can depend on the ambient temperature and on the number of impurities. Well, for some materials, like CdS, it also depends on the amount of light, because photons of light can give up their energy to electrons, thereby freeing them from their atomic nuclei. Having more free electrons means more current for the same applied voltage;

hence, the resistance of a photoresistor decreases with increasing light. How this property can be exploited in a detector for the hovercraft project is described in more detail later in this chapter.

4.6.4 Transistors

Transistors are semiconductor devices that can be used either as amplifiers or as high-speed electronic switches. Unlike the previous two-terminal components that we have discussed up to this point, transistors have three terminals. One of these is used to control current flow between the other two terminals. Bipolar-Junction Transistors (BJTs) use current in the control terminal (called the base) while Field-Effect transistors (FETs) use a potential (voltage) difference between the control terminal (called the gate) and one of the other terminals to control the current flow. Transistors are another example of a non-linear device.

In the following two subsections we give very brief explanations of how these devices work, concentrating on the switching aspects of these devices, since it is doubtful that you will find need in the hovercraft project for their amplification properties. Afterward, we discuss a few practical considerations regarding the use of these transistors. An important note for both types of transistors is that current is designed to go in only one direction and voltage is designed to be held off in only one direction. Thus, if you want to use transistors to run a fan backwards, you need more than one transistor to do the job. One possibility is the H-Bridge circuit described later in this chapter.

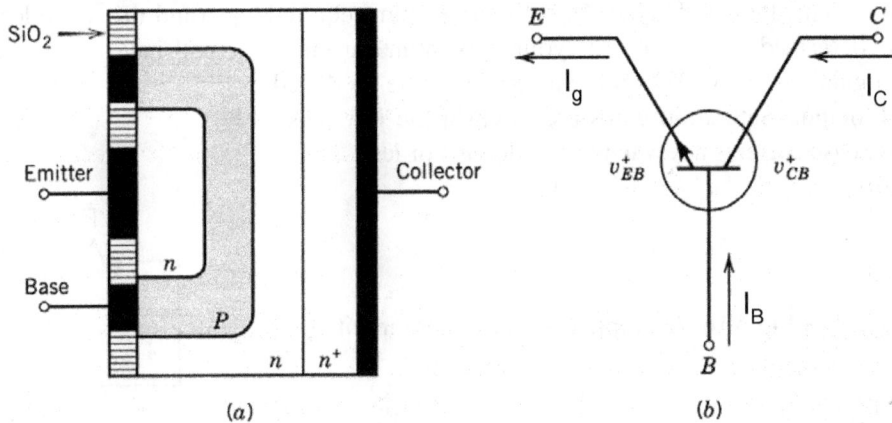

Fig. 4.15 Representation of a NPN bipolar transistor. (a) Planar structure in silicon.
(b) Circuit symbol.

4.6.4.1 BJTs

An NPN bipolar junction transistor (PJT) is illustrated in Fig. 4.15 (there is a similar PNP version). The devices are planar, therefore, they can be fabricated using lithographic methods in p-type and n-type doped silicon. The devices are extremely small with areas of 10^{-9} m^2.

The three-terminals of the device are labeled in the figure as follows: the base is represented by B, the collector by C, and the emitter by E. The main current flow in the device is from the collector to the emitter. The theory of operation of the bipolar transistor is beyond the scope of this book; however, the transistor will act as a current amplifier because relatively small base currents I_B produce large collector currents I_C. When $I_B=0$, there is no current flow into the collector ($I_C=0$), and the transistor is said to be in the "off" state. Like an open switch, no current flows from the collector to the emitter, and there can be a large voltage difference between those two terminals. Normally we label this voltage V_{CE}, for the voltage from the emitter to the collector. As such, V_{CE} is always greater than zero.

If the transistor is properly connected to a circuit, an example of which is given in Fig. 4.16, the collector increases linearly with the base current, at least for base currents that are "small enough." The small-signal transistor gain, β, is defined as the ratio of those two currents:

$$\beta = I_C / I_B \tag{4.30}$$

Typical values for β are in the tens and hundreds. As the base current becomes larger, the collector current will start to level off, and eventually the transistor becomes "saturated", or fully "on." Consider Fig. 4.16. As I_C increases, the voltage V_R across the resistor increases, and KVL around the mesh on the right tells us that V_{CE} must then decrease. The minimum V_{CE} is usually some small fraction of a volt, depending on the particular transistor (but it can be a few volts in high-current BJTs, so be careful), so this saturated state corresponds to the "on" state of a switch. The voltage is close to zero, and any current can flow from the collector to the base. It differs from a switch in that the maximum collector current is determined by the base current. To increase the maximum collector current, the circuit would need to be modified to produce a correspondingly larger base current. If this were a hovercraft application, R_C might represent your fan, V_B would be the NXT output, and V_S would be your battery pack. Your job would be to pick R_B small enough, so that it would allow for sufficient fan current without overloading the NXT or overheating the transistor. You also need to pick a transistor that can handle: (1) your maximum required V_{CE} (and this includes any fan motor noise, as described above) and (2) your maximum current I_C. It will have to have a large enough saturated gain, so that I_B does not exceed the value produced by the NXT output. Saturation gains are often in the low tens and saturation base currents a few tens of milliamps or less.

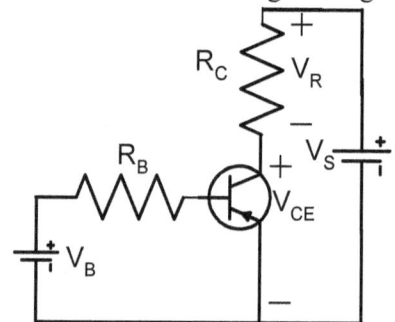

Fig 4.16 A simple BJT transistor circuit

4.6.4.2 MOSFETs

MOSFETs, a somewhat newer technology than BJTs (but still very mature devices), enjoy some benefits over BJTs with respect to the switching application. The symbol for an enhancement-mode, n-channel MOSFET is shown in Fig. 4.17, where the MOSFET is inserted into a simple switching circuit. The three terminals have different names for the MOSFET, but they have analogous functions to the BJT terminals. The gate controls the main current flow, which is from the drain to the source. The n-channel device is analogous to the NPN transistor in that the bulk of the current motion is through an n-type material. "Enhancement mode" means that there is no current flow from the drain to the source when there is zero voltage from the gate to the source. As the gate voltage is increased, the drain current I_D increases and the voltage V_{DS} decreases, Eventually V_{GS} is increased to the point where V_{DS} approaches zero and the MOSFET is saturated. As long as the V_{GS} required for saturation is less than the NXT output voltage, one simply needs to hook the gate and source directly to the NXT output in order to switch the transistor. *(NOTE: it may be necessary to put a resistor in parallel with the NXT outputs for the switch to work properly, but this is due to the NXT properties, not due to MOSFET performance characteristics.)*

Another possible advantage of the MOSFET is that they usually have smaller voltage drops from the drain to the source, compared to their BJT counterparts, so they act as better switches, with lower power loss and less heat generation. They are more sensitive to damage from electrostatic shocks, but if handled carefully, they are quite robust.

When buying MOSFETs, in addition to making sure it is the right type with adequate gate voltage properties, it must also have sufficient ability to hold off voltage in the off state and sufficient current capability when in the on state.

Fig. 4.17 A simple MOSFET circuit

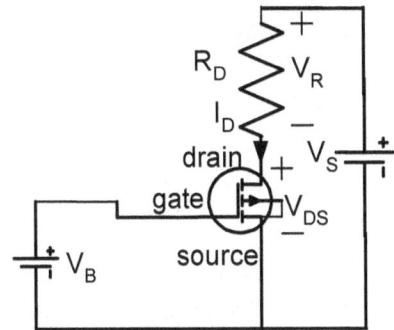

4.6.4.3 Transistor Packages and Heat Sinks

Another consideration when buying transistors of either type, BJT or MOSFET, is the housing, or packaging, that they come in. There are many, many housings for printed circuit board applications (PCBs), These packages don't have leads, but rather metal pads to connect to, and should be avoided at all costs. Many of the useful packages with leads have a package name that starts with TO – for Transistor Outline Package. The TO-92 package is very standard for low-power transistors. A better package for high power applications is the TO-220 package. This package comes in styles with two to seven connections (so the three connections needed for a transistor is just fine). They have a metal backing and a hole that can be used to attach the transistor to a heat sink – a metal block (usually with fins) that can be used to draw heat out of the transistor and decrease the operating temperature. Usually a paste-like heat sink compound is applied between the heat sink and the transistor to improve the heat flow. With a heat sink, this package can often dissipate up to 50 W without problem (though often becoming quite hot). There are many other styles of housings as well; the important considerations to reflect on are (1) if it will have the power capability you need and (2) if it will be compatible with your method for assembling your circuit, in terms of its size, connections, and heat sink requirements.

4.6.5 Voltage Regulators

Remember that in Section 4.4.7 we mentioned that battery voltages are anything but constant. When you buy a 12V battery pack, the voltage could be anywhere from about 11V to 14V and the battery pack would still be considered to be functioning properly. There are many devices, perhaps including your hovercraft, that need to have the battery voltage quite constant in order to function correctly, and the 30% variation typical in batteries is extremely inadequate. The hovercraft lift, for example, depends on the fan's flow rate, which depends on the motor's speed, which depends on the applied voltage. Many hovercrafts have worked well on preliminary trial runs, only to get stuck of their final test run after their batteries have "run down."

Voltage regulators are integrated circuits containing transistors and other electronic components that can maintain the voltage in a circuit extremely constant over the usable life of the battery that powers the circuit. Regulators are generally quite small – many of them come in the TO-220 package so that they can easily mount to heat sinks for temperature reduction.

DC voltage regulators come in two main varieties: linear regulators and switching regulators. Both are fairly cheap. Linear regulators simply "chop off" excess voltage. For example, if your fan needs 12 V at 2 A, but your battery produces 14.4 V, the linear regulator is placed in series with the circuit, and 2.4V is chopped off the battery voltage. This means that the regulator has to dissipate 4.8W all the time.

Linear regulators have only three terminals and require at most two small capacitors connected between the terminals to help reduce any voltage "ripple" that occurs with changing circuit conditions (See Fig. 4.18). Most voltage regulators have fixed output voltages, but some have voltages that can be varied via an adjustment to an external resistor.

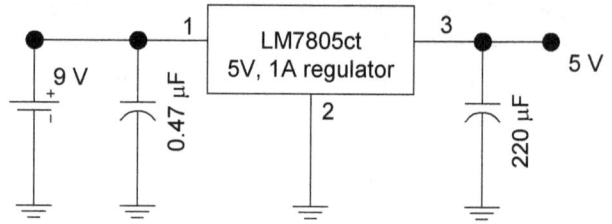

Fig. 4.18 A typical connection for a 5 V
output linear regulator.

When selecting a voltage regulator, there are a host of parameters that might be important. After the output voltage, the maximum current that the regulator can handle is critical. The regulator voltage will have a tolerance, which might not be too important, since it is voltage variations that can ruin a good hovercraft ride. The regulator will have limits on the maximum input voltage and the maximum variation in the output voltage with time and/or temperature. The regulator will have fluctuations specified as a function of how much the input voltage changes and how much the output current changes. Usually these variations will be small, but they should always be verified to be adequate. Some regulators will have current protection (so that you cannot draw too much current) and they may have a temperature sensor, which turns the regulator "off" if it overheats. The specifications will have this sort of information, and specifications for almost every electronic part sold can be found on line[1]. Finally, one last parameter that is very important is the dropout voltage. The dropout voltage specifies how much the input voltage has to be above the output voltage to work well. For low dropout voltage regulators, this voltage can be $0.5 - 1$ V. For other regulators, this voltage can be as much as two volts or more. Let's continue the example of a 12V fan powered by a battery that outputs 14.4 volts when fully charged and no current drawn. If you have a regulator that has a dropout voltage (like an LM1084) of $1.3 - 1.5$V, the output will be regulated at 12V only while the battery voltage is at least $13.3 - 13.5$V. When the battery voltage drops below this value, the regulator output will start to drop down and the specifications regarding the voltage regulation no longer apply.

Switching regulators are more complicated than linear regulators. They take in the constant input voltage, then the "chop the voltage" to make a time-varying voltage with a fairly high frequency (above the audible range). They increase or decrease the voltage of the periodic circuit as they convert the signal back to a constant dc voltage. They have several advantages. First, they are more efficient so they usually get less hot. Second, they can be used to lower or raise voltages and may not have a dropout voltage (or at least a smaller dropout voltage). They are somewhat more complicated to connect and usually require a few more extra components, but neither type of regulator is difficult to insert into a circuit and the product specification sheets always contain schematics of the circuits required to use the regulator.

4.6.6 Operational Amplifiers

Operational amplifiers, or *op-amps*, for short, form another class of integrated circuits, each one comprised of dozens of transistors and other passive components. Operational amplifiers are voltage amplifiers that convert supply (dc) power into signal power, which could be dc, sinusoidal, or even non-periodic in nature. They are very useful, very complex devices with a large list of operating characteristics. In this section we will talk only about a small subset of those properties for idealized op-amps.

---

[1] The authors are particularly fond of http://www.digi-key.com

Real op-amps have at least five connection pins, though some have seven or more. One pin is for the output voltage (V_o), two are for the inputs (V_+ and V_-), and two are for the power supply (or supplies). Op-amps can come in packages with anywhere from 5 to 8 pins. Sometimes op-amps get packaged together – you can fit four op-amps, for example, on a single chip with 14 pins, since all op-amps share the same power source(s). The standard symbol for an op-amp, with power supply connections, is given in Fig. 4.19a. If the op-amp has the rail-to-rail property, the output voltage can range anywhere from the minimum supply voltage to the maximum supply voltage. If an op-amp is NOT rail-to-rail, the minimum output voltage is a volt or so higher than the minimum supply voltage and the maximum output voltage is a volt or so lower than the maximum supply. Single-sided op-amps allow for one power supply with the negative supply terminal simply connected to ground. "Ground" is a node in a circuit (often connected to the negative terminal of a battery) that you agree to call zero volts. If it is not single-sided, the op-amp requires one positive supply and one negative supply.

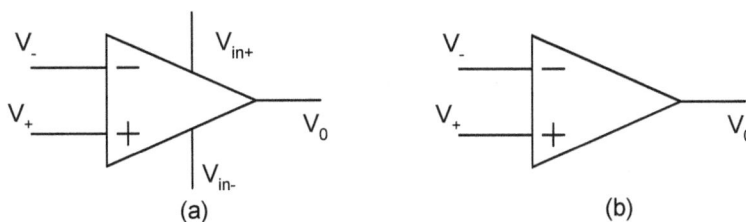

Fig. 4.19 The symbol for an op amp. All voltages are with respect to ground.

Ideal op-amps have three important properties that make it easy to design and analyze op-amp circuits, and at the same time, make op-amps very useful for many applications. When we consider ideal op-amps, we often ignore the power supply connections, as shown in Fig. 4.19b (don't forget that one must power a real op-amp for the circuit to work). The first ideal assumption is that there is infinite input resistance, which is to say that zero current flows into either input terminal. The second ideal assumption is that the output voltage, which is defined as:

$$V_o = A (V_+ - V_-),\qquad(4.31)$$

has an open-loop gain, A, which is nearly infinite: $A \rightarrow \infty$. Since the output voltage must be finite, this second assumption means that the two input voltages must be equal:

$$V_+ = V_-\qquad(4.32)$$

This assumption is not always true. As long as the output voltage does not try to exceed the power supply limitations, no more than one voltage source is directly connected to the inputs, and there is some feedback (which represents a piece of the output signal fed into the negative input terminal), this assumption is usually not violated. The final assumption is that the output voltage equation (Eq. 4.31) is true irrespective of any component connected to the op-amp output.

An example of a simple op-amp circuit is given in Fig. 4.20. This circuit is called a comparator. It is one of the few circuits that violates the second ideal assumption. If $V_1 > V_2$, then the output should ideally be positive infinity. In reality, the output voltage becomes as large as possible. If $V_1 < V_2$, the output should ideally be negative infinity. In reality, the output voltage becomes as negative or as small as possible. Thus, the op-amp compares the two inputs and makes a large difference in the output voltage even for small differences in the input signals. This is a simple analog-to-digital converter because there are an infinite number of input values and only two possible output values. One can think of the positive output as a logical one and the small negative output as a logical zero.

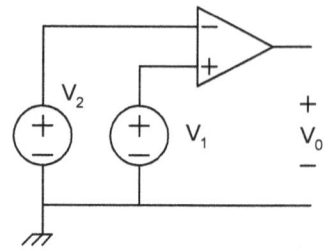

Fig. 4.20 A simple comparator circuit

Another example of an op-amp circuit is given in Fig. 4.21. The negative terminal is directly connected to the output, so the positive input terminal is at the output voltage. No current flows into the input, so it is easy to calculate the relation between input and output (see the following section):

$$V_o = V_1 R_B / (R_A + R_B) \tag{4.33}$$

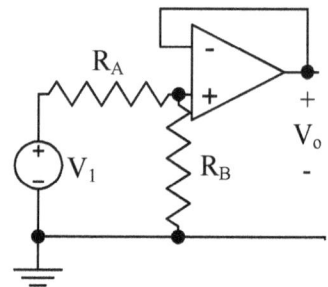

Fig. 4.21 A simple buffer circuit.

More complicated op-amp examples are given in Section 4.7.

4.7 APPLICATION EXAMPLES

4.7.1 Simple Series and Parallel Combinations

Assume there is a requirement for a particular sensor, that you must place a 5.40 kΩ resistor in series with your sensor in order to calibrate it. The sensor will not work correctly if the resistor is less than 5.395 kΩ or greater than 5.405 kΩ. How can this resistance be achieved? Standard resistors available at the local store only come in 5.1 kΩ and 5.6 kΩ packages. The resistors can vary by 5%, so if one has enough 5.6 kΩ resistors, perhaps an adequate resistor could be found. It's far better to build these with series or parallel combinations. With standard resistors, one could use: (a) two 2.7 kΩ resistors in series, (b) a 5.1 kΩ resistor and a 300 Ω resistor in series, (c) a 150 kΩ resistor in parallel with a 5.6 kΩ resistor, (d) a potentiometer, or (e) many, many other possibilities, depending on what component values happen to be available.

4.7.2 Voltage Divider

One consequence of resistors in series is voltage division. Consider the circuit in Fig. 4.22 and determine the voltage across R_2. From Ohm's law (see Eq. 4.17), we have:

$$I = V_s /(R_1 + R_2) \tag{4.34}$$

$$V_{d2} = IR_2 = V_s R_2/(R_1 + R_2) \tag{4.35}$$

Equation (4.35) shows that the voltage across R_2 is a fraction of V_s. The fraction is $R_2/(R_1 + R_2)$. Another way of expressing this result is:

$$V_{d2}/V_s = R_2/R_{eq} \qquad\qquad (4.36)$$

where R_{eq} is the equivalent series resistance.

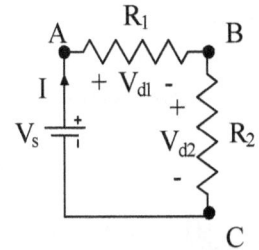

Fig. 4.22 A voltage division circuit.

The nice thing about Eq, 4.36 is that it remains correct even if there are more than two resistors in series, as long as it is the voltage across the second resistor that we wish to calculate.

Another consequence of Kirchhoff's laws is that voltage sources in series add just as resistors do as long as the orientation of the source is taken into account. That is why two 1.5 V flashlight batteries are used in an ordinary flashlight; the voltage of the two batteries adds to provide a total voltage source of 3 V. On the other hand, if one of the batteries is inserted upside down, it will act **against** the voltage of the first battery and give a net zero voltage. In that case, the flashlight will not work.

At this point, you should recognize that Ohm's law and Kirchhoff's two laws can be employed to analyze circuits. Also, you should be able to take a complex circuit and use equivalent resistances to reduce the circuit to its most elementary form.

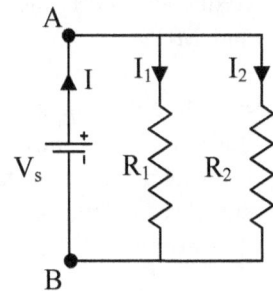

Fig.4.23 A current divider circuit.

4.7.3 Current Divider

One consequence of resistors in parallel is current division. Consider the circuit in Fig. 4.23 and determine the current through R_2. From Ohm's law [see Eqs. (4.24) and (4.26b)], we have:

$$I = V_s (R_1 + R_2)/ R_1 R_2 \qquad\qquad (4.37)$$

$$I_2 = V_s /R_2 = IR_1/(R_1 + R_2) \qquad\qquad (4.38)$$

Equation (4.38) shows that the current through R_2 is a fraction of I. The fraction is $R_1/(R_1 + R_2)$. If we want a formula that is good for more than two resistors in parallel, we need to write the formulas in terms of the conductances of the components:

$$G_1 = 1/R_1 \qquad G_2 = 1/R_2 \qquad G_{eq} = G_1 + G_2 \qquad\qquad (4.39)$$

For conductances, we get an expression that works for more than two resistors:

$$I_2/I = G_2/G_{eq} \qquad\qquad (4.40)$$

4.7.4 Reducing Motor Noise

There are at least three ways to control motors: Directly from the NXT, via a relay, or via a transistor. Relays are fairly insensitive to noise, but transistors may certainly be destroyed by high voltage noise. There are two ways to protect the transistors: by placing a capacitor in parallel with the transistor or in parallel with the motor. Normally these capacitors have high capacitance and are polarized. Thus, if the motor is to be reversed, the capacitors need to be in parallel with the transistors. It is difficult for a beginner to calculate how much capacitance is needed. Since they are relatively cheap, often the best practice is to try different values to see what works best. Unlike resistors, capacitors add in parallel, which is to say that if you stick n identical capacitors in parallel, the net capacitance is just n times the individual capacitance. The more capacitors you place in parallel, the lower the voltage spikes from the motor will be. These spikes can be measured on a device called an oscilloscope. When there is a good safety margin between the maximum voltage and the maximum voltage allowed by the transistor (maybe a factor of two), the transistors are protected. Increasing the capacitance too much may interfere with motor performance. If this were to happen, one simply could buy transistors with a higher breakdown voltage.

4.7.5 Controlling Motor Speed

A simple motor controller is shown in Fig. 4.24. The NXT uses PWM to control motor speed, at a frequency that is too high to switch the power on and off with a typical relay. However, both BJT and MOSFET transistors are more than adequate for the job. A MOSFET is shown in the figure since the connection is straightforward. The resistor R is there because the MOSFETs have nearly infinite gate resistance and the NXT needs a finite resistance to work properly. The exact resistance does not matter: anything from 1 kΩ to 10kΩ should certainly work just fine. The capacitor needed to protect the transistor from the motor noise is also shown in the figure. The battery pack which energizes the motor is shown as V_s; if a voltage regulator is used, Fig. 4.18 can be used to synthesize the correct circuit drawing.

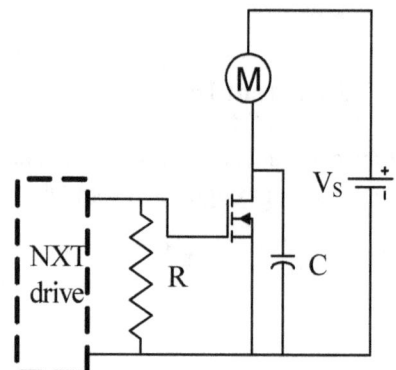

Fig. 4.24   A simple speed control

4.7.6 Changing the Motor Direction

In addition to adjusting the motor speed, the motor can be in one of three states: forward, off, reversed. If one simply wants to toggle between off and forward (slowly), the circuit shown in Fig. 4.25 can be used with a single SPST relay. When the NXT output is zero (motor off), the relay inductor is not energized, the switch is open, and the motor is not connected. When the output is fully on (motor speed five), the switch is closed and the motor runs.

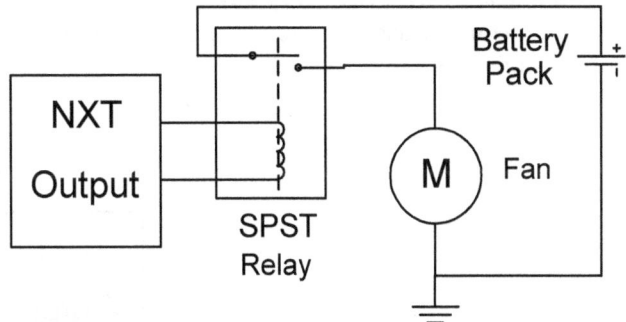

Fig. 4.25 A simple on-off circuit

A double-pole, double throw (DPDT) relay can be used if you want to toggle between forward and reverse (but never be off). This circuit is shown in Fig. 4.26. When the NXT output is zero, the center contact of each switch is in contact with its upper contact, causing the top of the motor to be connected to the positive battery terminal and the motor bottom to be connected to the negative battery terminal. When the NXT output is high, both switches change positions and the motor connections are reversed. Remember that this will not work if you have a motor that does not allow voltage reversals.

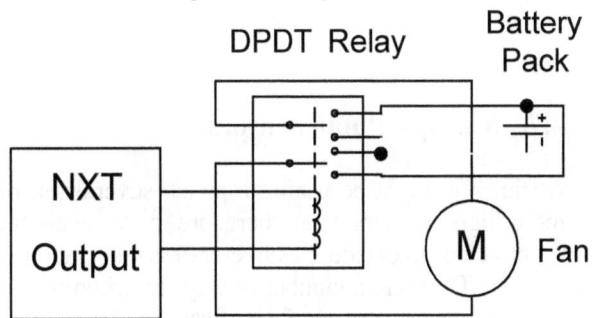

Fig. 4.26 A simple way to reverse fan direction

If you want to be able to access all three states (slowly), two DPST relays (and two diodes) can be used to accomplish this task, since the NXT output has three states: off, positive output, and negative output. The required circuit for this is shown in Fig. 4.27. When the NXT output is zero, the fan is not connected to anything and does not spin. When the NXT output power is on positive, the top NXT output (in the figure) is at a higher potential than the lower output. Thus, the upper diode is on and that energizes the upper relay, connecting the motor to the battery. Thus, the positive terminal of the battery is connected to the top of the motor. The lower relay is not energized because the lower diode is reverse-biased. When the NXT has the motor direction reversed, the lower NXT output is at a higher voltage than the upper output; the upper diode is off and the lower diode is on; the upper relay is off and the lower relay is energized, and the top of the motor is now connected to the negative terminal of the battery reversing the motor direction. The battery would be shorted out if both relays were on for some reason. If identical relays are used, the chance of that would be pretty small, but a good design practice is to always turn the motor off for a short period of time (a small fraction of a second) between motor direction changes.

A good way to both reverse fan direction and control motor speed with a PWM system is to use an H-bridge circuit. An H-bridge is a set of four transistors that are used to reverse the connections to the motor. When all four transistors are off, the fan receives no power. When one pair of transistors is on (and the other pair off), the fan is powered in the forward direction. When the transistor pair roles are reversed, the motor is powered in the reverse direction. The transistor pairs should never be turned on at the same time to prevent shorting the motors. One can certainly build an H-Bridge with individual transistors, but there are off-the-shelf H-Bridges that can handle up to 5A or more. Up to 5A, these chips are fairly cheap and readily available. They have some protections (like avoiding having all transistors on) and they are very compact and reliable. These chips, their wiring diagrams and explanations of circuit operation are all readily available on the web.

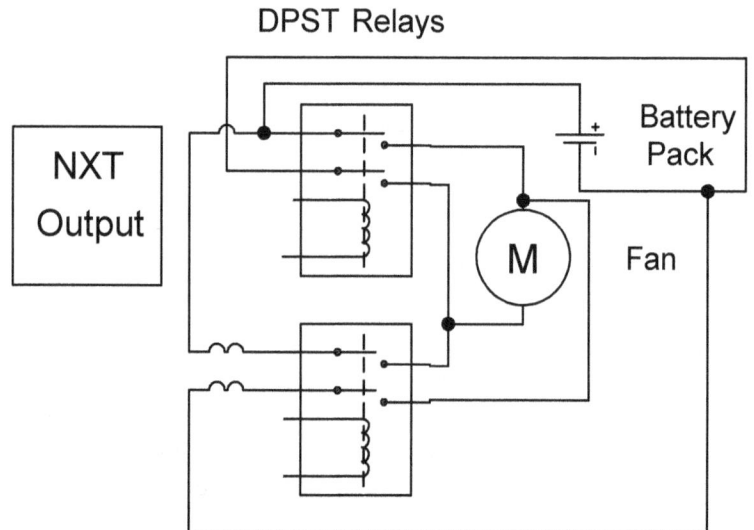

Fig. 4.27 A simple way to drive the motor forward, reverse and stop.

4.7.7 Multiple Photo Sensors

At times it might be useful to attach several sensors to the same NXT input port. Perhaps a design has more light sensors than there are ports available. Perhaps two proximity sensors are going to be "subtracted" to decide if a hovercraft is closer to one wall or another.

There are a number of ways to accomplish multiple sensor connection. Multiplexing is a complex way of looking at a number of sensors by slicing up time into fixed intervals and only looking at one particular sensor in each time slot. Here we will look at two examples that work even when the NXT is looking at all of the sensors all of the time.

The two examples demonstrate two ways of looking at two CdS photoresistors on the same input channel. The first uses only two photoresistors whereas the second uses op-amps as well.

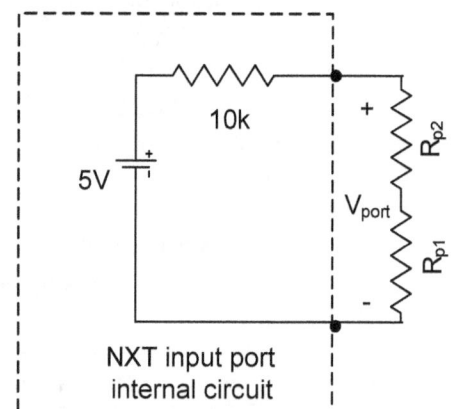

Fig. 4.28 Connecting two phtoresistors to one input port

Table 4.2.
Port voltages for two series CdS photoresistors

R_{p1} resistance (kΩ)		R_{p2} resistance (kΩ)		R_{p1} over:	R_{p2} over:	Vport (V)
Case I: individual identity unimportant (Fig. 4.28)						
WH: 5 k	BL: 12 k	WH: 5 k	BL: 12 k	White	White	2.50
WH: 5 k	BL: 12 k	WH: 5 k	BL: 12 k	White	Black	3.15
WH: 5 k	BL: 12 k	WH: 5 k	BL: 12 k	Black	White	3.15
WH: 5 k	BL: 12 k	WH: 5 k	BL: 12 k	Black	Black	3.53
Case II: individual identity important (Fig. 4.28)						
WH: 5 k	BL: 12 k	WH: 2 k	BL: 5 k	White	White	2.06
WH: 5 k	BL: 12 k	WH: 2 k	BL: 5 k	White	Black	2.50
WH: 5 k	BL: 12 k	WH: 2 k	BL: 5 k	Black	White	2.96
WH: 5 k	BL: 12 k	WH: 2 k	BL: 5 k	Black	Black	3.18
Case III: using the op-amp circuit (Fig. 4.29)						
WH: 5 k	BL: 12 k	WH: 5 k	BL: 12 k	White	White	0.63
WH: 5 k	BL: 12 k	WH: 5 k	BL: 12 k	White	Black	1.88
WH: 5 k	BL: 12 k	WH: 5 k	BL: 12 k	Black	White	3.13
WH: 5 k	BL: 12 k	WH: 5 k	BL: 12 k	Black	Black	4.38

Let's say that you want to know whether two sensors are situated over white floor, over black tape, or one over black and one over white. Take two photoresistors, called R_{p1} and R_{p2}, and put them in series and connect them to the NXT input as shown in Fig. 4.28. The 10 kΩ resistor in series with the 5V is an accurate model of the analog inputs of the NXT.

If you do not care *which one* of the sensors is over white or black, you can use the same type of CdS sensor. Let's say that you have a photoresistor that has a resistance of 5 kΩ when situated over white floor and 12 kΩ over black tape (these are representative values for one size of CdS sensors). When both cells are over white space, the combined series resistance is 10 kΩ, and by the voltage divider rule (Eq. 4.35), we can see that $V_{port} = 2.5$V, When one CdS sensor is over black tape and the over is over white space, the combined resistance is 17 kΩ, and Eq. 4.35 yields $V_{port} = 3.15$V. Finally, when both sensors are over black tape, the series resistance is 24 kW, and Eq. 4.35 yields $V_{port} = 3.53$V. These three voltages are pretty far apart (separated by over 1/3 V), and so it should be easy to decipher in a program for the NXT what the correct sensor status is. These results are summarized in Table 4.2. In the table, results are also shown for the case where is does matter which sensor is over white and which is over black. The smallest separation for the design values is over 0.2V, which is probably adequate to function well. However, both cases are well inferior to the third case in the table; those results correspond to the op-amp example discussed below.

The final example circuit is given in Fig. 4.29. This appears to be a very complicated circuit with three op-amps, two sensors, and seven additional resistors. The operation of the circuit is actually quite straightforward. The two op-amps to the left just function as comparators (see Fig. 4.20). The negative inputs are connected to 2.5 V. When either sensor is over white, its resistance is less than the average of the on and off resistances, so the positive terminal is at a higher voltage than the negative terminal, and the comparator output is at 5V. When the sensor is over black, the opposite is true and the comparator output is at zero volts (we assume that we have rail-to-rail op-amps).

The final op-amp circuit adds and subtracts input signals. The output voltage can be found from the input voltages by using Kirchhoff's Laws. We leave that as an exercise for the reader and only write the result below. We call V_B the output from the upper-left op-amp and we call V_A the output from the lower-left op-amp. The output voltage can be found to be:

$$V_{out} = (35 - 4\,V_A - 2\,V_B) / 8 \text{ Volts} \hspace{2cm} (4.41)$$

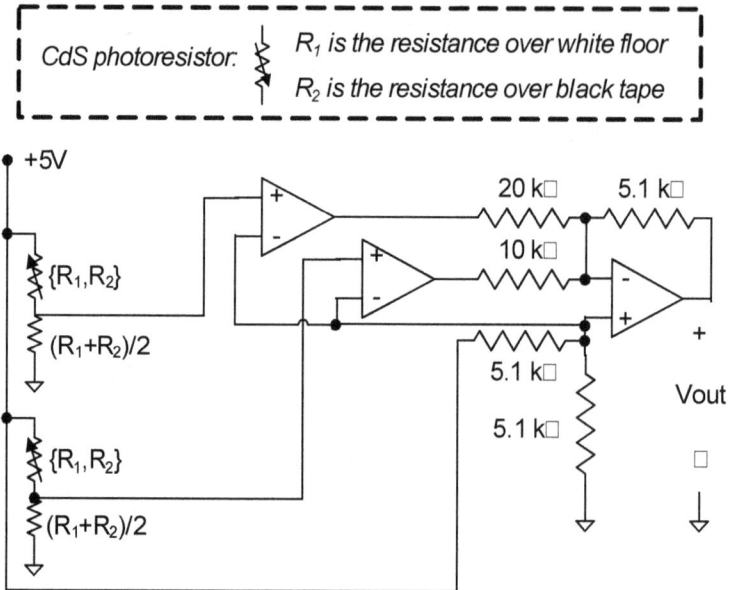

Fig. 4.29 Using one input to read two sensors

The values in Table 4.2 come from plugging in the digital possibilities for V_A and V_B. The op-amp circuit has a number of advantages over the other approach which enhances its robustness and flexibility: (1) much higher separation between output levels, (2) independence from exact photoresistance, (3) independence from moderate lighting variations, (4) automatic detection of specific sensor, and (5) adaptability to three (or perhaps four) sensors. It costs a little more and is a little more difficult to assemble.

4.7.8 System Requirements Calculation

Assume that you have two 12V, 10 Ω motors that you intend to power with a 14.4 V battery. There are also six sensors that take 100 mA each at 6 V. A 12V voltage regulator is in series between the battery and the motors to insure the correct operating voltage, but a series resistor will be used to drop the 12V regulator output to the 6V needed for the sensors. If you want the batteries to last for 5 trial runs of 15 minutes each, what is the minimum battery capacity (in mAH) that you need? What is the minimum acceptable current specification for the voltage regulator?

The problem may seem daunting, but a single figure will help organize the process (see Fig. 4.30. In Fig. 4.30a, the original layout is shown, with an "M" indicating a motor and an "S" indicating a sensor. The Resistance "R" has to be determined so that the voltage across the sensors is 6V. The capacitors need to be chosen so that the regulator works well, but they don't affect the calculations that we need to do. Likewise, we can replace the battery and regulator combination with an equivalent 12V battery. The current that this battery delivers to the circuit is the current that the voltage regulator must be able to handle. In Fig. 4.30b this simplification has been made, and the motors and sensors have been replaced by their equivalent resistances. The motor resistances were given in the problem definition and the sensor resistances were calculated by Ohm's Law (Eq. 4.4) given that they use 0.1A when energized to 6V. In Fig. 4.30c the two 10Ω parallel resistors have been replaced by their equivalent resistance and so have the six parallel 60Ω resistors. Using the voltage divider equation (Eq. 4.35), we can find that R = 10Ω. Combining the two 10Ω series resistors, after making one final parallel combination, the simplified circuit in Fig. 4.30d is achieved. Applying Ohm's Law to this circuit, we see that the required regulator current is 3A. Given that the batteries must last 1.25 hours, the battery needs a capacity of 3.75 AH or 3,750 mAH.

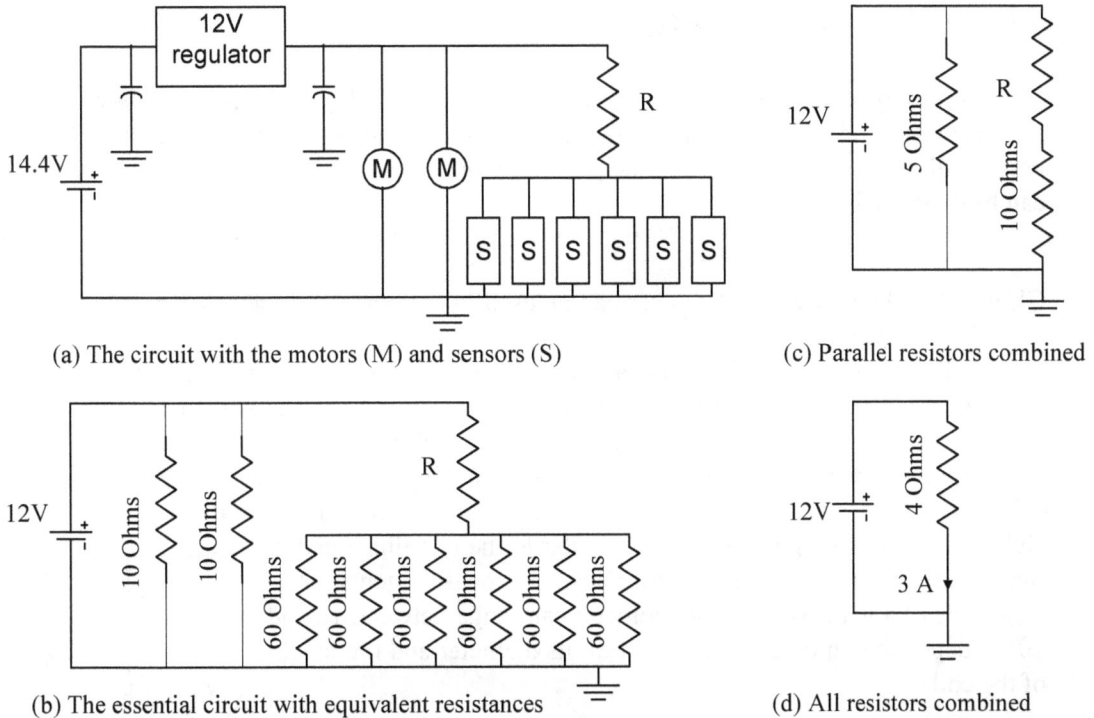

(a) The circuit with the motors (M) and sensors (S)

(c) Parallel resistors combined

(b) The essential circuit with equivalent resistances

(d) All resistors combined

Fig. 4.30 Schematic of the final circuit example

4.8 SENSORS AND TRANSDUCERS

Transducers are electromechanical devices that convert a mechanical change such as displacement or force into an electrical signal that can be monitored as a voltage after some degree of signal conditioning. A wide variety of transducers are commercially available for use in measuring mechanical quantities such as force, torque, strain, velocity, acceleration, etc. Transducer characteristics include: range, linearity, sensitivity and operating temperatures. The sensor that is incorporated into a transducer to produce an electrical output determines its characteristics. For example, a photodiode that detects reflected light intensity provides a transducer that senses the color or reflectivity of a nearby surface. A simple limit switch can also serve as a sensor to detect contact. The touch sensor marketed by LEGO for the NXT utilizes a short-throw, normally-open, contact switch as its sensor. When the switch is closed, the circuit resistance across its terminals goes from infinity to nearly zero providing an abrupt voltage change that is monitored by a circuit contained within the NXT. The sound sensor for the NXT incorporates a very small microphone that converts the pressure associated with a sound wave into an electrical signal. This signal is processed by circuits contained within the NXT and then can be used as a feedback signal for controlling one or more of the output ports on the NXT.

Sensors used in transducer design include switches, potentiometers, differential transformers, strain gages, capacitors, piezoelectric and piezoresistive crystals, thermistors, etc. Important features of a few of these sensors are described in this chapter.

4.9 POTENTIOMETERS

The simplest type of potentiometer, shown schematically in Fig. 4.31, is the slide-wire resistor. This sensor consists of a length L of resistance wire attached across voltage source V_s. The relationship between the output voltage V_o and the position x of a wiper, as it moves along the length of the wire, can be expressed as:

$$V_o = (x/L)V_s \qquad\qquad x = (V_o/V_s)L \qquad\qquad (4.42)$$

Clearly, the slide-wire potentiometer can be used to measure a displacement x if V_0 is measured and V_s and L are known.

Resistors fabricated from a short length of straight wire are not feasible for most applications, because the wire's resistance is too low. A sensor with a low resistance imposes excessive power drain on the voltage source. To alleviate this difficulty, high-resistance, wire-wound potentiometers are obtained by winding the wire around an insulating core, as shown in Fig. 4.32. The potentiometer illustrated in Fig. 4.32a is used for linear displacement measurements. Cylindrically shaped potentiometers, similar to the one illustrated in Fig. 4.32b, are used for angular displacement measurements. The resistance of a wire-wound potentiometer can range between 10 and 10^6 Ω depending upon the wire material, its diameter and the length of the coil.

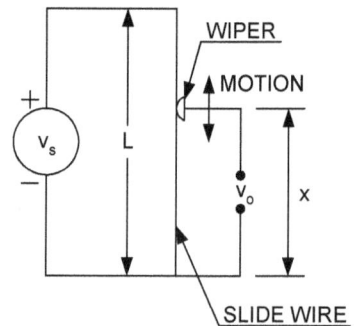

Fig. 4.31 Slide-wire resistance potentiometer.

The resistance of a wire-wound potentiometer increases in a stepwise manner as the wiper moves from one turn to the adjacent turn. This step change in resistance limits the resolution of the potentiometer to L/n, where n is the number of turns in the length L of the coil. Resolutions ranging from 0.05 to 1% are common, with the lower limit obtained by using many turns of very small diameter wire.

The active length L of the coil controls the range of the potentiometer. Linear potentiometers are available in many lengths up to about 1 m. Arranging the coil into a helix extends the range of the angular-displacement potentiometer. Helical potentiometers are commercially available with as many as 20 turns; therefore, angular displacements as large as 7,200° can be measured with relative ease.

In order to improve resolution, potentiometers have been introduced that utilize thin films of conductive plastic with controlled resistivity instead of wire-wound coils. The film resistance on an insulating substrate exhibits very high resolution together with lower noise and longer life. For example, a resistance of 50 to 100 Ω/mm can be obtained with conductive plastic films that are used for commercially available potentiometers with a resolution of 1 μm or less. The frictional force that must be overcome to move the wiper is depended on the construction details of the particular potentiometer; however, friction forces from 0.5 to 1.0 N are common. The life expectancy for potentiometers fabricated with conductive plastics is commonly specified at 20×10^6 strokes. An example of a linear motion potentiometer with a single end shaft is presented in Fig. 4.33. Note the three terminals used for electrical connections.

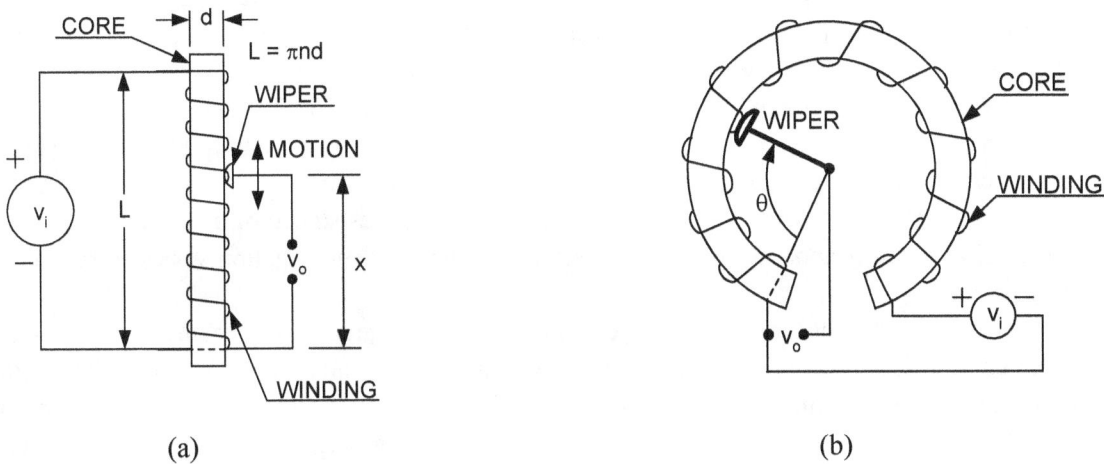

(a)   (b)

Fig. 4.32 Wire-wrapped resistance potentiometer for (a) axial displacements and (b) for angular displacements.

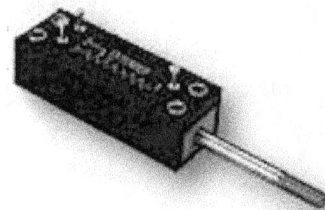

Fig. 4.33 A linear motion potentiometer.

The dynamic response of both the linear and the angular potentiometer is severely limited by the inertia of the shaft and wiper assembly. Because of its large inertia, the potentiometer is used only for static or quasi-static measurements where a high frequency response (bandwidth) is not required.

Potentiometers are used primarily to measure large displacements—10 mm or more for linear motion and 15° or more for angular motion. Potentiometers are relatively inexpensive yet accurate; however, their main advantage is simplicity of operation, because only a voltage source and a digital voltmeter (DVM) are required for a complete instrumentation system. The primary disadvantages of potentiometers are a limited frequency response that precludes their use for dynamic measurements and the force required to overcome friction of the wiper.

4.10 PHOTOELECTRIC SENSORS

In many applications where direct contact cannot be made with the object being examined, a photoelectric sensor can be employed to make physical measurements by monitoring changes in the intensity of light reflected off the object of interest. When light impinges on a photoelectric sensor, it either creates or modulates an electrical signal. Most photoelectric devices employ semiconductor materials and operate by either generating a current or by changing the semiconductor's resistivity. These devices are photodetectors that respond quickly to changes in light intensity.

4.10.1 Photoconducting Sensors

Photoconductive cells, illustrated in Fig. 4.34, are fabricated from semiconductor materials, such as cadmium sulfide (CdS) or cadmium selenide (CdSe), which exhibit a strong photoconductive response. When a photon with sufficient energy strikes a molecule of, say CdS, an electron is driven from the valence band to the conduction band and a hole remains in the valence band. This hole and the electron both serve as charge carriers and with continuous exposure to light, the concentration of charge carriers increases and the resistivity decreases. A circuit used to detect the resistance change ΔR of the photoconductor is also shown in Fig. 4.34. The resistance of a typical photoconducting sensor changes over about 3 orders of magnitude as the incident radiation varies from very dark to very bright.

When a photoconductor is placed in a dark environment, its resistance is high and only a small current flows. If the sensor is exposed to light, the resistance decreases significantly (the ratio of maximum to minimum resistance for R_d in Fig. 4.34 ranges from 100 to 10,000 in common commercial sensors); therefore, the output current can be quite large. The sensitivity depends on cell area, type of cell (CdS or CdSe) and the power limit for the cell. If the supply voltage (V_s) is set to supply the cell at its maximum power limit, then sensitivities of 0.2 mA/lx result for CdS type cells. A typical cadmium sulfide photoconductor exhibits a maximum dark resistance of about 1 to 2 MΩ. When subjected to a light intensity of about 2 foot-candles the resistance decreases to about 1 to 5 kΩ. The change of about three orders of magnitude is large, enabling many applications with very simple control circuits.

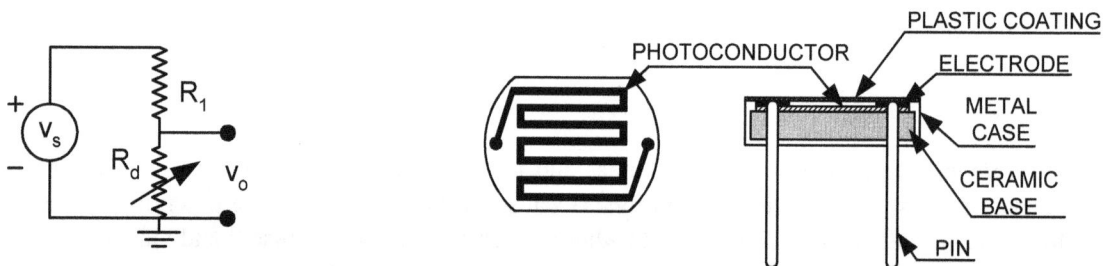

Fig. 4.34 When a photoconductor is used (modeled as R_d), the output voltage measured (shown as V_o) will vary with light intensity.

Photoconductors respond to radiation ranging from long thermal radiation through the infrared, visible and ultraviolet regions of the electromagnetic spectrum. The sensitivity, S, changes significantly with wavelength and drops sharply at both short and long wavelengths; consequently, photoconductive cells exhibit the same disadvantage as many other photodetectors. Their calibration depends on the wave length of the impinging light.

The photocurrent requires some time to develop after the excitation is applied and some time to decay after the excitation is removed. The rise and fall times for commercially available photoconductors are usually about a second. Because of these relatively long delays, the CdS and CdSe photoconductors are not suitable for dynamic measurements. Instead, their simplicity, high-sensitivity and low-cost lend them to applications involving sensing, counting and switching based on a slowly varying light intensity.

4.10.2 Photodiode Sensors

Photodiodes are semiconductor devices that respond to high-energy particles and photons. Photodiodes operate by absorption of photons or charged particles and generate a current that flows in an external circuit, which is proportional to the incident power. Photodiodes can be used to detect the presence or absence of minute quantities of light and can be calibrated for extremely accurate measurements from intensities below 1 pW/cm^2 to intensities above 100 mW/cm^2. Silicon is the most common semiconductor material used in fabricating planar diffused photodiodes. These photodiodes are employed in such diverse applications as spectroscopy, photography, analytical instrumentation, optical position sensors, beam alignment, surface characterization, laser range finders, optical communications and medical imaging instruments.

Planar diffused silicon photodiodes are P-N junction diodes. A P-N junction can be formed by diffusing either a P-type impurity (anode), such as boron, into a N-type bulk silicon wafer, or a N-type impurity, such as phosphorous, into a P-type bulk silicon wafer. The diffused area defines the photodiode active area. To form an ohmic contact it is necessary to diffuse another impurity into the backside of the wafer. The impurity is an N-type for a P-type active area and P-type for an N-type active area. Contact pads are deposited on defined areas of the front active area and on its backside. The active area is covered with an anti-reflection coating to reduce the reflection of the light for a specific predefined wavelength. The non-active area on the top is covered with a thick layer of silicon oxide. A schematic illustration of a planar diffused photodiode fabricated from N-type silicon is presented in Fig. 4.35.

By controlling the thickness of bulk substrate, the speed and responsivity of the photodiode can be controlled. When the photodiodes are biased, they are operated in the reverse bias mode, i.e. a negative voltage applied to anode and positive voltage to cathode.

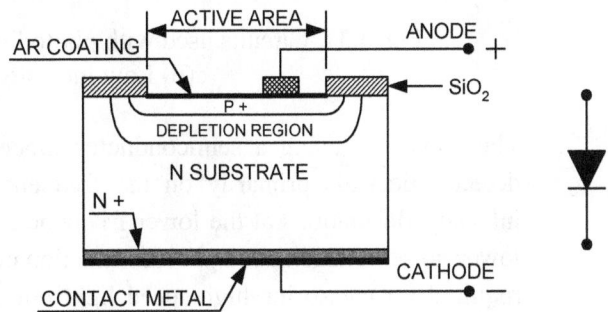

Fig. 4.35 Diagram of the diffusion areas used to create a P-N junction in a N-type silicon photodiode.

Electrical Characteristics of Photodiodes

A silicon photodiode can be represented by a current source in parallel with an ideal diode as shown in Fig. 4.36. The current source represents the current generated by the incident light, and the diode represents the P-N junction. In addition, a junction capacitance (C_J) and a shunt resistance (R_{SH}) are in parallel with the other components. A resistance (R_s) is in series in the circuit.

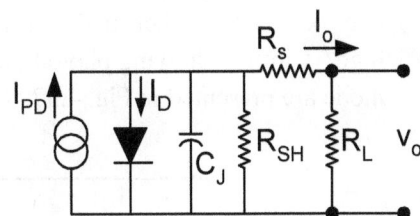

Fig. 4.36 Representative circuit for a photodiode.

Physically, the shunt resistance is the slope of the current-voltage curve of the photodiode at the origin, i.e. $V_0 = 0$. Although an ideal photodiode should have an infinite shunt resistance, actual values range from 10 MΩ to 1,000 MΩ. Experimentally the shunt resistance is determined by applying $\pm$ 10 mV across the diode, measuring the resulting current and calculating the resistance from Ohm's law. Shunt resistance is used to determine the noise current in the photodiode with no bias (photovoltaic mode). For superior photodiode performance, a very high shunt resistance is specified.

The series resistance of a photodiode arises from the resistance of the contacts and the resistance of the undepleted silicon shown in Fig. 4.35. It is used to determine the linearity of the photodiode in photovoltaic mode[2]. Although an ideal photodiode should have zero series resistance, typical values ranging from 10 Ω to as much as 1,000 Ω are measured.

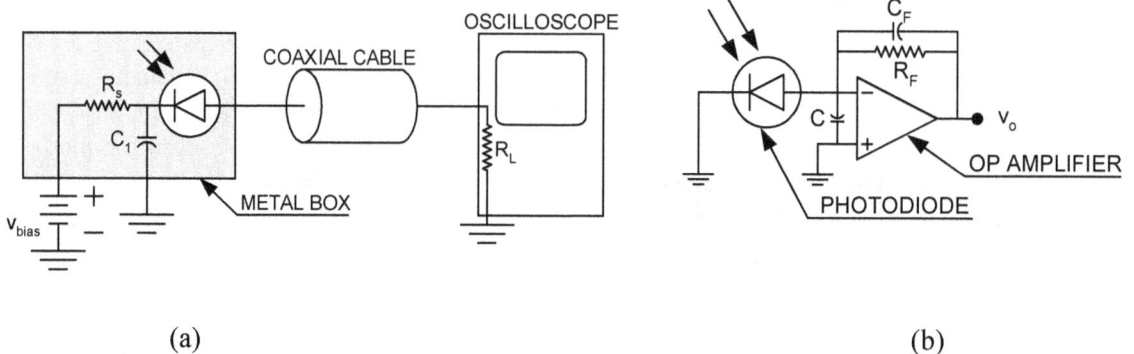

(a)

(b)

Fig. 4.37 Circuits used with photodiodes. (a) High intensity light at high frequency.
(b) Low intensity light with low frequency.

The choice of using a semiconductor diode as a photovoltaic detector or as a photoconductive detector depends primarily on the frequency response required in the measurement. For light intensity fluctuations at the lower frequencies (less than 100 kHz), a photovoltaic circuit exhibits a lower noise voltage than a photoconduction circuit with reverse bias voltage. For higher frequencies required for measuring high-speed light pulses or high frequency modulation of a continuous light beam, the reverse bias serves to accelerate the electron/hole transition times, which improves the frequency response. Photoconductive diodes operate over a frequency range from DC (0 Hz) to 100 MHz and are capable of measuring the intensity of light pulses with rise times in the 3 to 12 ns range.

Semiconductor photodiodes are small, rugged and inexpensive. Because of these advantages, they have replaced the vacuum tube detectors in most applications. The photodiodes may be used in either mode with operational amplifiers to give a responsivity that is exceptionally high. Circuits showing photodiodes operating in the photoconduction mode and in the photovoltaic mode are presented in Fig. 4.37.

---

[1] Photodiodes act as photovoltaic devices when no bias voltage is applied.

4.11   INTERFACING SENSORS WITH THE NXT CONTROLLER

The NXT is housed in a relatively small package with a monochrome LCD display and four buttons. In spite of its small size and simple interface, it is a powerful controller with two processors, four sensor ports, three motor ports, a USB port, and Bluetooth capability. A block diagram showing the primary components and their connections within the NXT controller is presented in Fig. 4.38.

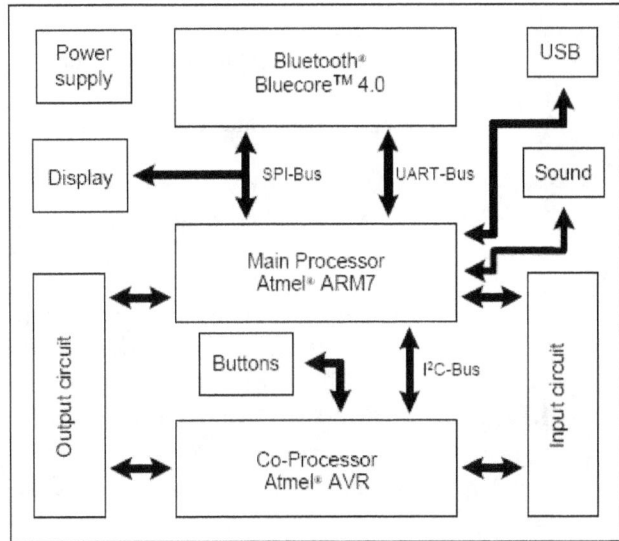

Fig. 4.38 Block diagram of the components within the NXT controller.

4.11.1 Component Specifications for the NXT Controller

The NXT controller employs a number of electronic devices to provide its broad functionality. A description of the hardware specifications for the NXT controller is given below:

Main processor: Atmel® 32-bit ARM® processor, AT91SAM7S256
- 256 kB FLASH
- 64 kB RAM
- 48 MHz

Co-processor: Atmel® 8-bit AVR coprocessor, ATmega48
- 4 KB FLASH
- 512 Byte RAM
- 8 MHz

Bluetooth wireless communication: CSR BlueCoreTM 4 v2.0 +EDR System
- Supporting the Serial Port Profile (SPP)
- Internal 47 Kbytes RAM
- External 8 Mbits FLASH
- 26 MHz

USB 2.0 communication port:
- Full speed port (12 Mbits/s)

Four input ports: 6-wire interface supporting both digital and analog interface

- A single high speed port, IEC 61158 Type 4/EN 50170 compliant

Three output ports:
  - 6-wire interface supporting input from encoders

Display: 100 x 64 pixel LCD black & white graphical display
  - View area: 26×40.6 mm

Sound output channel: 8-bit resolution
  - Supporting a sample rate of 2-16 kHz

Four user-interface buttons:

Power supply: 6 AA batteries
  - Alkaline batteries are recommended
  - A rechargeable Lithium-Ion battery (1400 mAh) may be employed

Connectors: 6-wire industry-standard connectors
  - RJ12 with right side adjustment

4.11.2 Input Ports on the NXT Controller

The NXT has four input ports that are used to connect sensors to the two processors with six wire leads. The six wire connector permits the acquisition of both analog and digital signals at each port, thus enabling the use of both analog and digital sensors with the NXT. A schematic diagram of the six pin connector is presented in Fig. 4.39. All of the ports have the same pin arrangement; however, the digital pins (5 and 6) are connected to a high speed RS485 controller to accommodate high frequency communications.

   Pin #1, the analog input, is connected to a 10 bit A/D converter that is incorporated within the AVR coprocessor. This pin is also connected to a current generator which supplies power to the sensors connected to the NXT. The analog signals are sampled and converted to digital format with the same sampling rate for all of the analog sensors—333 Hz. The sensors require power for 3 ms before they are capable of making measurements. Pin #2 is a ground connection. These two pins have the same functionality as the input terminals on the RCX, which permits the use of legacy LEGO sensors with the NXT using appropriate lead wires.

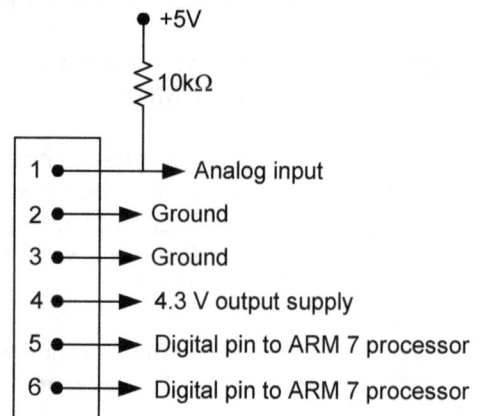

Fig. 4.39 Pin connections between the input ports and components within the NXT.

Output power is provided from Pin # 4 for both the input and output ports. The maximum current that can be drawn from the NXT's power supply is 180 mA, which is divided among the seven ports. Hence, the maximum current capability of an individual port is about 20 mA. If the total current required exceeds the 180 mA limit, the output from pin #4 is automatically decreased without warning. If the power supply is accidentally shorted to ground, the NXT resets.

Pins #5 and #6 are used for digital communications using what is known as I^2C protocol. The I^2C is a two-wire communication interface developed by Phillips Semiconductor many years ago to connect a CPU to peripheral chips in a television set. Today this two wire interface is used in many applications. In the NXT, digital communication requires two lines—Pin #5 controls the timing signals from its clock and Pin #6 transmits information in both directions between the NXT and the sensors attached to it. The communication speed is 9600 bits/s for each port. Pins #5 and #6 are connected directly to two I/O terminals on the ARM7 processor. Firmware in the NXT creates four independent buses, one for each input port.

4.11.3 Output Ports on the NXT Controller

The NXT has three output ports that are used to connect actuators to the power supply within the NXT. A six wire digital interface is used with the output ports permitting the actuators to provide information back to the NXT without having to employ an input port. A schematic diagram of the six pin connector for the output ports is presented in Fig. 4.40. Note that all three ports have the same pin arrangement.

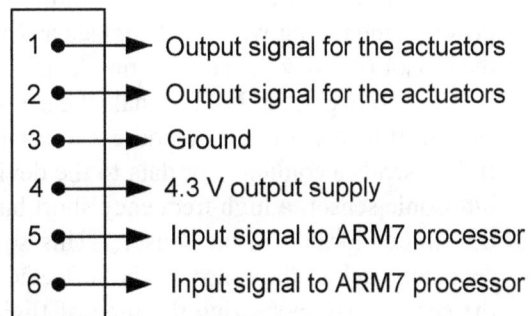

1	→	Output signal for the actuators
2	→	Output signal for the actuators
3	→	Ground
4	→	4.3 V output supply
5	→	Input signal to ARM7 processor
6	→	Input signal to ARM7 processor

Fig. 4.40 Pin connections between the output ports and components within the NXT.

Pins #1 and #2 provide power to the motors and actuators. A second internal power supply within the NXT can provide a continuous current of 700 mA to each of the three output ports and a peak current of 1 A. The output from this power supply is pulse width modulated that can be controlled by software to brake or float. This internal power supply incorporates a built-in thermal protection system that automatically lowers the current output when overheating of the supply occurs.

Pin #3 is employed to provide a ground connection for the output power supply associated with Pin #4. Output power is provided from Pin # 4 for both the input and output ports. The maximum current that can be draw from the NXT's power supply is 180 mA, as described above.

Pin #5 and Pin #6 provide connections to input pins on the ARM7 processor through a Schmidt trigger. Tachometer pulses from the encoders in the LEGO servo motors are counted using this circuit and the firmware stored in the NXT. The signals from the servomotor are also used to establish the direction of rotation of the servo motors.

4.11.4 LEGO Sensors and the NXT Controller

Ports 1, 2 and 3 on the NXT are used to interface with both active and passive sensors. Recall that passive sensors do not require a separate power lead from the NXT to function. Moreover, the output from a passive sensor is a voltage drop across the sensor terminals due to a resistance or a voltage source placed across the terminals of the NXT. For example, the voltage source can be from an externally powered Wheatstone bridge, a solar cell or a thermocouple. The resistive loads may be from switches, potentiometers, thermistors or fixed value resistors.

Active sensors require power to function. The light sensor from a LEGO Mindstorms Invention kit is an example of an active sensor because it requires power drawn from the NXT to operate the light emitting diode (LED) that illuminates the scene and activates the circuit containing the photodiode. The light reflected from the scene passes through a lens and impinges on a photodiode producing an output voltage that is monitored by the NXT and converted to a reading of light intensity. The supply voltage to the light sensor is provided through pin #1. A power generator connected to pin operates in a multiplexed mode. It is switched on to supply power to the light sensor for 3 ms and then off for 0.1s, which is sufficient to measure the output voltage from the photodiode. This design feature was added to the NXT to provide backwards compatibility with the RCX sensors.

The passive sensors that interface with the NXT controller include: the touch, the sound and the temperature sensors. These sensors do not need a separate power lead to function. The sound and temperature sensors generate a voltage that is related to the quantity they are measuring. The touch sensor generates a voltage within the NXT when its switch is closed. The output voltage from these sensors is sampled every 3 ms using the A/D converter in the AVR coprocessor.

The ultrasonic sensor included in the LEGO Mindstorm NXT kit is a digital sensor that requires an I^2C communication channel to function. The I^2C communication channel is available only on port #4 of the NXT. The communication is effected with two lines. The first line (Pin #5) provides the timing with clock signals and the second line (Pin #6) transmits information to and from the sensor (the slave). The communication speed is 9,600 bits/s.

To begin, the NXT signals the sensor that it is about to transmit a message by sending a start signal. It then sends a 7 bit address to establish a connection to the specific sensor it is monitoring[3]. It then sends a command or data to the device (sensor) and waits for information to return. With the ultrasonic sensor, a high frequency short burst of sound (ping) is generated by a piezoelectric crystal transmitter located in the sensor. This sound wave propagates to some nearby object where it is reflected back to the sensor. The reflected sound wave (echo) is monitored by a receiver located in the sensor. By measuring the time of flight between the ping and echo, it is easy to determine the distance between the target and the sensor. Sound waves travel at about 333 m/s (depending on temperature and pressure); hence, propagation over a meter takes 3 ms. At the maximum range of 2.5 m for the LEGO sonar sensor, the transit maximum time to and from the target is $2 \times 2.5 \times 3 = 15$ ms.

4.11.5 Additional Features within the NXT Controller

Bluetooth® Communication within the NXT Controller

Wireless communication is possible with the NXT using Bluetooth® technology. Wireless communications between four NXTs are possible (one master unit and three slaves), but the master NXT can only communicate with one other NXT (slave) at a time. This capability is enabled by using Serial Port Profile (SPP), which is essentially a wireless serial port that may be used for incoming or outgoing communications. Wireless communications are possible between the NXT and personal computers or other devices equipped with Bluetooth technology. It is possible to send

---

[3] The fact that a 7 bit address is used to specify the sensor implies that up to 128 devices could be arranged in parallel and monitored in sequence with the NXT.

and receive messages between NXT controllers and between the computer and NXT as a program is executing. To conserve power used by Bluetooth, its range has been limited to about 10 m. Communications to other Bluetooth units occur through four communication channels. Channel 0 is used by one or more of the slave NXTs sending information back to the master NXT. Channels 1, 2 and 3 are used by the master NXT for outgoing communications to the slave devices.

The Display on the NXT Controller

A monochrome liquid crystal display (LCD) with a resolution of 100×64 pixels is provided to enable readout and feedback to the operator. The viewing area on the display is 26×40.6 mm. There is an interface connecting the display to the ARM7 processor that operates at a frequency of 2 MHz. The display is updated in a continuous line by line replacement scheme that requires 17 ms for a total display update.

Power Management in the NXT Controller

Power is drawn from the six AA batteries housed in the NXT (either alkaline or lithium ion rechargeable). The power supply within the NXT provides three voltages—9 V from the batteries, 5 V and 3.3 volts for the ARM processor and the BlueCore® chip. The power supply is protected with a poly switch that is rated at 1.85 A for a continuous current and 3.3 A for the trip current. The current and power requirements placed on the power supplies depend upon whether the motors are loaded or not. These requirements are presented in Table 4.3.

Table 4.3
Current and power requirements for the NXT

Supply Voltage (V)	Current		Power(9 Volt Battery)	
	Maximum (mA)	Normal (mA)	Maximum (mW)	Normal (mW)
No Load on Motors				
9	339	114	5,184	1,422
5	271	112	1,744	448
3.3	72	38	410	216
Load on Motors				
9	2,901	848	26,109	7,632
5	271	112	1,142	307
3.3	72	38	410	137

4.12 DESIGNING AND BUILDING SENSORS

There are a number of different sensors commercially available from LEGO—touch, sound, light, temperature and sonar (distance). In addition the angle of rotation of the servo motors can be monitored using the internal encoders in the servo motors. These are excellent sensors designed specifically to interface with the NXT. The touch, sound and temperature sensors are passive and do not require separate power leads to operate. However, the light sensor is active because the LED and circuit containing the photodiode both require power to function. In addition to the LEGO sensors available, a number of companies provide sensors that are compatible with the NXT to measure other quantities. Also, it is possible to design and build your own sensors using components that are available from local electronic supply stores. In fact, with Internet orders, it is possible to obtain almost any electronic device or component from regional supply houses in a few days.

A Touch Sensor

Let's begin with the touch sensor because it is the simplest of all the sensors. It consists of a momentary switch, a LEGO support block to facilitate mounting in a LEGO design environment, and a connector that is compatible with the NXT. A schematic diagram of a momentary switch[4] that can be used for the touch sensor is shown in Fig. 4.41. As shown in Fig. 4.41, the momentary switch is in the normally open mode because a spring keeps the switch open until the push button is pressed, bringing the switch bar down on the contacts and closing the switch. When the button is released, the switch returns to the open position.

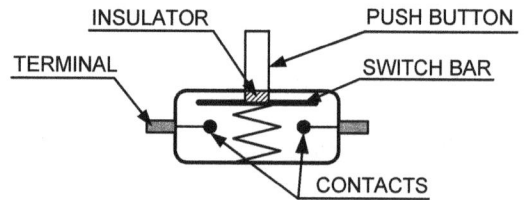

Fig. 4.41 A normally open momentary switch used in the design of a touch sensor.

If the touch sensor is to be employed with the NXT, the electrical connections must be compatible with the ports on the NXT. The easiest way of connecting the switch to the NXT is to take a LEGO lead wire with port compatible connectors on both ends and cut it into two pieces. Take one of the two pieces and strip the insulation for about 6 to 8 mm from the ends of the two wires connected to pins #1 and #2. Then use a soldering iron to solder the two LEGO lead wires to the appropriate terminals on the contact switch. Next insert the connector into Port 1 on the NXT.

At this stage of the development of the touch sensor, check if the NXT responds when the push button on the switch is depressed. With the touch sensor connected to one of the input ports, depress the gray triangular buttons on the NXT until you have activated the View icon. Select the view icon and continue to press the gray triangular buttons until the touch sensor icon is displayed on the LCD. Press the square orange button to show the reading from the touch sensor. The display should read 0 indicating an open switch. Press the button closing the switch and note that the display reads 1. This reading indicates that the switch is functioning properly and will act as a touch sensor. For this simple sensor no calibration is required—it is either open with a zero reading on the NXT or closed with a reading of 1.

If the touch sensor is to be used in a LEGO design environment, it should be mounted on a LEGO plate or brick. The selection of either a plate or a brick will depend on the size of the switch and the method you choose to attach the switch to the brick or plate.

A Light Sensor

A relatively complex light sensor is available from LEGO that interfaces with the NXT to measure light intensity on a scale of 0 to 100. This sensor incorporates a light emitting diode (LED) to illuminate the scene, a photodiode to measure either the direct light intensity or the reflected light, and circuitry to power both of these components. This is an active sensor that requires a separate power lead from the NXT to operate.

---

[4] Momentary switches are available from a large number of companies through electronic suppliers. For this example, we selected a Panasonic detector switch part number ESE-11HS1 that was available from Digi-Key for less than $1. This switch, which is rated at 5 Vdc and 10 mA, is actuated with a force of 35 g applied to its push button.

In many applications, less complex and lower cost light sensors fabricated from Cadmium Sulfide (CdS) photoconductors are sufficient. The characteristics of CdS photoconductors were discussed previously in Section 4.11.1. They are variable resistors with the resistance changing from very high values when the sensor is dark to nearly zero resistance when the sensor is exposed to bright light and the CdS becomes a good conductor. We have discussed the design of angle of rotation sensor using a potentiometer that is a variable resistor. The development of a CdS light sensor is similar.

Only two parts are required to build this sensor: a CdS photoconductor (photocell) and a lead wire that is compatible with the NXT. We purchased a package of five photocells from Radio Shack. One of these photocells is illustrated in Fig. 4.42.

Fig. 4.42 A photocell and its two lead wires anchored with a dime.

After soldering the lead wires from the photocell to the wires for pins #1 and #2 on a NXT lead wire, we check to determine some of the characteristics of our light sensor. Using the same procedure developed in calibrating the angle sensor, we explore the readings from our light sensor that is connected to port 1. The raw value is displayed for this light sensor. In normal room light, the sensor gives a raw value that ranges from about 200 to 400 depending on its orientation. When exposed directly to a bright light source, the raw value decreases to about 15. If we cover the sensor with a black cloth, the raw value increases to about 900.

Typically, a low cost photocell is employed as a sensor to turn lights on or off and to detect significant changes in light intensity. To check if the photocell is effective for this application, write a suitable program for the NXT, connect the sensor to port 1, run the program and view the output from the light sensor on the LCD display. In normal room light, the reading is 1. Shield the sensor from the light and the reading goes to 0. The sensor is acting like a switch, turning on when exposed to light and turning off when shielded from light.

It is possible to use the CdS sensor to measure light intensity; however, the sensor is affected by light coming from many different directions. To improve its performance for measuring intensity, the sensing element should be placed in a short, small-diameter tube to shield it from stray light. Additional improvement in performance is achieved if a suitable lens is fitted to the front of this tube to focus the incoming light on the sensing element.

4.13 SUMMARY

In this chapter, we have tried to give a basic, yet thorough, introduction to key electric circuit theory concepts. We started with a discussion of the basic physical quantities and later described the origin of the equations that are needed to solve for all circuit variables. We introduced the material classes that are used to make electronic parts before giving a basic description of a number of different components, such as resistors, transistors, op-amps, regulators, etc. The list of components included the majority of the electronic devices that might prove useful for the hovercraft project. In most instances, we developed the terminal equations necessary to analyze the circuit or gave examples of how the devices can be used.

We also introduced Kirchhoff's Voltage and Current Laws, the two equations which describe how components are connected together to form functioning circuits. We gave a number of examples of circuit analysis, including a hovercraft control and detector circuits, as well as classic examples that introduce the useful concepts of equivalent circuits for various resistor combinations, and in particular, series and parallel-connected resistors.

A few basic sensors have been described in this chapter. These sensors can be used to measure an unknown quantity—such as the use of a photo cell to measure light intensity. Important characteristics of each sensor that must be considered in the selection process include:

1. **Size** - with smaller being better because of enhanced dynamic response and minimum interference with the process or event.
2. **Range**—with extended range being preferred to increase the latitude of operation.
3. **Sensitivity**—with the advantage to higher output signals that require less amplification.
4. **Accuracy**—with the advantage to devices exhibiting errors of 1% or less after considering zero shift, linearity and hysteresis.
5. **Frequency response**—with preference for wide-bandwidth sensors that permit application in both static and dynamic loading situations.
6. **Stability**—with very low drift in output over extended periods of time and with very small output signal changes with variations in temperature and humidity preferred.
7. **Temperature limits**—with the ability to operate from cryogenic to elevated temperatures preferred.
8. **Economy**—with reasonable costs preferred.
9. **Ease of application**—with reliability and simplicity always preferred.

The methods employed to design and build touch and light sensors were described in considerable detail.

REFERENCES

1. Irwin, J. D., Basic Engineering Circuit Analysis, 8th ed., John Wiley & Sons, New York, 2004.
2. Irwin, J. D. and Kerns, D. V., Jr., Introduction to Electrical Engineering, Prentice-Hall, 1996.
3. Horowitz, P. and Hill, W., The Art of Electronics, 2nd ed., Cambridge University Press, New York, 1989.
4. Dorf, R. C. and Svoboda, J. A., Introduction to Electric Circuits, 6th ed., John Wiley & Sons, New York, 2003.
5. Brindley, K.: Sensors and Transducers, Heinemann, London, 1988.
6. Sedra, A. S. and K. C. Smith: Microelectronic Circuits, 3rd edition, Holt, Rinehart, and Winston, New York, 1991.
7. Dally, J. W., W. F. Riley and K. G. McConnell, Instrumentation for Engineering Measurements, 2nd edition, John Wiley, New York, NY, 1993.

EXERCISES

4.1. Draw the symbols used on drawings to represent a resistor, a capacitor and an inductor. Also, write the relations for the voltage across these components in terms of the current flow.

4.2. If you design a circuit containing an equivalent resistance of 2,400 Ω with a 9-V battery power supply, find the current drained from the battery. If the battery has a capacity of 2A-h (ampere-hours), determine the life of the battery. Assume that it cannot be recharged. What is the power dissipated by the resistor?

4.3. A size D, 1.5 V, alkaline battery has a capacity of 1,200 hours when the battery is discharged through a 120 Ω resistor. Determine the current flow through this resistor, and state the capacity in terms of ampere-hours (Ah).

4.4. A 9 V alkaline battery has a capacity of 7 hours when the battery is discharged through a 120 Ω resistor. Determine the current flow through this resistor, and state the capacity in terms of ampere-hours (Ah).

4.5. Write Kirchhoff's two laws. Use circuit diagrams to illustrate these two laws.

4.6. Draw a circuit with four series-connected resistors and give the expression for the equivalent resistance. If these resistors have values of 800, 600, 1,300 and 12,000 Ω, find the equivalent resistance. Draw a circuit diagram that contains the equivalent resistor, which is also equivalent to your initial circuit diagram.

4.7. Draw a circuit with four parallel-connected resistors, and give the expression for the equivalent resistance. If these resistors have values of 800, 600, 1,300 and 12,000 Ω, find the equivalent resistance. Draw a circuit diagram that contains the equivalent resistor, which is also equivalent to your initial circuit diagram.

4.8. You have a relay that runs off 5V DC. When energized, the current through the electromagnet wire is 100 mA. Assume that this wire can be modeled by a simple resistor.
   a. What is the resistance of the electromagnet wire?
   b. If the electromagnet wire can dissipate no more than 2 Watts, what is the maximum voltage that you can place across the electromagnet terminals?

4.9. Assume that you have a motor that can be modeled as a 0.35 Ω resistor. You have a FET transistor that can take up to 10A DC. This transistor will be used to energize the fan (in a series configuration).
   a. What is the maximum battery voltage that you can use if you want to make sure that you don't exceed the transistor current limit?
   b. Let's say that you set the FET gate so that only 6.0 A flows through the fan/transistor combination when the circuit is energized to the maximum battery voltage. How much power (on average) is being dissipated by the transistor?

4.10. Assume that you have three photo-sensors, all of which have resistances of 10 kΩ when they are not reflecting light. Assuming that is the case (lights are off):
   a. What is the effective (combined) resistance of the three sensors together, if they are connected in series?
   b. What is the effective (combined) resistance of the three sensors together, if they are connected in parallel?

4.11. Hint: You should use the voltage divider rule for both parts below:
   a. If you place a 4.2 kΩ resistor across the terminals of one of the NXT analog sensor inputs (pins 1 and 2), what voltage will you measure across that resistor (when the NXT is turned on)? Note that a figure of the internal circuit of the NXT is given in Fig. Ex4.11.
   b. If you want to measure 3.3V across one of the NXT sensor inputs, what value of resistance should you connect across those two sensor inputs?

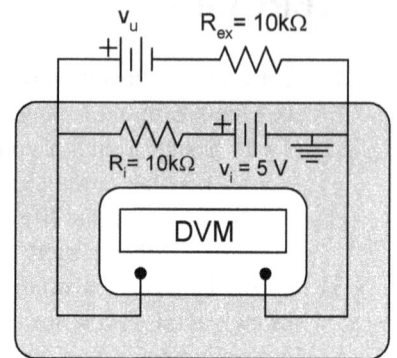

Fig. Ex4.11

4.12. Consider connecting various resistors to the NXT analog sensor inputs (terminals 1 and 2). Hint: You may want to use the voltage divider rule for all parts.
   a. If you place a 7.7 kΩ resistor across the NXT analog terminals, what voltage will you measure across that resistor (when the NXT is turned on)? Note that a figure of the internal circuit of the NXT is given in Fig. Ex4.11.
   b. If you want to measure 1.8 V across one of the NXT sensor inputs, what value of resistance should you connect across those two sensor inputs?
   c. If you want to measure 4.5 V across one of the NXT sensor inputs, what value of resistance should you connect across those two sensor inputs?

4.13. You have a relay that typically operates with an actuation voltage of 12V DC. When energized to this voltage, the current through the electromagnet wire is 80 mA. Assume that this wire can be modeled by a simple resistor.
   a. What is the resistance of the electromagnet wire?
   b. Let's say that you want to use the NXT output to energize this relay. What would be the power dissipated by the relay coil if the NXT output were 9V?
   c. Look up online the characteristics of a real 12V relay. Explain whether or not this relay would actually work with the NXT (in part b) and justify your answer with information from the relay's data sheet! Don't forget to give the part number for the relay!

4.14. You have three fans powering your hovercraft. Two propulsion fans run at 9.6V, 1A when running continuously and a 12V levitation fan that puts out a pressure of 0.15 psig when the flow rate is 10 CFM (ft^3/min). The conversion from electrical energy to output energy is 45%. For each of the batteries / battery combinations below, state whether or not they could be used to power your hovercraft for THREE trial runs on the ENES 100 course (worst case) and explain your reasoning clearly, supported by calculations.
   a. one 9.6 V, 2,200 mAh Lithium-ion battery
   b. one 12 V, 1,500 mAh Ni-Cad battery
   c. one 12 V, 700 mAh Ni-Cad battery and one 9.6V, 700 mAh Lithium-ion battery

4.15. A slide-wire potentiometer having a length of 100 mm is fabricated by winding wire with a diameter of 0.12 mm around a cylindrical insulating core. Determine the resolution limit of this potentiometer.

4.16. If the potentiometer of Exercise 4.15 has a resistance of 2,000 Ω and can dissipate 4 W of power, determine the voltage required to maximize its sensitivity. What voltage change corresponds to the resolution limit?

4.17. A 20-turn potentiometer with a calibrated dial (100 division per turn) is used as a balance resistor in a Wheatstone bridge. If the potentiometer has a resistance of 40 kΩ and a resolution of 0.05%, what is the minimum incremental change in resistance ΔR that can be read from its calibrated dial?

4.18. Why are potentiometers limited to static or quasi-static applications?

4.19. List several advantages of the conductive-film type of potentiometer.

4.20. A new elevator must be tested to determine its performance characteristics. Design a displacement transducer that utilizes a 10-turn potentiometer to monitor the position of the elevator over its 120-m range of travel.

4.21. Design a circuit to turn on outside lights at your home as it begins to get dark. Use a photoconduction cell in the circuit and provide for an adjustment to control the intensity level for switch activation.

4.22. Sketch the circuit for a photovoltaic cell and write a paragraph explaining its operation. Write another paragraph stating the advantages and disadvantages of this light sensor.

4.23. Sketch the circuit for a photodiode used in the photoconduction mode and write a paragraph explaining its operation. Write another paragraph stating the advantages and disadvantages of this light sensor.

Notes:

CHAPTER 6

BUILDING A HOVERCRAFT CONTROL SYSTEM IN ROBOTC

Bruce Jacob

6.1 INTRODUCTION: PROGRAMMING CONCEPTS

The computer is basically a calculator, just without the push buttons. There is little (in most cases, nothing) that a computer can do that a calculator cannot; the one thing that separates the two is the way that they are controlled. A push-button calculator is controlled when a person presses its buttons, which causes it to perform various functions—in particular, the functions that the person wanted, acting in the order that the person pushed the buttons, on whatever data the person entered into the calculator — by pushing buttons.

The computer is fundamentally identical to this picture: a person gets a computer to perform specified functions in a specified order, on data that is either entered into the computer by the person or gathered by the computer from its environment at run time. The only significant difference is that there is no button-pushing; the person controlling the computer must decide ahead of time what functions to perform, and, because the data values might not be known until run time, the person must anticipate the possible values and provide contingency plans. For example, a person using a calculator to perform a complex calculation will stop halfway through if it is clear that the divisor of a term is zero or so near to zero as to produce an essentially infinite result. There is no such human oversight at the time of running a computer program, so the possibilities must be written explicitly into the sequence of steps, in this case by checking the value of a divisor against zero before using it in a division operation.

This, then, is the main point to remember when writing control systems for computers: that they are like very fast, very small (and trusting) children. They will do whatever you tell them to, even divide by zero, a billion times a second, so it is up to you to be quite certain that what you *tell* them to do is exactly what you *want* them to do, because 999,999 times out of a million, any perceived misbehavior on the part of a computer is due to a design error in the program—i.e., it is doing precisely what you told it to do.

What makes autonomous control interesting (and extremely challenging) is the interplay between the computer program and the real-world environment in which the computer finds itself. When you drive a car, you are well aware of your senses, and you respond to what you see and hear by changing the car's controls (steering wheel, accelerator, brakes, etc.). You also anticipate the behavior of the car; for instance, you probably prepare for a right-hand turn well before you reach the street at which you intend to take the turn. When you control an RC hovercraft, you do essentially the same thing: you watch what the hovercraft is doing, how much it drifts when you turn off the fans; you also anticipate where it will be and how much to correct for over steer and tailspin in making turns.

What you will find the most challenging is how to operate in an environment where your senses are taken from you and replaced with far simpler and far more limiting ones. Compared to your senses, the hovercraft you build will have extremely rudimentary sensory inputs—such as binary tape detectors (light sensors detecting the presence or absence of a black tape underneath), distance sensors, and/or gyroscopes. Unless you explicitly create it in your program, your hovercraft will have no ability to remember past behavior (e.g., where it was a moment ago and what it was doing) or anticipate future stimuli and prepare for appropriate reactions.

What is Programming All About?

You have a set of resources at your disposal, for example:

- Computer/s (one or more microprocessors)
- Memory (temporary and/or permanent storage)
- Input devices (sensors)
- Output devices (actuators)

You also have a set of desired behaviors, for example:

- Follow a path
- Avoid hitting obstacles
- Turn on/off lamp when handclap detected
- Play brief salsa motif when simple task accomplished

A program is the connection between these two sets of things; it is your way to tell the system (e.g., hovercraft) how to behave. For instance, here is a program that will cause a walking robot to follow a path:

```
do the following forever:
    if on the path and the path is in front of you
        take a step forward
    otherwise
        turn slightly to the right
```

It should be relatively clear what this does, and how it does it. The robot is instructed to take a single step at a time and only when it is guaranteed that the step will keep the robot on the path (the "take a step forward" action is only allowed if the path is directly ahead). When the path is not directly ahead, the robot turns, a little bit at a time, until the path *is* directly ahead.

But what else does the robot do? What would be expected behavior of the robot, and what would be unexpected? Is walking the wrong way down the path an expected behavior? How about spinning in one place, forever? How about walking in circles or in a zig-zag manner? Should we expect the robot to exhibit any of these behaviors?

One should be able to look at code and imagine all its possible behaviors. For instance, at the end of the path, we should expect the robot not to stop, but to make a 180-degree turn, little by little, and return the way it came, so walking backward down the path is expected. How about doing this before we reach the end of the path—is that possible? Consider cases in which the path turns to the left, or the robot is walking at an angle a little to the right of the path—the robot will reach the edge of the path, at which point there will be no path directly ahead, and the robot will turn to the right until it is facing the way it came, at which point it will walk the wrong way down the path.

Can the robot spin in one place forever? What happens if it starts off the path, or if it accidentally steps off the path? In those circumstances, you should expect the robot to spin forever, because the code will not allow it to take a step unless the robot is already standing on the path. In addition, if the "slight" amount the robot turns is about 90 degrees and the robot is pointing 45 degrees away from the direction of the path, no matter how many times the robot turns, it will never face down the length of the path, and so it will never take a step forward. How about walking in circles? Is that ever possible? Yes, in scenarios similar to the one above (if the degree of turning and width of the path are just right). How about a zig-zag motion? Since the code does not ever have the robot turn left, this would be an unexpected behavior.

These are the types of questions that it is well worth asking of your code, as even the simplest code can produce complex behaviors. Will the code work if the robot starts out off the path, for instance next to it but facing it? Or how about starting off next to it but facing away? Is the code robust enough to handle unexpected situations, such as walking in a line that is just slightly askew, so that after some number of steps the robot finds itself off the path? Or what if the steps that the robot takes are so large that it could step off the path when close to the edge?

Here is a more robust version of the code, one that handles many of these odd conditions:

```
do the following forever:
    if the path is in front of you
        take a step forward
    otherwise
        turn slightly to the right
```

Can this spin forever? Yes, but only if the path is out of sensory range. Can this walk in circles? Just like before, it depends on the degree of turning and the width of the path relative to the step size. Unlike the previous code, this handles situations where the robot is not on the path, for instance at start or if the robot accidentally steps off the path; it also handles cases where the robot is pointing slightly askew relative to the path. Zig-zag is unlikely to happen, but the robot can still accidentally turn around and walk the wrong way down the path if it hits the right edge of the path before reaching the end of the path. The following would be an obvious solution to that problem:

```
do the following forever:
    if the path is in front of you
        take a step forward
    otherwise
        turn slightly toward the path
```

This avoids making 180-degree turns before the end of the path, but this code presupposes the ability to know just where the path is. Previously, all code worked with simple binary detectors: i.e., the path is or is not in front of us. This last piece of code requires more information: the path is in front of us, or, if not, we can tell whether it is to the left or to the right of us. Depending on the types of sensors you have at your disposal, this information may be trivial to get, or it may be impossible to get, and so understanding the abilities of your sensor devices is crucial in the development of your system's capabilities.

Just as important is understanding the abilities of your actuators: for instance, how is it that you cause the robot to turn? How do you know how far it turns each time? Do you need to know whether you are pointing straight or not, and, if so, how do you verify it? What actuator "takes a step forward" for you? How does it work? How far forward do you travel, and does the robot halt immediately after a step, or come to a "rolling stop," or drift significantly (e.g. if the robot is actually on wheels or rollers or floating off the ground)?

What is the NXT?

We will describe the Lego NXT controller in more detail a few pages from now, but here it is worthwhile to look at it in purely functional terms, including the sensory inputs that it accepts and the actuators it can control.

The NXT is a small computer, and so it can be programmed like any other computer. Its advantages are firstly that it is lightweight and battery powered and thus makes a good hovercraft controller (as opposed to a desktop or even laptop computer, which would weigh too much for a small hovercraft), and secondly that it has a diverse modular array of sensors and actuators designed to work with it, for instance:

- **light sensor** — a sensor returning a value proportional to the amount of light it detects, which can be used to determine the lightness/darkness of the terrain that you are traveling over or to head toward a bright light in the darkness
- **accelerometer** — a sensor returning values proportional to the amount of acceleration it detects along 3 axes, which can be used to measure a system's movement in 3 dimensions (acceleration is the 1st derivative of velocity and the 2nd derivative of position) ... note that the accelerometer measures both its movements and the effects of gravity, the sensor's constant acceleration in the up/down direction
- **gyroscope** — a sensor returning values proportional to the angular rotation it detects, which can be used to measure a system's movement in three dimensions, much like an accelerometer; its advantage is that, unlike the accelerometer, it does not also lump together the effects of movement and the effects of gravity
- **compass** — a sensor returning a heading (which way it is pointing relative to Magnetic North), which can be used to keep a system pointing in a known direction
- **proximity sensor** — an ultrasonic sensor returning a value proportional to the distance to the nearest object along the line of sight, which can be used to sense when approaching obstacles such as people, walls, traffic cones, etc.
- **servo** — an actuator that can be told to rotate any number of degrees, which can be used to implement wheels for propulsion, wheels for steering, doors that open and close, robotic arms, rudders and vanes, pulley systems, etc.
- **motor** — an actuator that can be told to spin at different speeds (i.e., run at different power levels), which can be used to implement wheels for propulsion, fans/propellers, etc.
- **home-grown actuators**— such as relays and transistors, which can be connected to the NXT by customizing a Lego output cable (i.e., cutting and splicing) and that enable the NXT to control actuators requiring much more power than the NXT is capable of delivering, such as high-power lift fans and thrust fans—in particular, the kinds of fans you will need to power your hovercraft.

These are extremely powerful sensors and actuators. At any time, the NXT can read the values of four different sensors connected to its inputs, and it can control the behavior of three different actuators connected to its outputs. Your job in the development of your hovercraft's control system is to determine which of these sensors and actuators make the most sense for you to work with, and then to develop a control program that periodically reads the input sensors values and controls the various actuators accordingly.

Examples of How To Get the NXT To Do Stuff

The following are some code samples of how to write ROBOTC code for the NXT. Do not worry if the syntax is confusing; we will discuss C-language specifics in detail in the "C Programming Basics" section of the chapter. This section is meant to serve as a preview.

The first example is a basic "hello, world" program, which just means that this is one of the simplest possible complete programs that you can imagine.

```
task main()
{
    nxtDisplayTextLine(3, "Hello");
    nxtDisplayTextLine(4, "World");
}
```

This prints a message on the NXT's LCD screen, taking up two separate lines of the display. Several things to note:

- The `task main` statement indicates two things: first, that it is a "task," a unit of code to execute together that is delimited by the curly braces, and second that it is the "main" task, which means that it will get executed first, before any other group of code that might be found in the program.
- As mentioned above, the curly braces delimit the code—like parentheses in mathematics, curly braces in C indicate what things should be grouped together.
- The print-out statements are function calls, wherein the stuff in the parentheses is the function's *arguments*—the instructions to the print-out mechanism. The first argument tells the NXT which line on the LCD display to print on, and the second is a *string* (a bunch of characters beginning and ending with double quotes) that tells the NXT what to print.
- All statements end in semicolons; this is a C-language convention.

Note also that the display function `nxtDisplayTextLine` can take additional arguments, allowing it to print out variables (a tool that is quite useful for debugging). The following example demonstrates this ability:

```
// Port to which the sensor is connected
const tSensors LightSensor = (tSensors) S1;
task main()
{
    int light_reading;
    /*
    this is another way to specify comments -- everything between the
slash- asterisk and the asterisk-slash is ignored, within a line or multi-
line
    */
    while (true) {
        light_reading = SensorValue(LightSensor);
        nxtDisplayTextLine(2, "Light = %d", light_reading);
    }
}
```

This example has a bit more to it:

- The very first line begins with two '/' slash characters. This is a *comment:* the double-slash indicates that the rest of the line should be ignored by the NXT, as it only contains information useful to the programmer. This is where you make notes to yourself in the code.

- The second line is actually generated by the ROBOTC programming environment (we'll get to that in class), and it declares that the object `LightSensor` is a sensor object that corresponds to a light sensor plugged into input 1 of the NXT.
- The first line of the main task is a variable declaration. It specifies that the object `light_reading` is to be used as an integer.
- The `while (true)` statement indicates that the subsequent code block (the set of statements delimited by the curly braces) should get executed repeatedly, as long as the value in parentheses evaluates to *true*. Since it is already declared to be true, that means the code block will loop forever. This is how you implement the `do the following forever:` statements in the pseudocode examples at the beginning of the chapter.
- The `nxtDisplayTextLine` function now has an additional parameter at the end: the variable that we wish to print to the screen. Also in the string is the odd-looking "%d" which indicates that a decimal-value integer will be printed out at that point.

Thus far we have shown how to print textual information to the NXT's display and how to get read data values from the NXT's sensor-input ports. The following example shows how to drive values to the NXT's output ports:

```
// Port to which the sensor is connected
const tSensors LightSensor = (tSensors) S1;

task main()
{
    motor[motorA] = 50;
    motor[motorB] = 50;

    wait1Msec(1000);

    while (SensorValue(LightSensor) > 50) {
        // do nothing, i.e. keep going
    }

    motor[motorA] = 0;
    motor[motorB] = 0;
}
```

This starts up the NXT's motors at a 50% power level (presumably driving wheels on the left- and right-hand sides), which moves the NXT forward at about half speed. After a 1-second delay, the NXT starts reading its input sensor value: it keeps reading the value over and over again while the value is greater than 50 (a percentage value, indicating the presence of light). Once the light reading is less than 50%, indicating the presence of a shadow or dark object, the `while` loop exits, and the motor power levels are set to 0, turning them off. Some things to note:

- `motor`, `motorA`, `motorB`, etc. are all reserved keywords. This means that the NXT expects them to be used in a certain way (as shown). So, for example, you can't have a variable called "motor" or "motorB" that you read and write the way the variable `light_reading` was used in the previous code example.
- The square brackets indicate an *array*, a C-language data structure that gathers together a set of similar objects. The name of this particular array is "motor", and the number in the brackets is the index into the array—so in this example `motor` indicates an array of things (probably motors), `motorA` being one and `motorB` being another.

- The value used to set a motor value is a number between 0 and 100, indicating a power level corresponding to a percentage of maximum.
- The `wait1Msec` function does as one would expect: it stalls for the specified number of milliseconds and then goes on. This stalls the execution of code, not the NXT, so while the code is stalling, the motors are still running along at power level 50.

Returning to the earlier theme of fully understanding one's code, how could this example produce odd behavior? For instance, note that it does *not* have the behavior of moving forward until detecting a dark spot and then stopping: it moves forward *for a minimum of 1 second* before it even looks for lightness/darkness, so if the dark spot is small and less than one second of travel away, the NXT could drive right over it without ever stopping—however, there are times when you might want exactly this behavior.

The following starts detecting the lightness/darkness immediately and also gives an example of time measurement:

```
// Port to which the sensor is connected
const tSensors LightSensor = (tSensors) S1;

task main()
{
    int timeTaken;

    ClearTimer(T1);
    motor[motorA] = 50;
    motor[motorB] = 50;

    while (SensorValue(LightSensor) > 50) {
        // do nothing, i.e. keep going
    }

    // found dark - stop & print results
    timeTaken = time1[T1]; // stop timing
    motor[motorA] = 0;      // shut down motors
    motor[motorB] = 0;

    nxtDisplayTextLine(2, "Time = %d", timeTaken);
}
```

This starts moving immediately and stops as soon as darkness is detected; in particular, if the NXT starts on a dark spot, it will halt before the motors have much time to get moving. Like the "motor" keywords, `time1` and `T1` are reserved keywords that must be used as shown: `time1` is an array of timers, `T1` being one of those timers. The call to `ClearTimer` sets timer `T1` to 0, so that the reading of the timer later on (in the statement `timeTaken = time1[T1];`) gets the time in milliseconds from the time of the clearing to the time of the reading. This gives the amount of time the NXT spent driving around, searching for darkness.

Pseudocode as a First Draft

In the beginning of the chapter, some code was offered as example control programs, but the code was written in English and probably would not run on any computer as-is. This is *pseudocode*, the most commonly used form of sketching out one's programming ideas.

In general, before writing a complex piece of software, one should formulate a reasonable picture of what it is the program should do: just as an artist draws sketches before starting to paint, so should an engineer sketch out the overall program before starting to write actual code. Pseudocode is a popular medium because it tends to be self-explanatory, and, despite its simplicity, it can still help discover design flaws, as we saw in the previous examples. For example, here is a rudimentary vehicle-control algorithm for following a dark line on a light floor, assuming that the vehicle has two downward-pointing light sensors (lightL and lightR) at the front, each one positioned on either side of the line, and two motors (motorL and motorR) that provide forward thrust:

```
do forever:
    read forward light sensors (lightL, lightR)
    respond to the 4 possible combinations of values:
        1.  (light, light) -- sensors are straddling line
            set motor speeds as follows:
                motorL = 100%
                motorR = 100%
        2.  (light, dark) -- lightR is on the tape, so turn right
            set motor speeds as follows:
                motorL = 100%
                motorR = 50%
        3.  (dark, light) -- lightL is on the tape, so turn left
            set motor speeds as follows:
                motorL = 50%
                motorR = 100%
        4.  (dark, dark) -- we are perpendicular to the tape, so spin slowly
                in a clockwise motion until we are not perpendicular to tape
            set motor speeds as follows:
                motorL = 20%
                motorR = -20%
```

This is not a particularly robust program, but it does specify reasonably well what is intended and how it will happen. Because the "code" is mostly human-readable English, it is self-documenting; any engineer reading it should be able to understand what it is intended to do as well as how it expects to go about it.

Note in particular the indentation; here, as in regular programming languages, the nesting of statements indicates their "scope" ... e.g., everything below and indented further right than the do forever statement is subservient to that statement (it comprises a loop that will execute until the computer is shut off); the numbered statements that are indented more than the respond to the 4 possible combos statement are the four possible sensor-value combinations, and one would expect that only one could be true, so for any given sensor reading only one will execute; the motorX = ... statements are indented further right than the set motor speeds statements above them, and so both motorL and motorR will be set if the previous set motor speeds statement is executed.

6.2   C PROGRAMMING BASICS

The ROBOTC environment does not implement the entire C programming language; it implements only a subset, but the subset it implements is more than powerful enough to create just about any control system you will need. This section offers a primer on the topic … in general, when a complete valid statement is shown, it is terminated with a semicolon; code snippets without semicolons at the end represent code that would be part of a larger statement.

Data Types

The following ROBOTC data types are at your disposal, not all of which are classic C types:

- **Boolean values** —Booleans are useful for specifying variables that indicate "true" or "false" and nothing else in between. For instance, your code could use these to indicate light/darkness detected by various sensors. Ultimately, they are represented by integers having the value 0 (false) or 1 (true).

```
bool right_sensor_onTape;
bool left_sensor_onTape;

if (right_sensor_onTape) {
    // do something
}
```

- **Integers** —We have seen integer values in the previous code examples; integers are declared as "int" and hold positive or negative integers. In the NXT, they are 16-bit quantities, so they hold values between –32768 and 32767. Any number outside of this range does not exist, as far as the NXT is concerned; for instance, if you have the value 32767 in a variable and you add the value 27 to it, the result will "wrap around" the number scale, and you will be left with a large negative number.

- **Floating point numbers** —Floating point numbers are used to represent fractions, which are otherwise not convenient to do with integers. For instance, if you did the following:

```
int a = 3;
int b = 5;
int c = a / b;

nxtDisplayTextLine(2, "Value = %d", c);
```

you would get garbage results. The variable "a" would get the results of the division, rounded down to the nearest integer, because *a* is incapable of holding a non-integer value.

```
float a = 3.0;
float b = 5.0;
float c = a / b;

nxtDisplayTextLine(2, "Value = %f", c);
```

This works as expected. Note the "%f" in the display function … this indicates that the variable is a floating-point number and should be interpreted and printed out as such.

- **Characters** —Characters are a subtype of integer; they are 8-bit quantities that range from 0 to 256 and are used to represent ASCII alphanumeric codes. They are declared as follows:

```
char a, b;
```

One thing to note about characters is that, while it is possible to use them like integers, it is more convenient to use them as printable ASCII codes, as follows:

```
char a = 37; // this is valid, but only used in small computers
a = a + 1;   // this is valid, but only used in small computers

char b = 'G'; // this is more often used
```

Note that when you use an ASCII value directly (like the letter G in the example above), you need to put single quotes around it.

- **Arrays** —An array is an aggregate data type, in particular a linear set of similar data objects. So, for example, you could have a bunch of integers packed together as follows:

```
int array[25];
```

This produces a grouping of 25 integers that are all packed together right next to one another. They are accessed as follows:

```
array[0]    // the first item in the array
array[1]    // the second item in the array
array[i]    // the "ith" item in the array (valid for 0 <= i <= 24)
array[24]   // the last item in the array
array[25]   // a bug in your code just waiting to wreak havoc
```

Note that the index into the array has to be an integer, and in particular it has to be a positive integer value that is less than the size of the array, otherwise you get weirdness. If you index into an array using an index value that is negative or equal to/larger than the size of the array, you will be reading data, but not data that is part of the array; you'll get garbage.

- **Strings** —A string is a character array, declared as follows:

```
string robotName = "wallE";
```

Note the double quotes at the beginning and end. In particular, the double quotes at the end come *before* the terminating semicolon. This will drive you nuts if you are American and have spent the last two decades of your life being told by your teachers always to put the punctuation *inside* the quotes. The rest of the English-speaking world is a bit more logical in this regard.

- **Structures** —Like arrays, structs are another aggregate data type. While arrays are great for collecting sets of similar data items, sometimes you want to create a set of potentially heterogeneous objects. The C "struct" type allows just that. The following gives an example of how to declare a struct:

```
typedef struct {
    int grade;
    string firstname;
    string lastname;
    char letterGrade;
    float GPA;
    int examplearray[12];
} myStruct_t;
```

This example creates a new user-defined data type called a "myStruct_t" which can be used to create variables in the same way that you can create variables of type `bool`, `int`, `float`, and `char`. Note that not all the fields need to be of different type; for instance, one could have a struct that is entirely integers. Also note that a struct can contain aggregate data types like arrays (or even other structs).

The members of the struct are not indexed as in an array; they are referenced by name. The following code snippet shows how to create an object of this type, as well as how the various fields of the struct are referenced.

```
task main()
{
    myStruct_t example;

    example.grade = 10;
    example.firstname = "Brockton";
    example.lastname = "Veendorp";
    example.letterGrade = 'A';
    example.GPA = 3.5;

    if (/* student is graduating */) {
        example.grade = example.grade + 1;
    }
}
```

- **Arrays of Structs** — Often times, the main reason to use a struct is to create a whole array of data records. The following is an example, which could be used within a hovercraft to keep track of the previous 50 sensor readings.

```
struct {
    int lightL;
    int lightR;
    int fanL;
    int fanR;
} history[50];
```

For example, if sensor readings and corresponding actions were taken once every tenth of a second, this would keep track of the last five seconds' worth of light-sensor readings and the power levels given to the two thrust fans. Individual components in this aggregate data type would be referenced as follows:

```
history[10].fanL
history[i].lightR
```

One thing to note about the invocation above is that the syntax is slightly different than shown in the previous section; specifically, there is no "typedef" keyword. Thus, this statement creates an array of structs called *history*, but it does not create a type that can be instantiated. The more complex method would be as follows:

```
typedef struct {
    int lightL;
    int lightR;
    int fanL;
    int fanR;
} state_t;
state_t history[50];
```

For the relatively small control programs you will write in this class, the simpler facility (making a single struct, not a whole type class of them) is probably more than sufficient.

- **Void** —the non-type. For purposes of basic programming in ROBOTC, this is generally only useful for declaring a function that does not return a value, i.e.

```
void turn_off_motor(int id)
{
    motor[id] = 0;
    return;
}
```

as opposed to the following, which returns an integer value:

```
int add_two_ints(int a, int b)
{
    return a + b;
}
```

We'll get into more detail on functions in the *Control Structures* section below.

Operations on Data

You can read data, you can write data, and you can perform various arithmetic and logical operations on data. Examples are given below.

- Reading/writing data

```
a = 3;        // assignment operation; puts the integer 3 into variable a
a = 3.5;      // puts the floating-point value 3.5 into the variable a
a = b;        // reads variable b and puts whatever is found into variable a
x = list[2]   // reads the 3rd item in array "list" & writes it to variable x
list[0] = y;  // reads y and writes it to the first item in the array "list"
list[i] = y;  // puts y into the ith slot of the array, e.g.:
              // if i=1, y is written into the second slot of the array;
              // if i=32, y is written into the 33rd slot of the array; etc.
a[x] = b[y];  // yes, this is possible, and can be very useful
a[f(x)] = 1;  // works if we know that function f returns an integer
              // of the proper size
```

- Arithmetic and logic operations — these operations work on integers and floating point numbers and produce a result of the same type as the input variables. So — add an integer to an integer, and you get an integer; add a float to a float, and you get a float. Be careful when operating on variables of different types; while it can be very useful, it creates an unintended bug 99 times out of 100. The following are examples:

```
a + b           // addition
a - b           // subtraction
a * b           // multiplication
a / b           // division
a++;            // increments a by 1
a--;            // decrements a by 1
a = b++;        // reads b, puts it into a, and then increments b by 1

a & b           // bit-wise AND function (e.g. 0010 AND 0110 equals 0010)
a | b           // bit-wise OR function (e.g. 0010 OR 0110 equals 0110)
!a              // logical negation (turns a non-zero number into zero, and
                // turns a zero into a 1 - e.g. if a==0110, !a equals 0)
~a              // bit-wise negation (e.g., if a==0110, ~a equals 1001)

a += b;         // short-hand for a = a + b;
a -= b;         // short-hand for a = a - b;
a *= b;         // short-hand for a = a * b;
a /= b;         // short-hand for a = a / b;
a &= b;         // short-hand for a = a & b;
a |= b;         // short-hand for a = a | b;
```

- Boolean operations — note that Boolean values are either 0 or 1 (false or true)

```
a == b          // equivalence operation; produces a Boolean result
a != b          // true if a does not equal b
a < b           // true if a is less than b
a <= b          // true if a less than or equal to b
a > b           // greater than
a >= b          // greater than or equal to
a = !b;         // puts the Boolean negation of b into a
a = ~b;         // puts the bit-wise negation of b into a
a && b          // true if a is true (or non-zero) AND b is true (or nonzero)
a || b          // true if a is true (or non-zero) OR b is true (or nonzero)
```

Control Structures

ROBOTC offers the following control structures for your code:

- **Task** — Perhaps the most fundamental control structure is that of the *task*, a ROBOTC-specific mechanism that corresponds roughly to a running program. We have already seen the main task; your program can have up to ten (10) others as well. Every task in your program is almost like a separate program that can run alongside the original main task, just as you can have Microsoft Word running alongside your browser, your email client, and your session of Quake III Arena, all at the same time, all on the same computer. Usually splitting out a program into separate tasks is done for convenience of logic and/or readability of code. Here is an example that starts up a separate task to pulse the motors on and off, something that you would need to do if you were to use relays instead of transistors to control your motors.

```
// global variables - necessary since we can't pass args directly to tasks
int motorA_power = 0;
int motorB_power = 0;
```

```
// reads motorA_power power value and "stutters" the output to motorA
// to approximate that power value -- loop 10 times per second
task stutter_motorA()
{
    int mSec_on = motorA_power;
    int mSec_off = 100 - mSec_on;

// loops ten times a second
while (true) {
        motor[motorA] = 100;
        wait1Msec(mSec_on);
        motor[motorA] = 0;
        wait1Msec(mSec_off);
    }
}


// reads motorB_power power value and "stutters" the output to motorB
// to approximate that power value -- loop 10 times per second
task stutter_motorB()
{
    int mSec_on = motorB_power;
    int mSec_off = 100 - mSec_on;

    // loops ten times a second
    while (true) {
        motor[motorB] = 100;
        wait1Msec(mSec_on);
        motor[motorB] = 0;
        wait1Msec(mSec_off);
    }
}


// assumes motorA is on left and motorB is on right
task main()
{
    // start them up just so we know it is safe to stop them later
    motorA_power = 0;
    motorB_power = 0;
    StartTask(stutter_motorA);
    StartTask(stutter_motorB);

        while (true) {
          if (/* sensor readings tell us we should turn right*/) {
              StopTask(stutter_motorA);
              StopTask(stutter_motorB);
              motorA_power = 100;
              motorB_power = 50;
              StartTask(stutter_motorA);
              StartTask(stutter_motorB);
        } else if (/* sensor readings tell us we should turn left */) {
              StopTask(stutter_motorA);
              StopTask(stutter_motorB);
              motorA_power = 50;
              motorB_power = 100;
              StartTask(stutter_motorA);
              StartTask(stutter_motorB);
        } else if (/* sensor readings tell us we should stop */) {
              StopTask(stutter_motorA);
              StopTask(stutter_motorB);
              motorA_power = 0;
              motorB_power = 0;
```

```
                    StartTask(stutter_motorA);
                    StartTask(stutter_motorB);
          } else {        // go forward
                    StopTask(stutter_motorA);
                    StopTask(stutter_motorB);
                    motorA_power = 75;
                    motorB_power = 75;
                    StartTask(stutter_motorA);
                    StartTask(stutter_motorB);
          }
     }
}
```

The main task starts up the tasks whenever it wants the thrust fans to go; the global values are used to indicate the desired power levels for the fans.

The responsibility of the tasks is to buffer the main task from having to stutter the motors (this is one way to effectively achieve a lower power level, which would be useful if you are using relays and not transistors to control your fans). You can imagine how difficult it would be if you had to intertwine into the main code loop the while(true) loops in the stutter tasks that implement the timed on/off behavior. It would be nasty-looking code, prone to exhibiting weird bugs, etc. The task allows the behavior to be consolidated in one place, where it can be shut off at will and from outside the task. Unfortunately, ROBOTC does not allow one to pass arguments directly to a task, thus the necessity of the motorX_power global variables (which are a bit hackish).

We start the tasks with zero power levels at the beginning of the program (first actions of the main task); this is done in case trying to stop a non-existent task turns out to cause a run-time error (it may or may not; you might want to experiment to find out for yourself).

- **Function** — We have already seen functions in action. For instance, the StartTask and StopTask primitives are functions predefined by ROBOTC. You can also declare your own functions. When invoked, functions are unlike tasks in that they do not become independent entities. That is, when you call a function, it is as if you executed the function's code in place. The following erroneous code example illustrates:

```
// takes as arguments the motor identifier and a power value and "stutters"
// the output to approximate that power value -- loop 10 times per second
void stutter_motor(int whichMotor, int powerLevel)
{
    int mSec_on = powerLevel;
    int mSec_off = 100 - mSec_on;

    // loops ten times a second
    while (true) {
        motor[whichMotor] = 100;
        wait1Msec(mSec_on);
        motor[whichMotor] = 0;
        wait1Msec(mSec_off);
    }
    return;
}

// assumes motorA is on left and motorB is on right
task main()
{
        while (true) {
```

```
                    if (/* sensor readings tell us we should turn right*/) {
                        stutter_motor(motorA, 100);
                        stutter_motor(motorB, 50);
                    } else if (/* sensor readings tell us to turn left */) {
                        stutter_motor(motorA, 50);
                        stutter_motor(motorB, 100);
                    } else if (/* sensor readings tell us we should stop */) {
                        motor[motorA] = 0;
                        motor[motorB] = 0;
                    } else {       // go forward
                        stutter_motor(motorA, 75);
                        stutter_motor(motorB, 75);
                    }
            }
    }
```

This is an attempt to simplistically re-do the previous task-based example using functions; the stuttering behavior is implemented in functions which are called by the main task, instead of tasks that are started & stopped by the main task. This would be a more elegant way, for example, to pass arguments from the main task to the subroutines that actually perform the stuttering.

Though it might appear to be a more elegant way to write the previous code example, this code fails to work: as soon as the `stutter_motor` function is called, it will go into an endless loop and never return. This is not meant to disparage functions, merely an illustration of how they differ from tasks. Functions are extremely convenient, as they allow the passing of arguments (as shown) as well as return values. Because of this they allow a programmer to use a modular approach to program design, wherein the low-level details of how things are done are kept in very low-level functions, and the high-level functions (the main task and the handful of functions it calls) can be written in such a way as to hide all of those low-level implementation details.

Taking a modular approach tends to facilitate reading, understanding, and thus debugging one's code. For instance, the following code is a rewrite of the earlier example that uses a function to initialize the global variables and start/stop tasks.

```
// global variables - necessary since we can't pass args directly to tasks
int motorA_power = 0;
int motorB_power = 0;

task stutter_motorA()
{
    // as before
}

task stutter_motorB()
{
    // as before
}

// takes as input power levels for motorA and motorB
void run_motors(int mA, int mB)
{
        StopTask(stutter_motorA);
        StopTask(stutter_motorB);
        motorA_power = mA;
        motorB_power = mB;
        StartTask(stutter_motorA);
```

```
            StartTask(stutter_motorB);
            return;
    }

    // assumes motorA is on left and motorB is on right
    task main()
    {
        // start them up just so we know it is safe to stop them later
        motorA_power = 0;
        motorB_power = 0;
        StartTask(stutter_motorA);
        StartTask(stutter_motorB);

        while (true) {
            if (/* sensor readings tell us we should turn right*/) {
                run_motors(100, 50);
            } else if (/* sensor readings tell us we should turn left */) {
                run_motors(50, 100);
            } else if (/* sensor readings tell us we should stop */) {
                run_motors(0, 0);
            } else {   // go forward
                run_motors(75, 75);
            }
        }
    }
```

This code should be much easier to read than the original example upon which it is based. This is the power of functions; whenever you see the same block of code happening over and over again, with only minor variations, which is a sure bet that you should encapsulate the code in a function call.

As mentioned, another aspect of functions is that they can return values, as in the following code example.

```
    const tSensors light1 = (tSensors) S1;
    const tSensors light2 = (tSensors) S2;
    const tSensors light3 = (tSensors) S3;
    const tSensors light4 = (tSensors) S4;

    float get_sensor_avg()
    {
        float avg;
        int sum = SensorValue(light1);

        sum += SensorValue(light2);
        sum += SensorValue(light3);
        sum += SensorValue(light4);

        avg = (float)sum;
        return avg / 4.0;
    }

// prints out a new sensor reading every 0.1 second
task main()
{
    while (true) {
        nxtDisplayTextLine(2, "Avg = %f", get_sensor_avg());
        wait1Msec(100);      // wait a tenth of a second
    }
}
```

Among other things, this code example illustrates the C-language notation for casting data types into other types, in particular how to create a floating-point number out of an integer. The `(float)` notation in front of the variable name `sum` near the end of the function tells the system to read the integer value of `sum` and then turn it into a floating-point number before putting its value into the variable `avg` (which is a floating-point number).

- **While loop** — We have already seen the while loop in action. Formally, it takes as an argument a Boolean statement (something that evaluates to either true or false) and conditionally executes whatever code block comes after it. The important thing to remember is this ordering: first the evaluation of the condition statement and then the execution of the following code block. This happens repeatedly until the condition statement evaluates to false, at which point the NXT moves on to execute the instructions that come after the while loop.

```
while (condition) {
    // code block to be executed (one or more instructions),
    // often called the "loop body"
}
// following instructions
```

This produces a loop format, because the code block will be executed repeatedly until the condition becomes false (the integer value 0). As soon as the condition evaluates to false, the NXT jumps to the "following instructions" after the loop body.

Note again that the condition test happens before the loop body; this means that it is possible for the loop body to be skipped, if for example on the first test of the condition it evaluates to false. Sometimes, this is not desired; sometimes a programmer might wish for the loop body to be executed at least once before evaluating the conditional value. This is what the do-while loop is for, described next.

- **Do-while loop** — As mentioned above, sometimes a programmer needs a loop structure in which the loop body is executed at least once before evaluating the conditional value. For this, the do-while loop is used, as shown below.

```
do {
    // code block to be executed (one or more instructions),
    // often called the "loop body"
} while (condition);
// following instructions
```

This structure is also a loop; the loop body will be executed repeatedly until the NXT evaluates the Boolean condition to false. The difference is that the loop body is guaranteed to be executed at least once, even if we know ahead of time that the condition is false, for instance in the following (contrived) case:

```
do {
    // loop body
} while (false);
```

This will execute the loop body once, then test the conditional which is hard-coded to be false; the condition will fail, and the loop will terminate, passing control to the following instructions.

One thing to note: the semicolon at the end of the `while();` statement is required for a do-while loop.

- **For loop** — The for loop is generally used for iteration; it takes the following form:

```
for ( /*initialization*/ ; /*condition*/ ; /*update*/ ) {
    // loop body
}
```

The semicolons between the three components (initialization, condition, update) are required, even if a particular component is left blank. The components function as follows:

- **initialization** — This happens once, at the start of the loop, before the condition evaluation. It is a good way to set up the variables that will be in use.

- **condition** — This is identical to the while-loop condition; it is evaluated before the loop body, so if it evaluates to false on the first pass, the loop body is never executed.

- **update** — This is a statement that is executed after the loop body and before the next evaluation of the condition statement.

Here is a typical use of the for loop:

```
int i, array[1024];

for (i=0; i<1024; i++) {
    array[i] = 0;
}
```

Note that the *initialization* and *update* components can have multiple statements, each separated by commas.

- **If/then/else** — This control-flow structure was shown in an earlier code example. It allows a programmer to specify one or more code blocks, at most one of which will be executed, based on one or more conditionals.

A simple if-then example that executes the code body only if the sensor value satisfies some conditional, so the code body may or may not be executed:

```
if (SensorValue(mySensor) < 50) {
    // code body -- react to the sensor result
}
```

The following if-then-else example provides two options, one of which is guaranteed to be executed:

```
if (SensorValue(mySensor) < 50) {
    // turn left
} else {
    // turn right
}
```

Having a simple `else` statement at the end of an "if" construct is the default case; it specifies code that will be executed if all other conditionals fail. The following is a more complex example:

```
int value = SensorValue(mySensor);
if (value < 25) {
   // turn left
} else if (value < 50) {
   // turn right
} else if (value < 75) {
   // go back
} else {
   // stop
}
```

Note that later conditions in an if-else control block are tested only if all earlier ones fail, and later conditions are not tested if an earlier test succeeds, so only one code block will execute, in particular the first to satisfy its corresponding condition. So the following would be an error:

```
int value = SensorValue(mySensor);
if (value < 25) {
   // turn left
} else if (value < 75) {
   // go back
} else if (value < 50) {
   // turn right
} else {
   // stop
}
```

The third condition will never succeed; any value greater than 25 and less than 50 will be caught by the second conditional. Note also that we read the sensor value only once and not multiple times, which would have been the case, at least potentially, if we had placed the SensorValue(mySensor) function call in each of the conditionals.

- **Switch** — If you have a control structure similar to the following, where your conditionals are, for the most part, testing for equality as opposed to greater than/less than, you might want to use a switch statement:

```
int value = SensorValue(mySensor);
if (value == 1) {
   // turn left
} else if (value == 2) {
   // go back
} else if (value == 3) {
   // turn right
} else if (value == 4) {
   // spin around
} else if (value == 5) {
   // stop
} else if (value == 6) {
   // spin clockwise
} else if (value == 7) {
   // spin counterclockwise
} else if (value == 0) {
   // shut down
} else {
   // blink lights
}
```

This type of construct can be difficult to read, and because it is difficult to read, it can harbor latent bugs easily. A better construct is the switch statement, which would implement the construct above as follows:

```
int value = SensorValue(mySensor);
switch (value) {
     case 1:
          // turn left
          break;
     case 2:
          // go back
          break;
     case 3:
          // turn right
          break;
     case 4:
          // spin around
          break;
     case 5:
          // stop
          break;
     case 6:
          // spin clockwise
          break;
     case 7:
          // spin counterclockwise
          break;
     case 0:
          // shut down
          break;
     default:
          // blink lights
}
```

At the switch (value) statement, the value will be read once, and the NXT will immediately jump to which ever code block has the corresponding value expressed with its case statement. The default case is not required but is an extremely good idea to use, even if it is left empty by putting no code after it (perhaps include a comment to the effect that the omission of corresponding code is intentional).

The various break; statements are required, except for the very last case in the list. If a particular break; statement is omitted, the program will compile successfully, but at run-time the control flow will "fall through" from one case to the next.

- **Continue** and **break** in loops — There are two statements that provide extremely powerful ways to augment the behavior of loops. The *continue* statement causes a loop to halt mid-loop and go directly to the top of the loop. For example, here is its use in a while loop:

```
while (true) {
     int value = SensorValue(mySensor);
     if (value == 1) {
          // turn left
     } else if (value == 2) {
          // go back
     } else {
          continue;
     }
     // blink lights
     // update state
```

```
        // etc.
    }
```

In this example, the *continue* statement causes the execution of the code to skip over the stuff at the end of the loop (blink lights, update state, etc.) and go straight to the top of the loop, at which point the conditional is re-evaluated, and the sensor value is read.

A *break* statement similarly disrupts control flow, but it instead causes control to exit the loop entirely.

```
while (true) {
    int value = SensorValue(mySensor);
    if (value == 1) {
        // turn left
    } else if (value == 2) {
        // go back
    } else {
        break;
    }
    // blink lights
    // update state
    // etc.
}

nxtDisplayTextLine(3, "Error situation");
```

This example is just like the previous one, except that the *continue* statement is now a *break* statement. This is one way to exit a loop that would otherwise execute forever. It is usually used in loops to detect anomalous or erroneous situations; normal events should be reflected in the while loop's conditional.

An additional note on loop and if/then conditionals: they are usually Boolean statements, statements that evaluate to either 0 (false) or 1 (true). Because a computer only understands numbers, these are represented with integers, which can have the value 2 or –37, or 1023, and so on. Thus, while it is best practice to use code similar to this:

```
char x, array[1024];
int i;

for (i=1023; i > 0; i--) {
    x = array[i];
}
```

it is just as valid to say the following (the only change to the code is that `i > 0` has been replaced by simply `i`).

```
char x, array[1024];
int i;

for (i=1023; i; i--) {
    x = array[i];
}
```

Why does this work? Because the NXT checks to see if the conditional is zero or non-zero; it does not do a strict test of 0 vs. 1. So any number other than 1 is considered equivalent to 1 for the purposes of determining a Boolean result. As soon as the variable *i* reaches the value zero, both of these loops will terminate.

6.3 THE LEGO® NXT® CONTROLLER

The NXT incorporates a microprocessor capable of storing programs that provide instructions for controlling the fans and other actuators used in your team's hovercraft. In this application, the control of the fans was complicated by the need to steer the hovercraft and to control its speed particularly around curves in the track. Moreover, the requirement is for autonomous control, which implies that the NXT must be programmed so that it provides command signals to the lift and propulsion fans that enable the hovercraft to negotiate the track in the minimum possible time.

Decision based control requires sensors in the system that provide feedback signals, which give information essential in controlling the system. When the system is in operation, the sensors provide continuous signals that are monitored by the NXT. When the level of a feedback signal deviates sufficiently from a specified level, the power to one or more actuators (fans) is adjusted to maintain the sensor signal within a control band centered about the control level. For decision-based control to be effective, the microprocessor must be provided with instructions that establish the upper and lower levels of the control band. Sensor signals that are outside the control band cause the controller to energize actuators (fans), which adjust the parameters affecting the system so as to maintain control. Programming for decision-based control is more challenging as the program entails a more structure involving forks, jumps and loops and sometimes loops within loops.

The NXT, shown in Fig. 6.7, is a controller that contains an Atmel ARM processor (AT91SAM7S256), which incorporates a high performance 32 bit RISC architecture. This processor contains 256 Kbytes of flash memory (non-volatile) with single cycle access at 30MHz. It also contains an additional 64 Kbytes of SRAM (static random access memory). The ARM processor includes three different clocks to provide timing signals with frequencies that range from 22 to 42 kHz for the slow clock and 80 to 220 MHz for the PLL clock.

The NXT also contains an 8-bit Atmel co-processor (ATmega48) that is used with the main processor for measuring voltage and controlling actuators. This processor contains 4 Kbytes of flash memory (non-volatile) with single cycle access at 30MHz. It also contains an additional 512 bytes of SRAM and 256 bytes of EEPROM memory. It has two 8-bit timer/counters, one 16-bit timer/counter and a real time counter. Voltage is measured with an eight channel 10 bit analog to digital converter (ADC). The 10-bit ADC in the NXT can acquire and store voltage measurements at frequencies up to 200 Hz. Six pulse width modulated (PWM) channels are available in the co-processor for supplying power to the motors and actuators, although the NXT only uses three of these channels.

Communication between the NXT and a computer is accomplished with either a USB cable or by Bluetooth wireless connection (radio). A CSR BlueCore™ 4 single chip radio and base band integrated circuit is employed to implement a Bluetooth 2.4GHz system, which provides enhanced data rates up to 3Mbps. The chip interfaces to 8Mbit of external flash memory and contains 47 Kbytes of RAM. When used with the CSR Bluetooth software stack, it provides a fully compliant Bluetooth system for data and voice communications. The BlueCore™ transmitter is powered by a 6-bit digital-to-analog-converter with a dynamic range greater than 30 dB. Its receiver incorporates channel filters and a digital demodulator for improved sensitivity and co-channel rejection. The USB 2.0 communication port is capable of 12 Mbits/s.

The number of programs you can store in the NXT is only limited by the memory available. The NXT microprocessor can perform both integer and floating point (decimal) mathematics. You can program the NXT to read and write files to its memory.

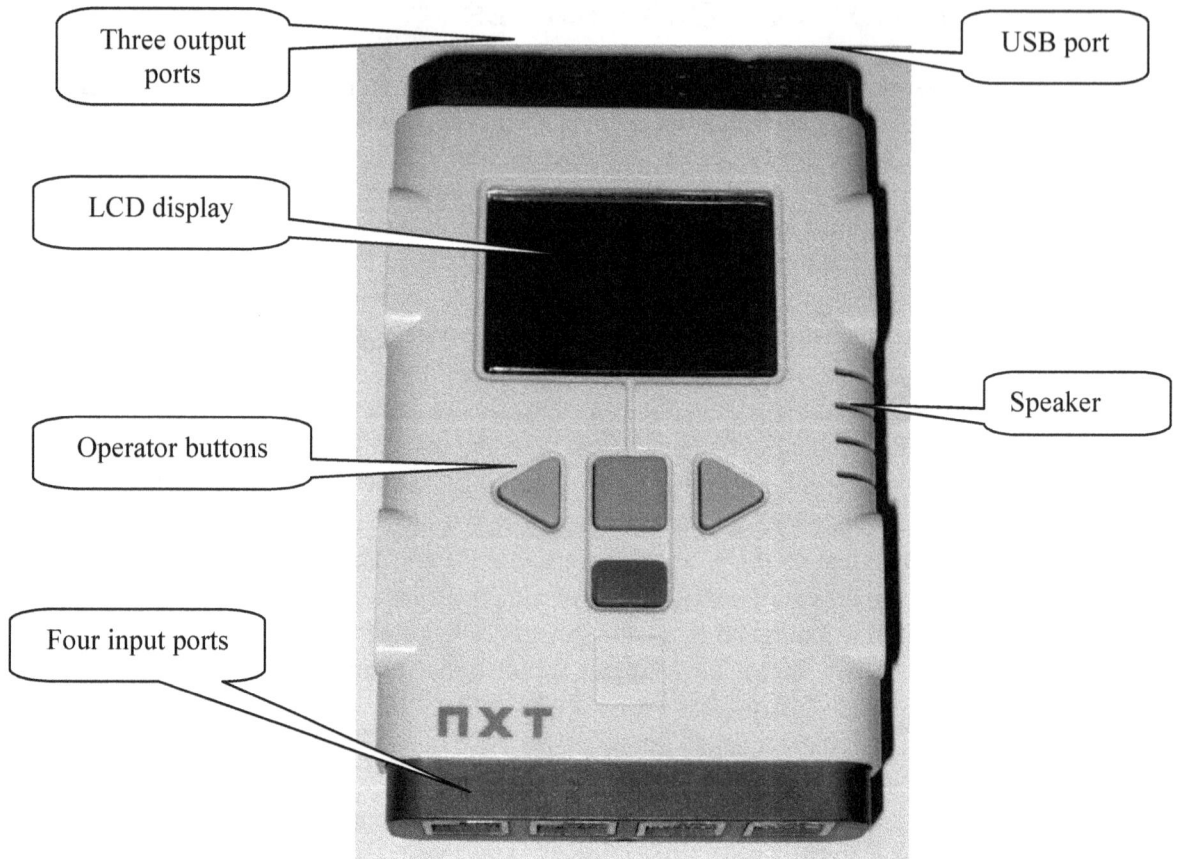

Fig 6.1 The NXT controls the actions of autonomous robots and vehicles.

6.4   SUMMARY

The basic programming concepts were described in by introducing you to the simplicity and speed of a computer. By performing at high speeds it can accomplish a wide variety of tasks using two numbers (0 and 1) and very clever circuit and software design. As you begin to design a hovercraft control system you will be using your computer and the NXT with their memories, sensors and output devices (fans). Using the signals generated by the sensors as feedback, you will design a program in ROBOTC to control the behavior of your hovercraft to follow a path and avoid obstacles found along the route.

Snippets of code are introduced early in the chapter to show you how to program the NXT to perform specific task. These are short examples that illustrate the use of `task main()`, curly and square brackets and semicolon when programming in ROBOTC. Other more substantial examples show the techniques for adding comments, to read and write sensor readings and to transmit power to the output ports. Pseudocode is defined and its use as a first draft for documenting a program is illustrated with an example employing a `do forever` loop.

Programming basics are covered in considerable detail with definition of data types including:

1. Boolean values
2. Integers
3. Floating point numbers
4. Characters
5. Arrays
6. Strings
7. Structures
8. Voids

A large number of different operations on data which included: reading and writing data, arithmetic and logic operations and Boolean operations were illustrated with code symbols. Finally control structures were described which included: task (main and auxiliary), functions, while loop, do-while loop, for loop, if/then/else and switch structures. Extensive sections of code were written to illustrate the structures and the concepts.

Finally a brief description of the electronic components found in the NXT microcontroller was given. The NXT contains two microprocessors, one of which runs on a 32 bit RISC architecture. It employs a 10 bit ADC that can acquire and store voltage measurements from the sensors at frequencies up to 200 Hz. It contains flash, SRAM and EEPROM memory for storing programs and data. A Bluetooth system is available for data and voice communications. A monochromic LCD is used for displaying data and other readout. Communication with a PC is accomplished with either Bluetooth or with a cable with USB connectors.

EXERCISES

6.1 Describe the difference between a calculator and a computer.

6.2 What makes autonomous control of your hovercraft interesting.

6.3 Describe some behaviors you anticipate in attempting to control your hovercraft.

6.4 Would it be easier to have a wheeled robot run the track instead of a hovercraft? If so why?

6.5 What is a binary detector? Will you be dealing with binary detectors in controlling your hovercraft?

6.6 What sensors are you planning to use in designing your control system?

6.7 What actuators do you plan to use in designing your control system?

6.8 Write a program to display ENES 100 on line 2 of the NXT's LCD and your team's name on line 4.

6.9 Is punctuation important in writing code in ROBOTC? Give explanations for { } ( ) ; [ ]

6.10 Why is important to pay attention to numbers designated as integers or floating point?

6.11 Write a single line of code that would introduce a delay of 4.5 seconds in a program.

6.12 Write a snippet of code that would record the time of operation of actuators if the output of a light sensor exceeding 55%.

6.13 Describe the role of indentation when writing code in ROBOTC.

6.14 What are the Boolean values? Why are they important in designing you hovercraft control system.

6.15 Prepare a listing of ASCII alphanumeric codes.

6.16 Define the data types used in ROBOTC.

6.17 Explain the following read and writing commands:

```
a = 3;
a = 3.5;
a = b;
x = list[2]
list[0] = y;
list[i] = y;
```

6.18 Explain the following Boolean operations:

```
a == b
a != b
a < b
a <= b
a > b
a >= b
a = !b;
a = ~b;
a && b
a || b
```

6.19 How many task can you program in ROBOTC? Why would you write a program containing more than he mail task?

6.20 Write a snippet of code containing a `while` loop.

6.21 Write a snippet of code containing a `do-while` loop.

6.22 Write a snippet of code containing a `for` loop.

6.23 Write a snippet of code containing a `if/then/else` control structure.

6.24 What happens when a `break` statement is inserted in a loop?

6.25 What happens when a `continue` statement is inserted in a loop?

6.26 Why does the NXT contain a analog to digital converter (ADC)? How many counts are possible with this 10 bit converter? Why is necessary to incorporate eight channels of ADC?

6.27 Describe the capabilities of the LCD on the NXT.

6.28 Describe the function of each button on the NXT.

6.29 What sensors are commercially available for use with the NXT?

6.30 What actuators are commercially available for use with the NXT.

6.31 What is Bluetooth?

PART II

ENGINEERING GRAPHICS

CHAPTER 7

COMPUTER AIDED DESIGN (CAD)

7.1 INTRODUCTION TO CAD

A product consists of components. In the design process individual components are designed first, afterwards the designed components are assembled. In this chapter, we introduce a CAD software system to create the individual components and their assemblies: Pro/ENGINEER Wildfire, Version 5.0.

There are a variety of hovercraft models, which have been designed and constructed. The following picture shows one of the hovercraft models. The 3 maximum dimensions, or the dimensions of the total height, the total width and the total length, are 206, 280 and 540 mm, respectively.

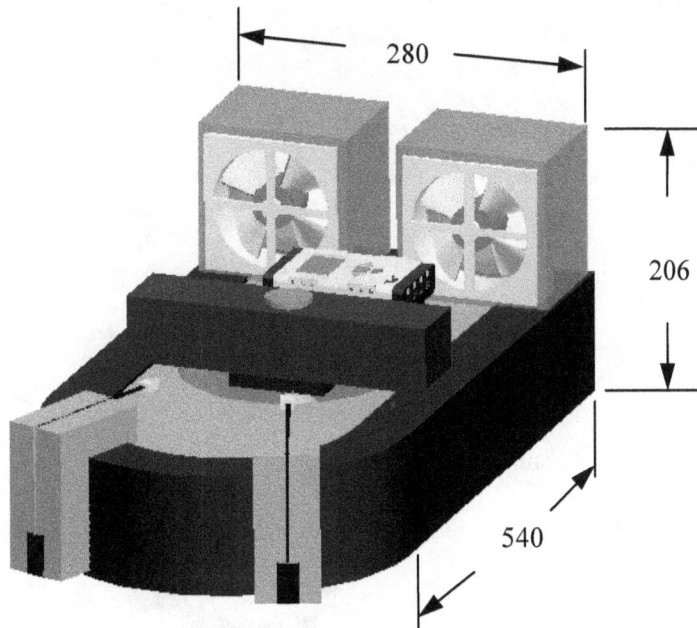

As illustrated in the above picture, the hovercraft model has a lift fan which is located on the central part. There are 2 propulsion fans located on the rear part. The lift fan and propulsion fans are all attached to the base structure. To autonomously control the motion of the hovercraft, 2 sensors and a micro-computer or a NXT box are used. The sensors and the NXT box are also attached to the base structure.

In the following, we will present the steps and procedures for designing this hovercraft model using Pro/ENGINEER. To facilitate the presentation, we decompose this hovercraft model into 4 sub-systems. These 4 sub-systems are:

1.   The base structural system

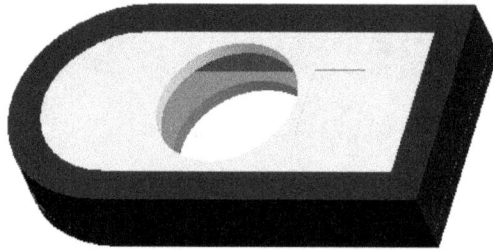

2.   The lift fan system where an axial fan or a centrifugal fan may be used.

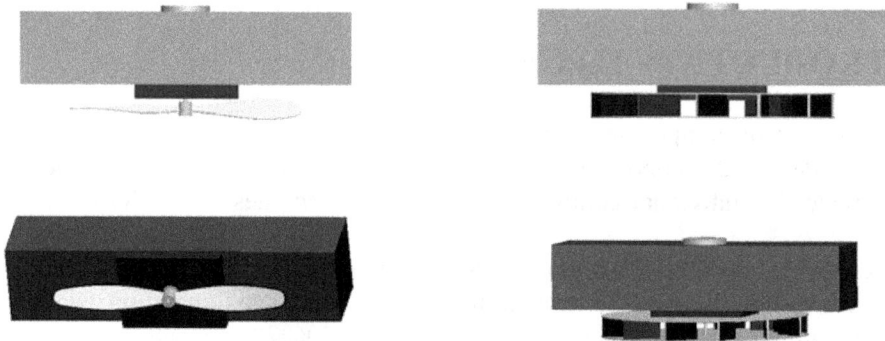

3.   The propulsion fan system.

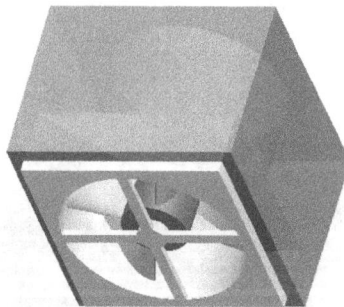

4.   The sensor-based control system.

In the following sections, we present the procedure to create each of the 4 subsystems, starting with the base structural system. For each sub-system, we present an exploded view first, illustrating all components used in the sub-system. Afterwards, we present a step-by-step procedure to create a 3D solid model for each component. Finally, we assemble the created 3D solid models to construct a 3D solid model of the subsystem. When the 3D solid models of the 4 subsystems are available, we assemble the four (4) 3D solid models of subsystems to construct a 3D solid model of the hovercraft model.

7.2   CREATION OF THE BASE STRUCTURAL SYSTEM

Figure 7.1 is an exploded view of the base structural system. As illustrated, this subsystem has a top_plate component, a bottom_plate component, 3 support-bar components and a skirt component.

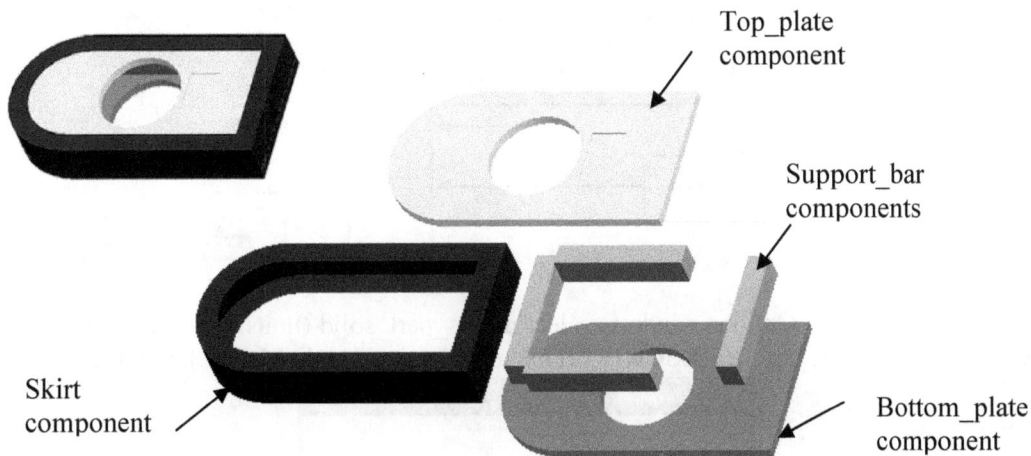

Fig. 7.1 An exploded view of the base structural system showing the 4 components.

We begin with creating the support_bar component to introduce the Pro/ENGINEER design system. Afterwards, we construct the top_plate component, the bottom_plate component and the skirt component. Finally, we assemble them together.

Creation of the Support_Bar Component

To start the design process using Pro/ENGINEER, you need to launch the CAD system. Click the **Start** menu > **Program** > **PTC** > **Pro ENGINEER**, or just click the displayed icon of **ProE**.

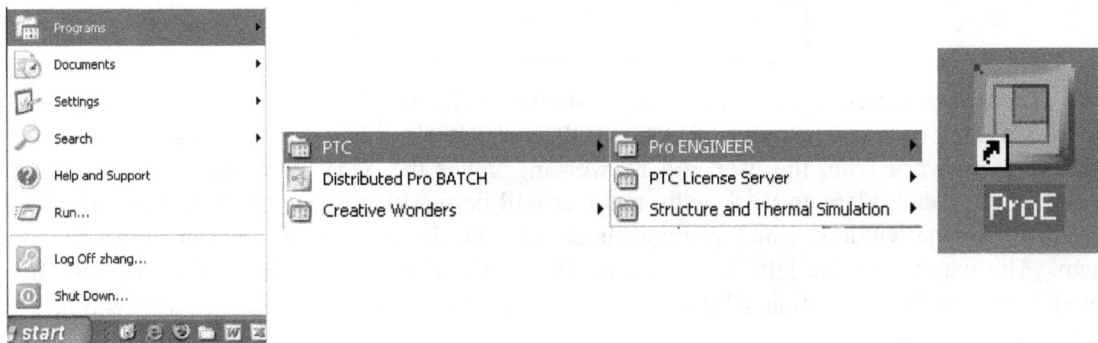

Step 1: select the icon, called "**Create a new file**", which is displayed on the menu bar. Make sure **Part** is selected. Type *support_bar* as the name of the file > clear the icon of **Use default template** > **OK**.

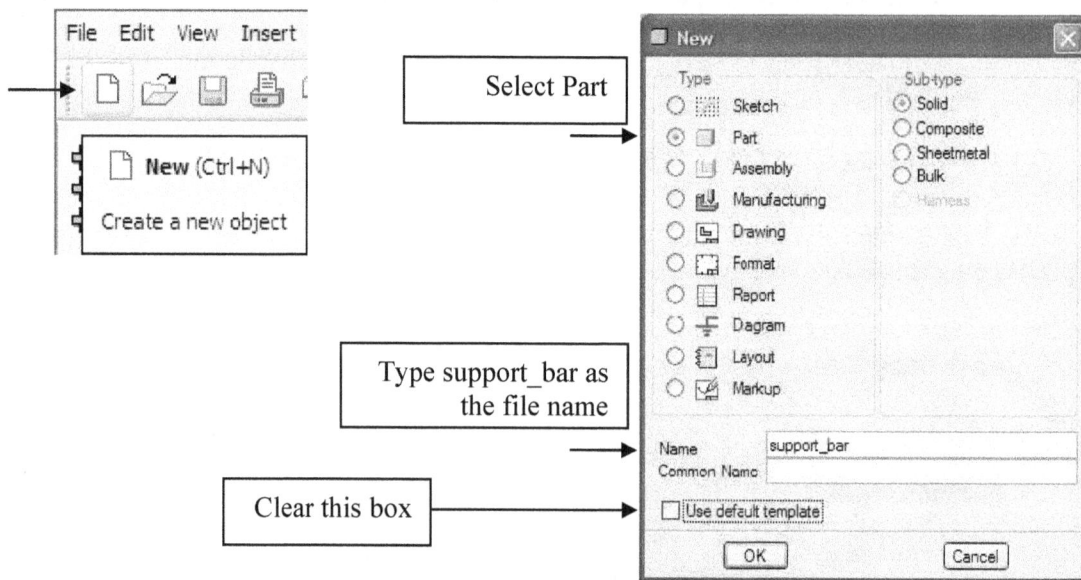

In the New File Options window, select mmns_part_solid (units: Millimeter, Newton, Second) as the unit, type *support_bar* in **DESCRIPTION**, and type s*tudent* in **MODELED_BY**, then click **OK**.

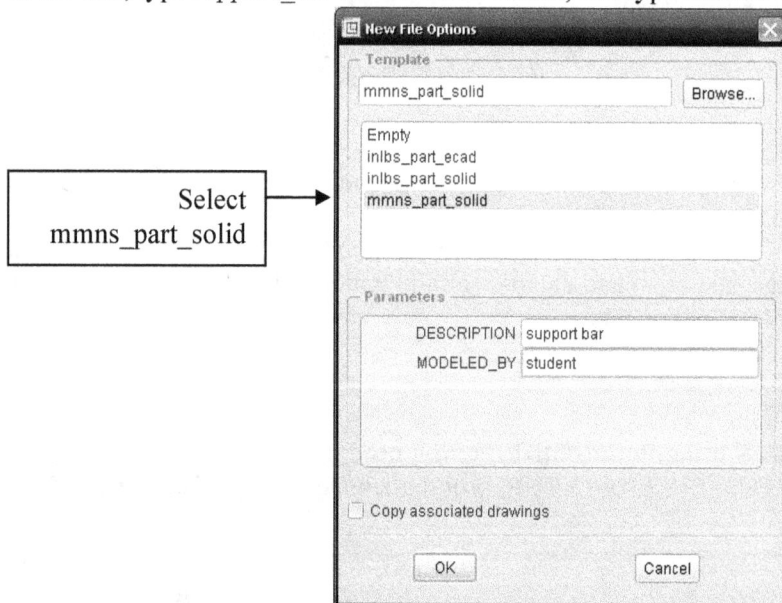

The design screen will be on display, as shown on the next page. The work screen has a display of three (3) datum planes, namely, FRONT, TOP, and RIGHT. When a user selects FRONT as the sketch plane to work with, the user will be working under the XOY coordinate system. If TOP is selected as the sketch plane to work with, the user will be working under the XOZ coordinate system. There is a message window, which provides users with the instruction how to work with the software system. The window on the left side is called Model Tree, which records the feature(s) the user has created in the process of creating a 3D solid model. On the right side of the screen, the Feature Creation Toolbar is on display.

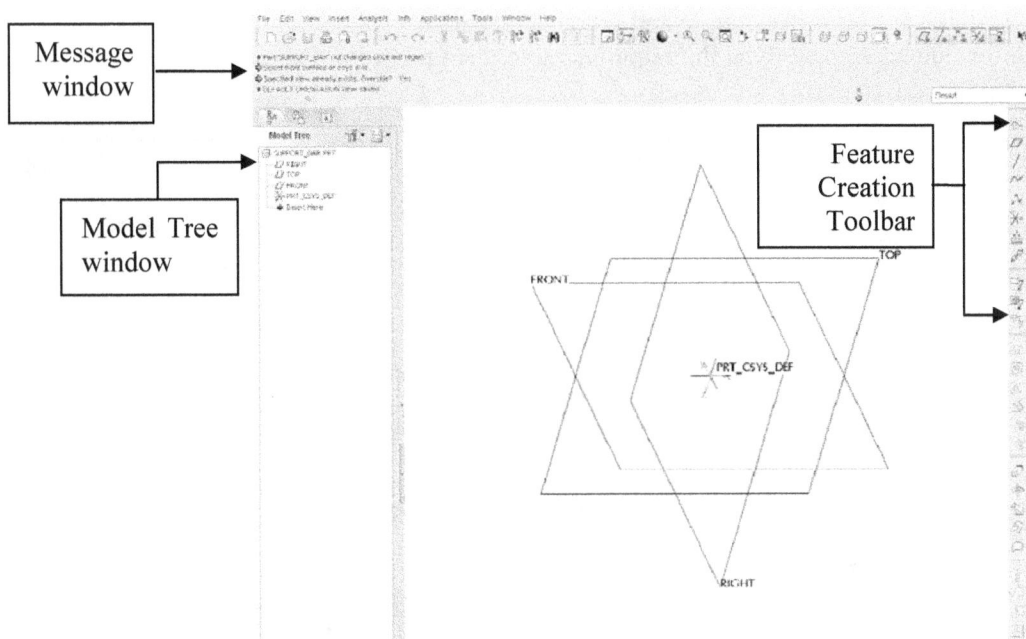

Message window

Model Tree window

Feature Creation Toolbar

Directly select the icon of **Extrude.** From the dashboard, set the thickness value to 40. Activate **Placement** to define a sketch plane > **Define.**

Thickness

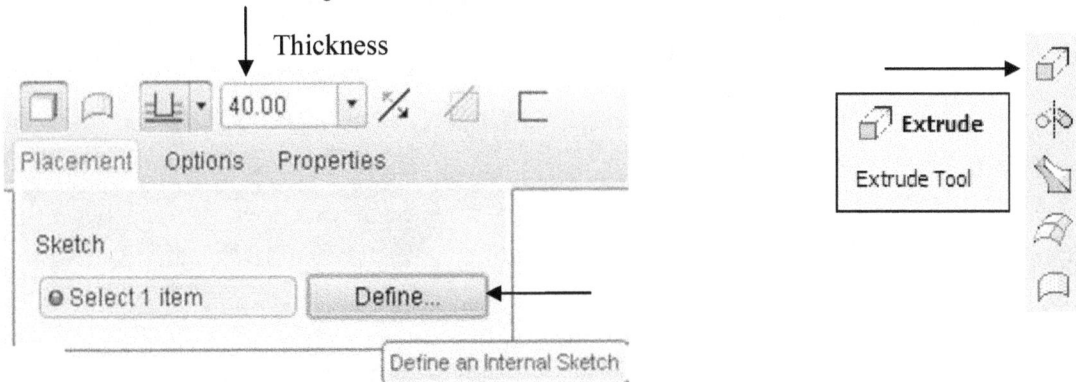

Placement   Options   Properties

Sketch

⊙ Select 1 item   Define...

Define an Internal Sketch

Extrude
Extrude Tool

From the screen, click **FRONT** to select the **FRONT** datum plane as the sketch plane, and click **Sketch** to accept the default orientation.

Sketch

Placement

Sketch Plane

Plane   FRONT:F3(Use Previous

Sketch Orientation

Sketch view direction   Flip

Reference   RIGHT:F1(DATUM PL...

Orientation   Right

Sketch   Cancel

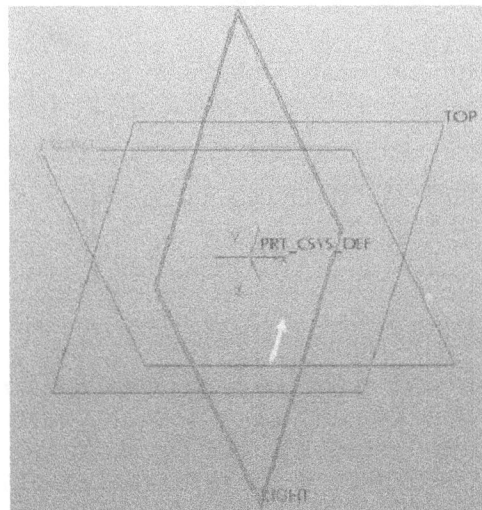

Click the icon of **Rectangle** and sketch a rectangle. To modify the value of a dimension, double click the value on display, type the required value, then press the **Enter** key. The 2 dimensions are 50 and 240, respectively.

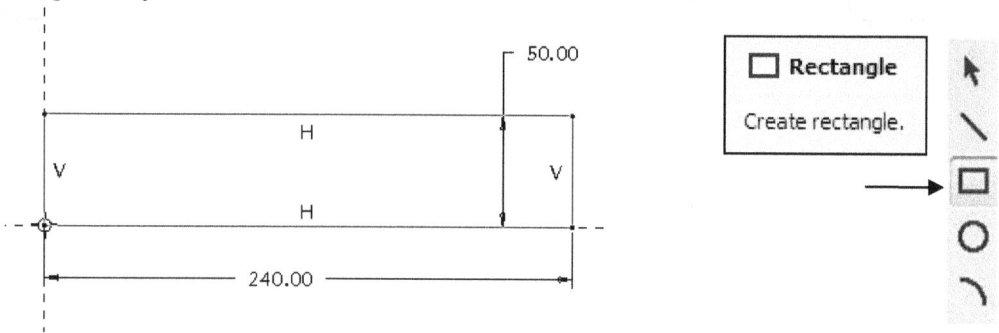

Upon completing this sketch, click the icon of **Done**. From the feature control panel, click the icon of **Apply and Save**.

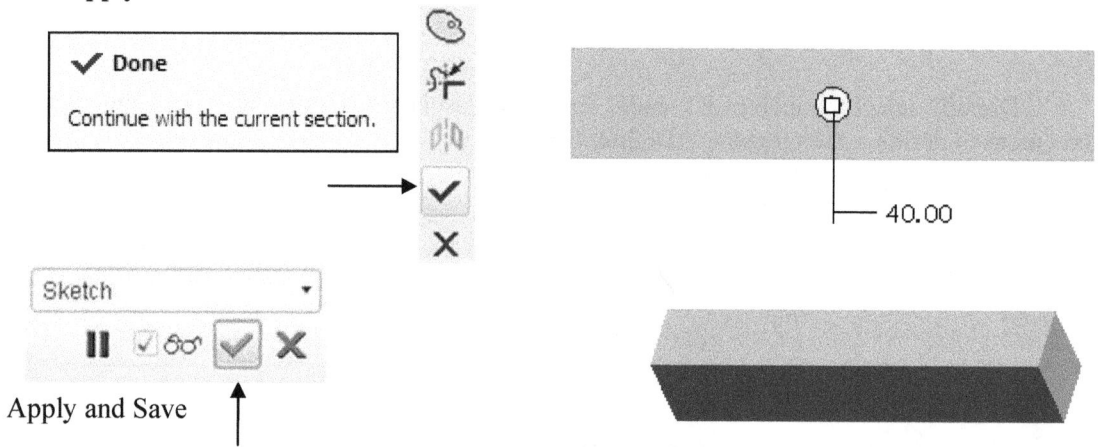

Apply and Save

Step 2: Save the created support_bar component.
Click the icon of **Save the active object** > **OK**.

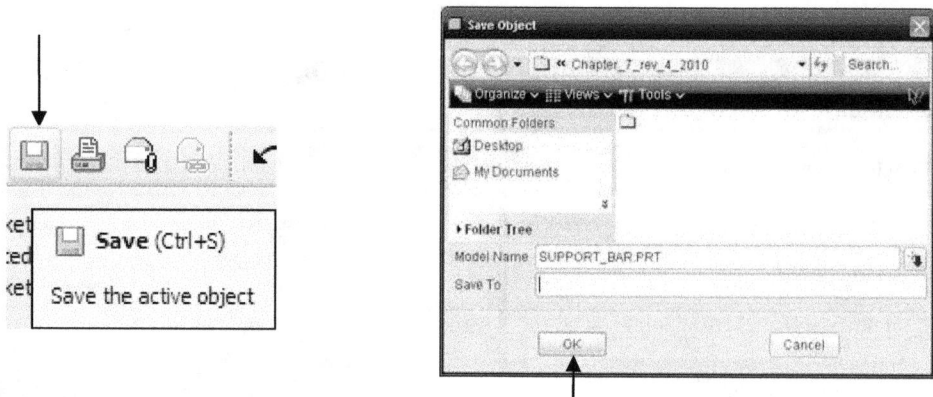

Creation of the Top_Plate Component

Step 1: Create a 3D solid model of the top plate component.
File > New > Part > type *top_plate* as the file name and clear the icon of **Use default template** > **OK**.

Select mmns_part_solid (units: Millimeter, Newton, Second) and type *top_plate* in **DESCRIPTION**, and s*tudent* in **MODELED_BY**, then click **OK.**

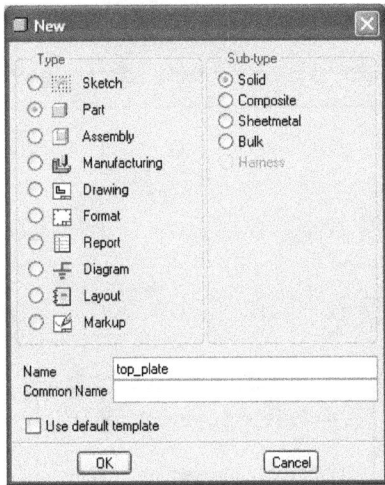

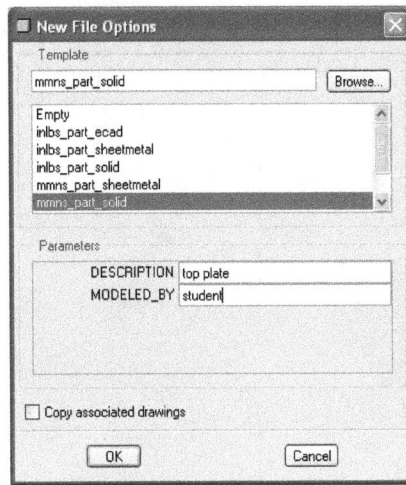

Directly select the icon of **Extrude.** From the dashboard, set the thickness value to 18. Activate **Placement** to define a sketch plane > **Define.**

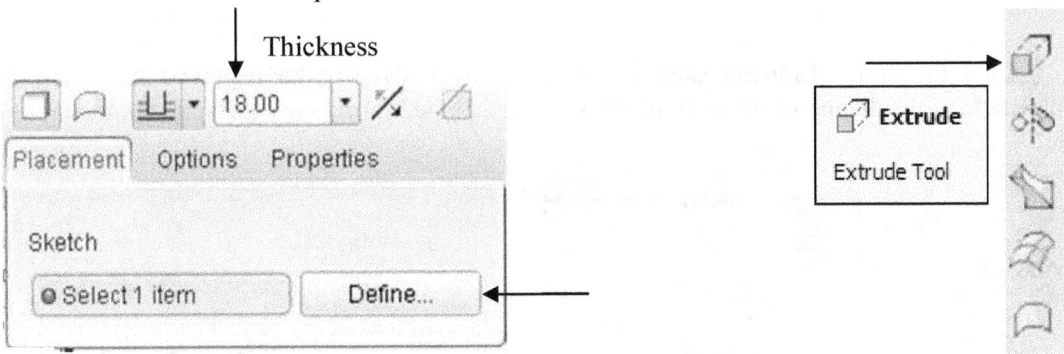

Select the **TOP** datum plane as the sketch plane, and accept the default orientation.

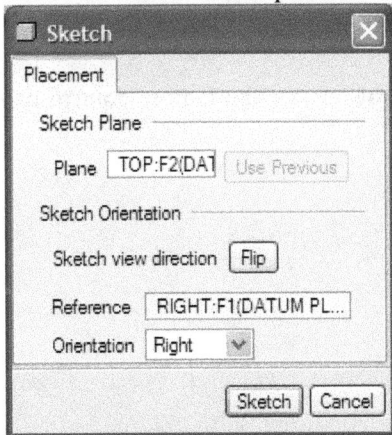

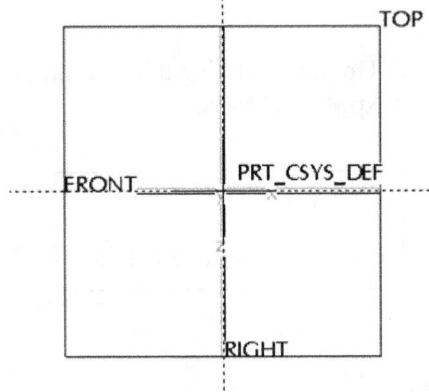

Click the icon of **Circle** and sketch a circle. Note the center of this circle is at the origin of the coordinate system. Click the icon of **Delete**, and click the half circle on the right side so that the half circle on the left side remains on display. Modify the displayed dimension to 140 by double clicking the displayed dimension.

Center and Point

Create circle by picking the center and a point on the circle.

PRT_CSYS_DEF

140.00

Delete Segment

Dynamically trim section entities.

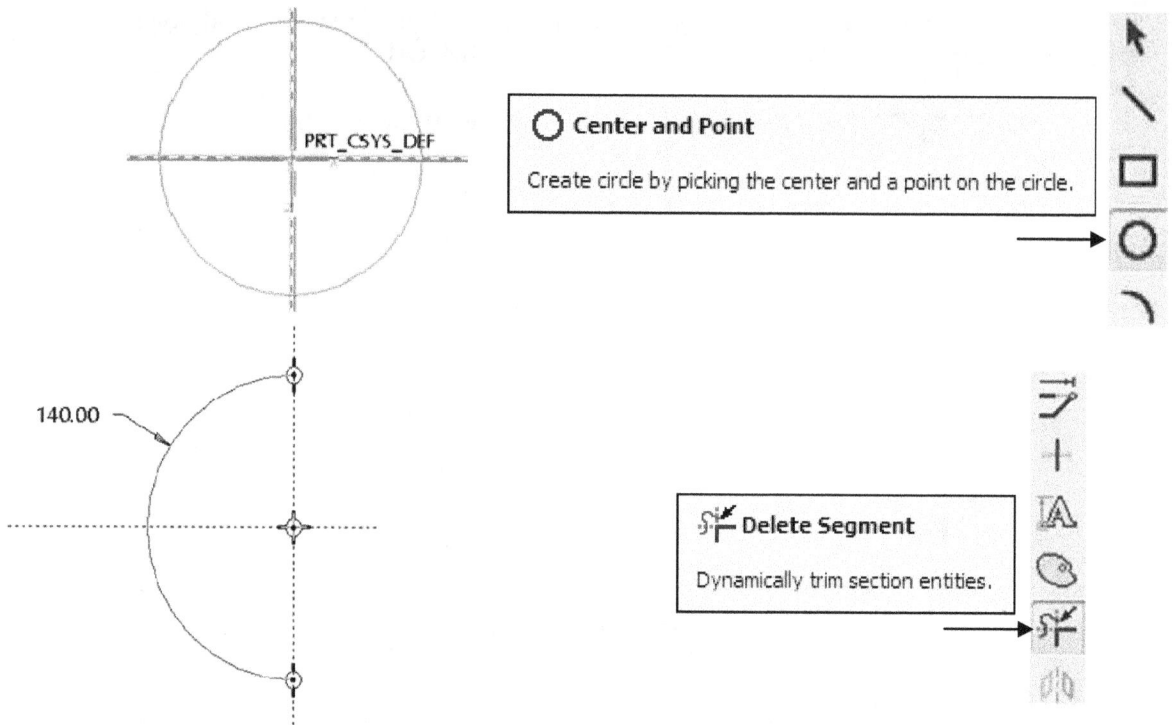

Click the icon of **Line** and sketch 3 lines, as shown. Modify the displayed dimension to 400 by double clicking the displayed dimension. Make sure that the 2 horizontal lines are tangential to the half circle.

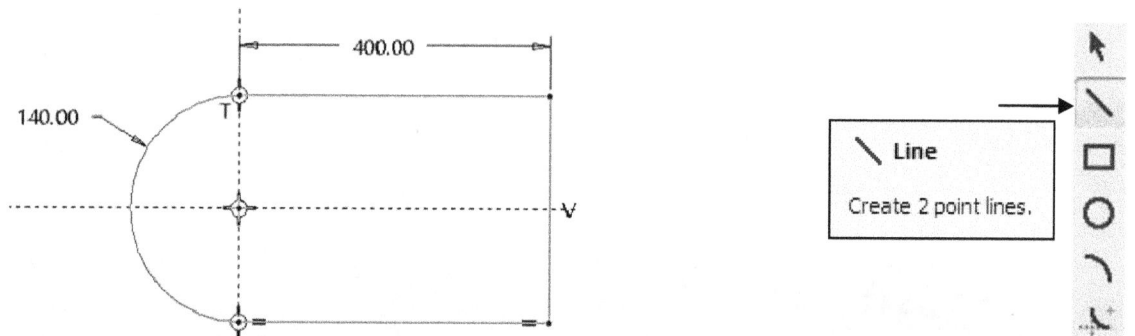

400.00

140.00

Line

Create 2 point lines.

Upon completing this sketch, click the icon of **Done**. From the feature control panel, click the icon of **Apply and Save**.

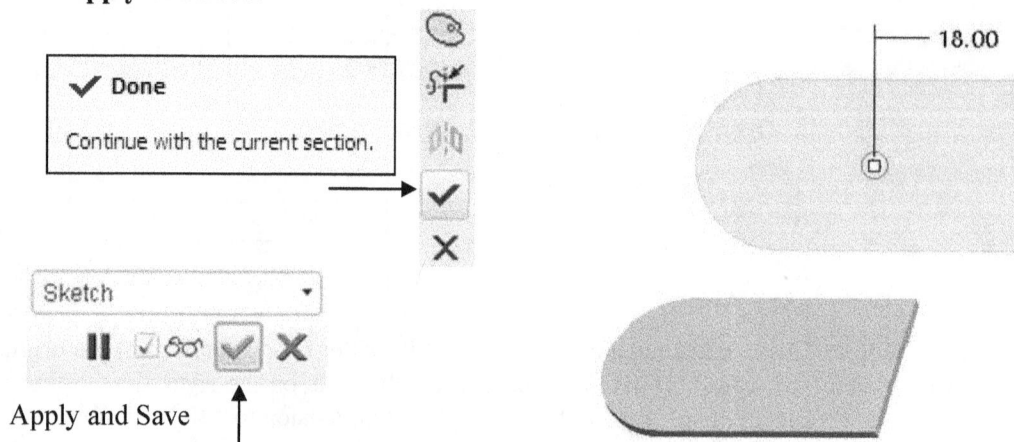

Done

Continue with the current section.

Sketch

Apply and Save

18.00

Step 2: Create a circular feature to accommodate the lift fan through a cut operation

Directly select the icon of **Extrude**. From the dashboard, activate the **Cut** operation, set **Thru All** as the depth choice. Activate **Placement** to define a sketch plane > **Define.**

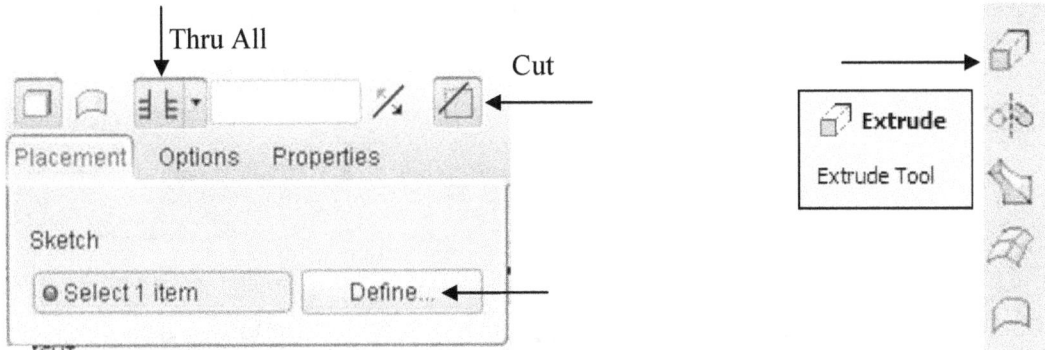

Select the top surface of the plate (do not select the **TOP** datum plane) as the sketch plane, and accept the default orientation.

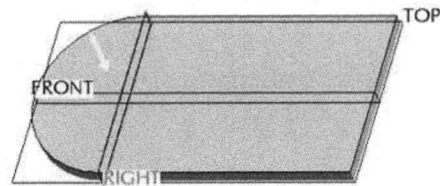

Click the icon of **Circle** and sketch a circle. The diameter value is *170* and the position dimension is *130*, as shown.

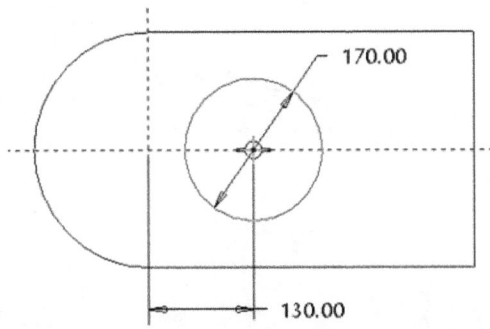

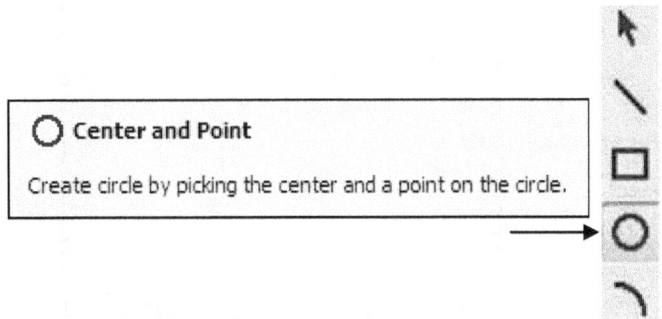

Upon completing this sketch, click the icon of **Done**. From the feature control panel, click the icon of **Apply and Save**.

Apply and Save

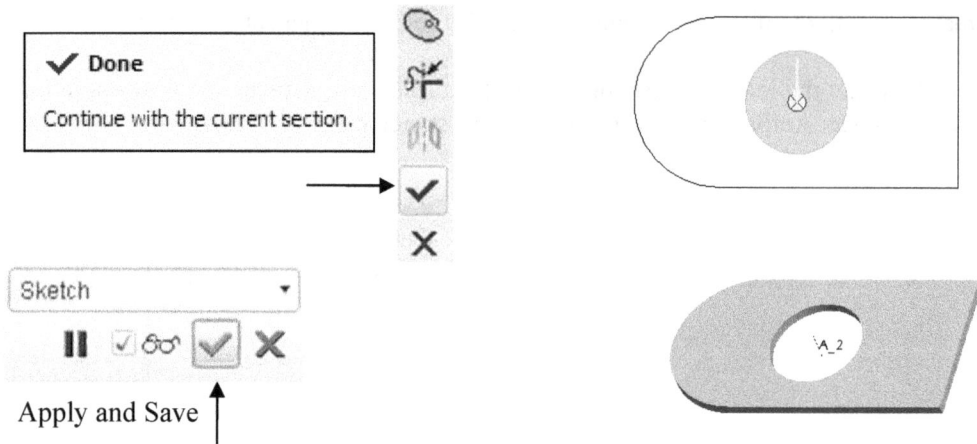

Step 3: Create a slot feature to accommodate the NXT, the size of which is 70 x 102 x 2 mm.

Directly select the icon of **Extrude.** From the dashboard, activate the **Cut** operation; specify 2 as the depth of cut. Activate **Placement** to define a sketch plane > **Define.**

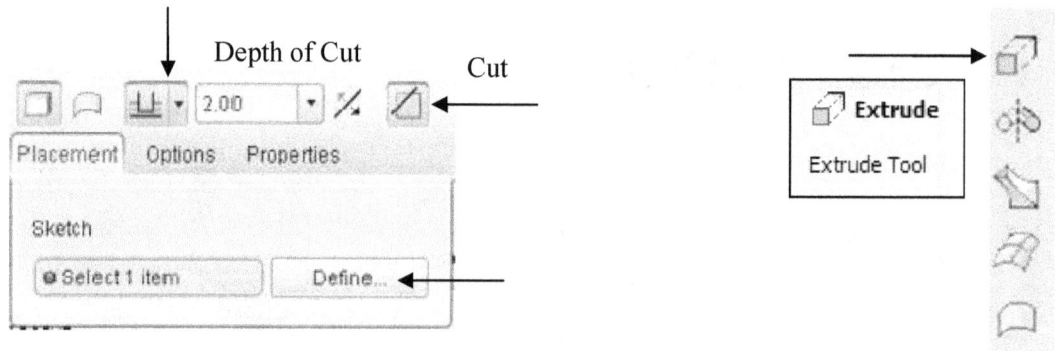

Depth of Cut

Cut

Extrude
Extrude Tool

Select the top surface of the plate (do not select the **TOP** datum plane) as the sketch plane, and accept the default orientation.

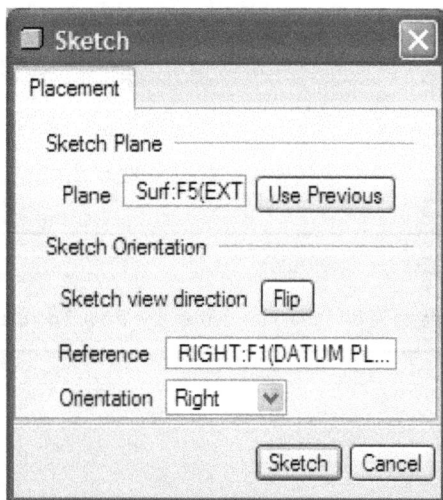

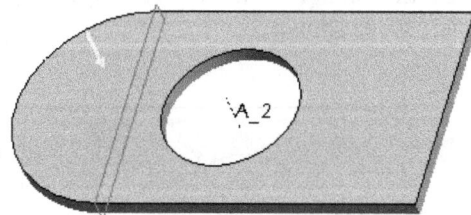

Right-click and hold, select **Centerline**.

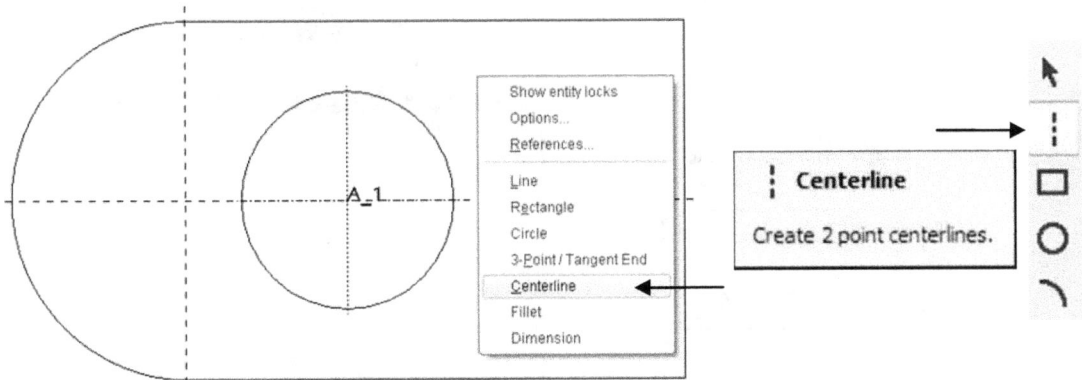

Sketch a horizontal centerline. Click the icon of **Rectangle** and sketch a rectangle, which is symmetric about the horizontal centerline. The 2 dimensions are 70 and 102, respectively. Note there is a dimension of 90 to position this rectangle with respect to the reference line on the right side, as shown.

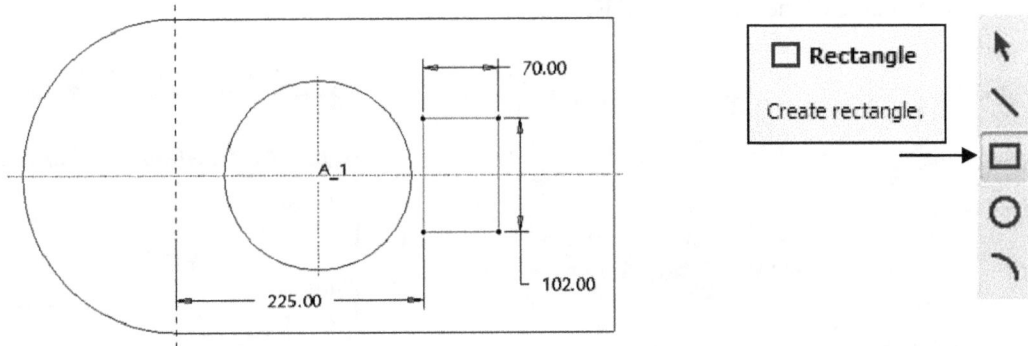

Upon completing this sketch, click the icon of **Done**. From the feature control panel, click the icon of **Apply and Save**.

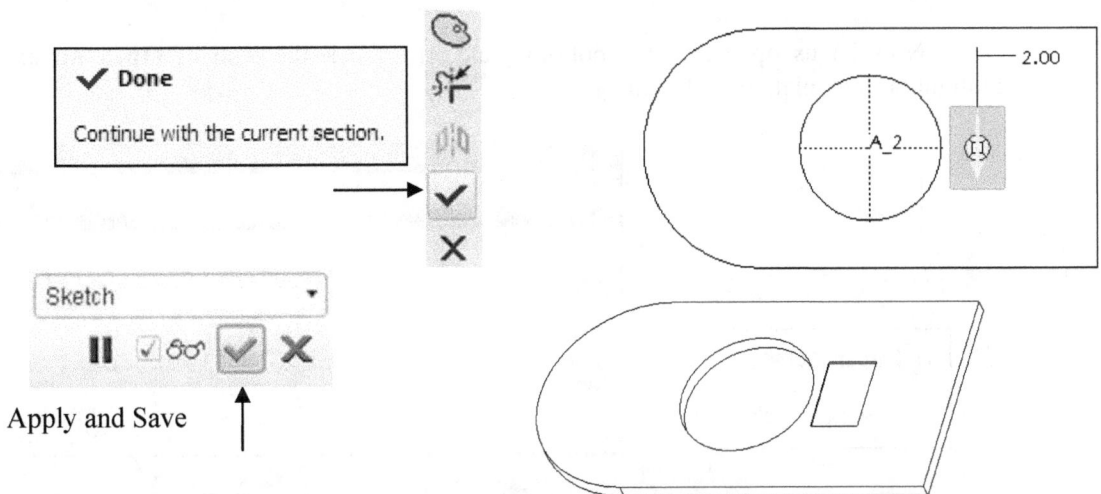

Apply and Save

Step 4: Save the created top_plate component.
Click the icon of **Save the active object** > **OK**.

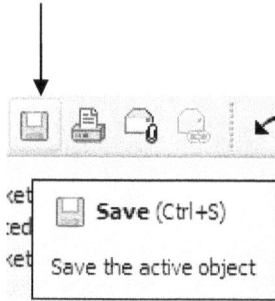

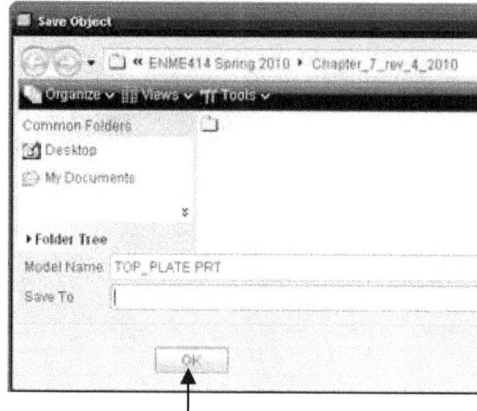

Creation of the Bottom_Plate Component

To create the bottom_plate component, click **File** > we use **Save a Copy** and type *bottom_plate* as the name of the new file > **OK**.

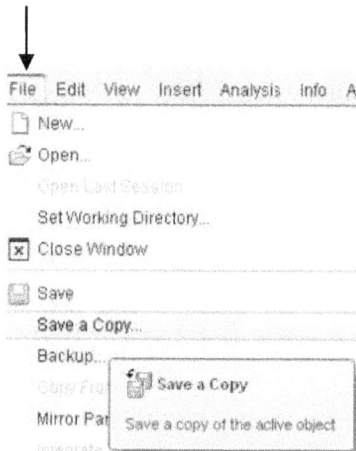

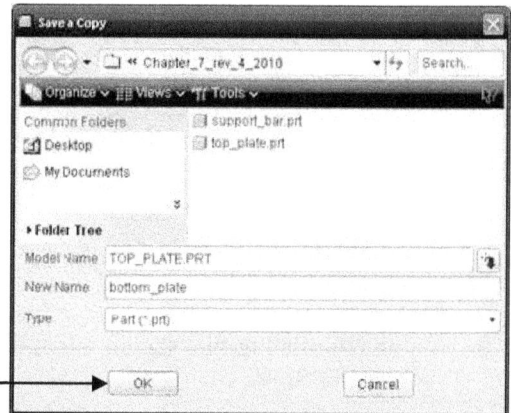

Now let us open the file: bottom_plate.prt. Click the icon of **Open an existing object** > highlight bottom_plate.prt > **Open**.

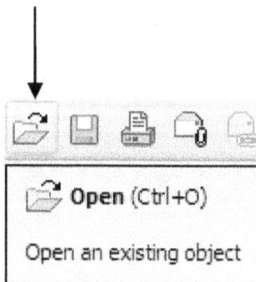

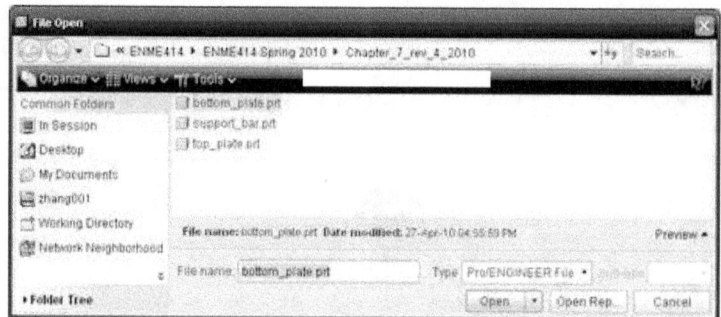

In the model tree, highlight the slot feature or highlight **Extrude 3** > right-click and hold, select **Delete** > **OK**. The slot feature does not appear on the display any more.

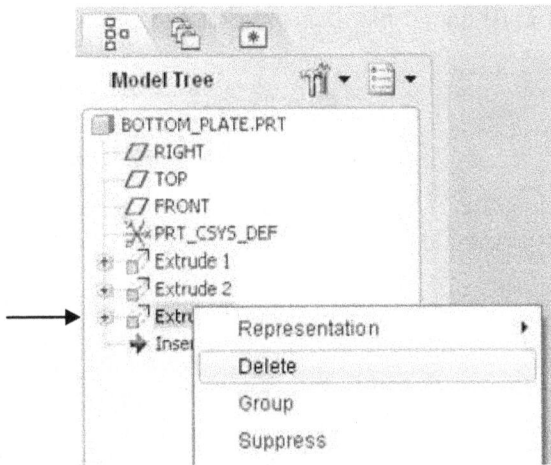

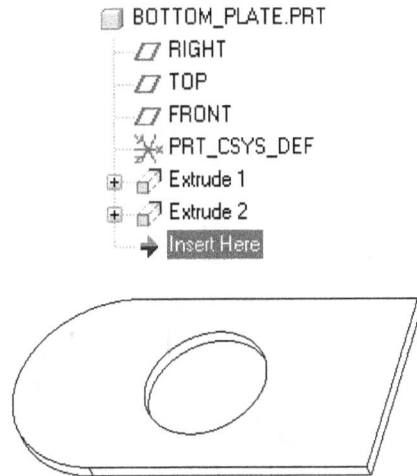

Now let us save the modified bottom_plate component. Click the icon of **Save the active object** > **OK**.

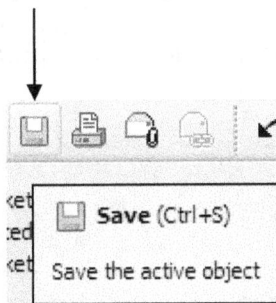

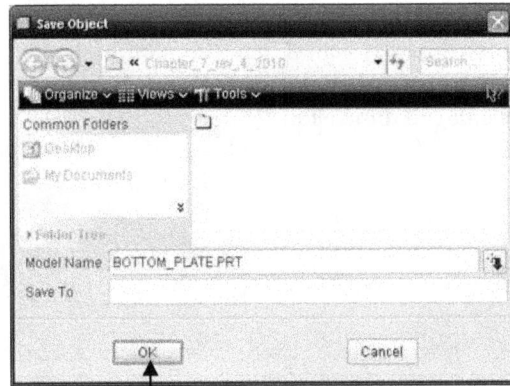

Creation of the Skirt Component

Step 1: Create a 3D solid model of the skirt component.

File > **New** > **Part** > type *skirt* as the file name and clear the icon of **Use default template** > **OK**.

Select mmns_part_solid (units: Millimeter, Newton, Second) and type *skirt* in **DESCRIPTION**, and s*tudent* in **MODELED_BY**, then **OK**.

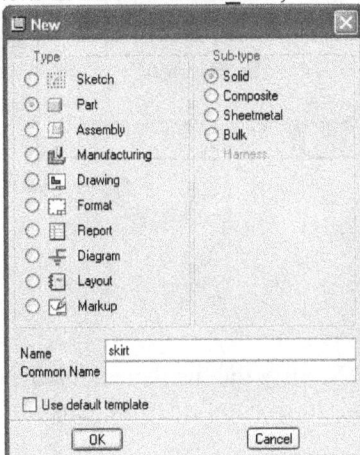

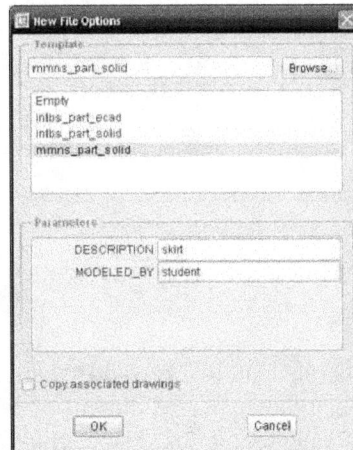

Click the icon of **Sketch Tool**. Pick the **TOP** datum plane as the sketch plane and accept the default setting for orientation > **Sketch.**

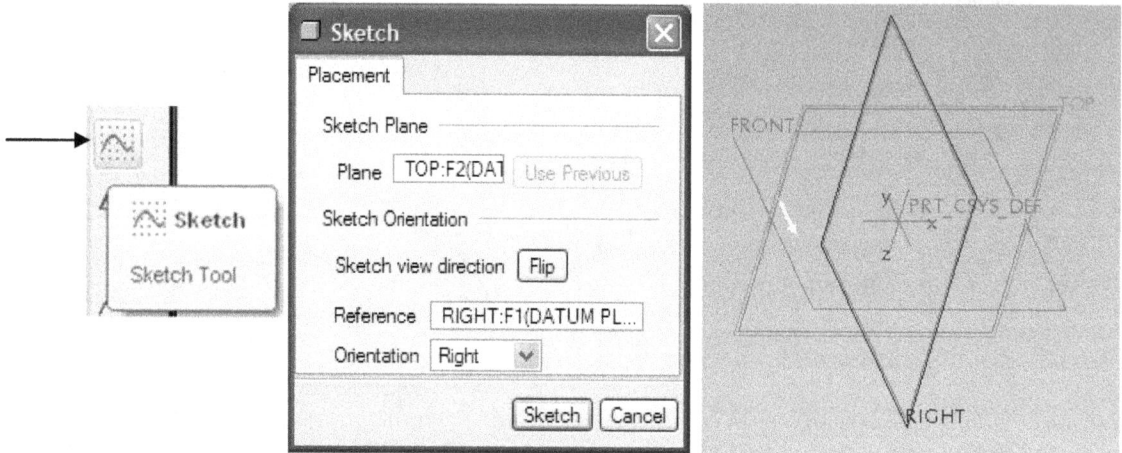

Click the icon of **Circle** and sketch a circle. Note the center of this circle at the origin of the coordinate system. Click the icon of **Delete**, and click the half circle on the right side so that the half circle on the left side remains on display. Modify the displayed dimension to 140 by double clicking the displayed dimension.

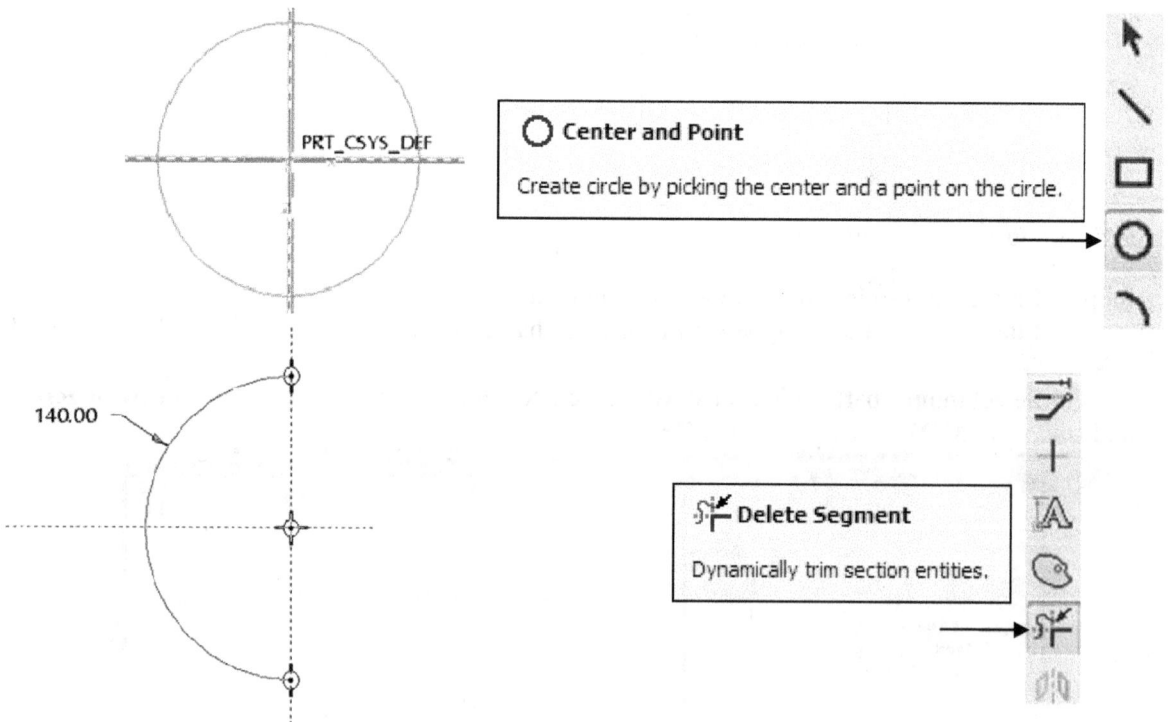

Click the icon of **Line** and sketch 3 lines, as shown. Modify the displayed dimension to 400 by double clicking the displayed dimension. Make sure that the 2 horizontal lines are tangential to the half circle. Upon completing this sketch, click the icon of **Done**.

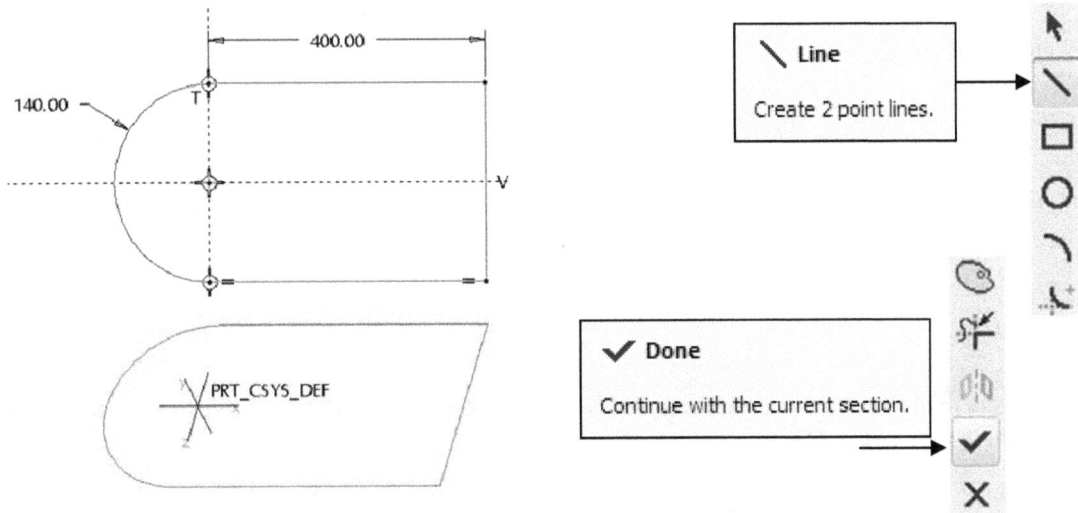

From the main menu, select Insert > Sweep > Thin Protrusion > Select Trajectory.
Select **Curve Chain** > pick the sketched trajectory > **Select All > Done**.

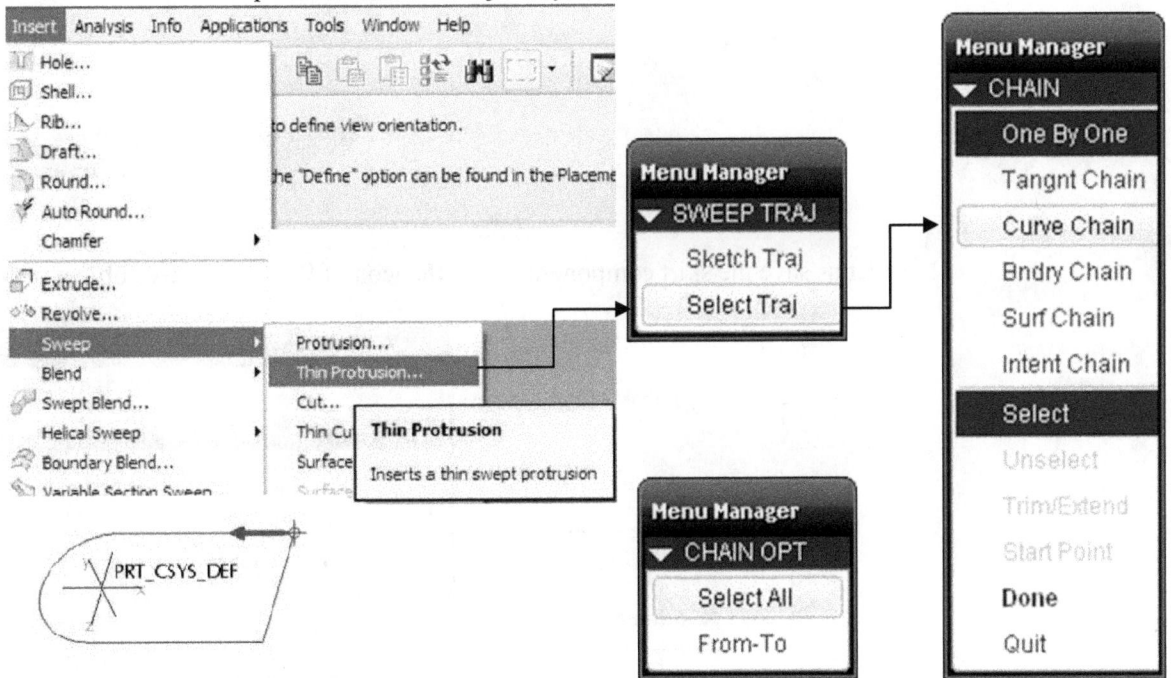

Click the icon of **Line** and sketch 3 straight lines, following the centerlines, as shown below. The 2 dimensions are 40 and 86, respectively.

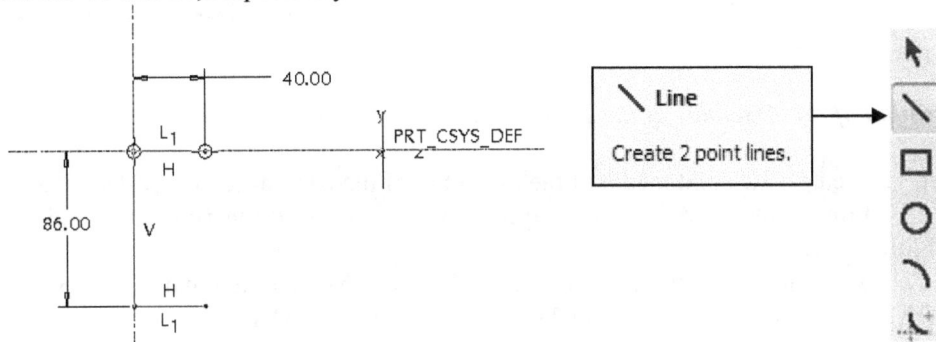

Upon completing, click the icon of **Done**.

For the side of material, **Flip > Okay** to ensure that the material is added to the outside of the sketch, as shown.

A menu for thickness automatically appears. For the thickness of the skirt material, type *1.0* and press the **Enter** key > **OK** to complete the creation.

Now let us save the skirt component. Click the icon of **Save the active object** > **OK**.

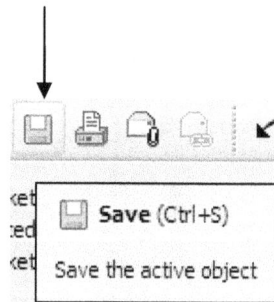

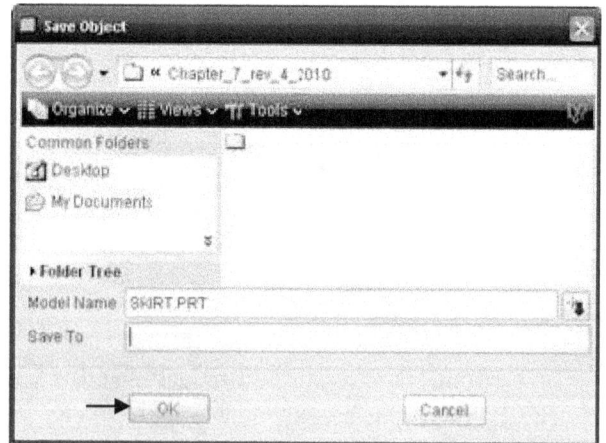

Creation of the Assembly of the Base Structural System

Step 1: Create a 3D solid model of the assembly of the base structural system.

File > **New** > **Assembly** > type *base_structure* as the file name and clear the icon of **Use default template** > **OK**.

Select mmns_asm_design (units: Millimeter, Newton, Second) and type *base_structure_system* in **DESCRIPTION**, and *student* in **MODELED_BY**, then **OK**.

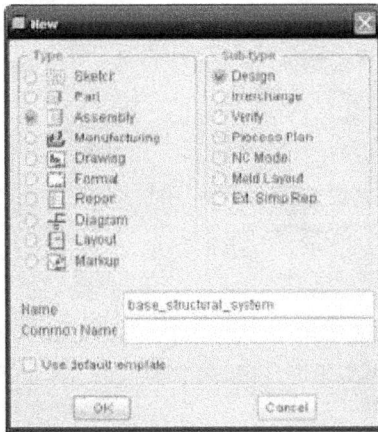

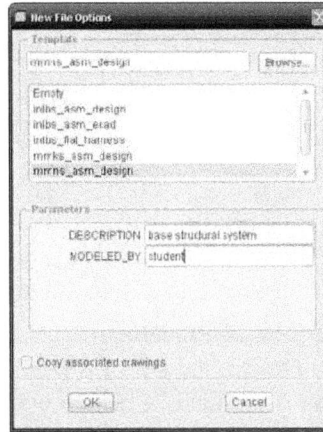

Click the icon of **Add component** > highlight bottom_plate.prt > **Open**.

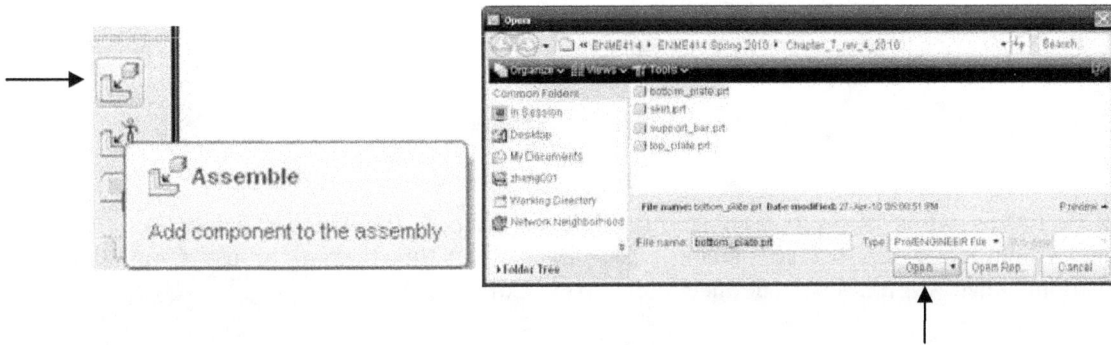

Select **Default** > click the icon of **Applies and Saves** from the constraint control panel.

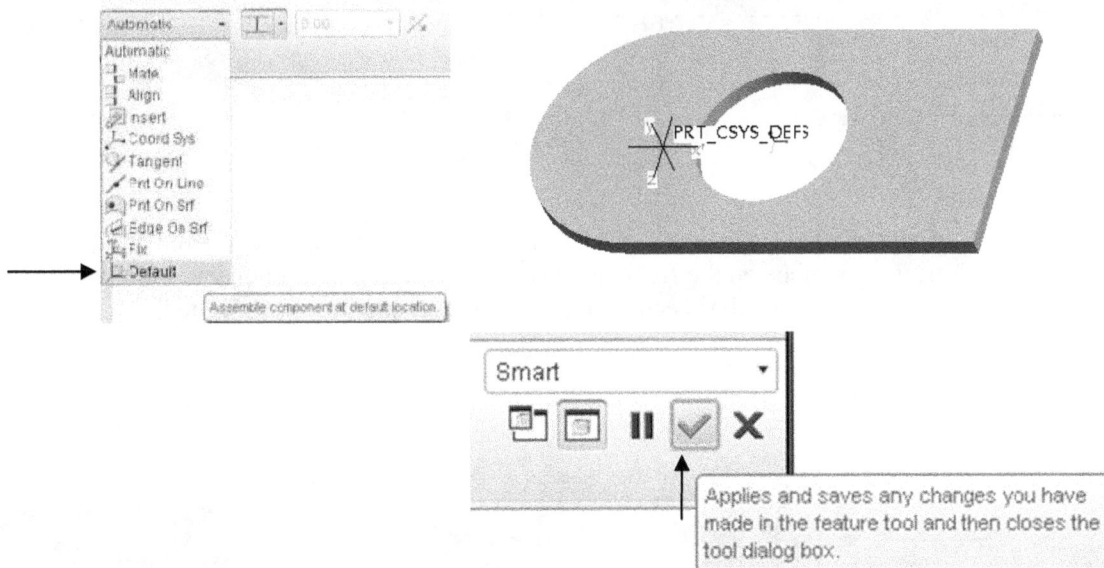

Click the icon of **Add component** > highlight support_bar.prt > **Open**.

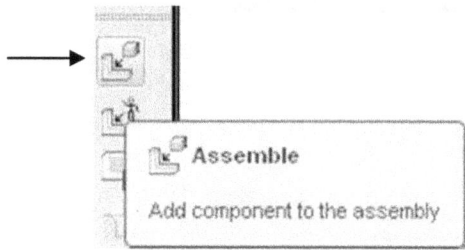

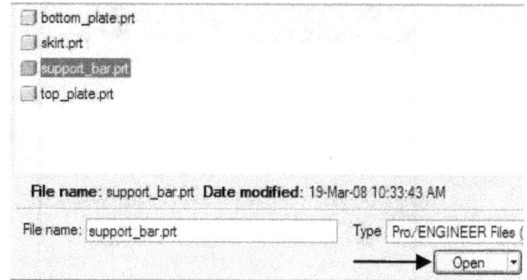

Click the icon of **Placement**. Let us define the first constraint. Select the top surface from the bottom_plate component and select the bottom surface from the support_bar component. A constraint called **Mate** has been defined. Click **New Constraint**.

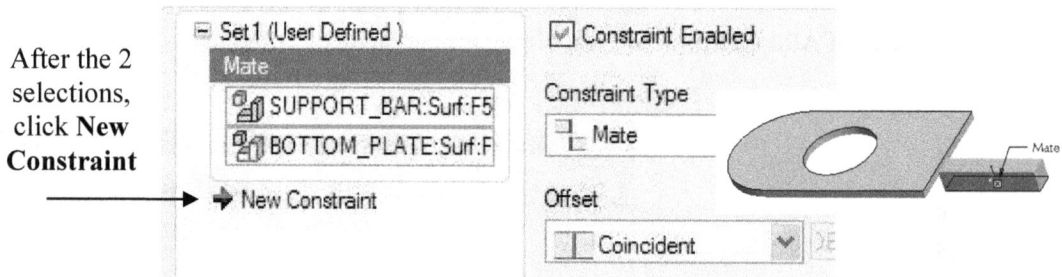

After the 2 selections, click **New Constraint**

Let us define the second constraint. Select the **ASM_RIGHT** and select the **RIGHT** datum plane from the support_bar component. Set the offset value to zero or use **Coincident**. A new constraint called **Align** has been defined. Click **New Constraint**.

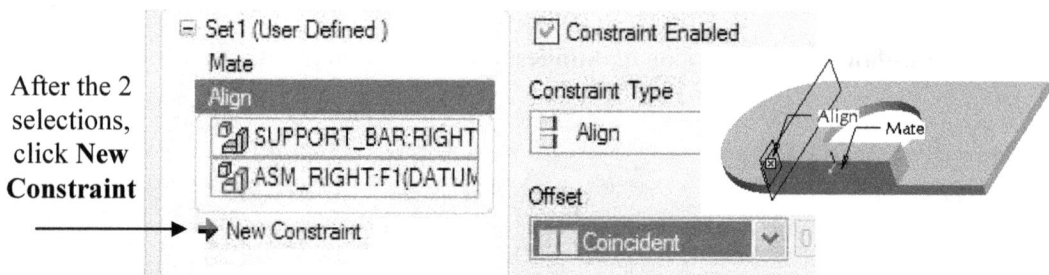

After the 2 selections, click **New Constraint**

Let us define the third constraint. Select the surface on the right side from the bottom plate component and select the surface on the right side from the support bar component. Set the offset value to 2. A new constraint called **Align** has been defined. "Fully constrained" is indicated.

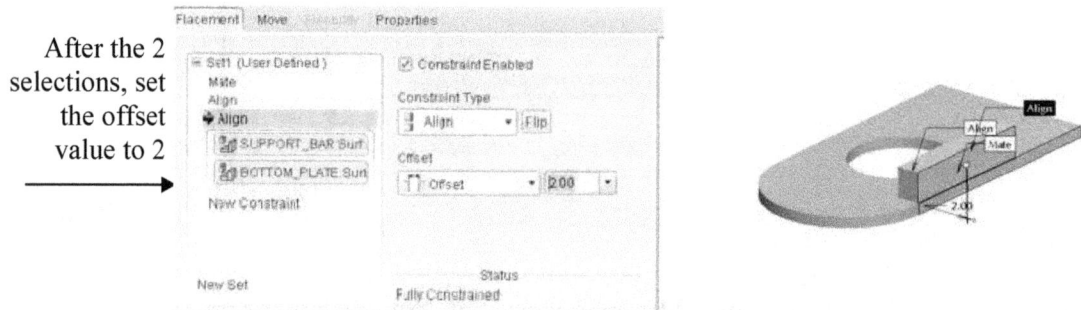

After the 2 selections, set the offset value to 2

At this moment, the support_bar component is fully constrained. Click the icon of **Applies and Saves** from the constraint control panel

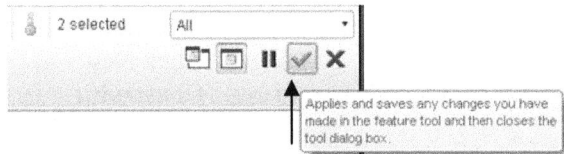

Now let us assemble the support_bar component on the other side. Click the icon of **Add component** > highlight support_bar.prt > **Open**.

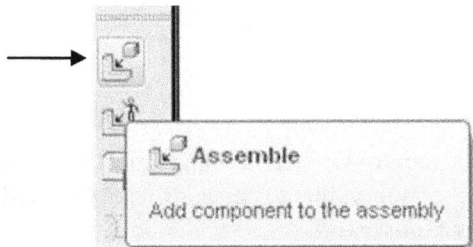

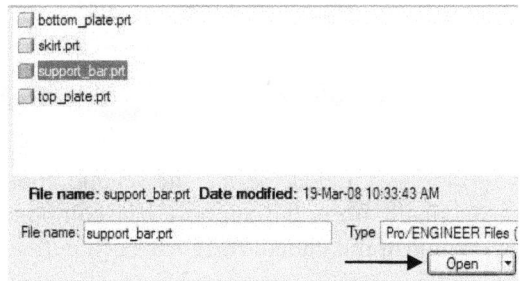

We follow the same procedure to define the 3 constraints as we did in the process of assembling the first support_bar component.

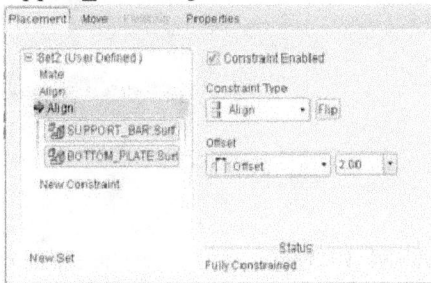

Now let us assemble the support_bar component at the front position. Click the icon of **Add component** > highlight support_bar.prt > **Open**.

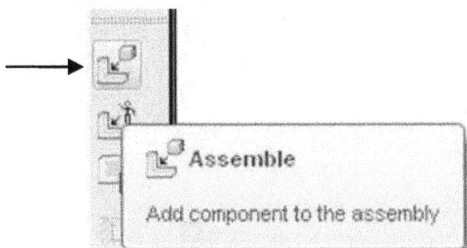

Let us define the first constraint. Select the top surface from the bottom_plate component and select the bottom surface from the support_bar component. A constraint called **Mate** has been defined. Click **New Constraint**.

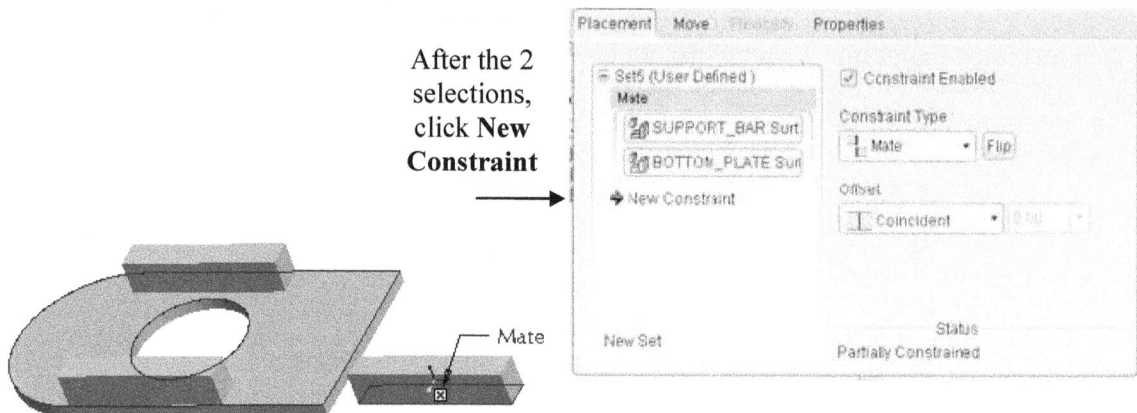

After the 2
selections,
click **New**
Constraint

Mate

Let us define the second constraint. Select the **ASM_RIGHT** and select the **FRONT** datum plane from the support_bar component. Set the offset value to zero or use **Coincident**. A new constraint called **Mate** has been defined. Click **New Constraint**.

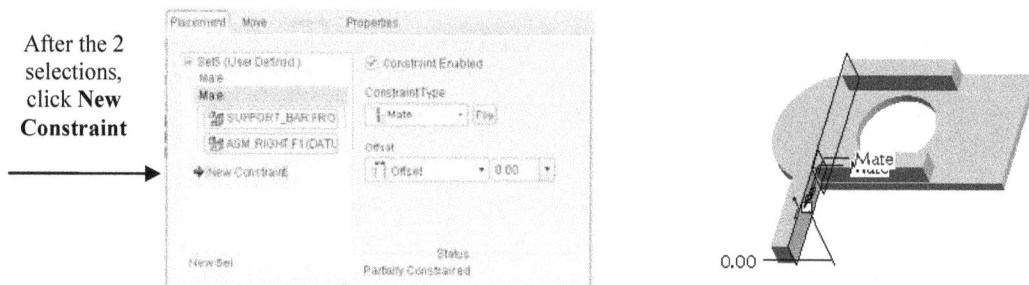

After the 2
selections,
click **New**
Constraint

0.00

Let us define the third constraint. Select the **ASM_FRONT** and select the end surface from the support_bar component. Set the offset value to 120. A new constraint called **Align** has been defined. At this moment, the support_bar component is fully constrained. Click the icon of **Applies and Saves** from the constraint control panel

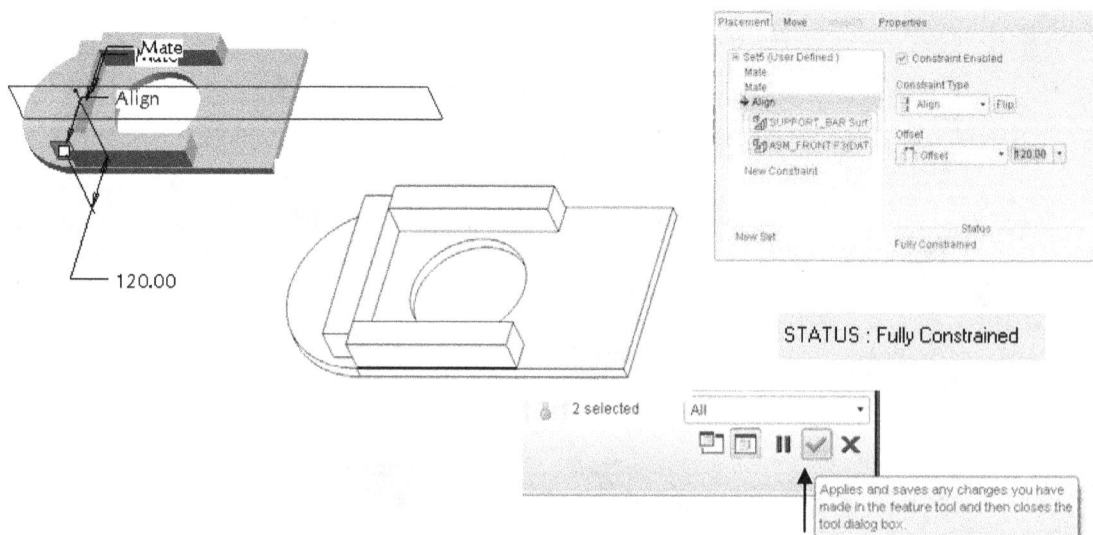

120.00

STATUS : Fully Constrained

2 selected   All

Applies and saves any changes you have
made in the feature tool and then closes the
tool dialog box.

Now let us assemble the support_bar component at the rear position. Click the icon of **Add component** > highlight support_bar.prt > **Open**.

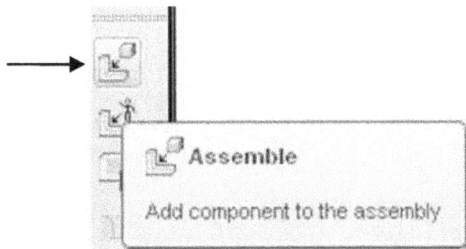

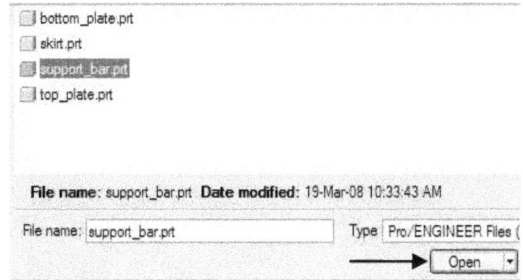

Let us define the first constraint. Select the top surface from the bottom_plate component and select the bottom surface from the support_bar component. A constraint called **Mate** has been defined. Click **New Constraint**.

After the 2 selections, click **New Constraint**

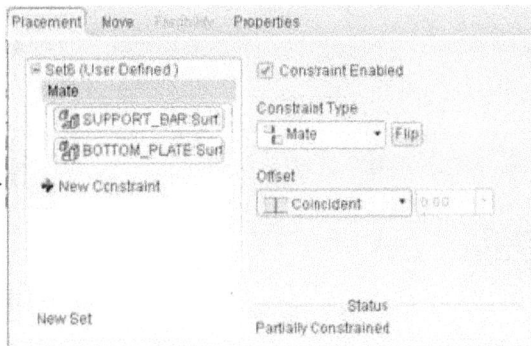

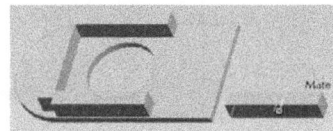

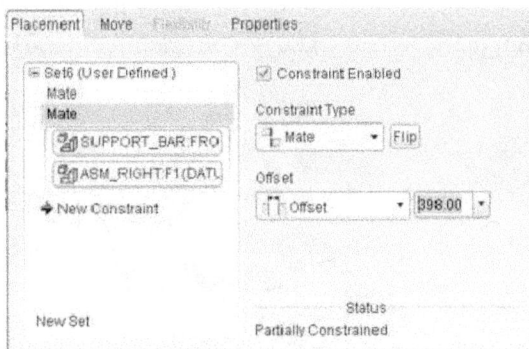

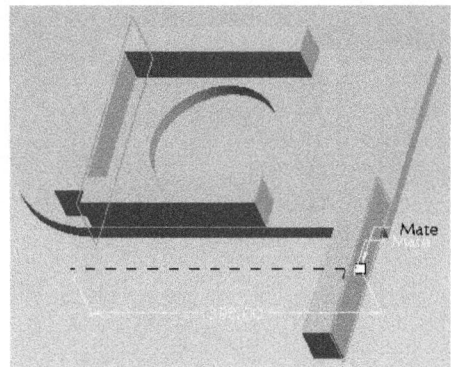

Let us define the second constraint. Select the **ASM_RIGHT** and select the **FRONT** datum plane from the support_bar component. Set the offset value to 398. A new constraint called **Mate** has been defined. Click **New Constraint**.

After the 2 selections, click **New Constraint**

Let us insert the third constraint. Select the **ASM_FRONT** and select the end surface from the support_bar component. Set the offset value to 120. A new constraint called **Align** has been defined. At this moment, the support_bar component is fully constrained. Click the icon of **Applies and Saves** from the constraint control panel

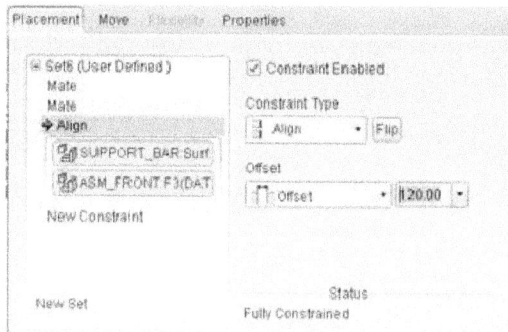

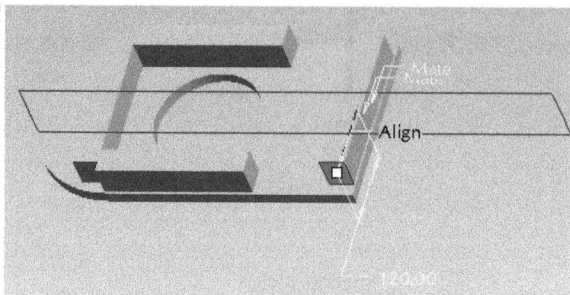

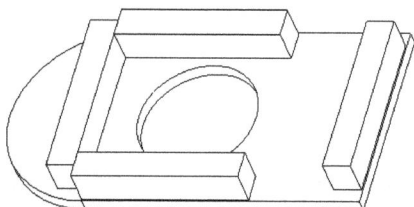

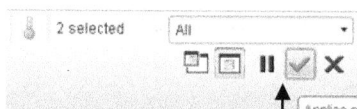

Now let us assemble the top_plate component. Click the icon of **Add component** > highlight top_plate.prt > **Open**.

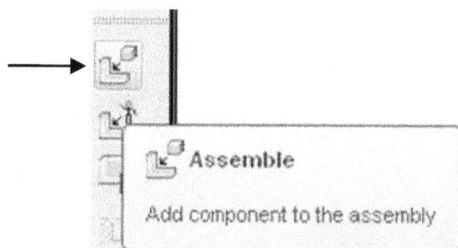

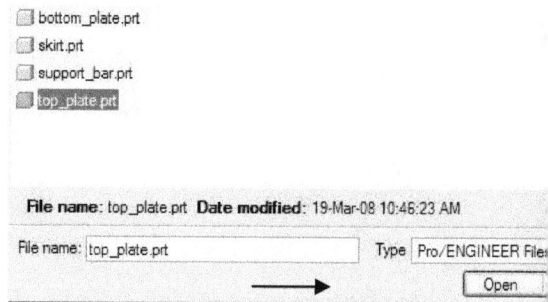

Let us define the first constraint. Select the top surface from the support_bar component and select the bottom surface from the top_plate component. A constraint called **Mate** has been defined. Click **New Constraint**.

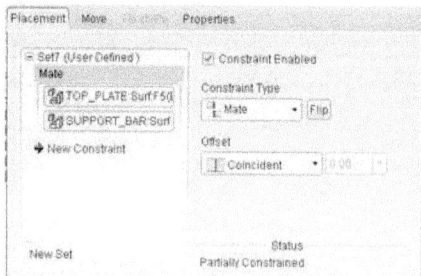

Let us create the second constraint. Select the axis from the bottom_plate component and select the axis from the top_plate component. A new constraint called **Align** has been defined. At this moment, the top_plate component is fully constrained. Click the icon of **Applies and Saves** from the constraint control panel.

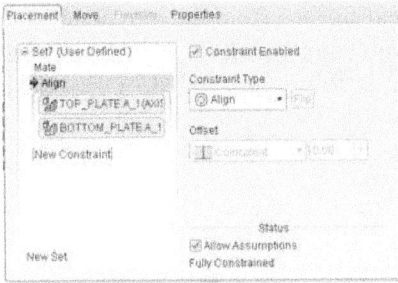

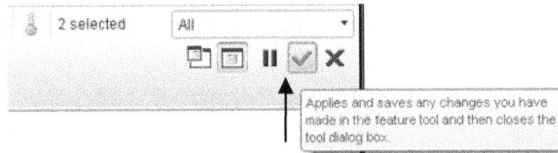

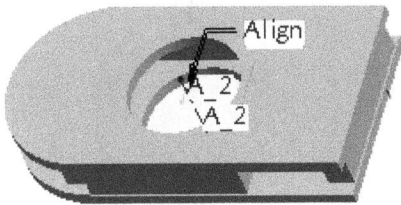

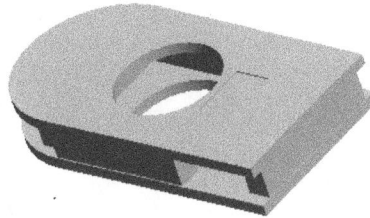

Now let us assemble the skirt component. Click the icon of **Add component** > highlight skirt.prt > **Open**.

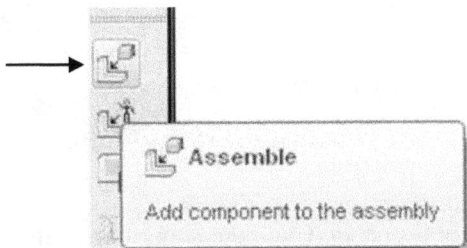

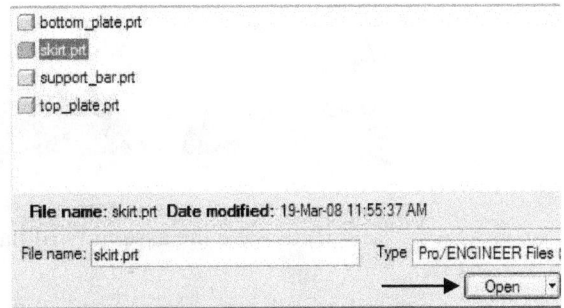

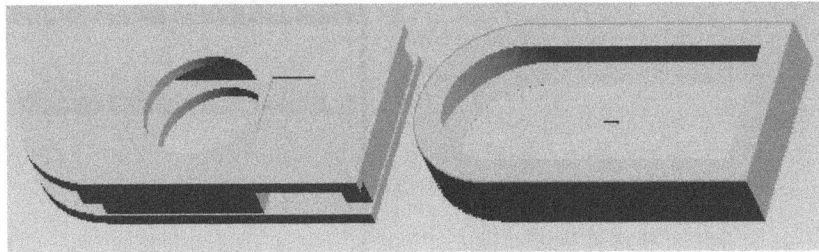

Let us define the first constraint. Select the top surface from the top_plate component and select the inner surface from the skirt component. A constraint called **Mate** has been defined. Click **New Constraint**.

After the 2 selections, click **New Constraint**

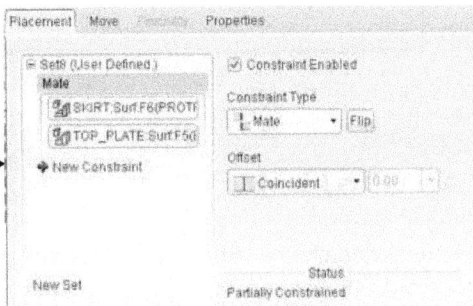

Let us define the second constraint. Select the cylindrical surface from the top_plate component and select the inner part of the cylindrical surface from the skirt component. A new constraint called **Insert** has been defined. At this moment, the skirt component is fully constrained. Click the icon of **Applies and Saves** from the constraint control panel.

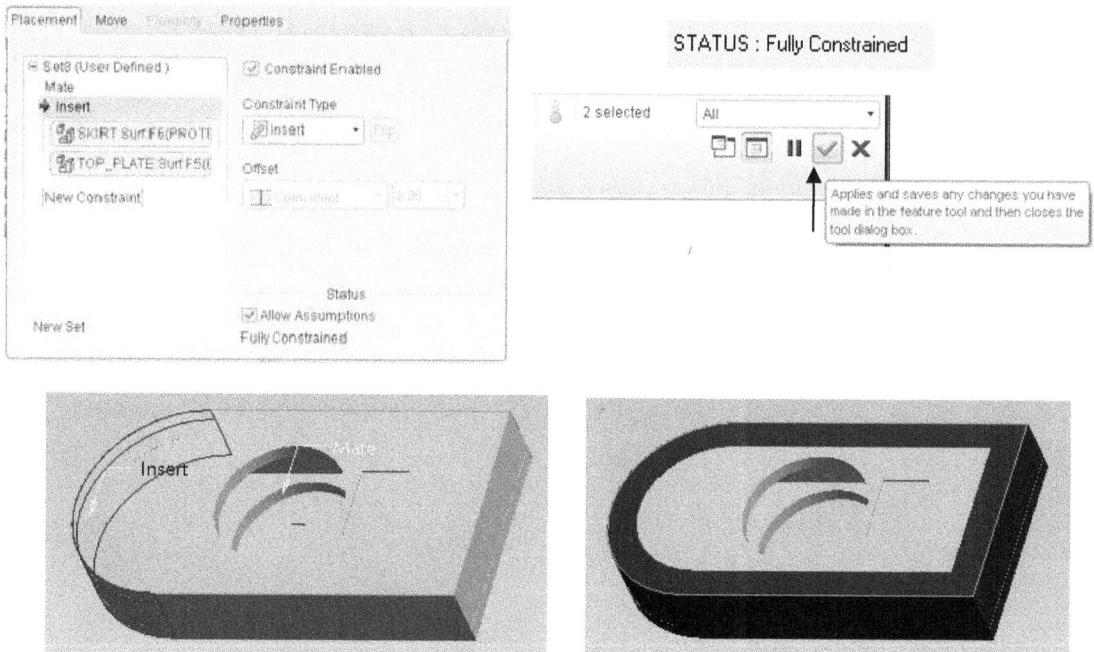

Now let us save the assembly of base structural system. Click the icon of **Save the active object > OK**.

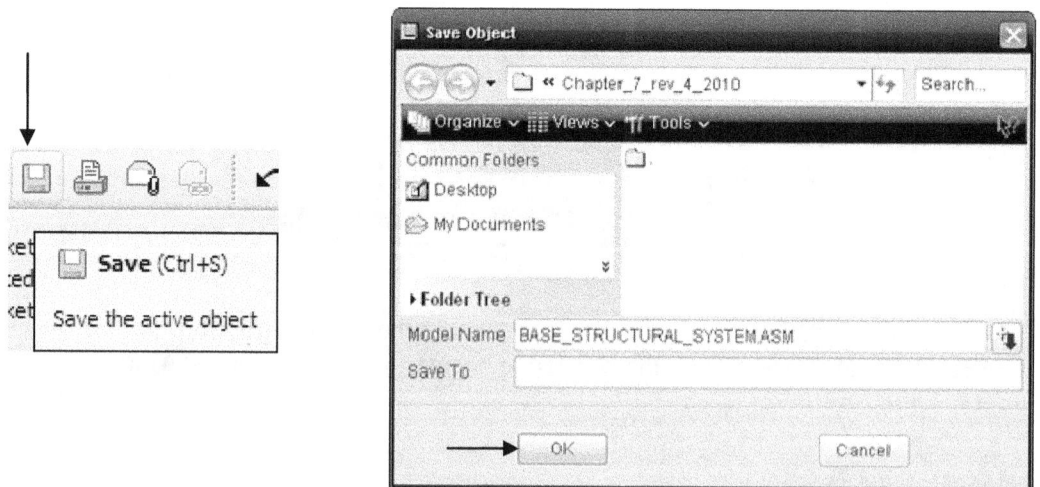

7.3 CREATION OF A LIFT FAN SYSTEM

Figure 7.2 is an exploded view of the lift fan system. As illustrated, this subsystem has a motor, a holding bar, a fan support and a blade for an axial fan or a turning wheel for a centrifugal fan.

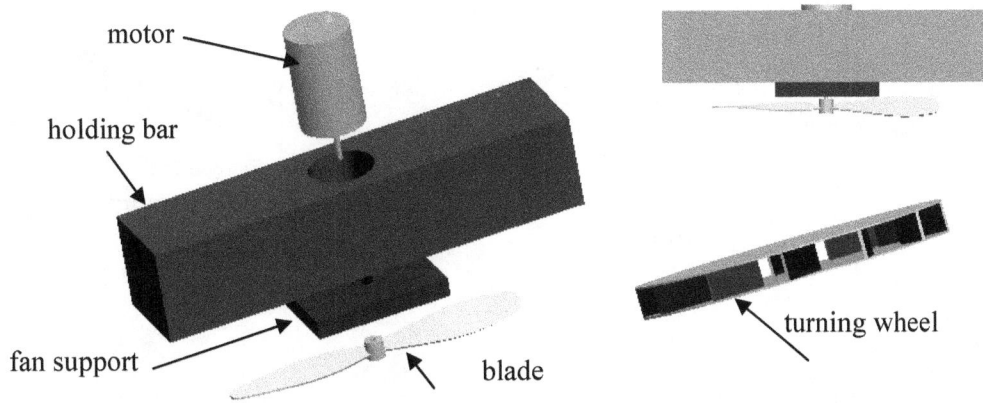

Fig. 7.2 An exploded view of the lift fan system showing the 4 components.

We begin with creating the holding_bar component. Afterwards, we construct the motor support component, the motor component, the blade component, and the turning wheel component. Finally, we assemble them together.

Creation of the Holding_Bar Component

Step 1: Create a 3D solid model of the holding bar component.
File > **New** > **Part** > type *holding_bar* as the file name and clear the icon of **Use default template** > **OK**.

Select mmns_part_solid (units: Millimeter, Newton, Second) and type *holding_bar* in **DESCRIPTION**, and s*tudent* in **MODELED_BY**, then **OK.**

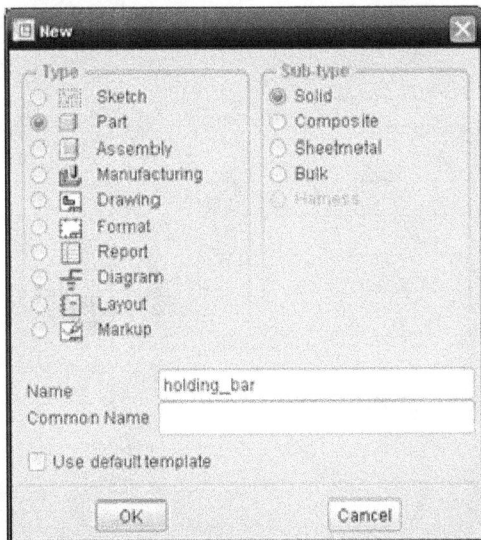

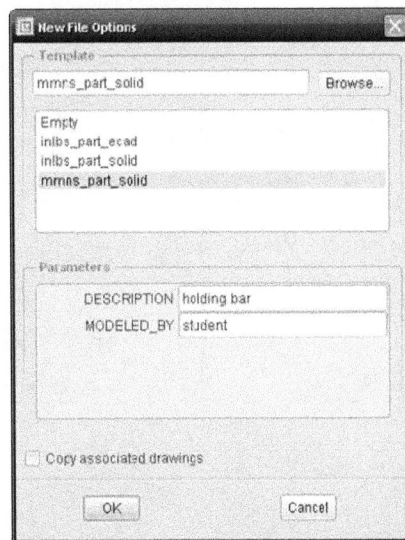

Directly select the icon of **Extrude.** From the dashboard, set the thickness value to 52. Activate **Placement** to define a sketch plane > **Define.**

Thickness

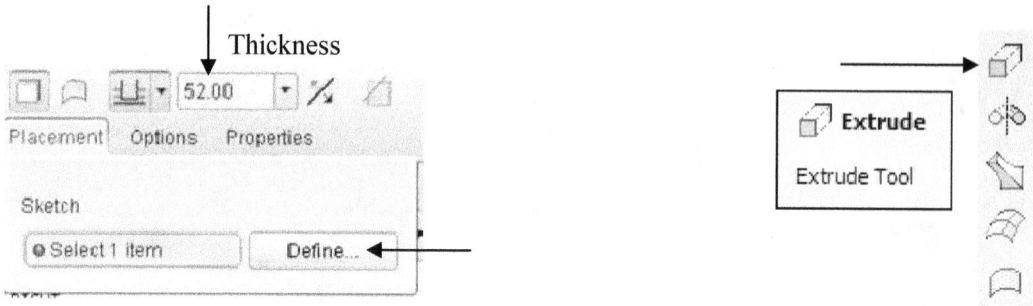

Select the **TOP** datum plane as the sketch plane, and accept the default orientation.

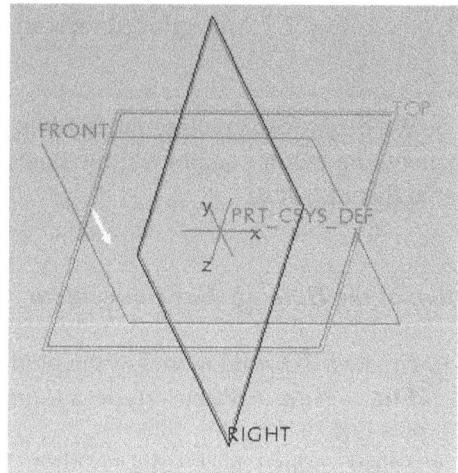

Right-click and hold, select **Centerline**. Sketch a vertical centerline and a horizontal centerline.

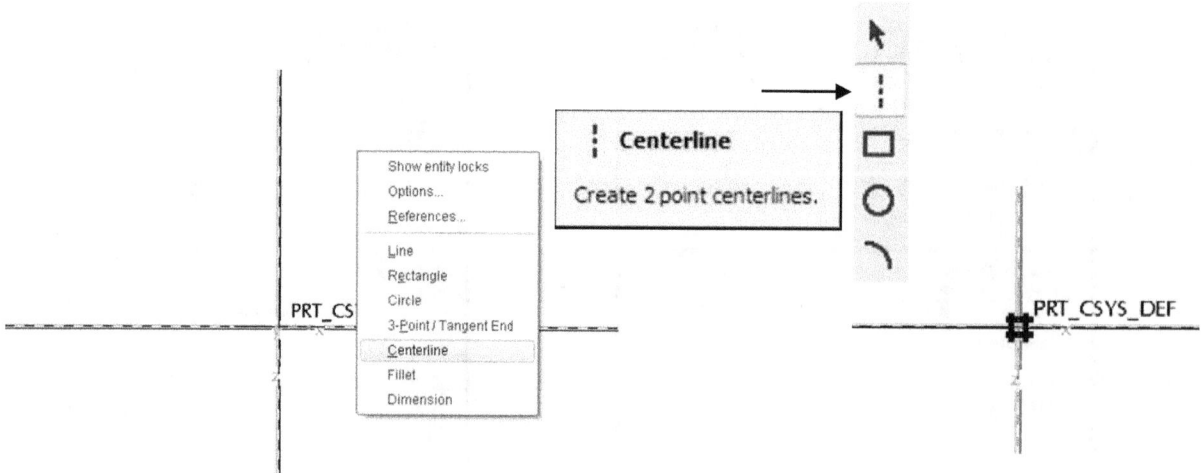

Click the icon of **Rectangle** and sketch a rectangle, which is symmetric about the 2 centerline. The 2 dimensions are 240 and 56, respectively.

240.00

56.00

☐ **Rectangle**

Create rectangle.

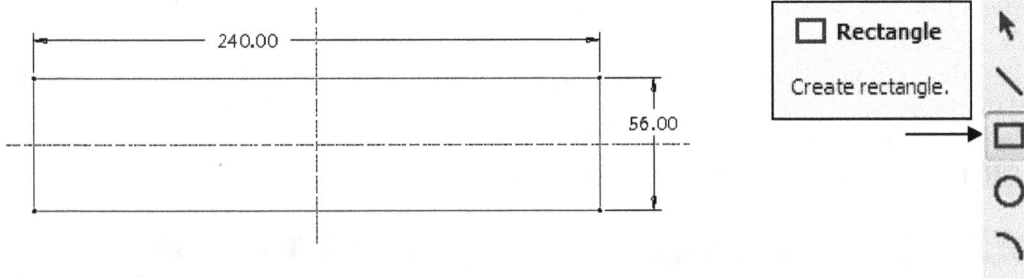

Click the icon of **Circle** to sketch a circle. The diameter value is 36, as shown.

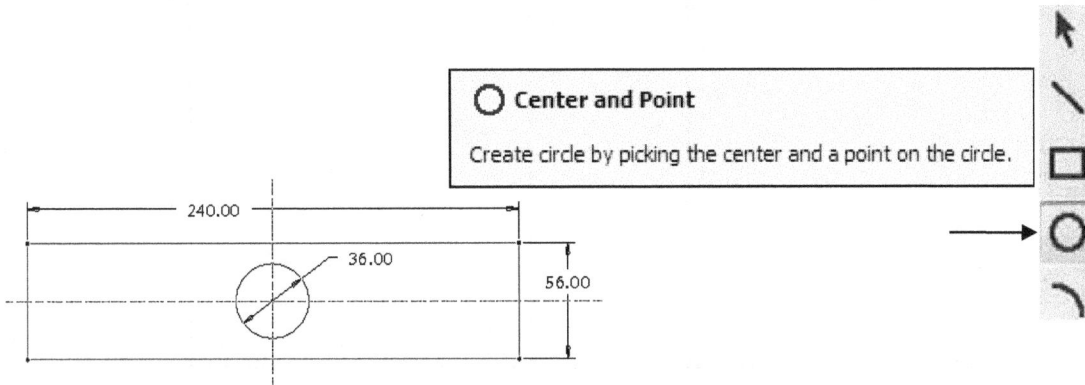

○ **Center and Point**

Create circle by picking the center and a point on the circle.

240.00

36.00

56.00

Upon completing this sketch, click the icon of **Done**. From the feature control panel, click the icon of **Apply and Save**.

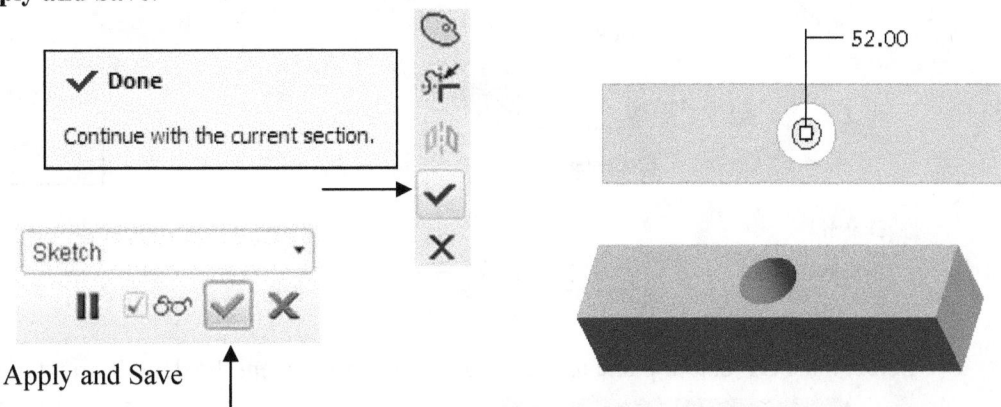

✓ **Done**

Continue with the current section.

Sketch

Apply and Save

52.00

Now let us save the holding bar component. Click the icon of **Save the active object** > **OK**.

🖫 **Save (Ctrl+S)**

Save the active object

Save Object

◄ Chapter_7_rev_4_2010   Search...

Organize ∨ Views ∨ Tools ∨

Common Folders
Desktop
My Documents

▶ Folder Tree

Model Name HOLDING_BAR.PRT

Save To

OK   Cancel

Creation of the Motor Support Component
Step 1: Create a 3D solid model of the motor support component.

File > **New** > **Part** > type *motor_support* as the file name and clear the icon of **Use default template** > **OK**.

Select mmns_part_solid (units: Millimeter, Newton, Second) and type *motor_support* in **DESCRIPTION**, and s*tudent* in **MODELED_BY**, then **OK.**

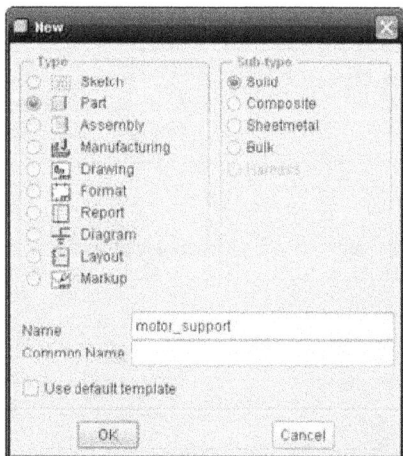

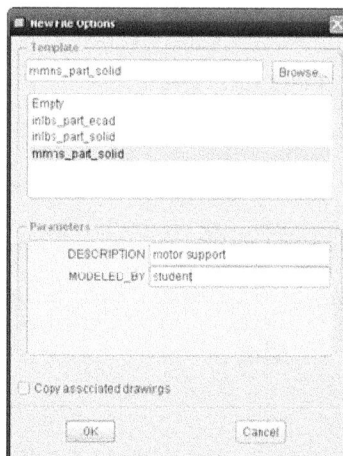

Directly select the icon of **Extrude.** From the dashboard, set the thickness value to 10. Activate **Placement** to define a sketch plane > **Define.**

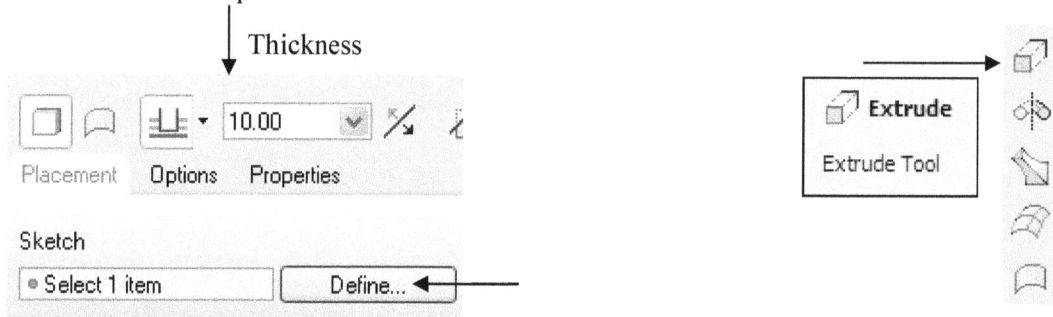

Select the **TOP** datum plane as the sketch plane, and accept the default orientation.

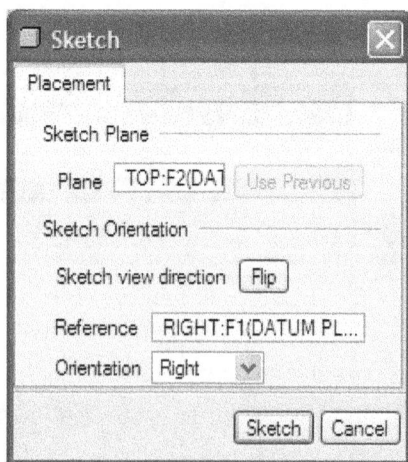

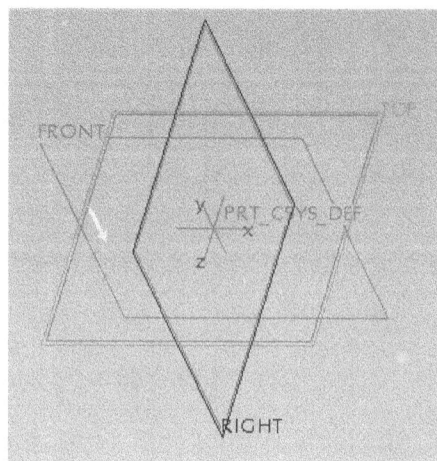

Right-click and hold, select **Centerline**. Sketch a vertical centerline and a horizontal centerline.

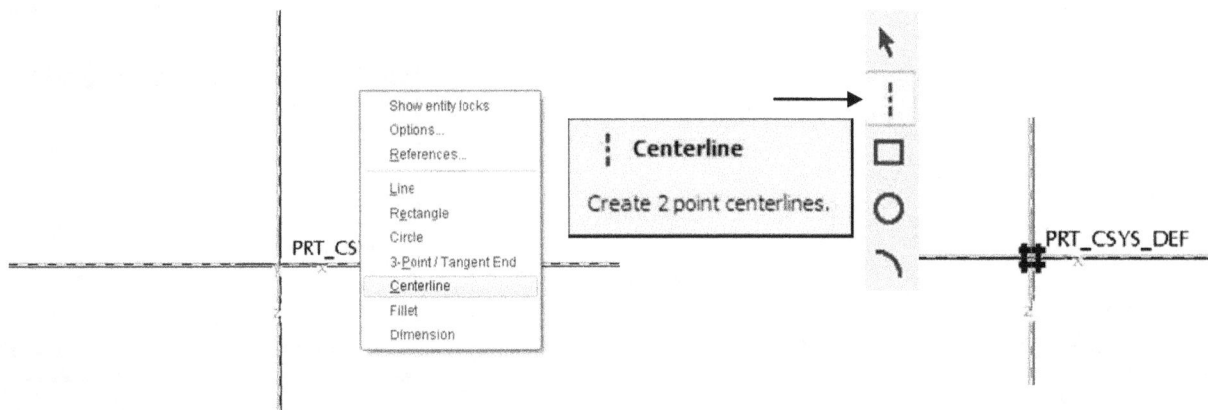

Click the icon of **Rectangle** and sketch a rectangle, which is symmetric about the 2 centerlines. The 2 dimensions are 76 and 56, respectively.

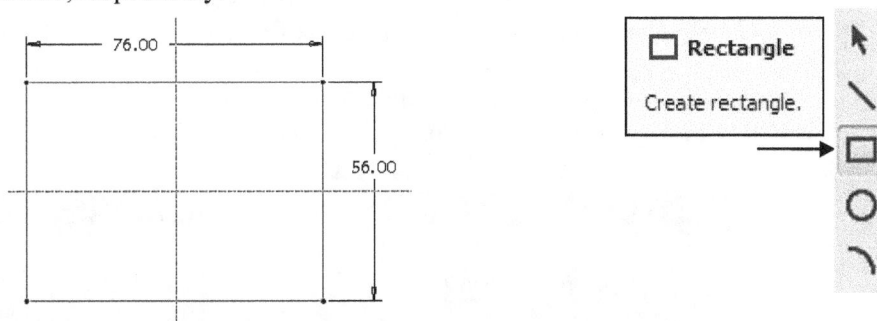

Click the icon of **Circle** to sketch a circle. The diameter value is 12, as shown.

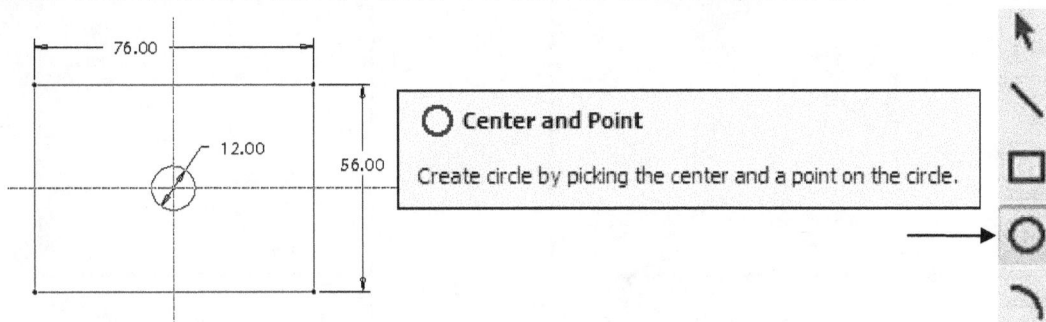

Upon completing this sketch, click the icon of **Done**. From the feature control panel, click the icon of **Apply and Save**.

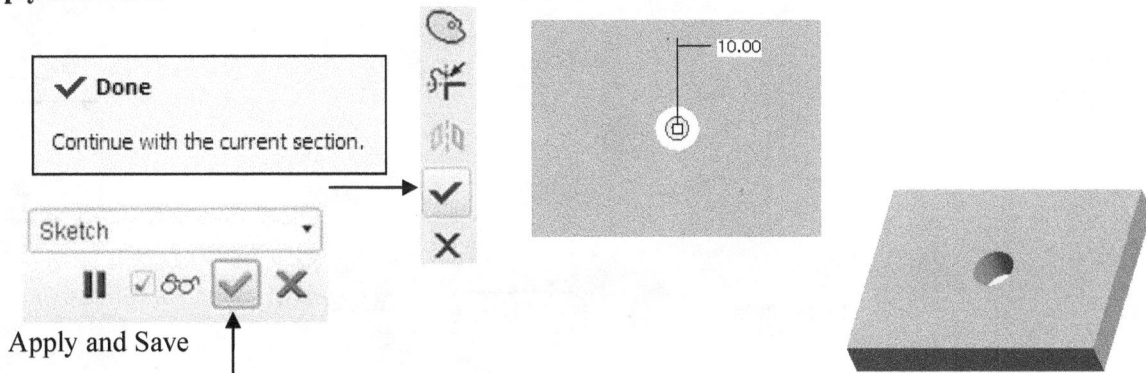

Apply and Save

Now let us save the motor support component. Click the icon of **Save the active object > OK**.

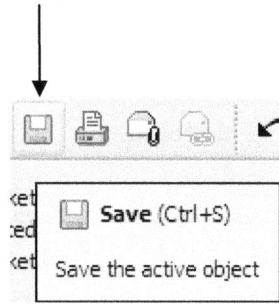

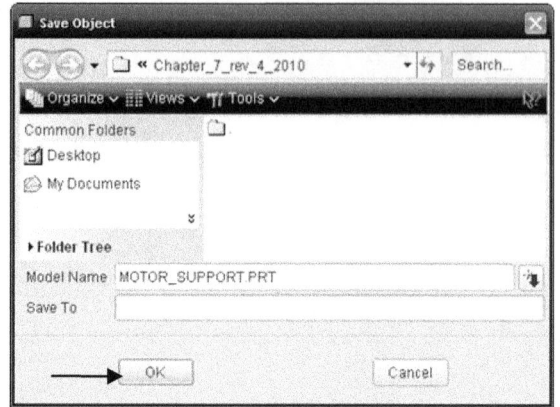

Creation of the Motor Component

Step 1: Create a 3D solid model of the motor component.

File > **New** > **Part** > type *motor* as the file name and clear the icon of **Use default template** > **OK**.

Select mmns_part_solid (units: Millimeter, Newton, Second) and type *motor* in **DESCRIPTION**, and s*tudent* in **MODELED_BY**, then **OK.**

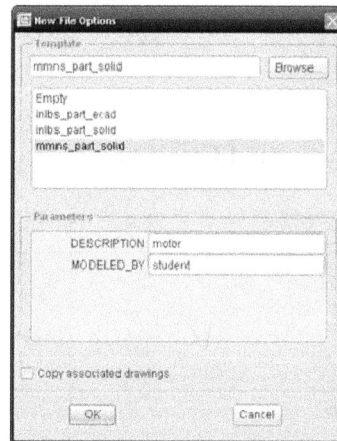

Directly select the icon of **Extrude.** From the dashboard, set the thickness value to 56. Activate **Placement** to define a sketch plane > **Define.**

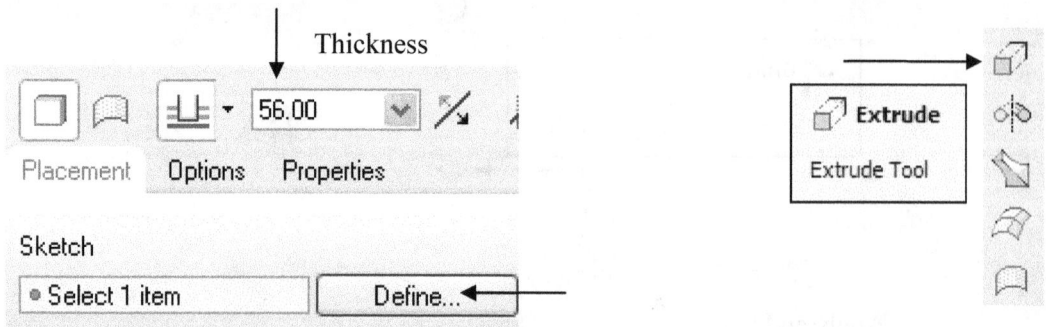

Select the **TOP** datum plane as the sketch plane, and accept the default orientation.

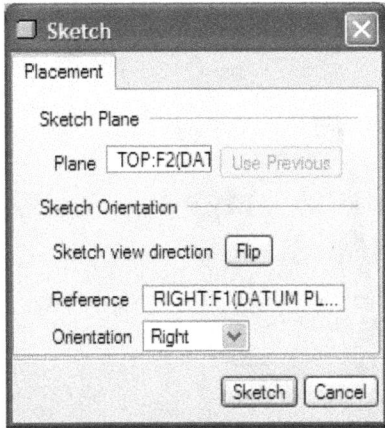

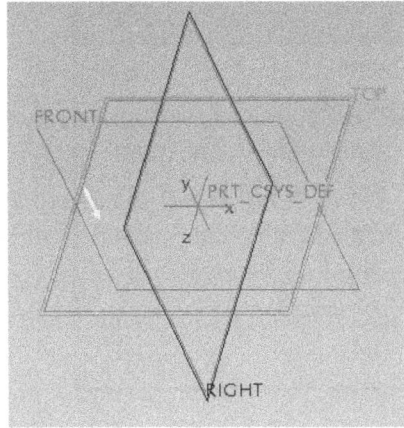

Click the icon of **Circle** to sketch a circle. The diameter value is 36, as shown.

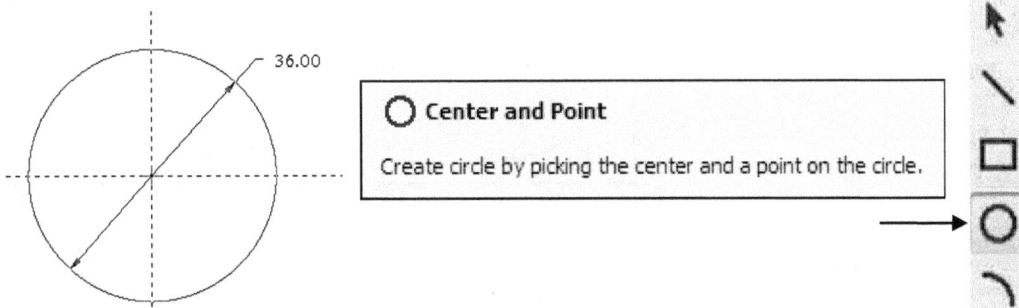

Upon completing this sketch, click the icon of **Done**. From the feature control panel, click the icon of **Applies and Saves**.

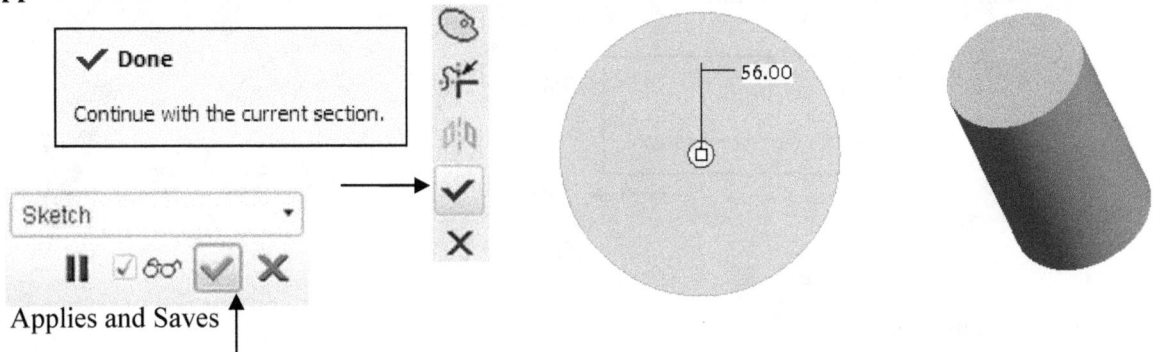

Applies and Saves

Step 2: Create a shoulder on the bottom surface of the created cylinder.

Select the icon of **Extrude**. From the dashboard, set the thickness value to 6. Activate **Placement** to define a sketch plane > **Define.**

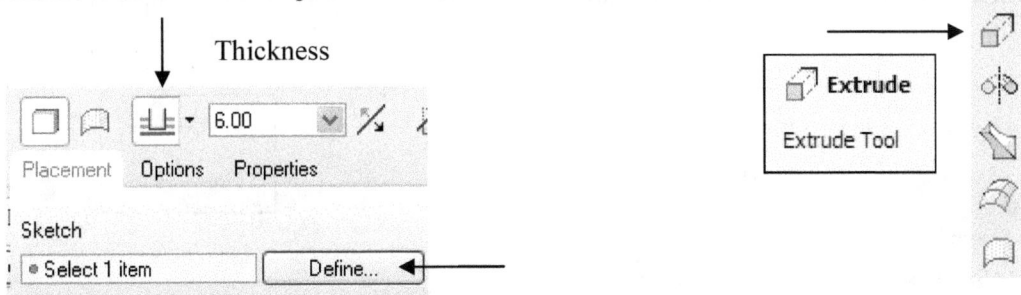

Select the bottom surface of the cylinder as the sketch plane, and accept the default orientation.

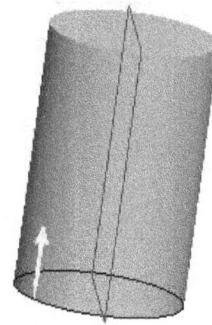

Click the icon of **Circle** to sketch a circle. The diameter value is 12, as shown.

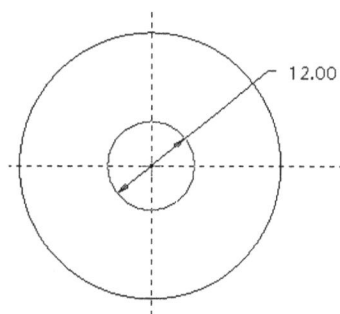

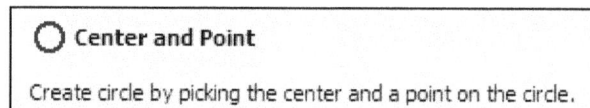

12.00

○ **Center and Point**

Create circle by picking the center and a point on the circle.

Upon completing this sketch, click the icon of **Done**. From the feature control panel, click the icon of **Applies and Saves**.

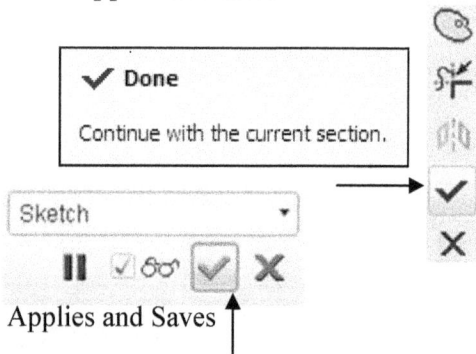

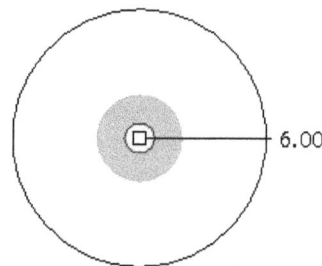

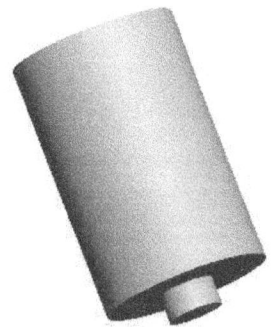

✔ **Done**

Continue with the current section.

Sketch

Applies and Saves

6.00

Step 3: Create the shaft of the motor.

Select the icon of **Extrude.** From the dashboard, set the thickness value to 16. Activate **Placement** to define a sketch plane > **Define.**

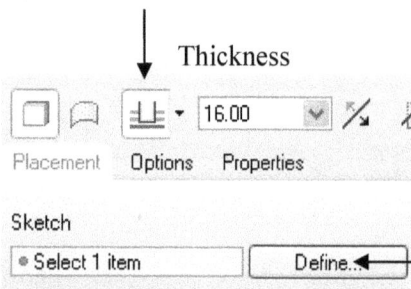

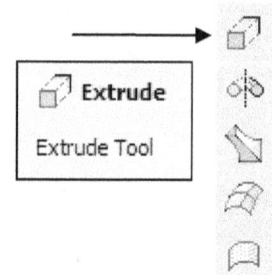

Thickness

16.00

Placement   Options   Properties

Sketch

• Select 1 item   Define..

⌐ **Extrude**

Extrude Tool

Select the bottom surface of the shoulder as the sketch plane, and accept the default orientation.

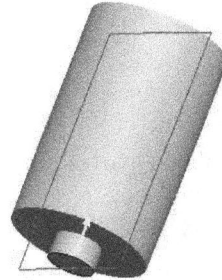

Click the icon of **Circle** to sketch a circle. The diameter value is 3, as shown.

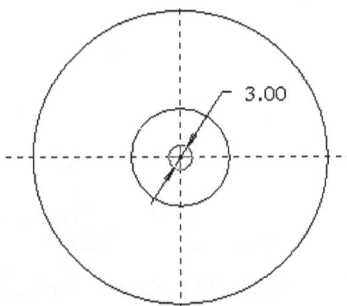

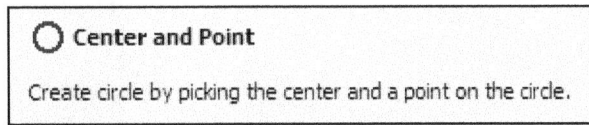

Upon completing this sketch, click the icon of **Done**. From the feature control panel, click the icon of **Applies and Saves**.

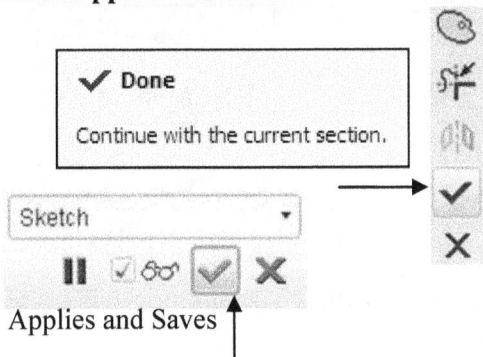

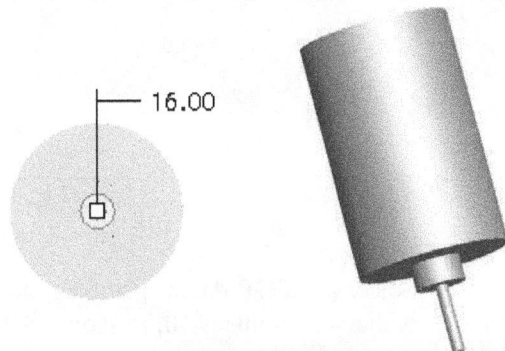

Now let us save the motor component. Click the icon of **Save the active object > OK**.

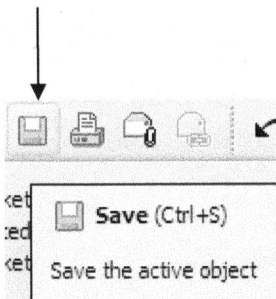

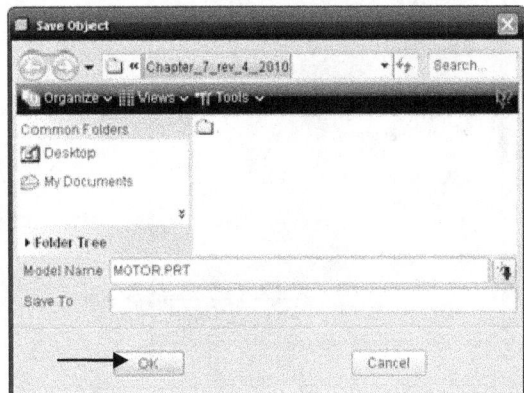

Creation of the Blade Component for an Axial Fan

Step 1: Create a 3D solid model of the blade component.
   File > **New** > **Part** > type *blade* as the file name and clear the icon of **Use default template** > **OK**.

   Select mmns_part_solid (units: Millimeter, Newton, Second) and type *blade* in **DESCRIPTION**, and s*tudent* in **MODELED_BY**, then **OK.**

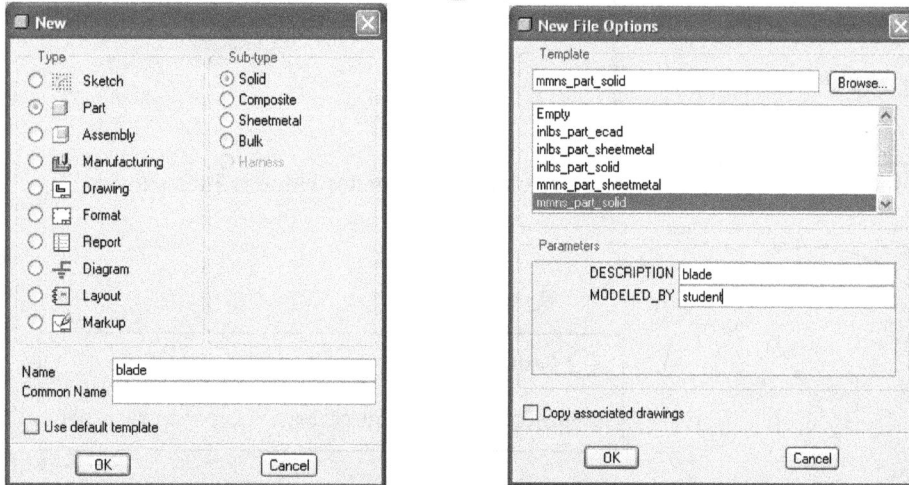

   Directly select the icon of **Extrude.** From the dashboard, use **Symmetry** and set the depth value to 1. Activate **Placement** to define a sketch plane > **Define.**

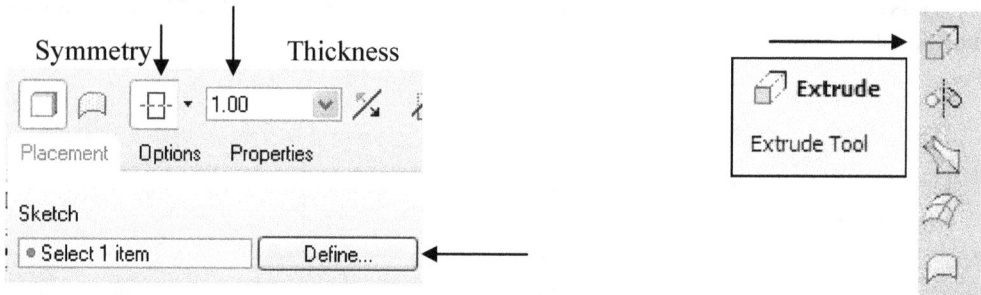

   Select the **TOP** datum plane as the sketch plane, and accept the default orientation. Sketch a circle. Its diameter value is 10, as shown below.

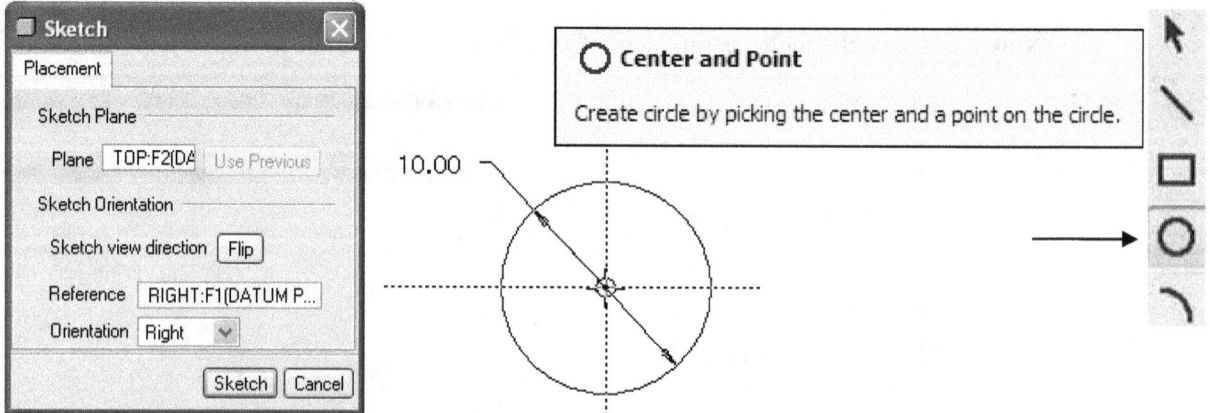

   Pick the sketched circle, right-click and hold, select **Construction** so that the circle in the solid line is converted to a circle in the dashed line.

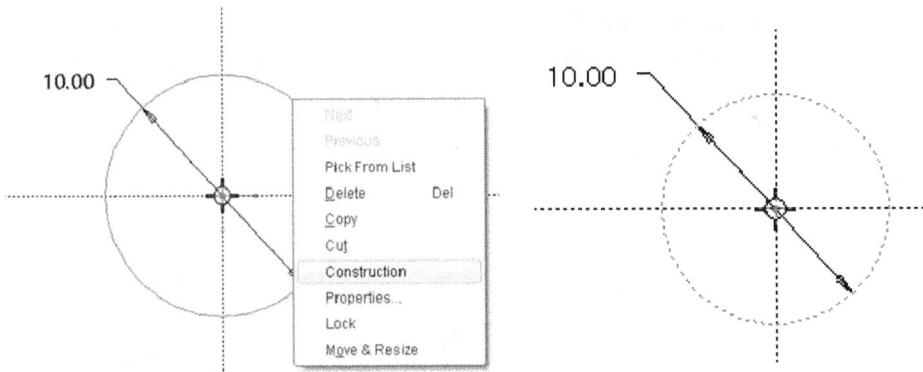

Click the icon of **Point** to define a point on the construction circle, as shown.

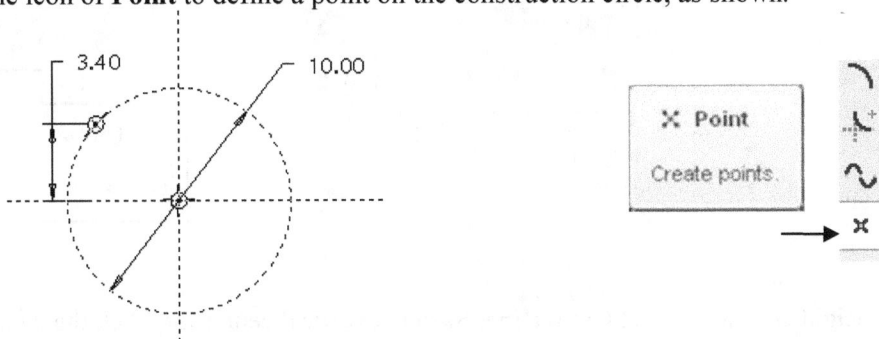

Follow the same procedure to define 4 more points, as shown.

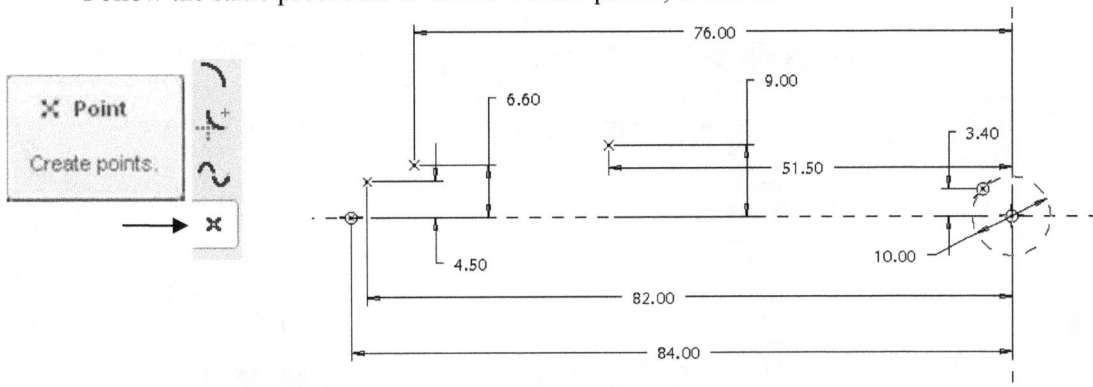

	point 1	point 2	point 3	point 4
x coordinate	51.5	76.0	82.0	84.0
z coordinate	9.0	6.6	4.5	0.0

Select the icon of **Spline** to sketch a curve by connecting the 5 points defined.

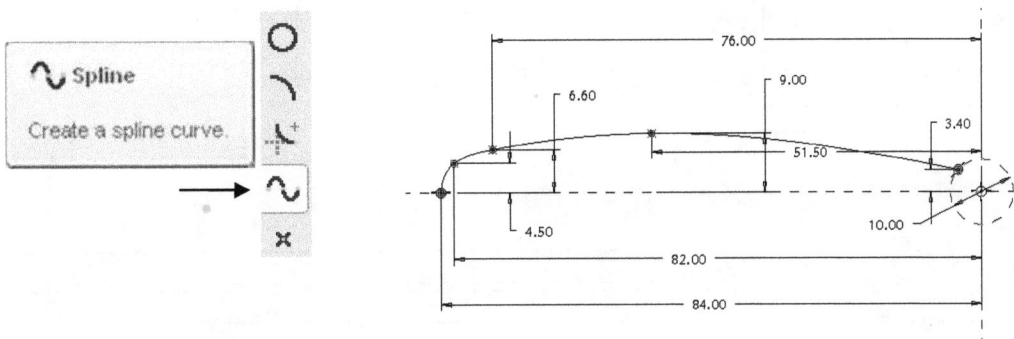

Right click and select **Centerline**. Sketch a horizontal centerline. Pick the sketched curve. Click the icon of **Mirror**, and pick the horizontal centerline to obtain the second curve. Sketch a vertical line to form a closed sketch.

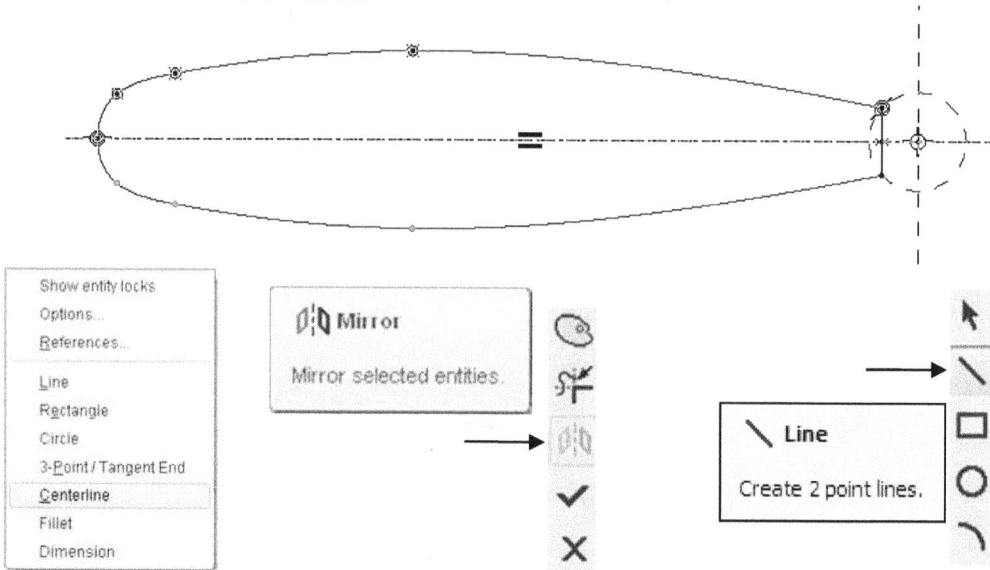

Right click and select **Centerline**. Sketch a vertical centerline. Pick the 2 curves and the vertical line. Click the icon of **Mirror**, and pick the vertical centerline to obtain the mapped curves and line.

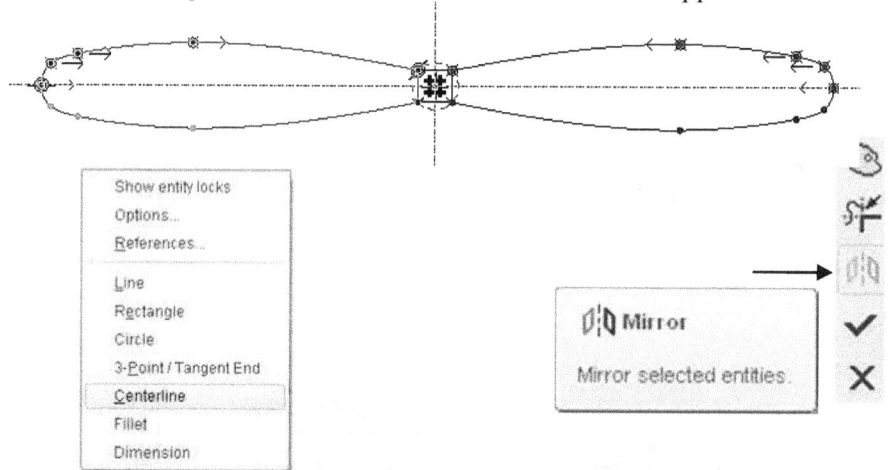

Upon completing this sketch, click the icon of **Done**. From the feature control panel, click the icon of **Apply and Save**.

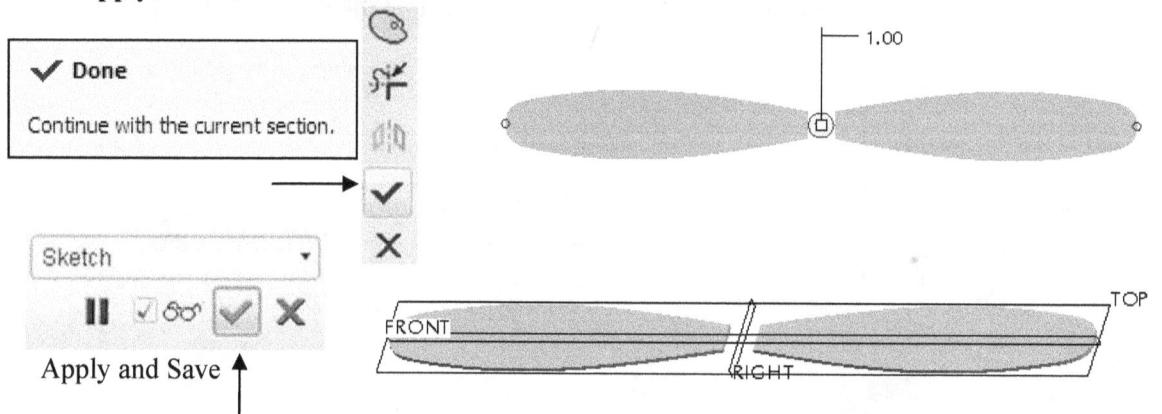

To create the twist feature, we use the function called **Warp**. Let us twist the right side of the blade first. To do so, from the main menu, select **Insert** > **Warp** > **References** > pick the created model. For **Direction**, pick the **RIGHT** datum plane. Click the icon of **Twist**. Specify 30 as the twist angle. Click the icon of **Apply and Save**.

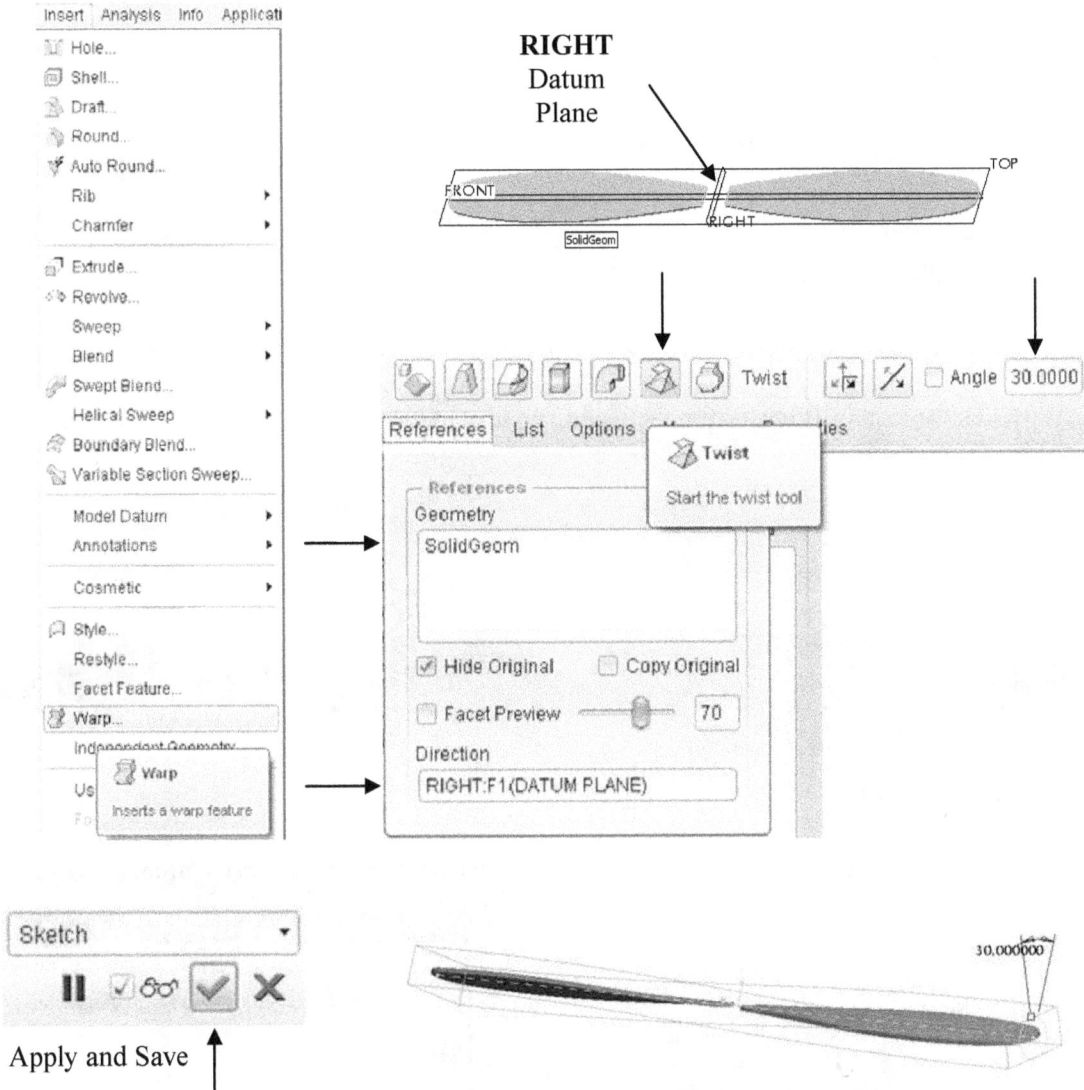

Now let us create the central part. Directly select the icon of **Extrude**. From the dashboard, use **Symmetry** and set the depth value to 10. Activate **Placement** to define a sketch plane > **Define.**

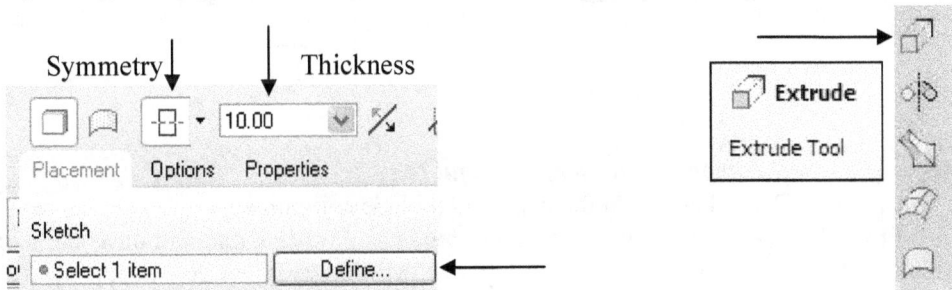

Select the **TOP** datum plane as the sketch plane, and accept the default orientation. Sketch 2 circles. Their diameter values are 10 and 3, respectively, as shown below.

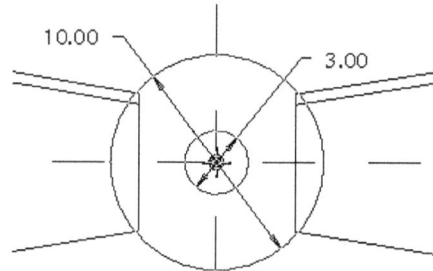

Click the icon of **Done**. From the feature control panel, click the icon of **Apply and Save.**

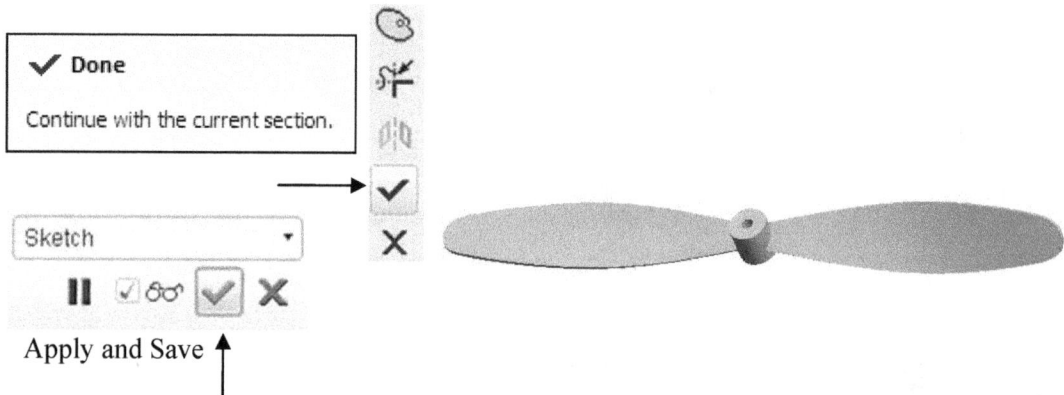

Apply and Save

Now let us save the blade component. Click the icon of **Save the active object** > **OK**.

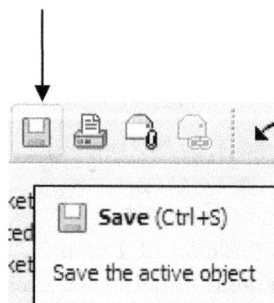

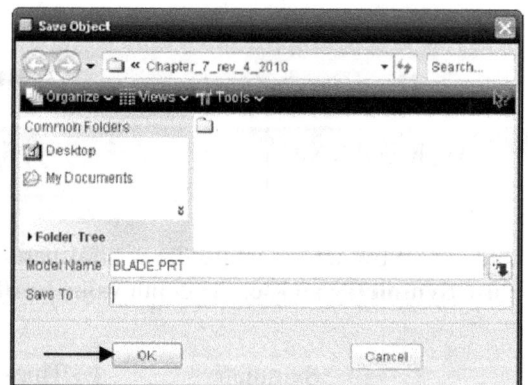

Creation of a wheel component for a centrifugal fan
Step 1: Create a 3D solid model of the centrifugal blade component.
    File > **New** > **Part** > type *centrifugal_blade* as the file name and clear the icon of **Use default template** > **OK**. Select mmns_part_solid (units: Millimeter, Newton, Second) and type *centrifugal blade* in **DESCRIPTION**, and *student* in **MODELED_BY**, then **OK**.

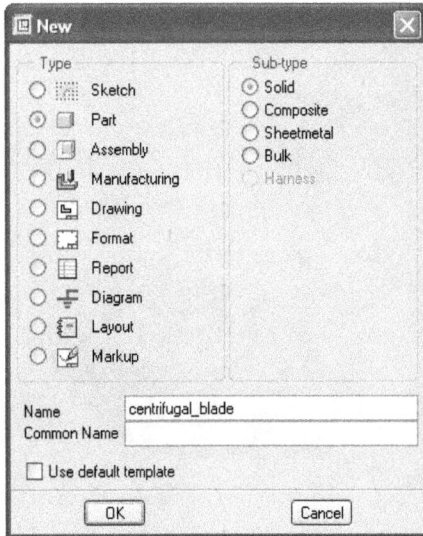

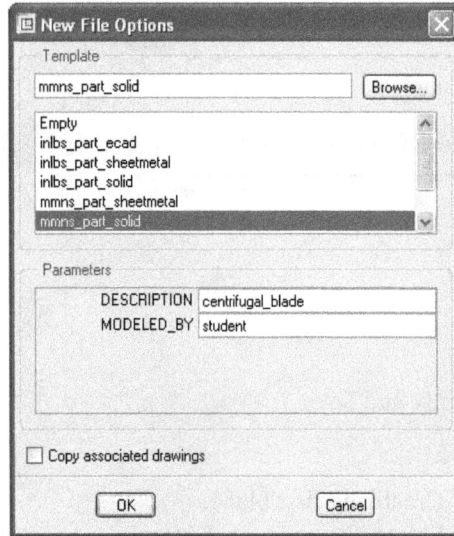

Select the icon of **Extrude.** From the dashboard, set the thickness value to 1. Activate **Placement** to define a sketch plane > **Define.**

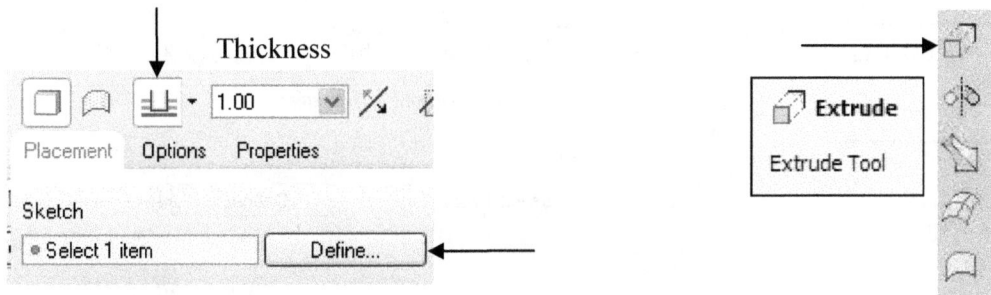

Select the **TOP** datum plane as the sketch plane, and accept the default orientation. Sketch a circle. Its diameter value is 168, as shown below.

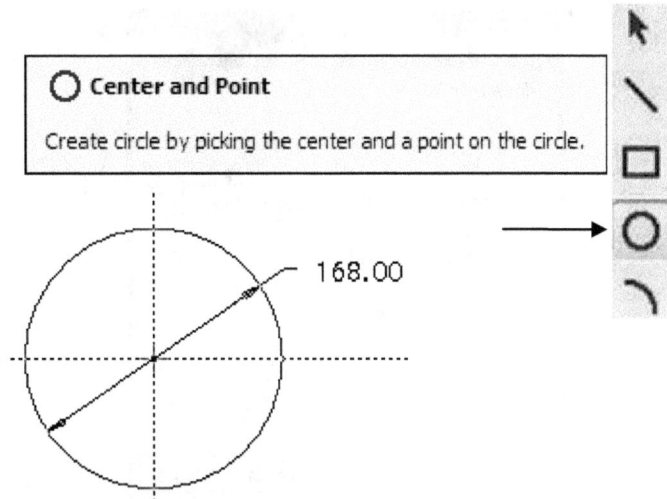

Upon completing this sketch, click the icon of **Done.** From the feature control panel, click the icon of **Apply and Save**.

Done

Continue with the current section.

Apply and Save

Sketch

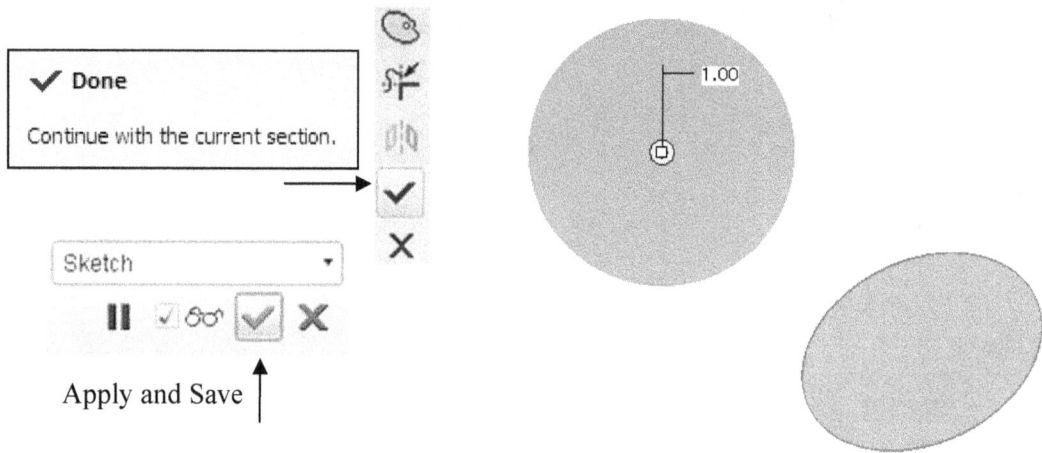

Step 2: Create the first blade.
Select the icon of **Extrude.** From the dashboard, set the thickness value to 15. Activate **Placement** to define a sketch plane > **Define.**

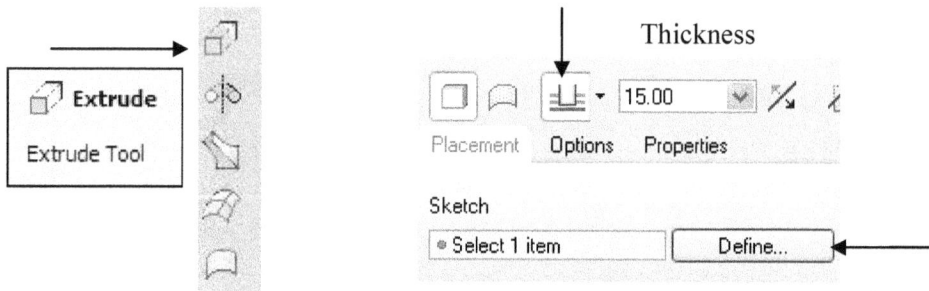

Extrude

Extrude Tool

Thickness

15.00

Placement   Options   Properties

Sketch

Select 1 item   Define...

Select the top surface of the cylindrical plate as the sketch plane, and accept the default orientation.

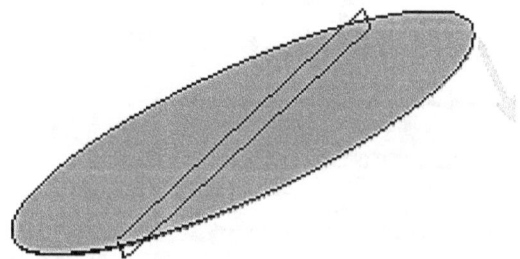

Sketch

Placement

Sketch Plane

Plane   Surf:F5(EX   Use Previous

Sketch Orientation

Sketch view direction   Flip

Reference   RIGHT:F1(DATUM P...

Orientation   Right

Sketch   Cancel

Click the icon of **Line.** Sketch a line inclined at angle equal to 50° with respect to the horizontal axis, as shown.

Line

Create 2 point lines.

50.00

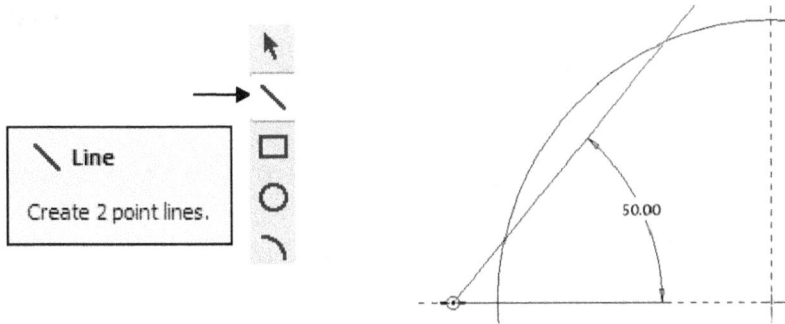

Pick the sketched line > right-click and hold, select Construction, as illustrated below. The solid line is converted to a dashed line.

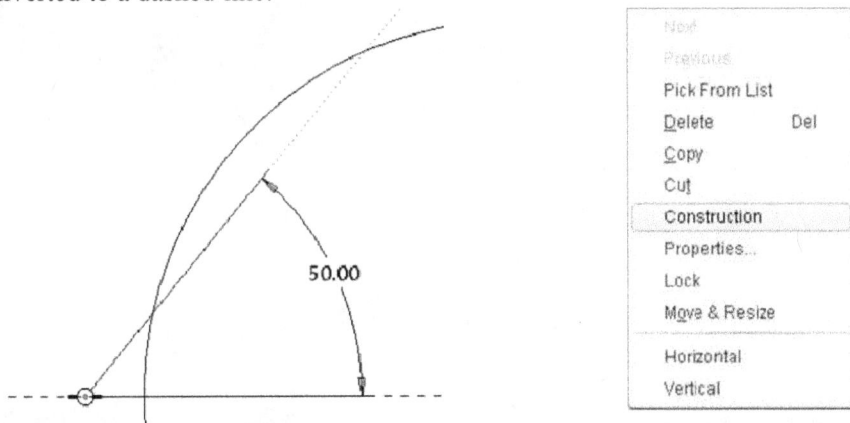

50.00

Next
Previous
Pick From List
Delete   Del
Copy
Cut
Construction
Properties...
Lock
Move & Resize

Horizontal
Vertical

Click the icon of **Point**, define a point at the intersection of the construction line and the circle. Specify the distance of 10 with respect to the horizontal axis.

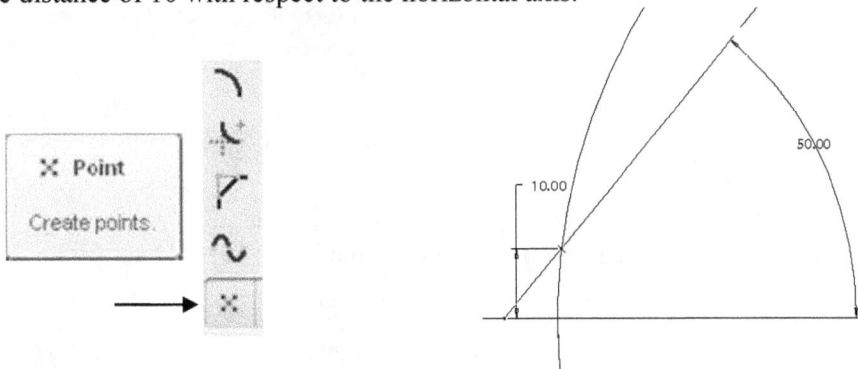

✕ **Point**

Create points.

50.00

10.00

Click the icon of **Line**. Sketch a rectangle, as shown. The 2 dimensions are 30 and 2, as shown.

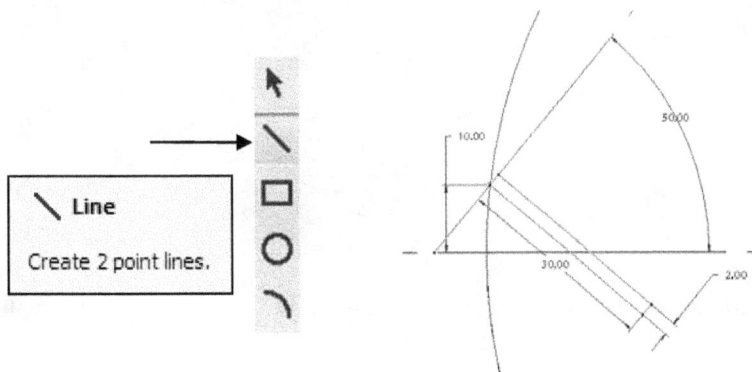

Line

Create 2 point lines.

10.00

50.00

30.00

2.00

Upon completing this sketch, click the icon of **Done**. From the feature control panel, click the icon of **Apply and Save**.

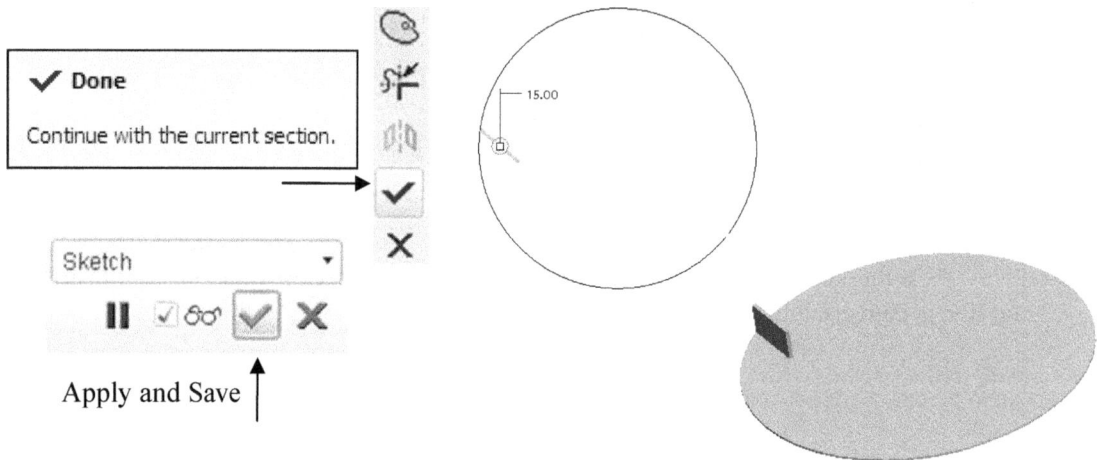

Apply and Save

Step 3: Use **Pattern** to create the other 11 blads.
From the model tree, highlight the first blade feature > right click and pick **Pattern** > **Axis**.

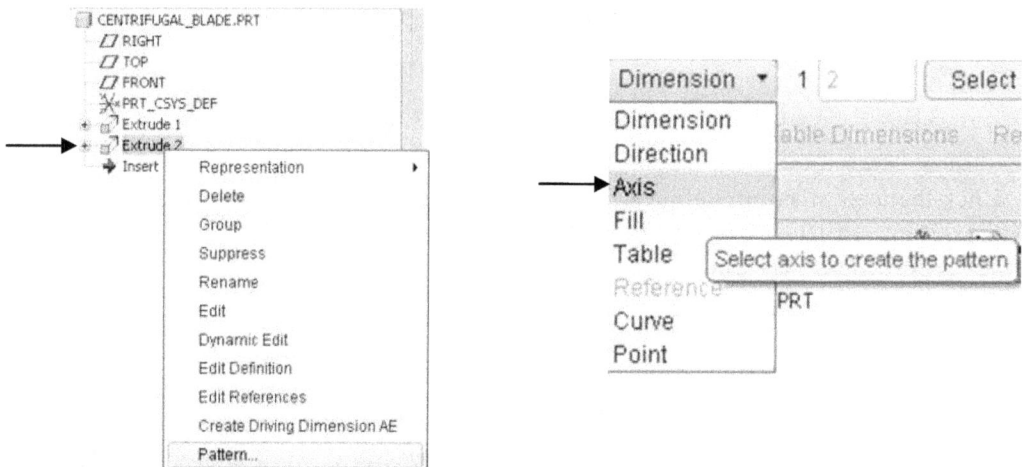

Pick the axis on display > type 12 and 30 > click the icon of **Complete**.

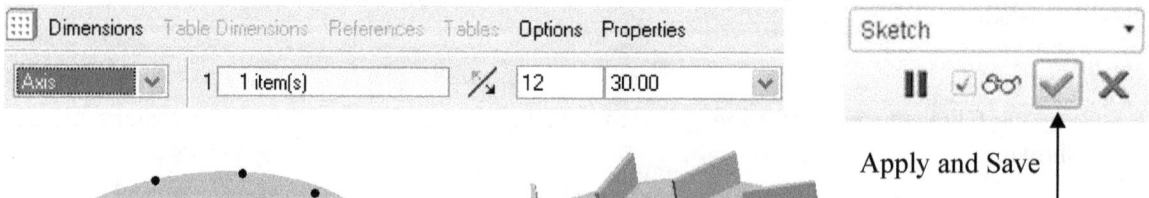

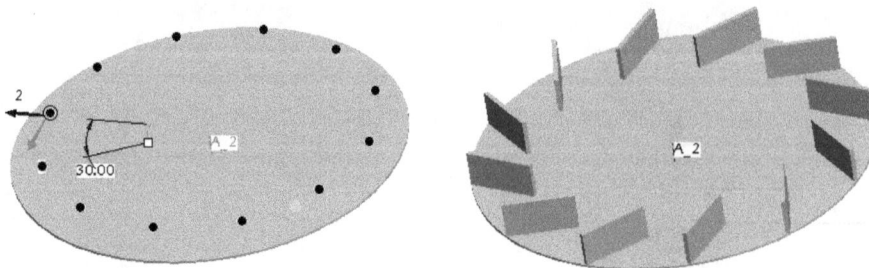

Apply and Save

Step 4: Add the top plate.
Select the icon of **Extrude**. From the dashboard, set the thickness value to 1. Activate **Placement** to define a sketch plane > **Define.**

Thickness

Extrude

Extrude Tool

Placement   Options   Properties

Sketch

Select 1 item   Define...

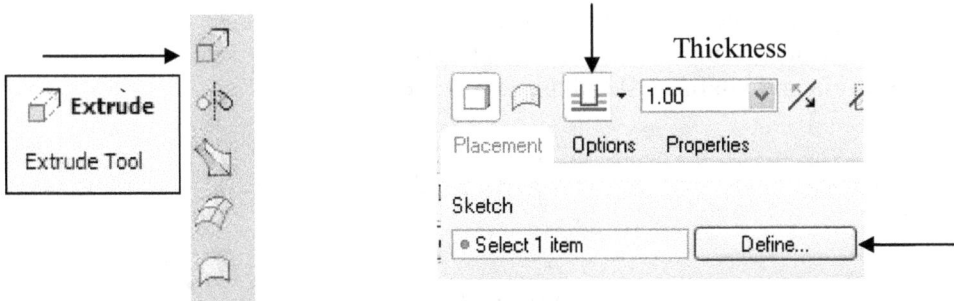

Select the top surface of the blades as the sketch plane, and accept the default orientation.

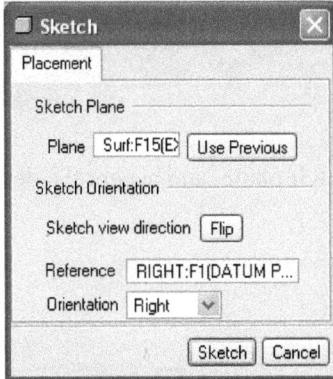

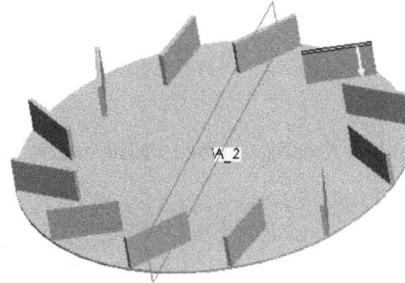

Sketch

Placement

Sketch Plane

Plane Surf:F15(E) Use Previous

Sketch Orientation

Sketch view direction Flip

Reference RIGHT:F1(DATUM P...

Orientation Right

Sketch   Cancel

Sketch a circle. Its diameter value is 168, as shown below.

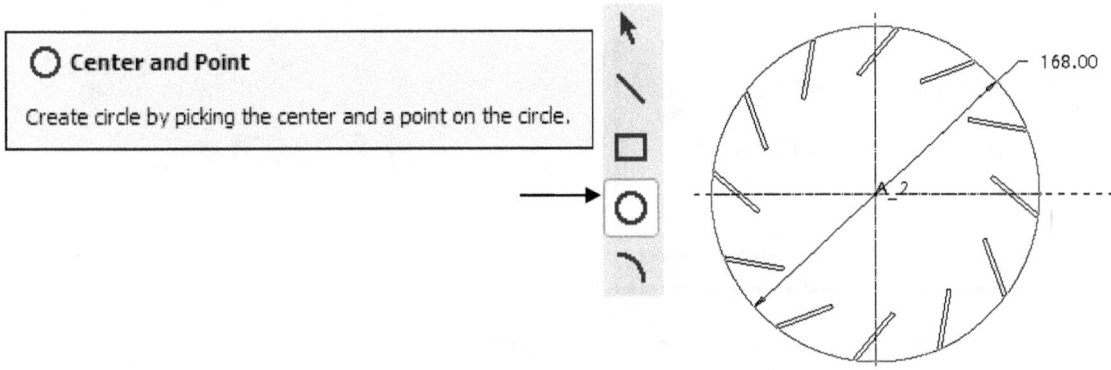

○ **Center and Point**

Create circle by picking the center and a point on the circle.

168.00

A_2

Upon completing this sketch, click the icon of **Done**. From the feature control panel, click the icon of **Apply and Save**.

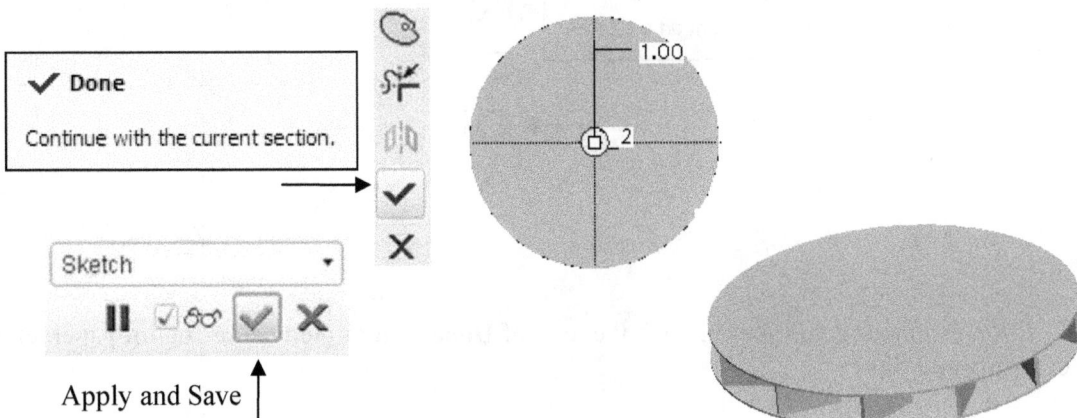

✓ **Done**

Continue with the current section.

1.00

2

Sketch

Apply and Save

Step 5: Create a through-all hole at the central location.
  Select the icon of **Extrude.** From the dashboard, select **Cut** > select **Thru All** as the depth choice. Activate **Placement** to define a sketch plane > **Define.**

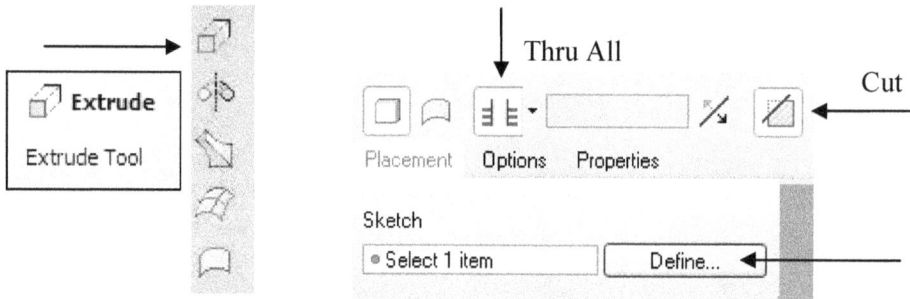

Select the top surface of the second blade as the sketch plane, and accept the default orientation.

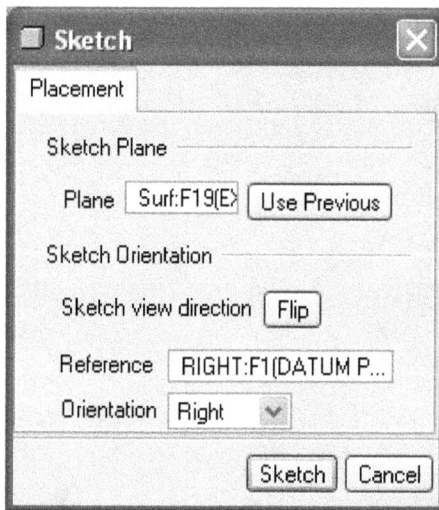

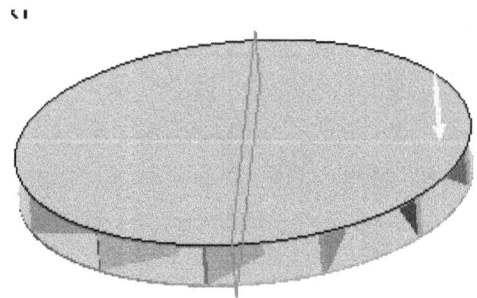

Sketch a circle. Its diameter value is 3, as shown below.

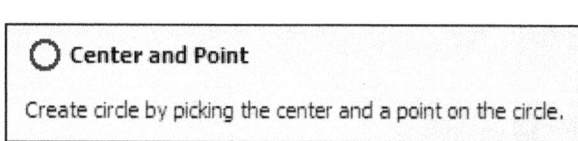

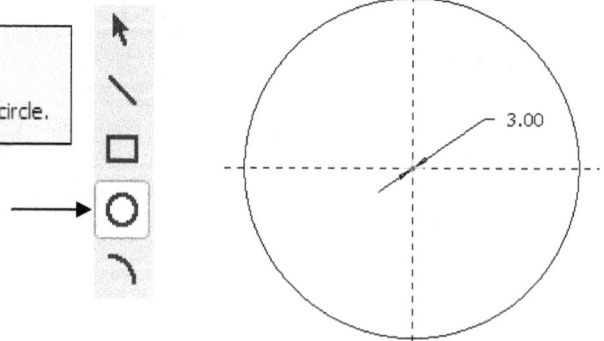

  Upon completing this sketch, click the icon of **Done**. From the feature control panel, click the icon of **Apply and Save**.

✓ **Done**

Continue with the current section.

✓

✗

Sketch ▾

❚❚ ☑ 👓 ✓ ✗

Apply and Save ↑

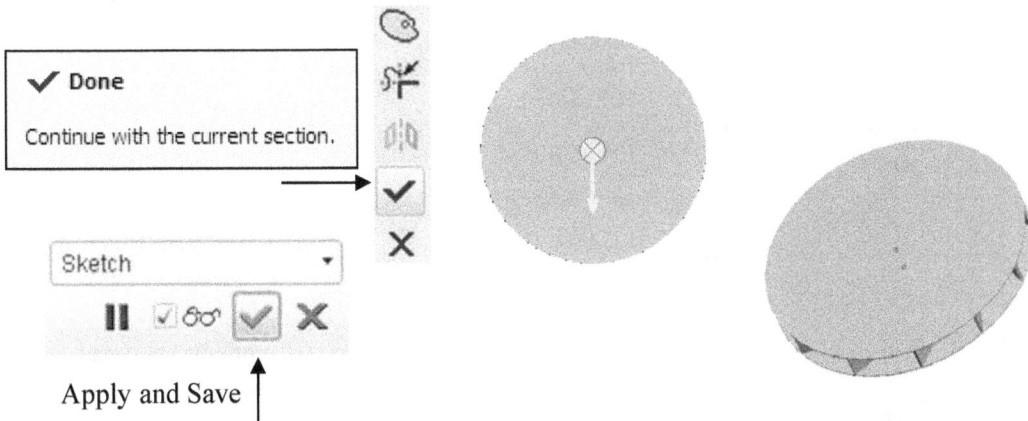

Now let us save the centrifugal blade component. Click the icon of **Save the active object** > **OK**.

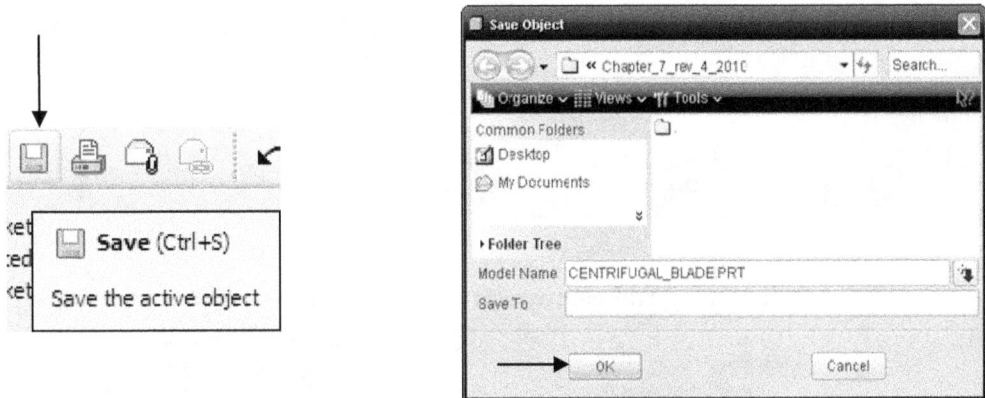

💾 Save (Ctrl+S)

Save the active object

Save Object

Chapter_7_rev_4_2010

Organize ∨ Views ∨ Tools ∨

Common Folders
Desktop
My Documents

▸ Folder Tree

Model Name CENTRIFUGAL_BLADE.PRT

Save To

OK   Cancel

Creation of the Assembly of the Lift Fan System

Step 1: Create a 3D solid model of the assembly of the base structural system.

File > New > Assembly > type *lift_fan_system* as the file name and clear the icon of **Use default template > OK**.

Select mmns_asm_design (units: Millimeter, Newton, Second) and type *lift_fan_system* in **DESCRIPTION**, and s*tudent* in **MODELED_BY**, then **OK.**

Click the icon of **Add component** > highlight holding_bar.prt > **Open**.

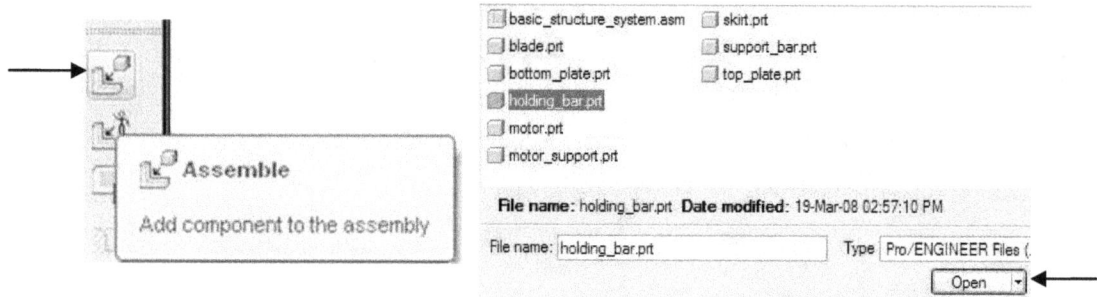

Select **Default** > click the icon of **Apply and Save** from the constraint control panel.

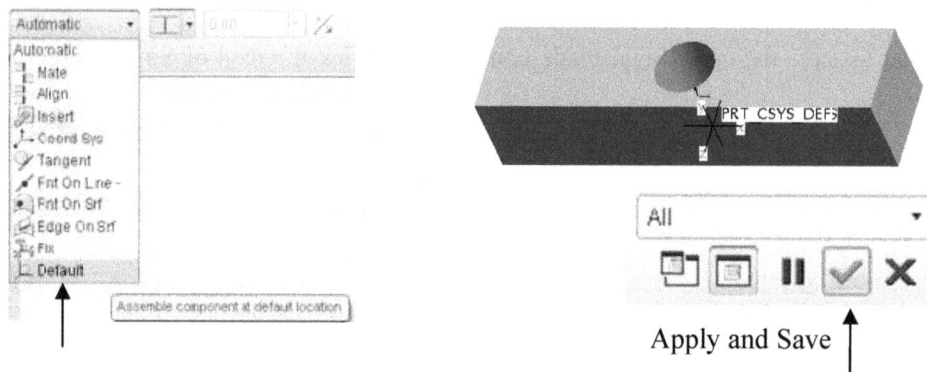

Apply and Save

Click the icon of **Add component** > highlight motor_support.prt > **Open**.

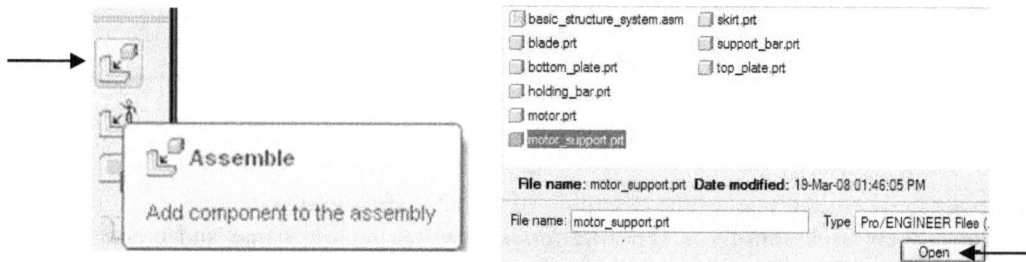

Click the icon of **Placement**. Let us define the first constraint. Select the top surface from the motor_support component and select the bottom surface from the holding_bar component. A constraint called **Mate** has been defined. Click **New Constraint**.

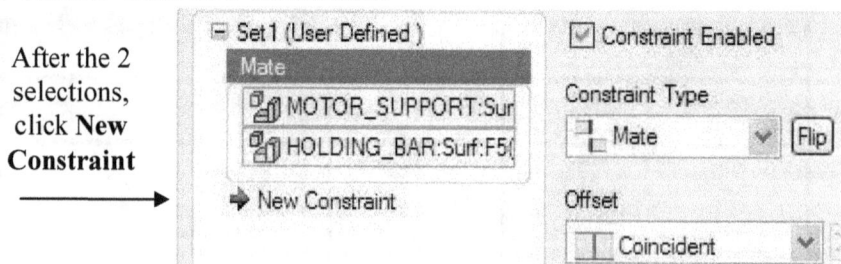

After the 2 selections, click **New Constraint**

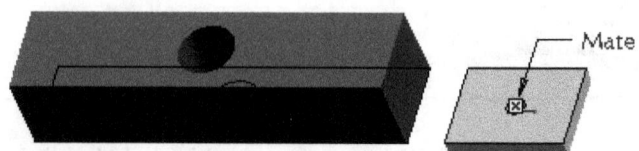

Let us define the second constraint. Select the axis from the motor_support component and the axis from the holding_bar component. Use **Coincident**. A new constraint called **Align** has been defined. At this moment, the support_bar component is fully constrained. Click the icon of **Apply and Save** from the constraint control panel.

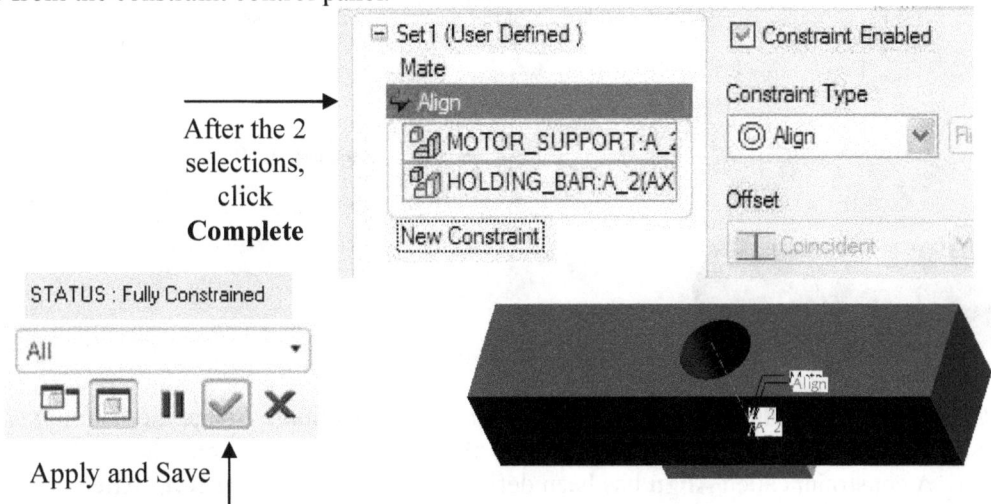

After the 2 selections, click **Complete**

Apply and Save

Now let us assemble the motor component on the other side. Click the icon of **Add component** > highlight motor.prt > **Open**.

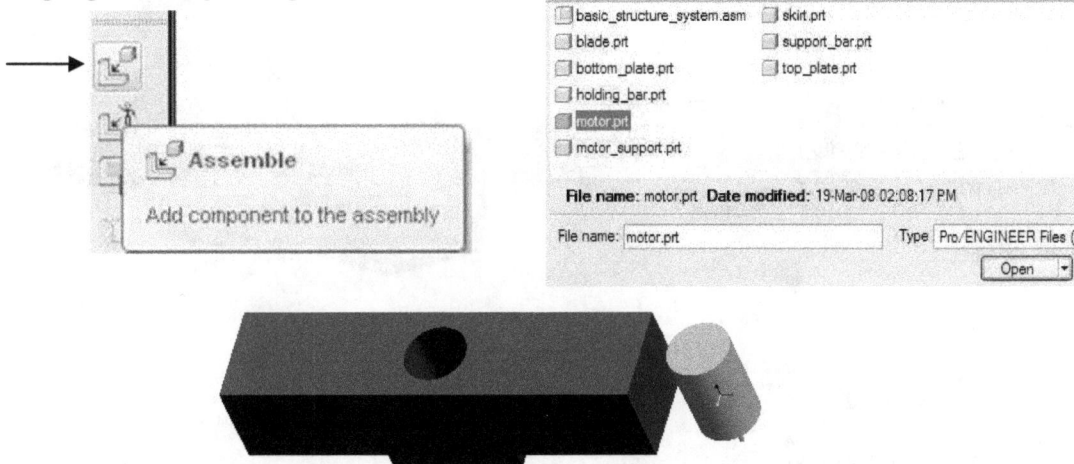

We follow the same procedure to insert 2 constraints as we did in the process of assembling the motor support component.

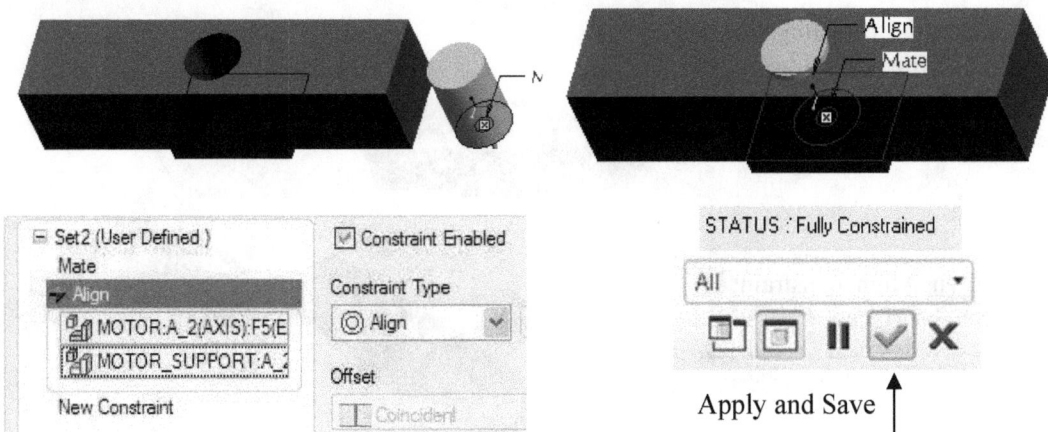

Apply and Save

Now let us assemble the blade component. Click the icon of **Add component** > highlight blade.prt > **Open**.

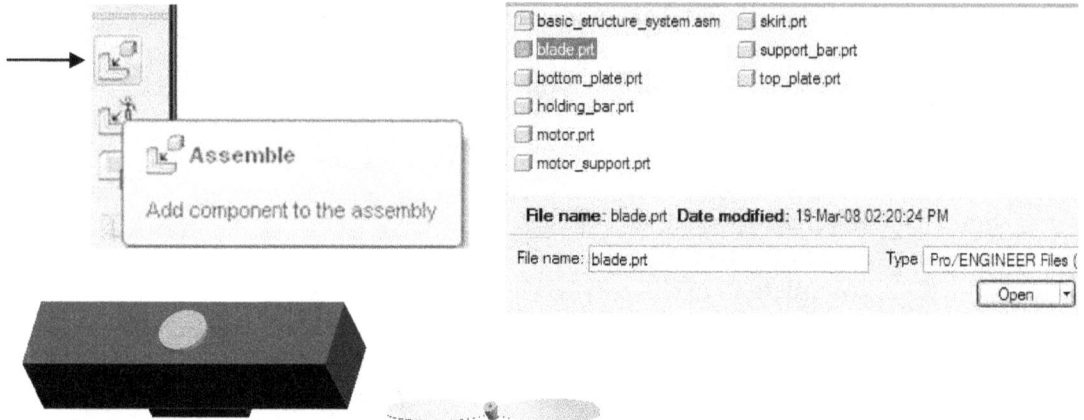

Click the icon of **Placement**. Let us define the first constraint. Select the bottom surface from the cylindrical part of the blade component and the bottom surface from the shaft of the motor component. A constraint called **Align** has been defined. Click **New Constraint**. Afterwards, select the axis from the blade component and the axis from the motor component. A new **Align** constraint is defined, thus completing the assembly process for the lift fan system.

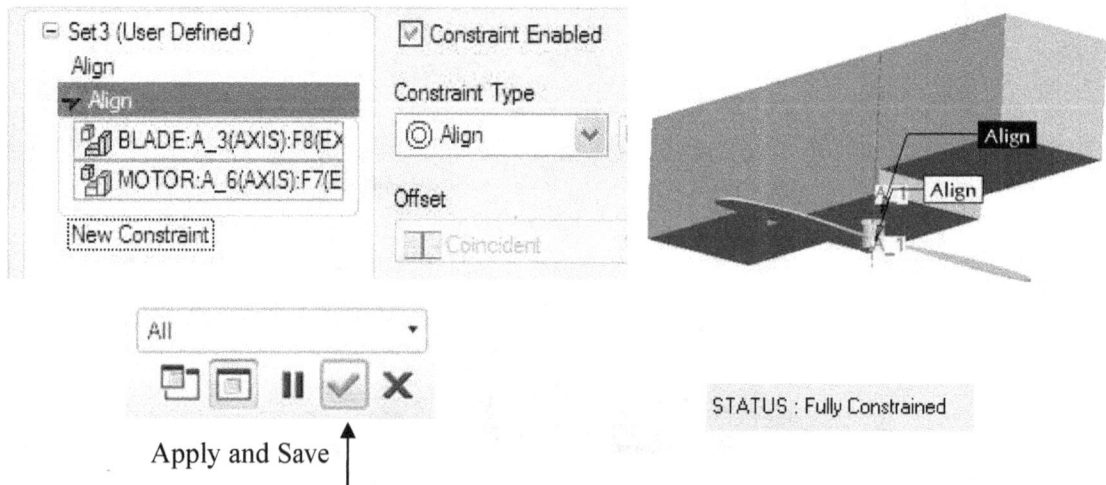

Apply and Save

If a user wants to assemble the centrifugal plate, he or she may complete the following:

Open the centrifugal blade file.

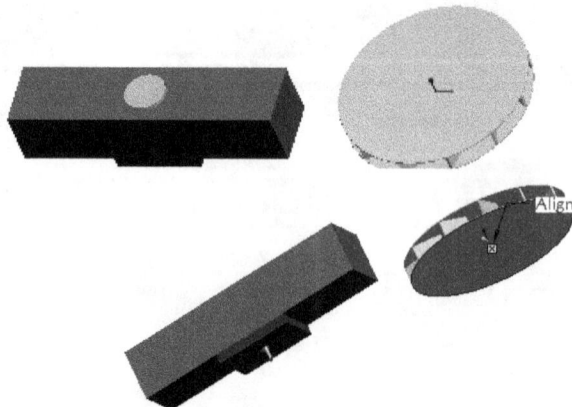

Define an Align constraint between 2 surfaces.

Define an Align constraint between
2 axes.

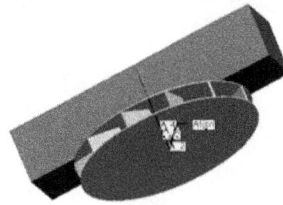

Now let us save the centrifugal blade component. Click the icon of **Save the active object** >
OK.

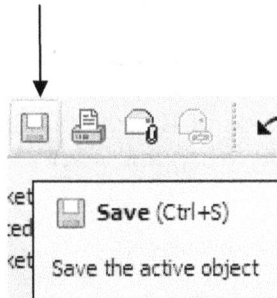

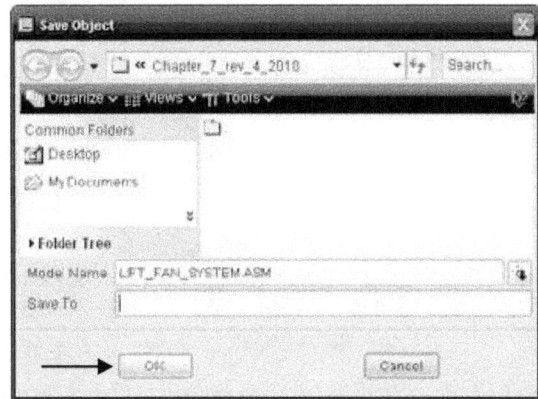

7.4 CREATION OF A PROPULSION FAN SYSTEM

Figure 7.3 is an exploded view of the propulsion fan system. As illustrated, this subsystem has a propulsion fan, a propulsion support component, and a propulsion fan base.

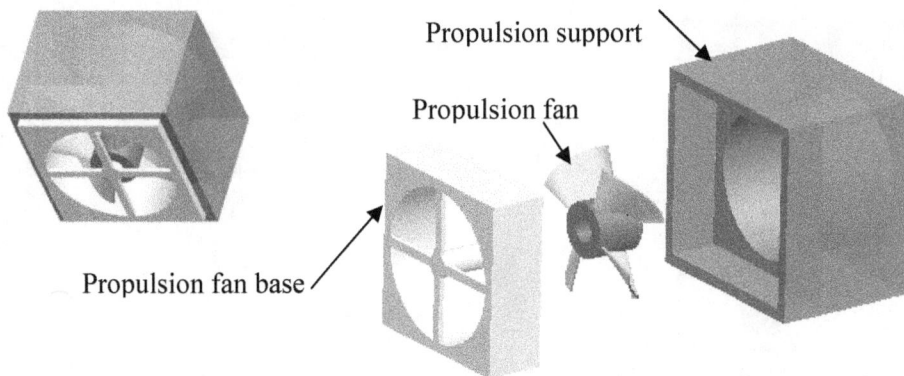

Fig. 7.3 An exploded view of the propulsion fan system showing the 3 components.

Creation of the Propulsion Fan Component

Step 1: Create a 3D solid model of the propulsion fan component.
      File > **New** > **Part** > type *propulsion_fan* as the file name and clear the icon of **Use default template** > **OK**. Select mmns_part_solid (units: Millimeter, Newton, Second) and type *propulsion_fan* in **DESCRIPTION**, and s*tudent* in **MODELED_BY**, then **OK**.

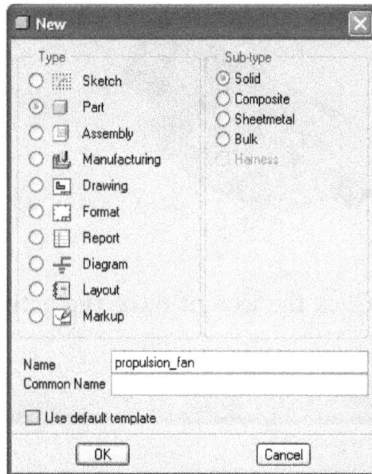

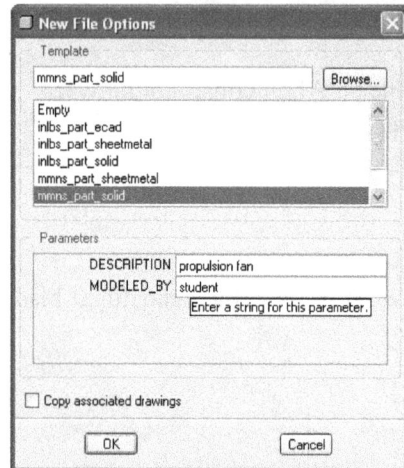

The middle part is a cylinder. Directly select the icon of **Extrude.** From the dashboard, use **Symmetry** and set the depth value to 30. Activate **Placement** to define a sketch plane > **Define.**

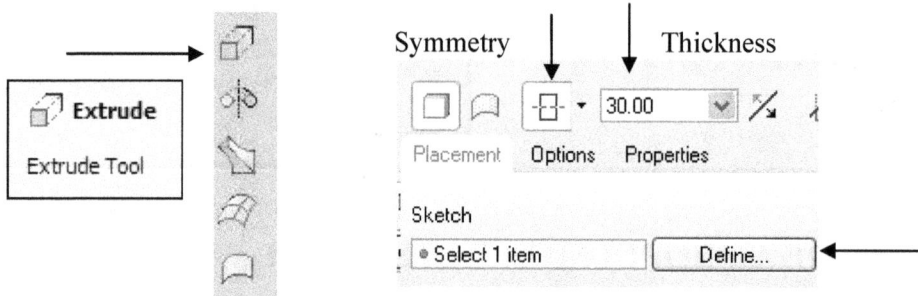

Select the **FRONT** datum plane as the sketch plane, and accept the default orientation. Sketch 2 circles. Their diameter values are 34 and 15, respectively. Click **Done** > **Apply and Save.**

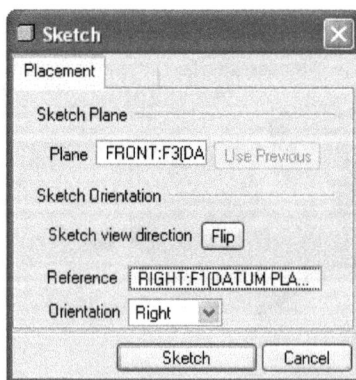

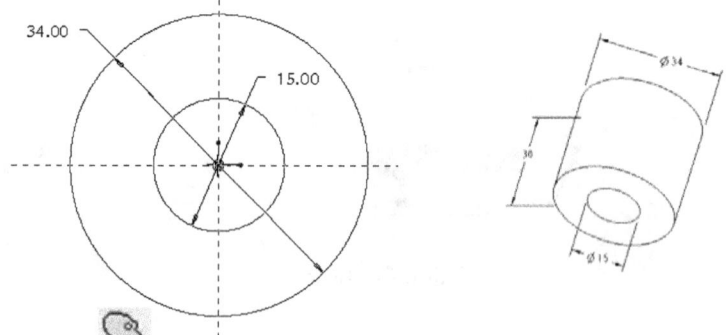

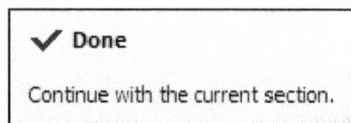

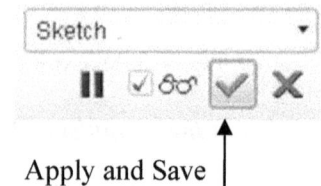

✓ **Done**

Continue with the current section.

Apply and Save

After creating the middle part, let us create the 4 blades. We will use the function called surface modeling, or **Boundary Blend** to create the blades, as shown below.

As illustrated below, a blade consists of 4 curves on the boundaries of the blade. These 4 curves form a closed form. Among the 4 curves, two are straight lines.

The first datum curve created

The second datum curve obtained through projection

The 2 straight lines are created to form a closed form for surface creation

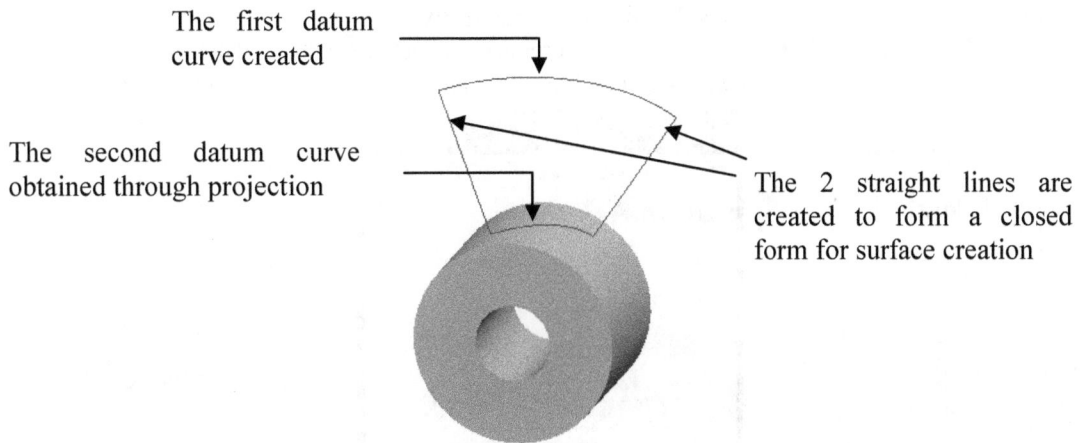

Step 2: Create 2 datum planes. The offset values are 12 and (-12) with respect to **FRONT** datum plane, respectively.

Click the icon of **Datum Plane Tool** > pick the **FRONT** datum plane > type *12* > **OK**. Repeat this process to create the second datum plane. Make sure that you select the **FRONT** Datum Plane and type *-12*.

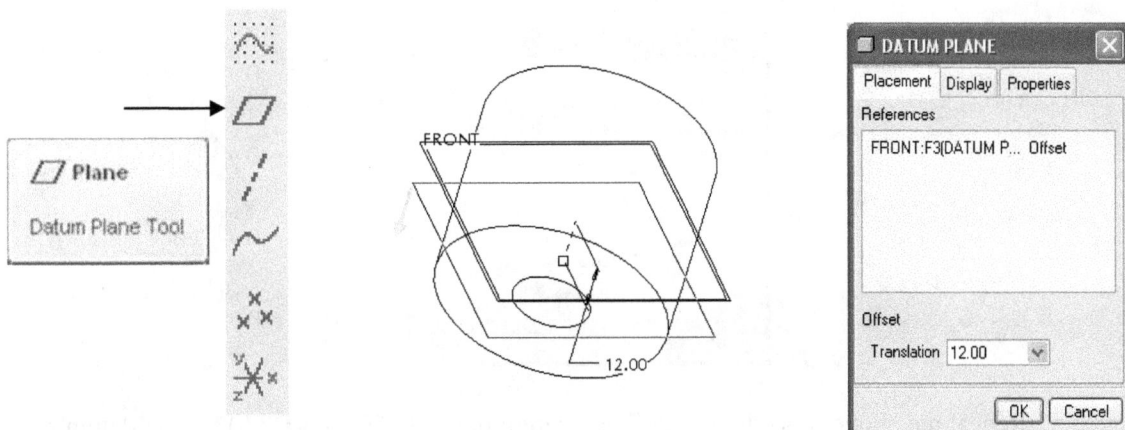

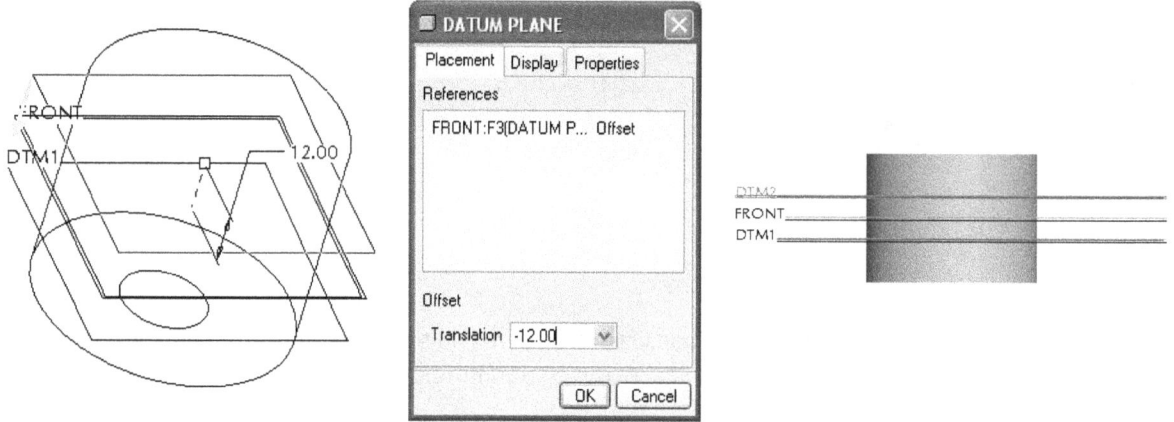

Step3: Sketch 2 circles on **DTM1** and **DTM2**, respectively.

From the toolbar of datum feature creation, select the icon of **Sketch Tool** > pick **DTM1** > **Sketch.** Click the icon of **Circle** and sketch a circle and specify *90* as the diameter value.

Repeat the above procedure to sketch a circle on **DTM2**. The diameter value is *90*.

Step 4: Define a datum point on the circle created on **DTM1** and define a datum point on the circle created on **DTM2**.

Click the icon of **Create Datum Point** > click a location on the sketch circle > change the ratio value to *0.40*, as shown > **OK**, completing the creation of **PNT0**.

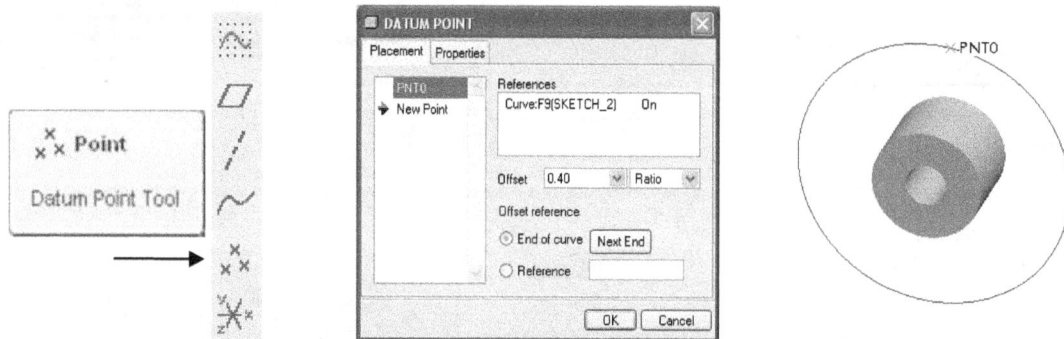

Repeat the above procedure to define a datum point, **PNT1**, on **DTM2**. The datum point is on the sketched circle. The length ratio value is 0.1, as shown.

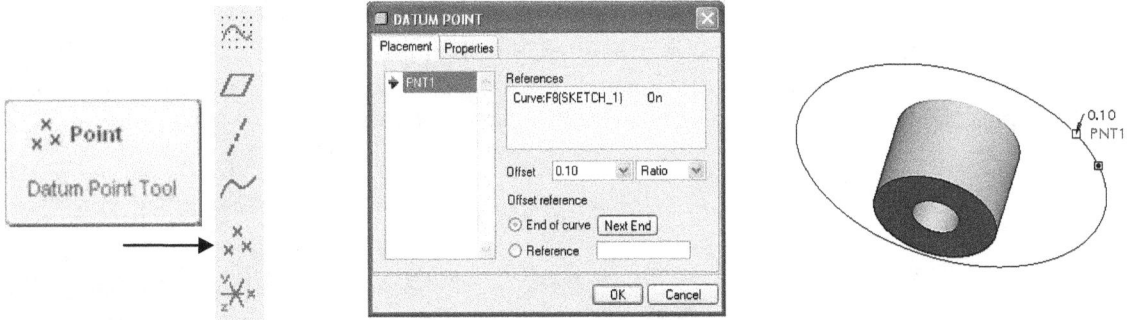

Step 5: Create a datum curve going through PNT0 and PNT1.
   Click the icon of **datum curve** > **Thru Points** > **Done**.
   Spline > Whole Array > Add Point > pick PNT1 first, then PNT0 > Done.

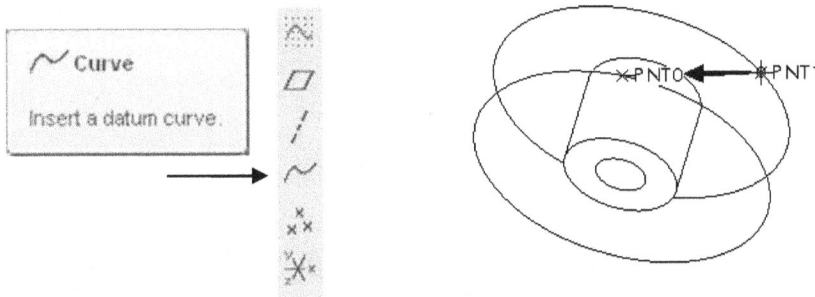

From the Curve window, click **Tangency** > **Define.** From the pop up window, **Start** > **Crv/Edge/Axis** > **Tangent** > pick the datum curve on **DTM2** > **Okay**.

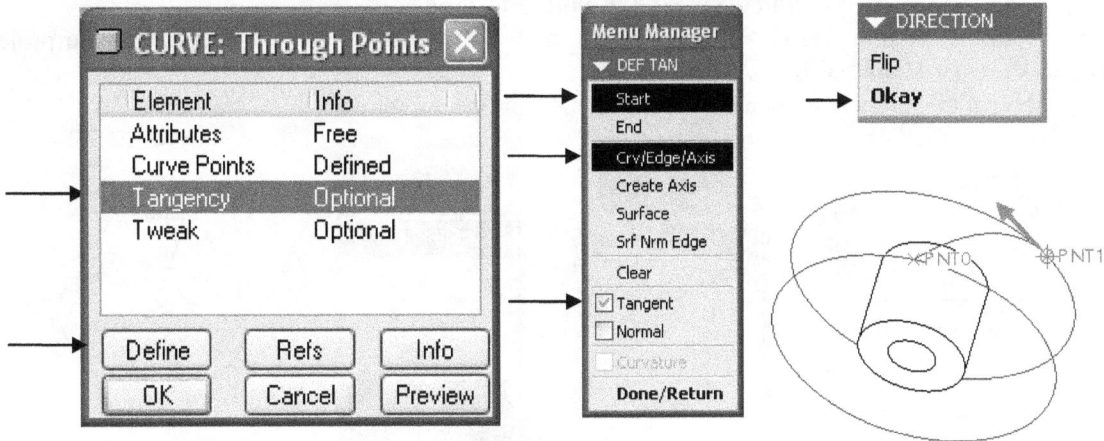

Repeat this process for the end point, **End** > **Crv/Edge/Axis** > **Tangent** > pick the datum curve on **DTM1** > **Okay** > **Done/Return** > **OK.**

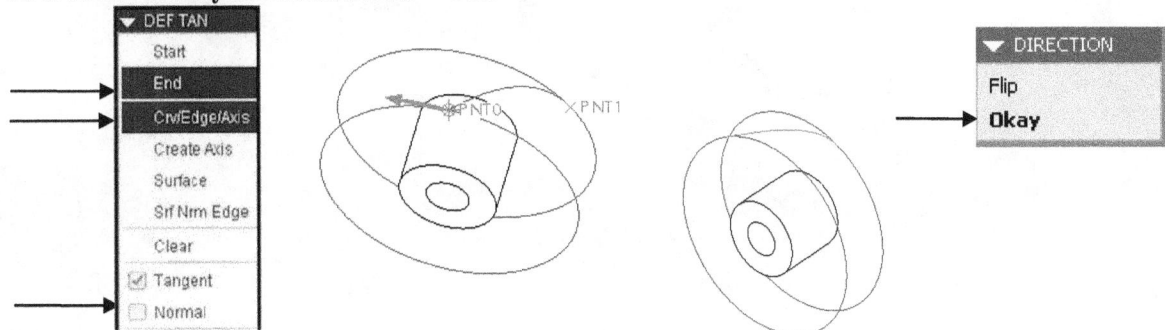

Step 6: To create the second datum curve, we use the function called **Project.**
 Highlight the datum curve just created. From the top menu, click> **Edit > Project > Normal to surface** > pick the cylindrical surface and a projected curve is created > **OK.**

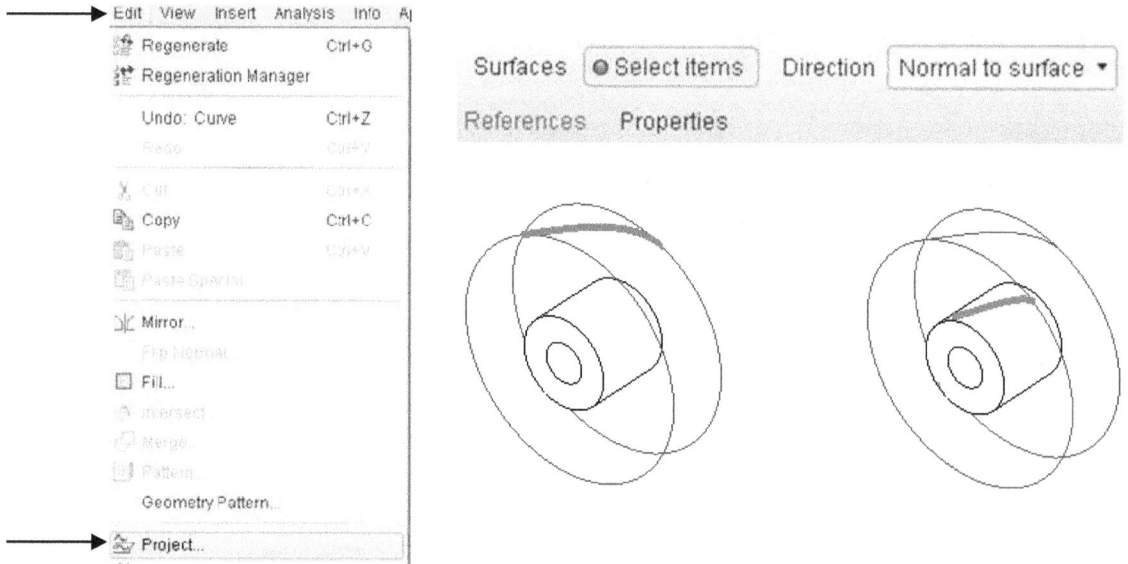

Step 7: Create the 2 straight lines, or 2 datum curves.

 Click the icon of **datum curve > Thru Points > Done.**
 Spline > Whole Array > Add Point > pick **PNT0** first, then pick the end of the projected curve, as shown > **Done > OK.**
 Repeat this process to create the second straight line, as shown.

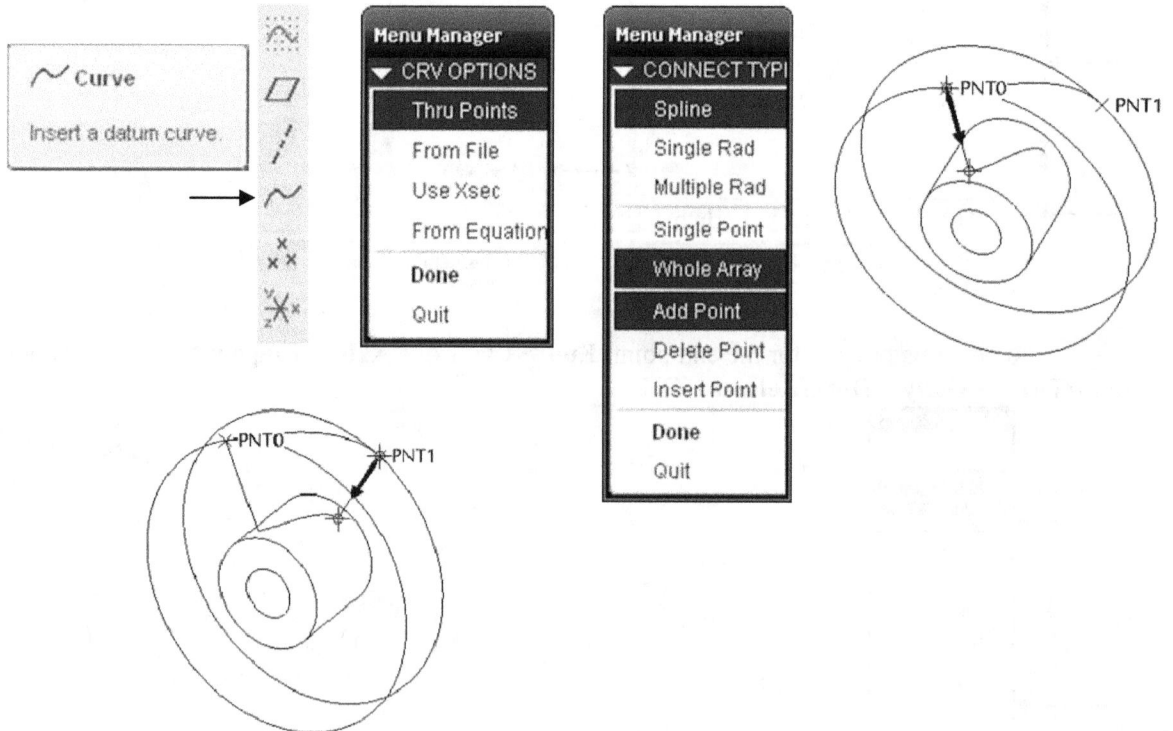

Step 8: Create a surface using the 4 datum curves created.
Click the icon of **Boundary Blend Tool**. For the first direction, pick the 2 straight lines.

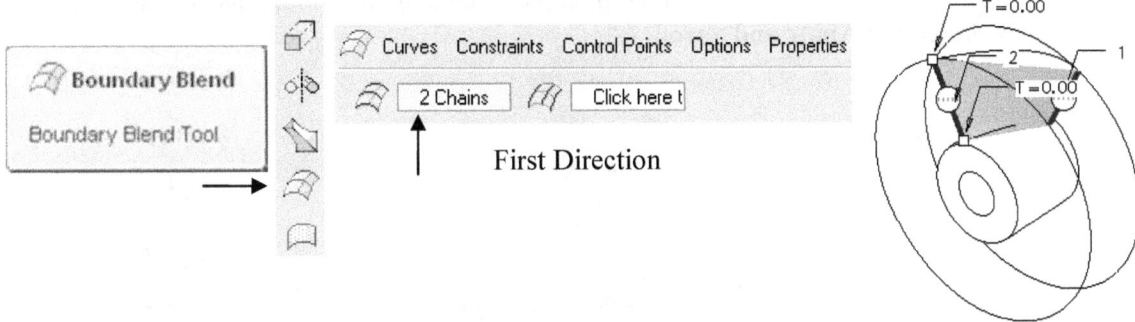

First Direction

For the second direction, pick the 2 datum curves.

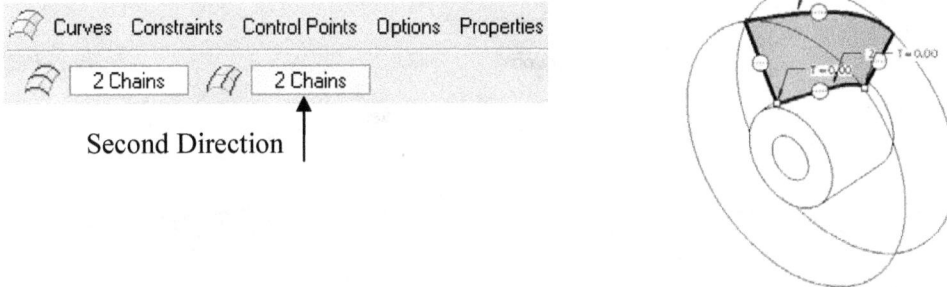

Second Direction

Step 9: Use the Thicken tool to convert the surface feature to a 3D solid feature.
Highlight the created surface > from the main toolbar, select **Edit > Thicken** > specify *1* as the thickness value . Click the arrow box a few times to make sure "middle-plane" is used. Click the check mark to apply and save.

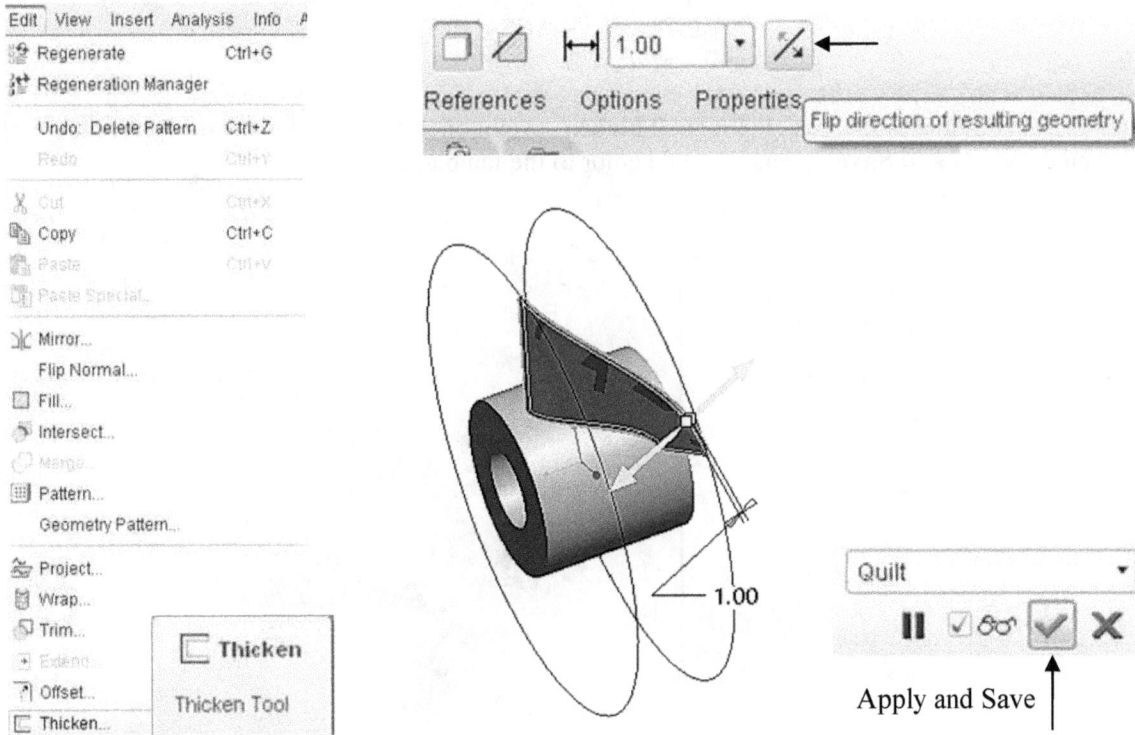

Flip direction of resulting geometry

1.00

Quilt

Apply and Save

Step 10: Use Pattern and Reference Pattern to create the 3 other fan blades
From the model tree, highlight **Boundary Blend 1** > right click and pick **Pattern** > **Axis** and pick the axis of the cylinder > set the number of copies including the current one to 4 and the increment to 90 > click the icon of **Apply and Save**.

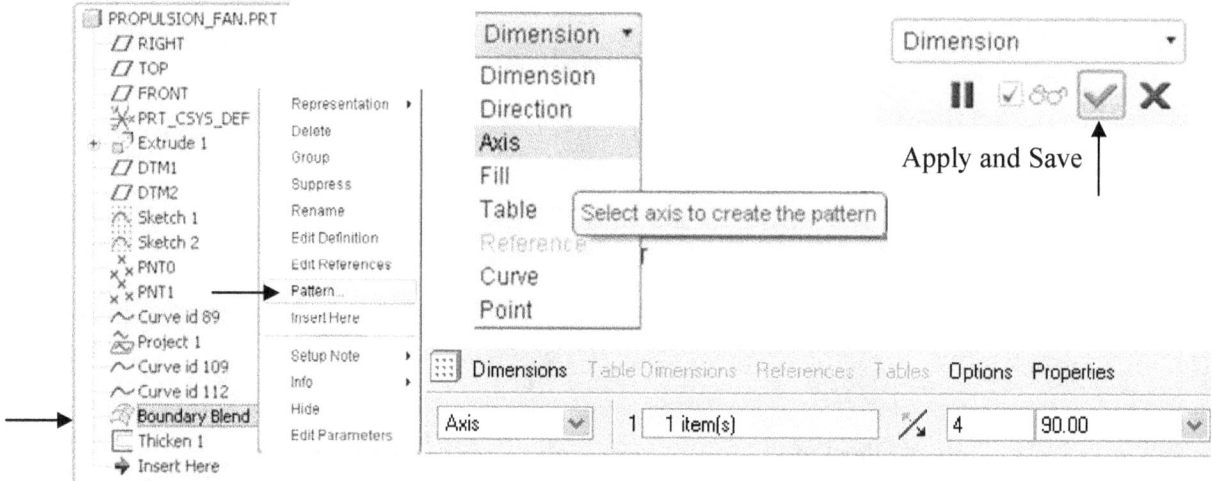

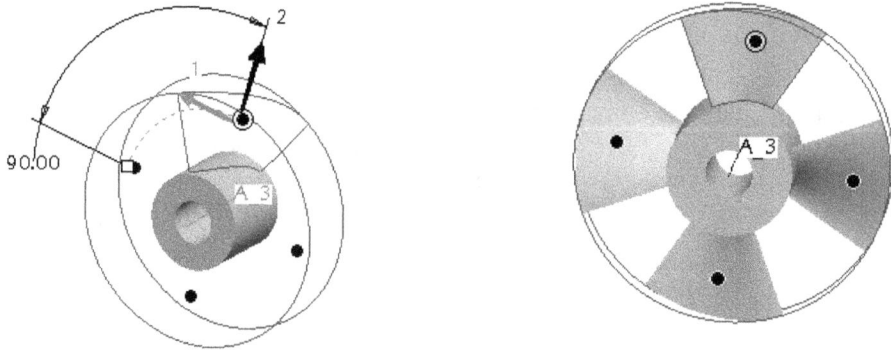

From the model tree, highlight **Thicken** > right click and pick **Pattern** > **Reference** > click the icon of **Apply and Save**. Users may add color to the fan blades.

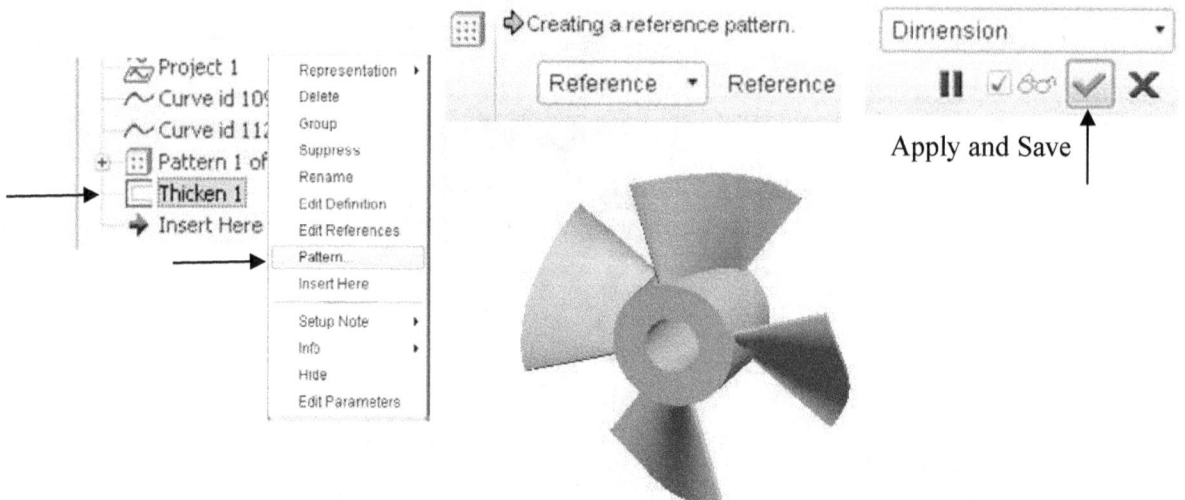

Now let us save the propulsion fan component. Click the icon of **Save the active object > OK**.

Creation of the Propulsion Fan Base Component

Step 1: Create a 3D solid model of the propulsion_fan_base component.

File > New > Part > type *propulsion_fan_base* as the file name and clear the icon of **Use default template > OK.**

Select mmns_part_solid (units: Millimeter, Newton, Second) and type *Propulsion_fan_base* in **DESCRIPTION**, and s*tudent* in **MODELED_BY**, then **OK.**

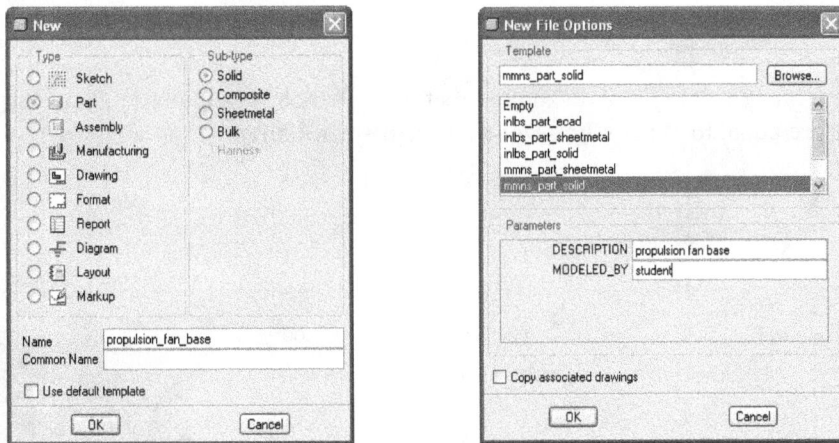

Directly select the icon of **Extrude.** From the dashboard, use **Symmetry** and set the depth value to 36. Activate **Placement** to define a sketch plane > **Define.**

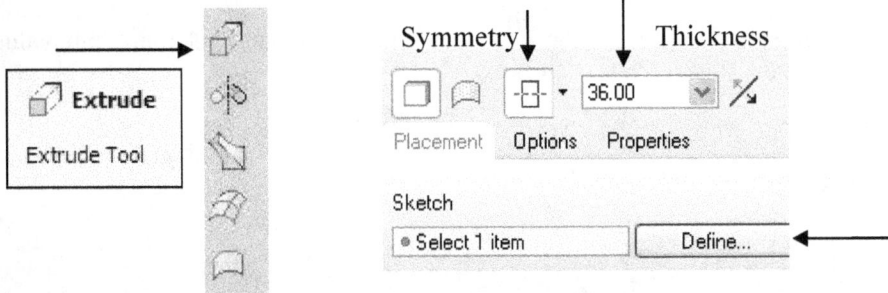

Select the **FRONT** datum plane as the sketch plane, and accept the default orientation.

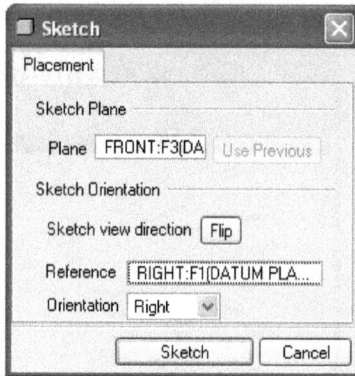

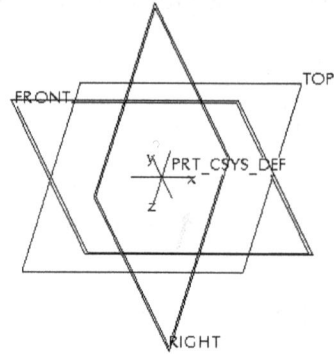

Right-click and hold, select **Centerline**. Sketch a vertical centerline and a horizontal centerline.

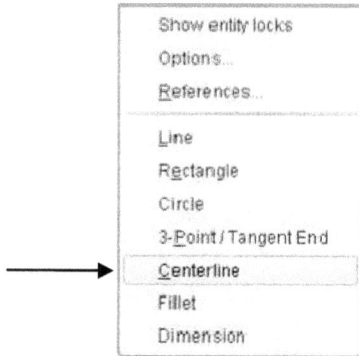

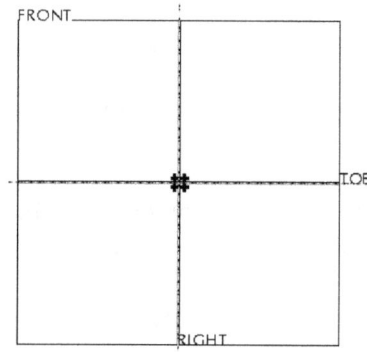

Click the icon of rectangle and sketch a rectangle, which is symmetric about the 2 centerlines. The 2 dimensions are equal to 106. Click **Done** and **Apply and Save**.

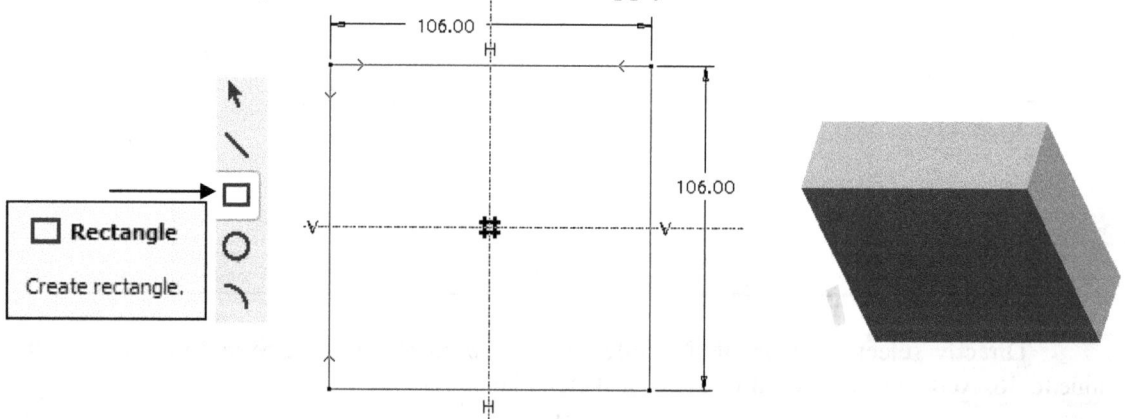

Select the icon of **Extrude.** From the dashboard, click **Cut** and set the depth value to *33.* Activate **Placement** to define a sketch plane > **Define.**

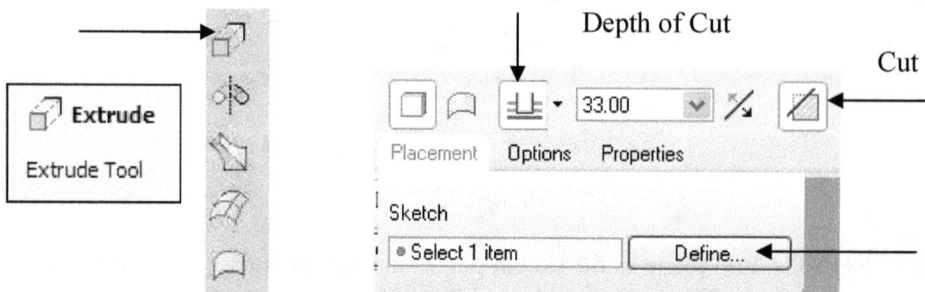

Depth of Cut

Cut

Select the front surface of the block as the sketch plane, and accept the default orientation.

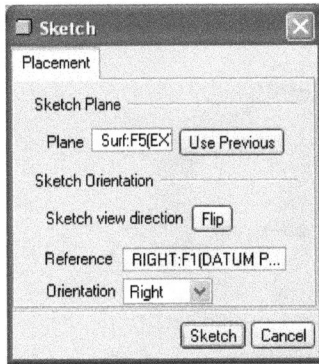

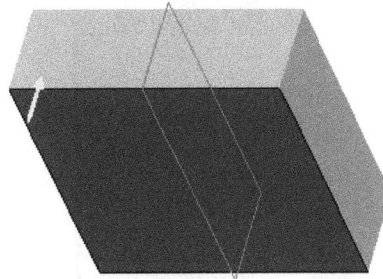

Click the icon of **circle** and sketch 2 circles. The dimensions of the 2 diameters are 94 and 15, respectively.

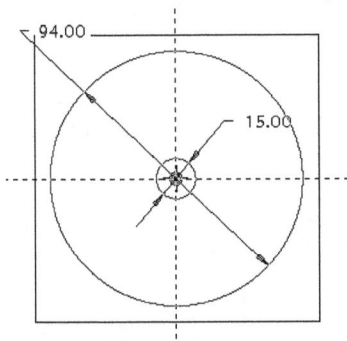

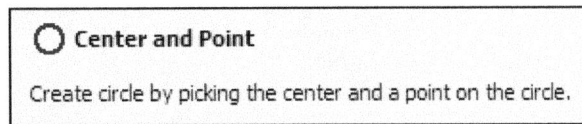

Upon completing the sketch, click the icon of **Done** and the icon of **Apply and Save** from the feature control panel.

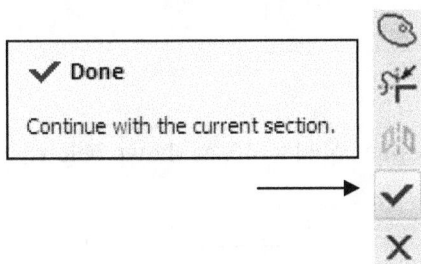

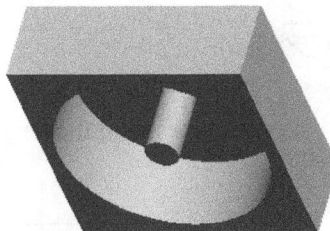

Select the icon of **Extrude.** From the dashboard, click **Cut** and select **Thru All** as the choice of depth. Activate **Placement** to define a sketch plane > **Define.**

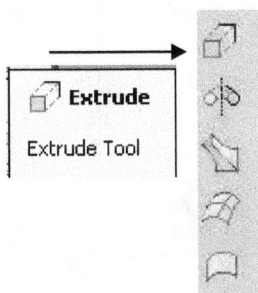

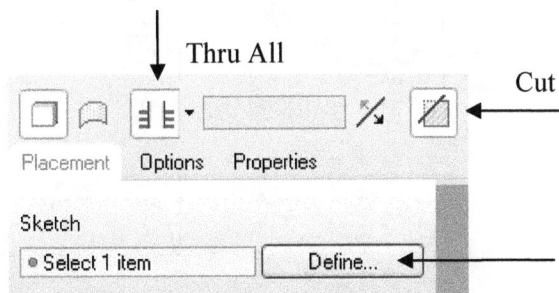

Select the front surface of the block as the sketch plane, and accept the default orientation > **Sketch**.

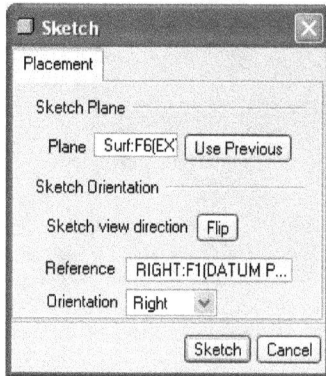

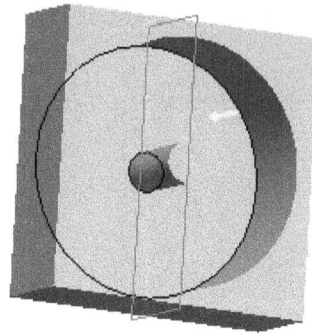

Click the icon of **Use Edge** > pick the 2 sketched circles, as shown below.

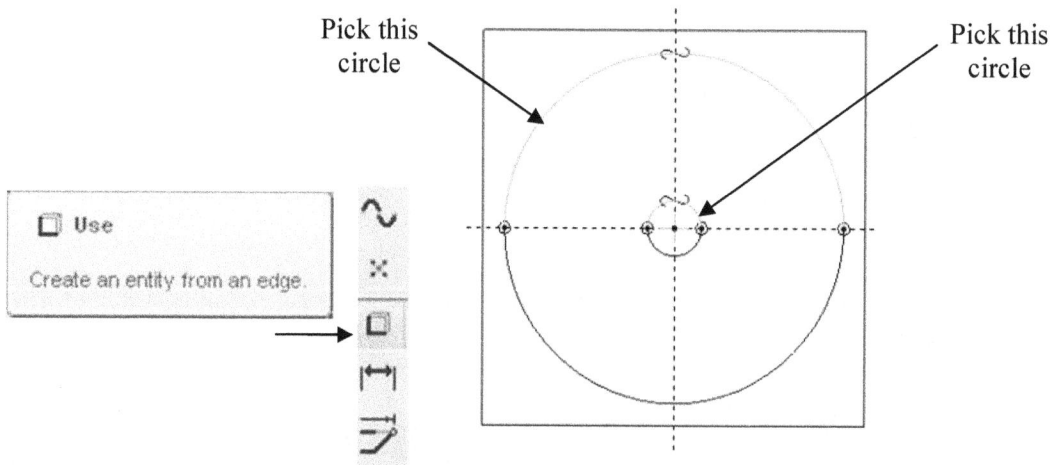

Pick this circle

Pick this circle

☐ Use

Create an entity from an edge.

Click the icon of **Line** > sketch 2 lines with the length value equal to 40, as shown below:

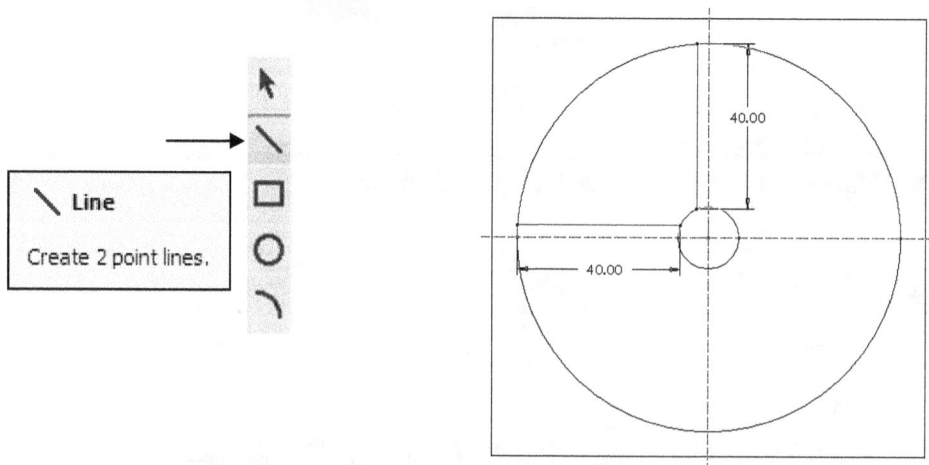

☐ Line

Create 2 point lines.

40.00

40.00

Click the icon of **Delete** so that a closed form of sketch is defined by deleting those sections, which do not belong to this closed form.

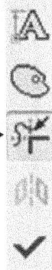

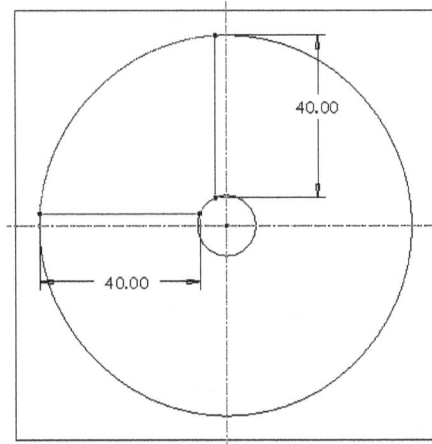

Upon completing this process, click the icon of **Done** and click the icon of **Apply and Save** from the feature control panel.

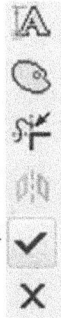

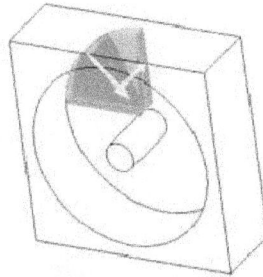

Apply and Save

Click the icon of **Mirror Tool** > click TOP to obtain a new cut feature, as shown. Click the check mark to Apply and Save. Repeat this process to obtain 2 more cut features, as shown.

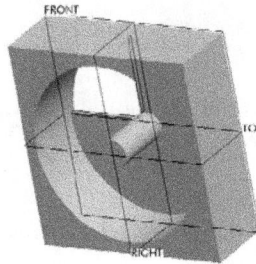

Pick TOP

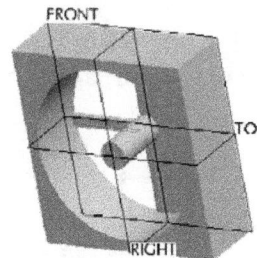

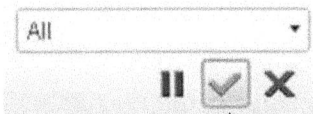

Apply and Save

Now let us save the propulsion fan base component. Click the icon of **Save the active object** > **OK**.

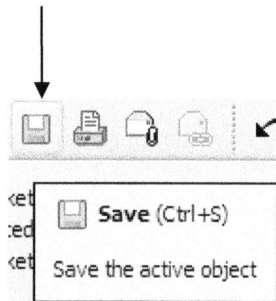

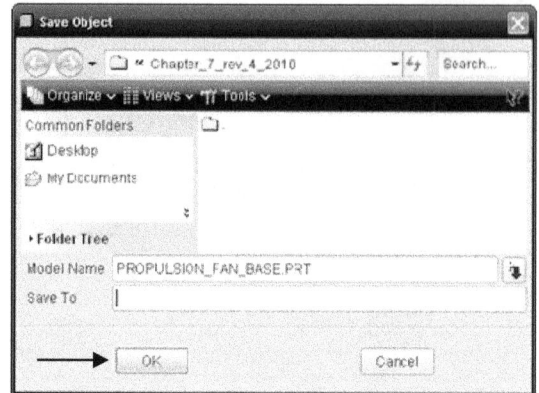

Creation of the Propulsion Support Component

Step 1: Create a 3D solid model of the propulsion_support component.
File > **New** > **Part** > type *propulsion_support* as the file name and clear the icon of **Use default template** > **OK**.

Select mmns_part_solid (units: Millimeter, Newton, Second) and type *propulsion_support* in **DESCRIPTION**, and *student* in **MODELED_BY**, then **OK.**

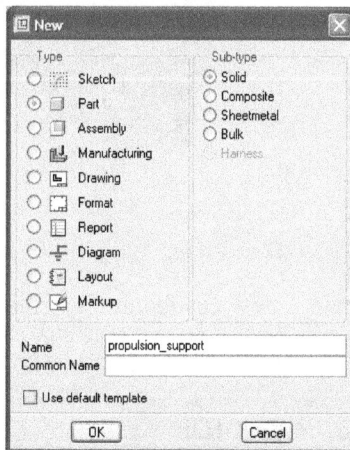

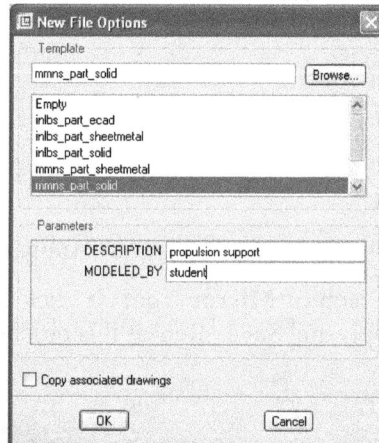

Directly select the icon of **Extrude.** From the dashboard, set the depth value to 90. Activate **Placement** to define a sketch plane > **Define.**

Select the **FRONT** datum plane as the sketch plane, and accept the default orientation.

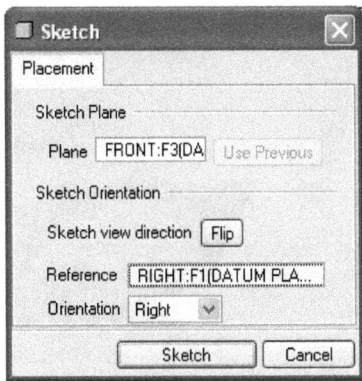

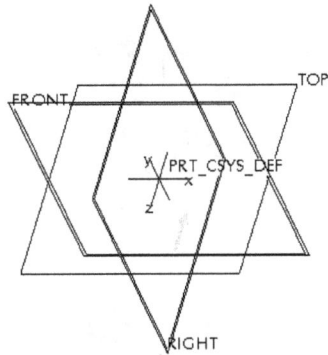

Right-click and hold, select **Centerline**. Sketch a vertical centerline and a horizontal centerline.

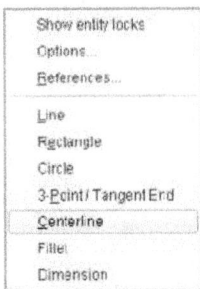

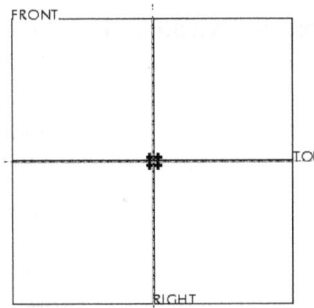

Click the icon of **rectangle** and sketch a rectangle, which is symmetric about the 2 centerlines. The 2 dimensions are equal with the value equal to 120.

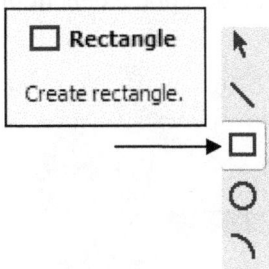

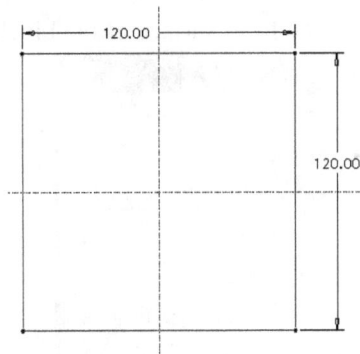

Click the icon of circle and the dimension of diameter is 95.

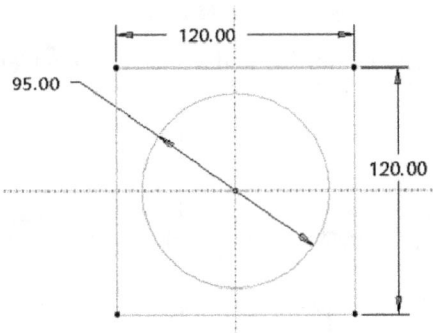

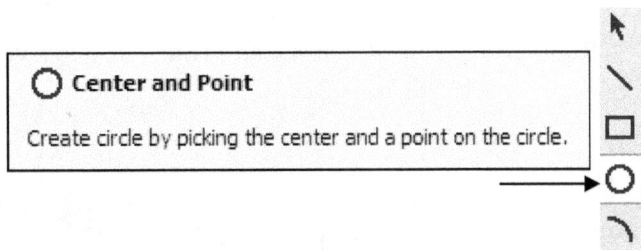

Upon completing this process, click the icon of **Done** and click the icon of **Apply and Save** from the feature control panel.

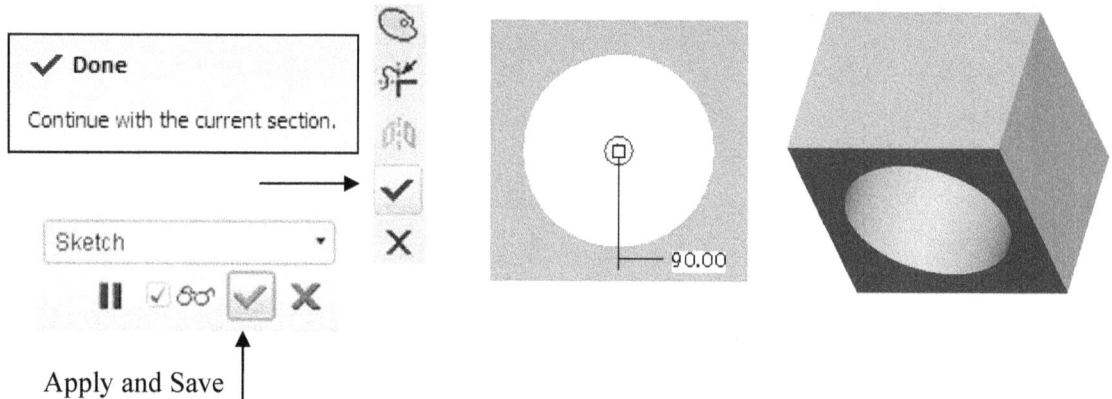

✔ **Done**

Continue with the current section.

Sketch

Apply and Save

90.00

Select the icon of **Extrude.** From the dashboard, click **Cut** and set the depth value to *30.* Activate **Placement** to define a sketch plane > **Define.**

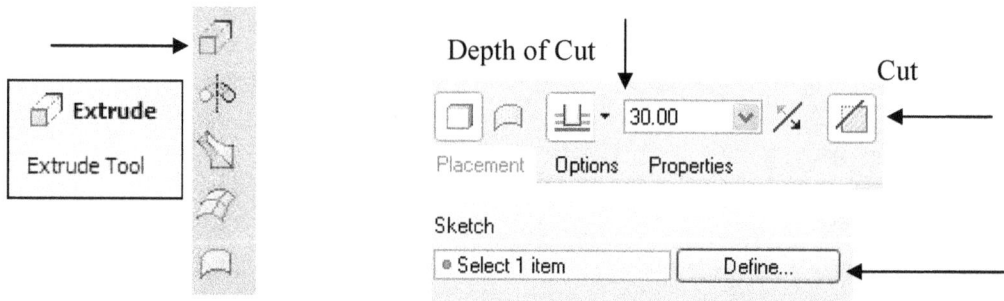

Extrude

Extrude Tool

Depth of Cut

Cut

30.00

Placement   Options   Properties

Sketch

Select 1 item   Define...

Select the front surface of the block as the sketch plane, and accept the default orientation.

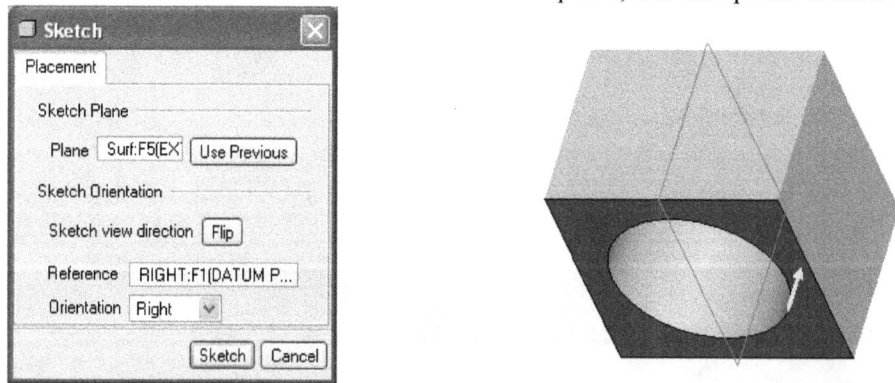

Sketch

Placement

Sketch Plane

Plane Surf:F5(EX Use Previous

Sketch Orientation

Sketch view direction Flip

Reference   RIGHT:F1(DATUM P...
Orientation Right

Sketch   Cancel

Right-click and hold, select **Centerline.** Sketch a vertical centerline and a horizontal centerline.

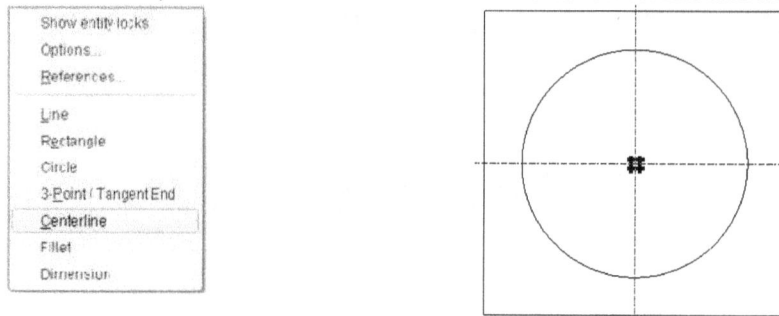

Show entity locks
Options...
References

Line
Rectangle
Circle
3-Point / Tangent End
Centerline
Fillet
Dimension

Click the icon of rectangle and sketch a rectangle, which is symmetric about the 2 centerlines. The 2 dimensions are equal with the value equal to 106.

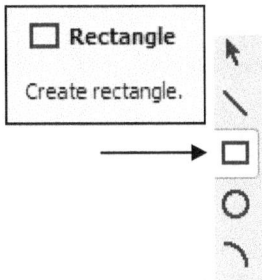

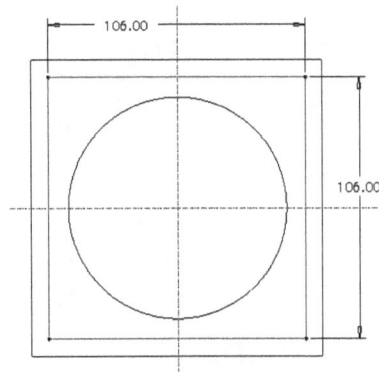

Upon completing this process, click the icon of **Done** and click the icon of **Apply and Save** from the feature control panel.

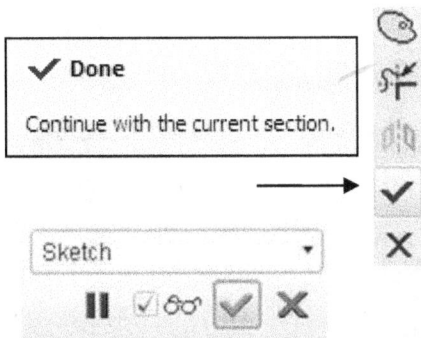

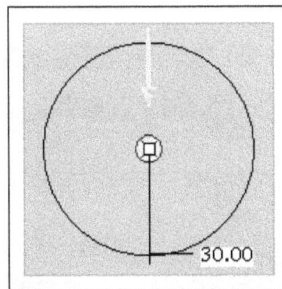

Apply and Save

Now let us save the propulsion fan base component. Click the icon of **Save the active object >** **OK**.

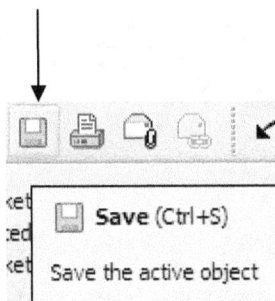

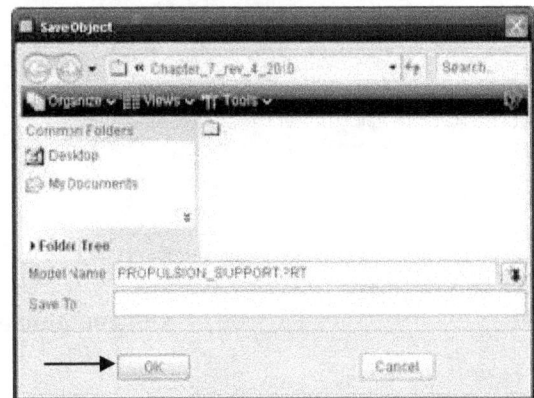

Creation of the Assembly of the Propulsion Fan System

Step 1: Create a 3D solid model of the assembly of the base structural system.
File > **New** > **Assembly** > type *propulsion_fan_system* as the file name and clear the icon of **Use default template** > **OK**.

Select mmns_asm_design (units: Millimeter, Newton, Second) and type *propulsion_fan_system* in **DESCRIPTION**, and *student* in **MODELED_BY**, then **OK.**

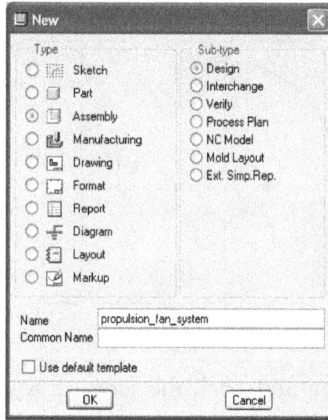

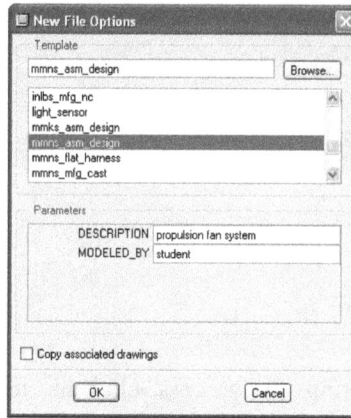

Click the icon of **Add component** > highlight propulsion_support.prt > **Open.**

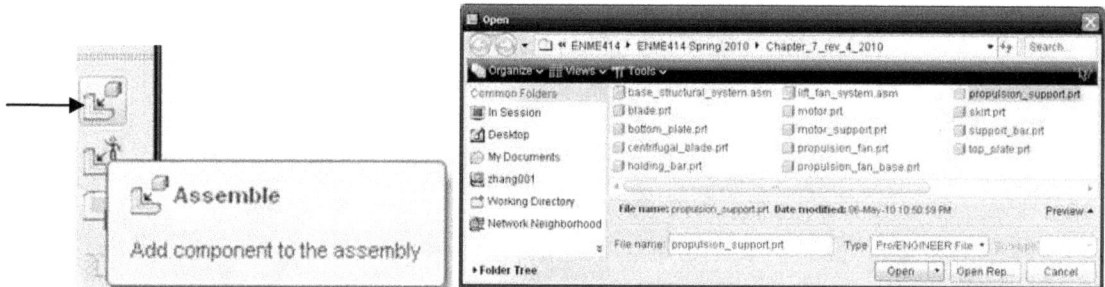

Select **Default** > click the icon of **Apply and Save** from the constraint control panel.

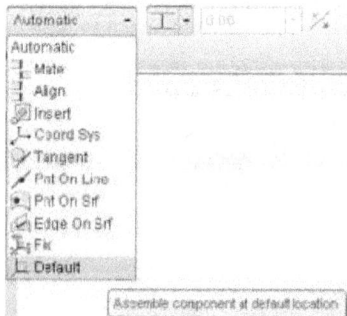

Apply and Save

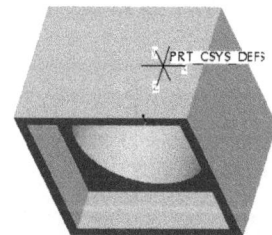

Click the icon of **Add component** > highlight propulsion_fan_base.prt > **Open.**

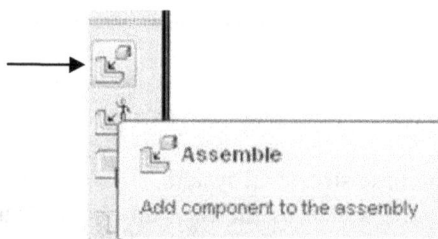

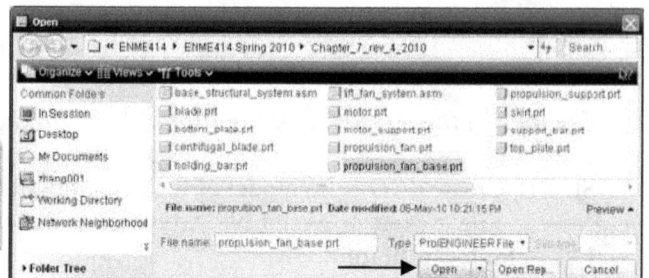

While holding down both the **Ctrl** key and the **Alt** key, press the middle button of the mouse and rotate the base component to facilitate the assembling process. If users want to move the base component, not rotating, they should hold down both the **Ctrl** key and the **Alt** key, press the right button of the mouse.

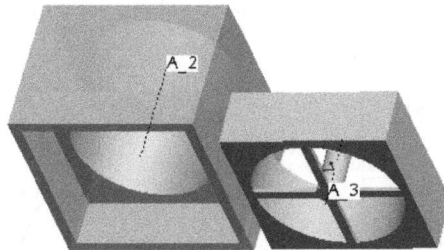

Click the icon of **Placement**. Let us define the first constraint. Select the axis from the propulsion_support component and the axis from the propulsion_fan_base component. A constraint called **Align** has been defined. Click **New Constraint** to define 2 Mate constraints, as shown below.

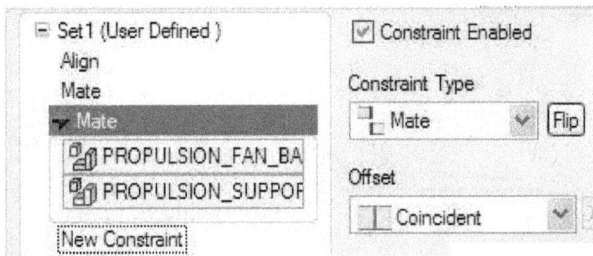

Align constraint

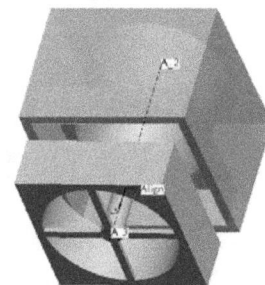

Mate constraint 2

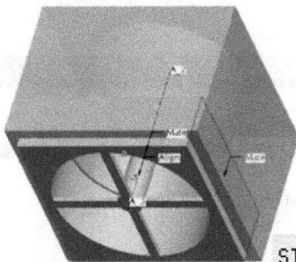

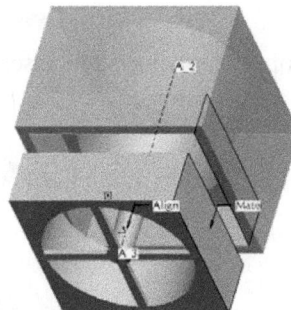

Mate constraint 1

Apply and Save

STATUS : Fully Constrained

At this moment, the propulsion_fan_base component is fully constrained. Click the icon of **Apply and Save** from the constraint control panel.

Now let us assemble the propulsion fan component on the other side. Click the icon of **Add component** > highlight propulsion_fan.prt > **Open**.

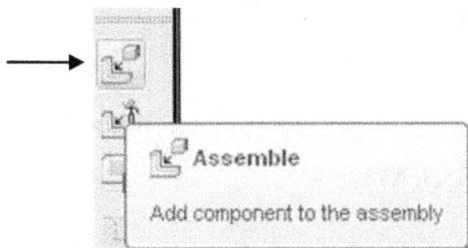

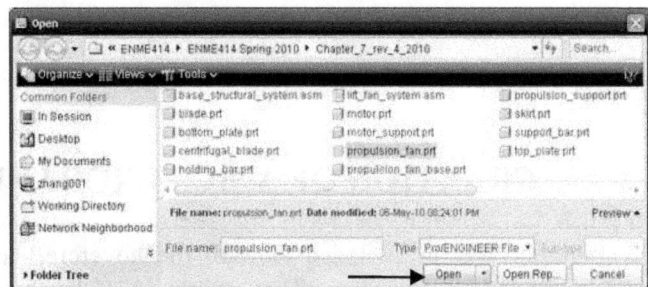

We follow the same procedure to define 2 constraints as we did in the process of assembling the propulsion_fan_base component.

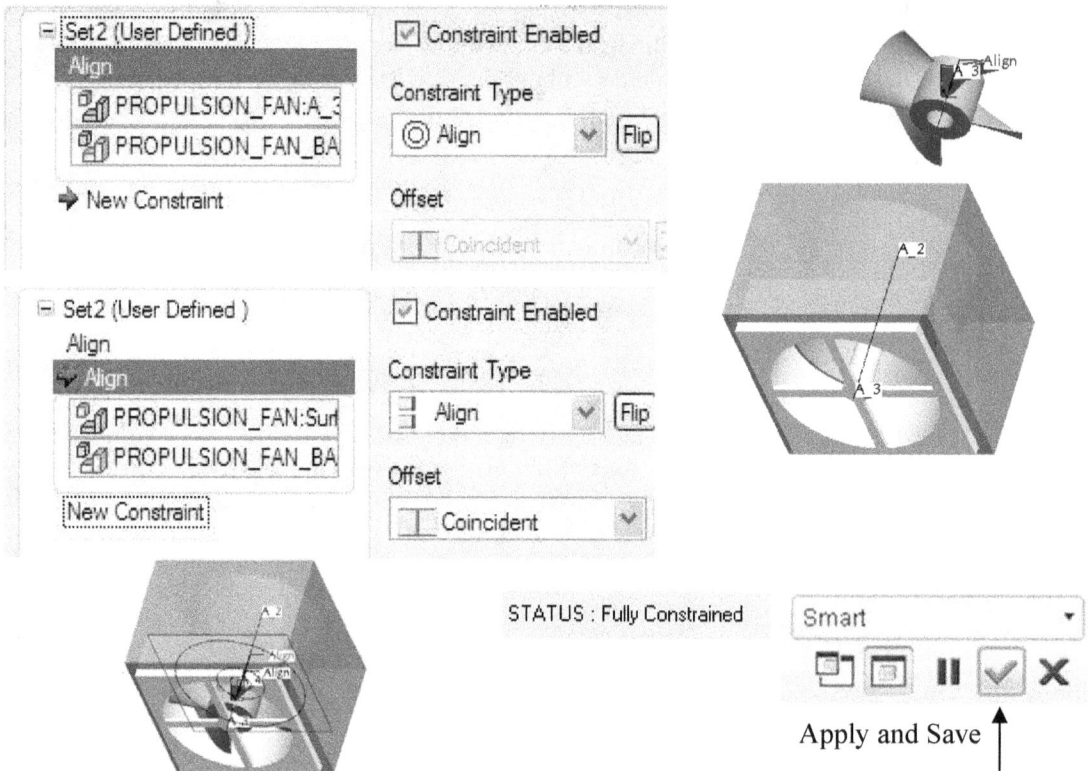

Now let us save the propulsion fan system assembly. Click the icon of **Save the active object** > **OK**.

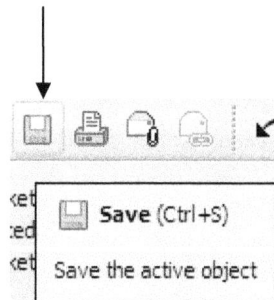

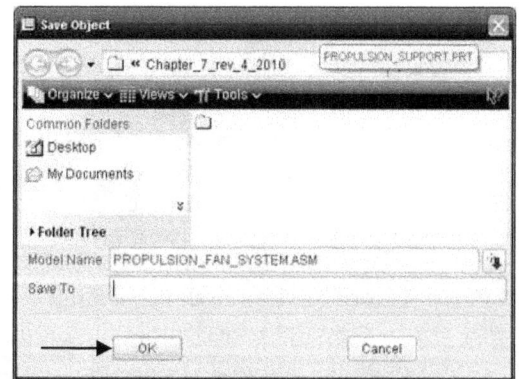

7.5   CREATION OF A HOVERCRAFT ASSEMBLY

Figure 7.4 is an exploded view of the hovercraft assembly. As illustrated, the hovercraft assembly consists of 4 subsystems. They are the base structure subsystem, the lift fan subsystem, the propulsion fan subsystem, and the sensor-based control system. Note that the sensor-based control system is not presented in this chapter.

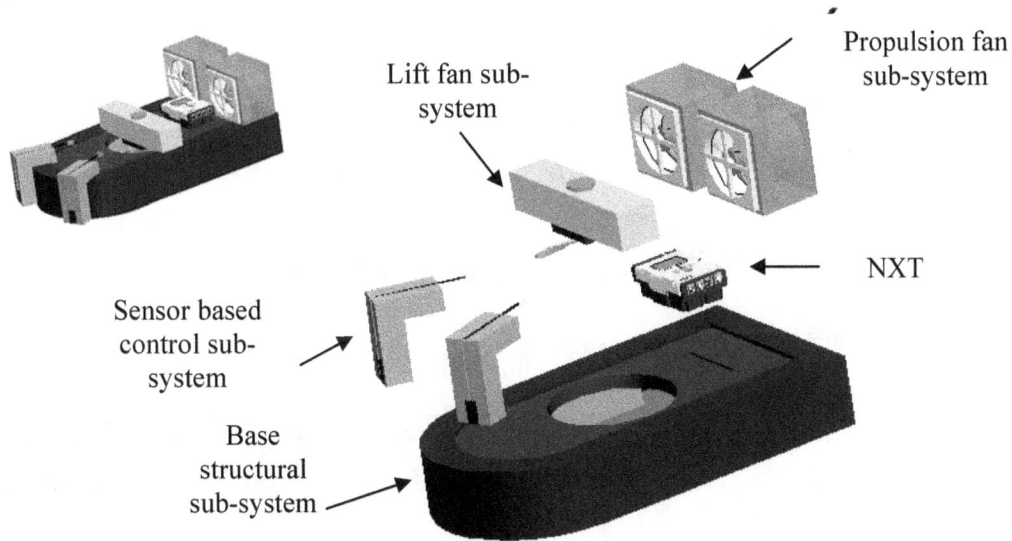

Fig. 7.4 An exploded view of the hovercraft assembly showing 4 subsystems

Step 1: Create a 3D solid model of the hovercraft assembly.

File > **New** > **Assembly** > type *hovercraft_assembly* as the file name and clear the icon of **Use default template** > **OK**.

Select mmns_asm_design (units: Millimeter, Newton, Second) and type *hovercraft_assembly* in **DESCRIPTION**, and s*tudent* in **MODELED_BY**, then **OK.**

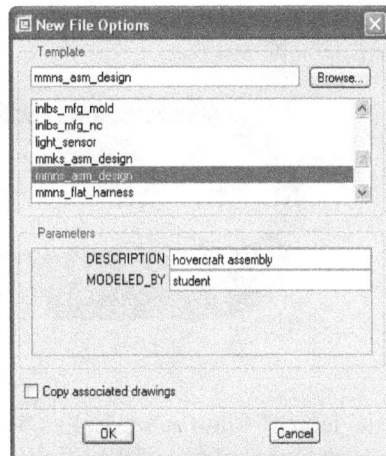

Click the icon of **Add component** > highlight base_structural_system.asm > **Open**.

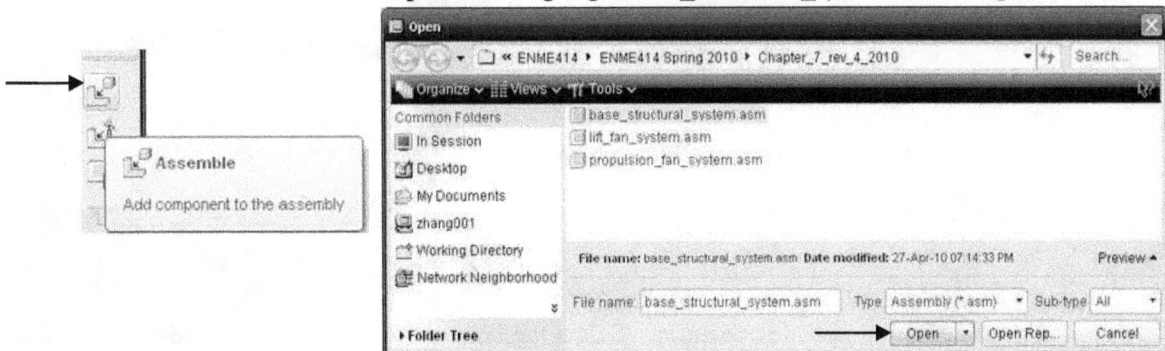

Select **Default** > click the icon of **Apply and Save** from the constraint control panel.

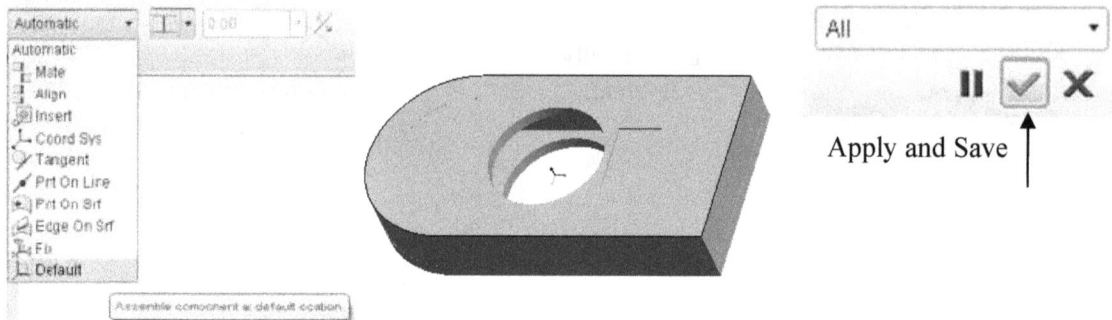

Apply and Save

Click the icon of **Add component** > highlight lift_fan_system.asm > **Open**.

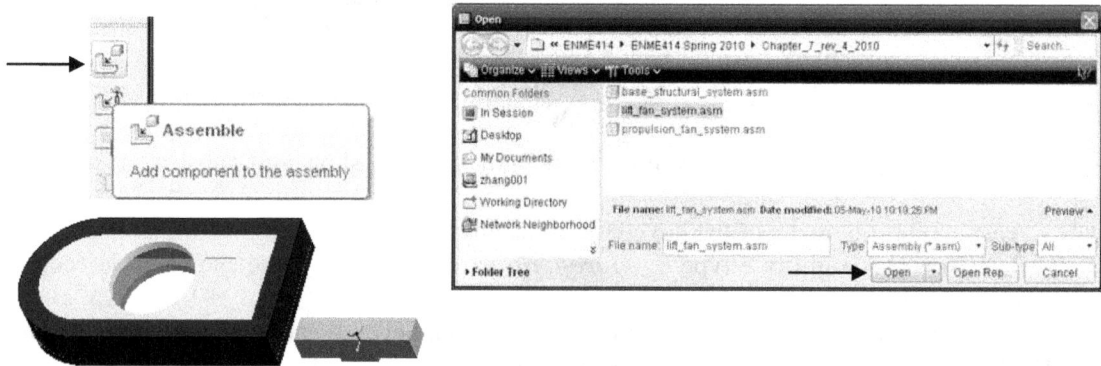

While holding down both the **Ctrl** key and the **Alt** key, press the middle button of the mouse and rotate the lift fan sub-system.

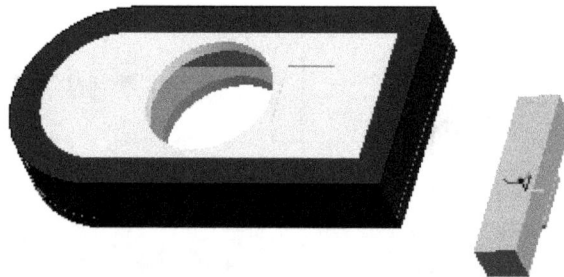

Click the icon of **Placement**. Let us define the first constraint. Select **ASM_FRONT** from the lift fan assembly and select **ASM_Right** from the base structural assembly and set the offset distance to *130*. A constraint called **Align** has been defined. Click **New Constraint** to insert a new constraint.

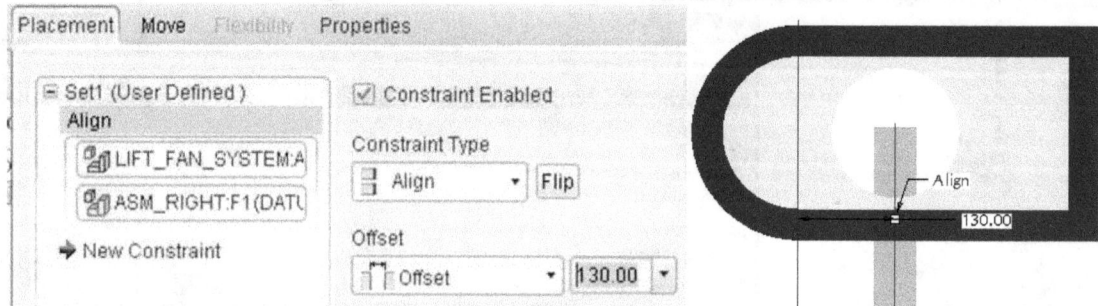

Let us define the second constraint. Select the bottom surface from the holding bar and the top surface from the skirt. A constraint called **Mate** has been defined. Click **New Constraint** to insert a new constraint.

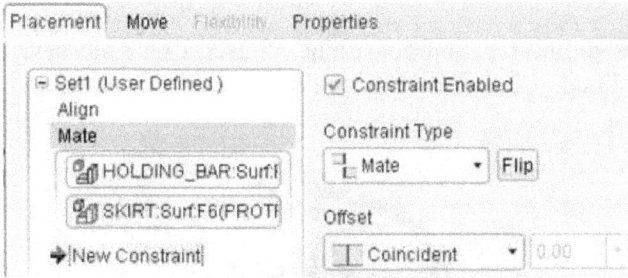

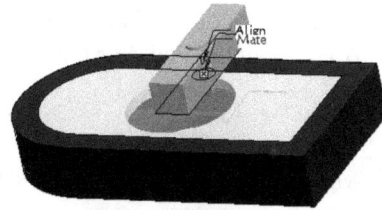

Let us define the third constraint. Select the axis from the motor and select the axis from the top plate. A constraint called **Align** has been defined. At this moment, the propulsion_fan_base component is fully constrained. Click the icon of **Complete** from the constraint control panel.

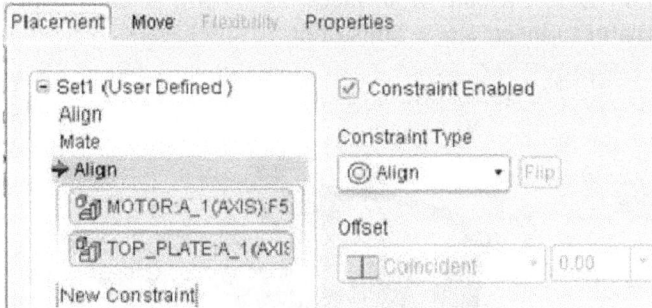

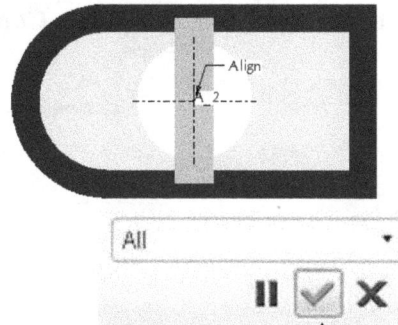

Apply and Save

Now let us assemble the propulsion_fan_system. Click the icon of **Add component** > highlight propulsion_fan_system > **Open**.

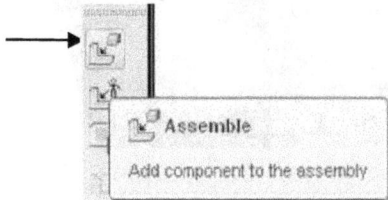

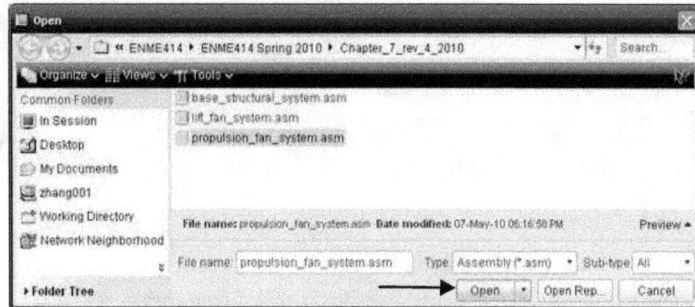

While holding down both the **Ctrl** key and the **Alt** key, press the middle button of the mouse and rotate the propulsion fan system.

Click the icon of **Placement**. Let us define the first constraint. Select a pair of surfaces, as shown. A constraint called **Align** has been defined. Click **New Constraint** to insert a new constraint.

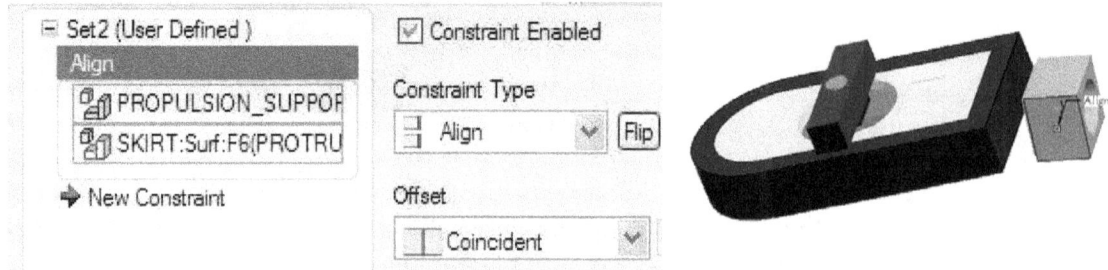

Let us define the second constraint. Select a pair of surfaces, as shown. A constraint called **Align** has been defined. Click **New Constraint** to insert a new constraint.

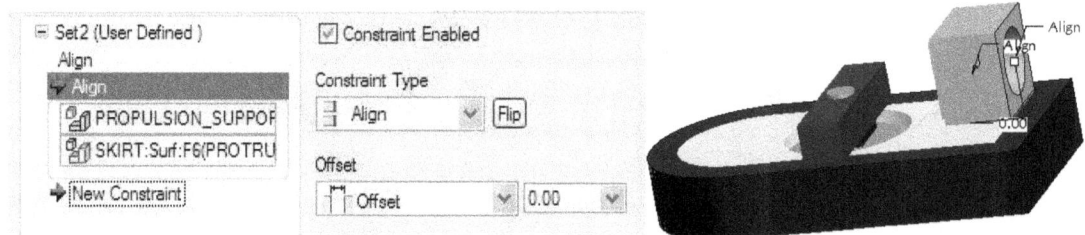

Let us define the third constraint. Select a pair of surfaces, as shown. A constraint called **Mate** has been defined. At this moment, the propulsion fan system is fully constrained. Click the icon of **Apply and Save** from the constraint control panel.

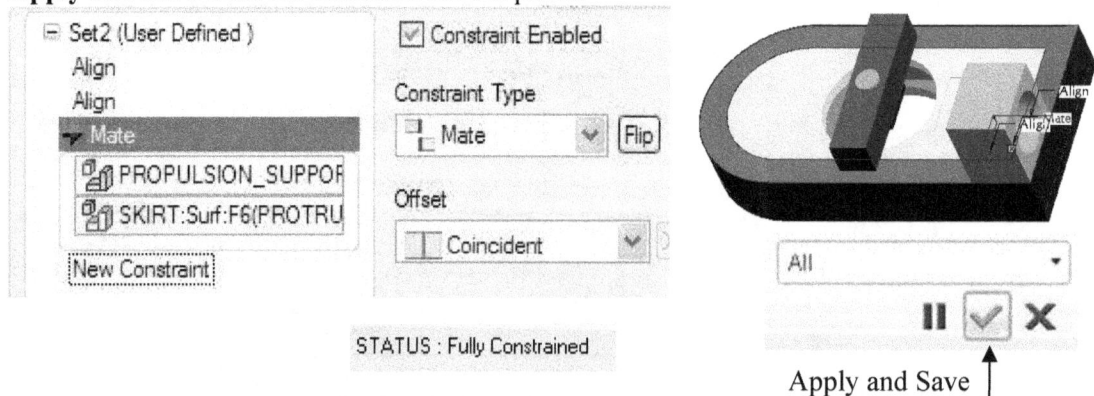

STATUS : Fully Constrained

Apply and Save

Repeat the above procedure to assemble another propulsion fan system on the other side, as shown. Click the icon of **Save the active object** > **OK**.

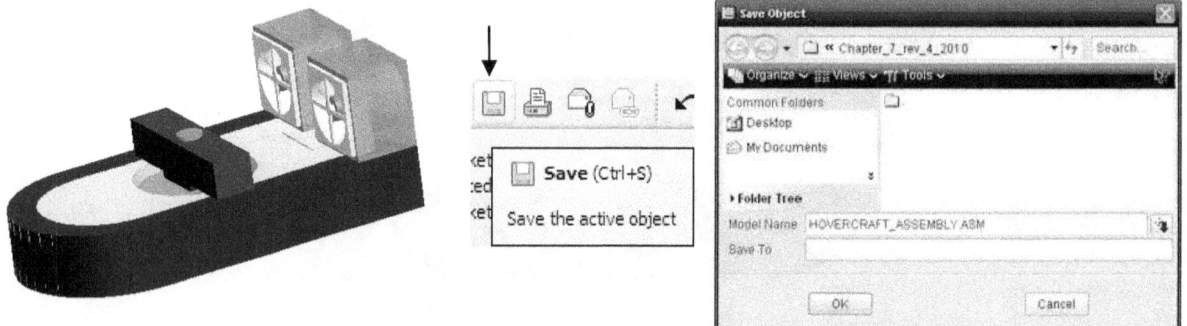

If a 3D solid model of the sensor_based_control sub-system is available, users may assemble it

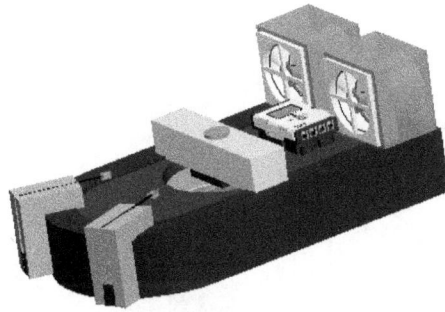

7.6 PREPARATION OF AN ENGINEERING DRAWING

In the section, we prepare an engineering drawing for the propulsion support component created in section 7.4. First, we select the icon, called "**Create a new object**", which is displayed on the menu bar. A **New** window appears, as illustrated below. This is the same step we used to open a new file when creating the 3D solid model of the propulsion support component. However, this time, "**Drawing**" mode, instead of "**Part**" mode, should be selected. Type propulsion_support as the name for the new drawing file. Clear the box of **Use default template** because we do not want to use the default setting for the drawing work. Afterwards, click the button of **OK**.

This brings up a new window called "**New Drawing**", as shown below. Make sure that the file of the 3D solid model of propulsion_support.prt is shown. Otherwise, use the "**Browse**" option to locate it. Select **Empty** under Specify Template, and select the paper size to be **A**. Afterwards, click the button of **OK**.

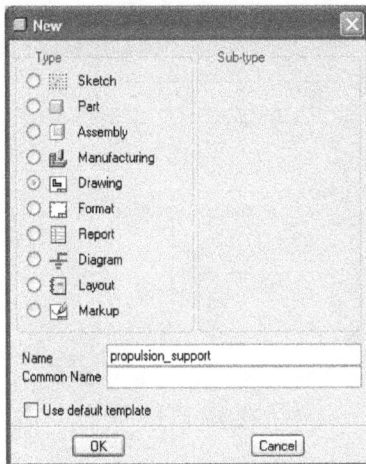

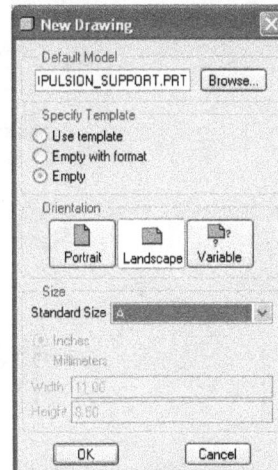

Click the icon of **Layout**. Select the icon of **General** which is displayed on the tool bar. Select a location on the drawing screen as the center point for the **Front View** (click the left button of mouse). A general view appears on the screen.

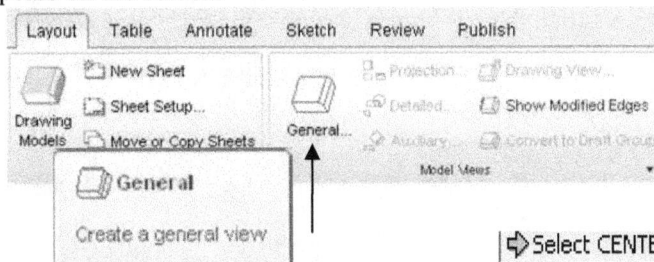

In the pop-up Drawing View window, select **FRONT** > **Apply** > **Close**, the construction of the Front View is completed.

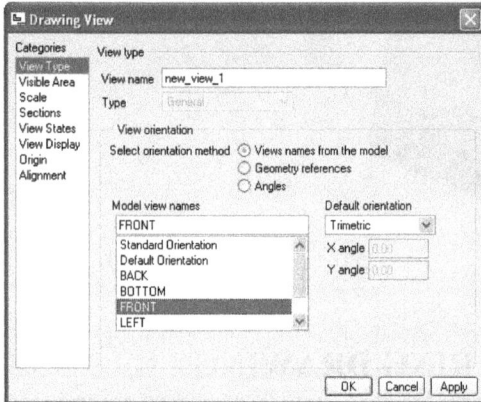

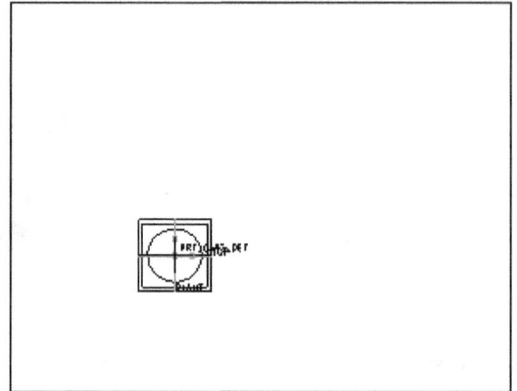

To insert the right side view through projection, first pick the **FRONT** View just created, right-click and hold, and then select **Insert Projection View** > move the curser to the right side and click the left button of mouse, and the construction of the Right-sided View is completed. Follow the same procedure to create the Top View, as shown below.

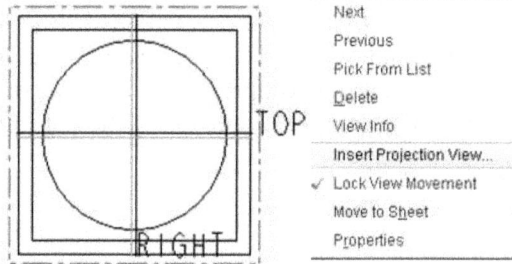

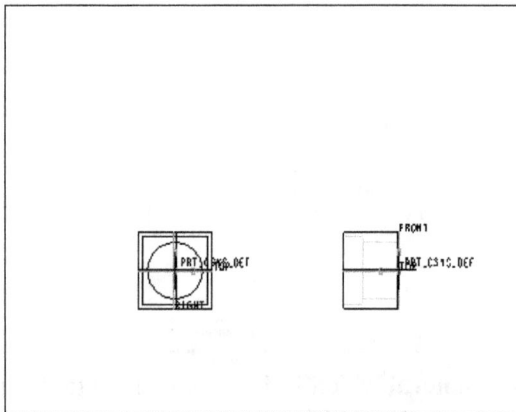

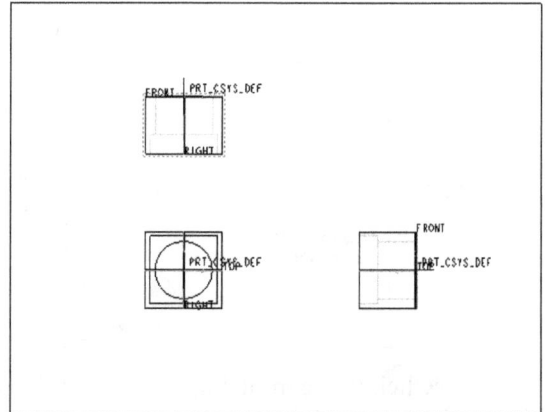

Select the icon of **General** to create a 3D view. Select a location on the drawing screen as the center point for the 3D view (click the left button of mouse). A general view appears on the screen.

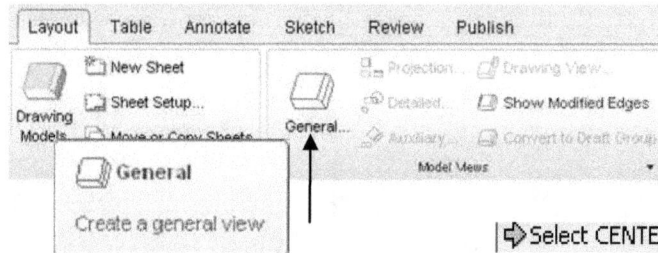

Select CENTER POINT for drawing view.

In the pop-up Drawing View window, select **Standard Orientation** > **Apply** > **Close**, the construction of a 3D view is completed.

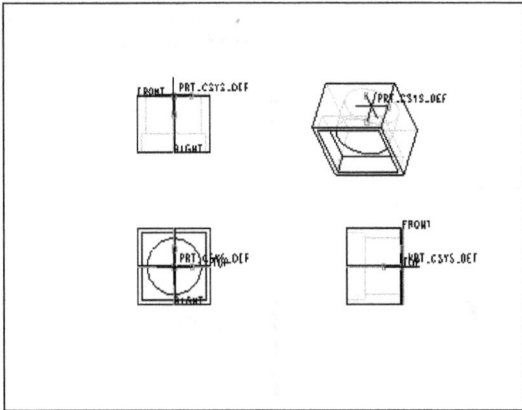

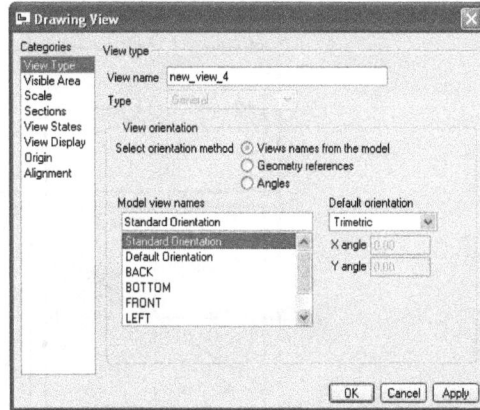

The user may notice that the names of the datum planes, such as FRONT, RIGHT and TOP, appear on the drawing. The name of coordinate system, such as PRT_CSYS_DEF, also appears. To clean the drawing screen, click the icon of **datum planes**, the icon of **datum axes**, the icon of **datum points**, and the icon of **coordinate system** from the main toolbar to disable their displays > click the icon of **repaint** (redraw the current views) from the main toolbar.

Disable their displays

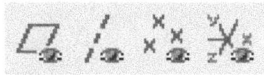

Enable their displays

Redraw or Repaint the current View

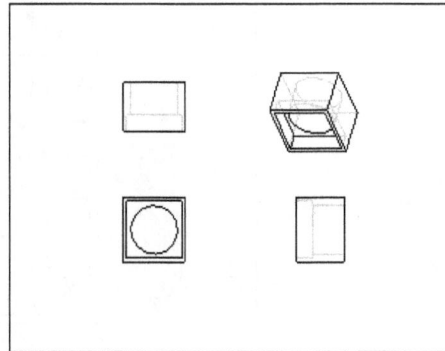

Upon completing the layout, we start adding dimensions. Click the icon of **Annotation**. Select the icon of **Show Model Annotation**, which is displayed on the toolbar.

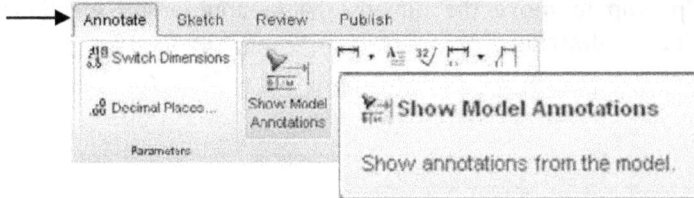

In the pop-up window, select the icon of **dimensions**, and click the button of **Show All** > **OK**.

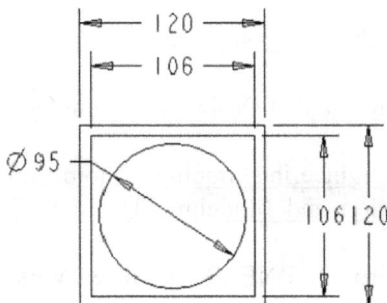

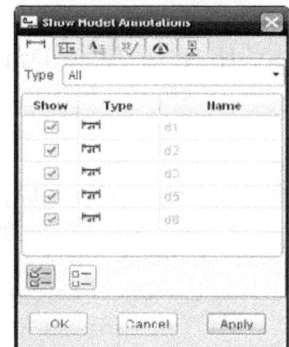

To add centerlines or axes to the drawing, click the icon of **Show Model Annotation**, again. Click the circular feature

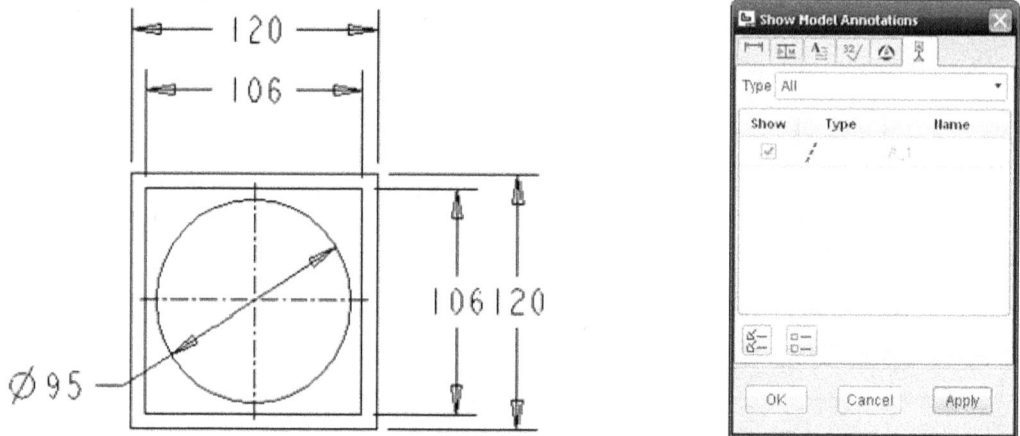

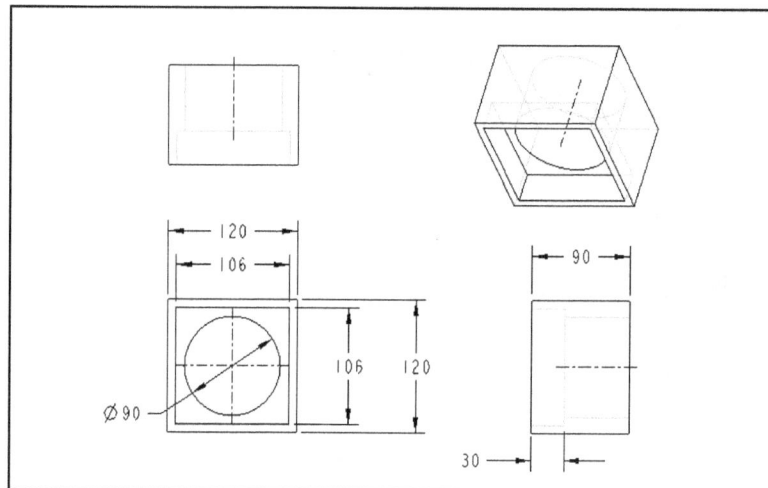

Users may use the icon of pickup to move the dimensions to appropriate locations, thus completing the preparation of an engineering drawing.

REFERENCES

1. Foley, J.D., Dam, A.V., Feiner, S., and Hughes, J., Computer Graphics, Principles and Practice, 2nd edition, McGraw-Hill, 1990.
2. Groover, M.P., and Zimmers, E.W., Computer-aided Design and Manufacturing, Englewood Cliffs, NJ, 1984.
3. Kalameja, A.J., AutoCAD 2006 Tutor for Engineering Graphics, Thomson Delmar Learning, 2006.
4. Qi, G., Engineering Design, Communication and Modeling: Using Unigraphics NX, Thomson, Delmare Learning, 2006.
5. Zhang, G.M., Engineering Design and Pro/ENGINEER Wildfire, Version 4.0, College House Enterprises, LLC., 2008.

EXERCISES

7.1.   An object is shown in a 3D space. Use Pro/ENGINEER to create a 3D solid model of it. Afterwards, prepare an engineering drawing, as shown.

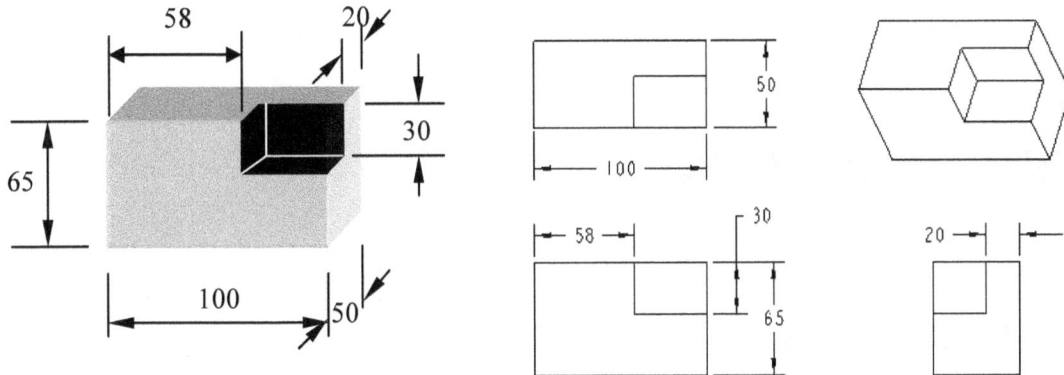

7.2.   An object is shown in a 3D space. Use Pro/ENGINEER to create a 3D solid model of it. Afterwards, prepare an engineering drawing, as shown.

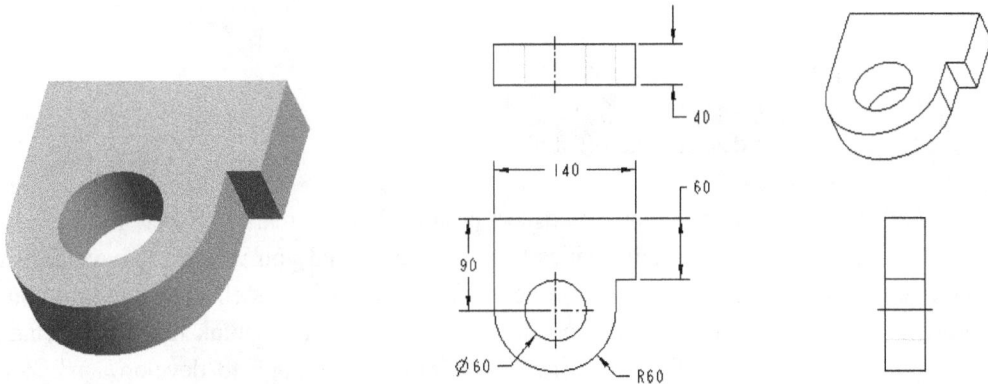

7.3.   An object is shown in a 3D space. Use Pro/ENGINEER to create a 3D solid model of it. Afterwards, prepare an engineering drawing, as shown.

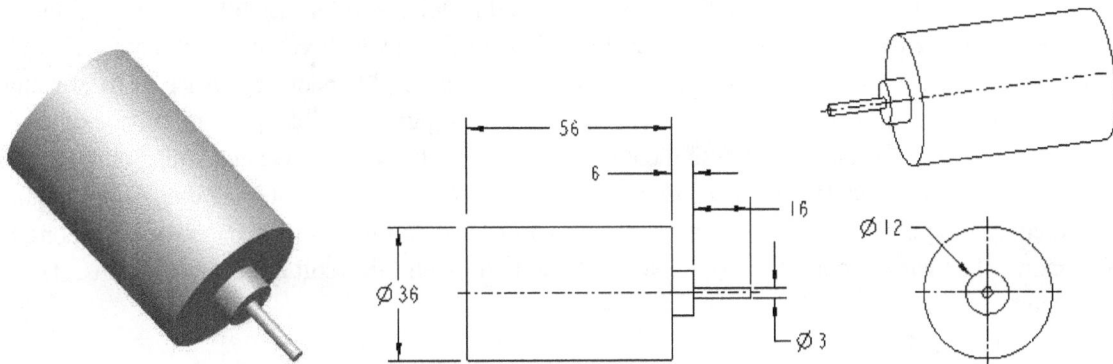

CHAPTER 8

MICROSOFT EXCEL

8.1 INTRODUCTION

Microsoft® Excel, a spreadsheet program, is an extremely powerful tool in engineering because it is useful in many different applications. In addition to providing an introduction for using the Excel spreadsheet, techniques for three important applications will be described in detail in this chapter. These applications include:

- Preparing tables and graphs
- Making calculations
- Conducting design trade-off studies

While there are several different spreadsheet programs on the market, Excel® has been selected because it is the most popular. Many universities have adopted it and you will likely have access to it in campus computer labs during your tenure in college. The content in this chapter is focused on both developing your entry-level skills in Excel and encouraging you to begin to think like a practicing design engineer. Hopefully, you will find the spreadsheet tool important enough to develop a much higher skill level through independent study.

8.2 EXCEL BASICS

The version of Excel described in this chapter is found in Microsoft Office 2007. However, much of the content in this chapter is applicable to earlier versions of Excel which you may still be using. Microsoft Office 2007 has been given a major facelift as compared to prior versions of Office. The key change is that the menu-based software has been replaced with an all-purpose "ribbon" along the top of the screen. The ribbon provides a tabbed browsing environment that contains all of the items found under the familiar Microsoft menu headers (i.e., File, Edit, View, Insert, etc.), with a minimal user effort required for searching drop down menus. It is advised that you take time to explore each of the tabs along the top ribbon to familiarize yourself with the new configuration for all Microsoft Office applications. This ribbon is shown in Fig. 8.1.

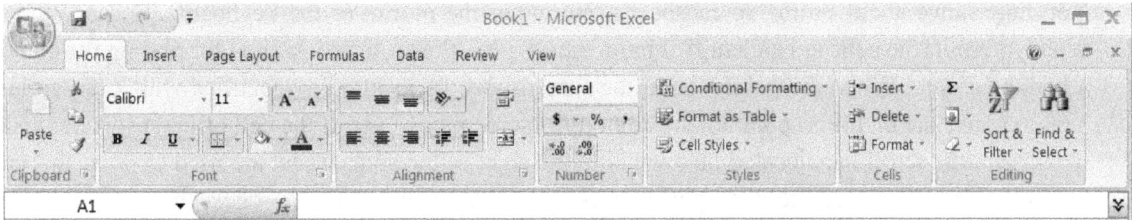

Fig. 8.1 The "ribbon" menu of Microsoft Office Excel 2007

Most of the screen is occupied by the spreadsheet, which is simply a large table with many columns and rows as shown in Fig. 8.2. The columns are labeled across the top of the table with letters A, B, C,… M and beyond. The rows are labeled down its left side with numbers 1, 2, 3,… 25 and beyond. Each small block, called a cell, is identified with its coordinates (e.g. in Fig. 8.2, the cell A1 is outlined with a border indicating it is active). The letter (column location) is placed before the number (row location). The cells are simply locations in a very large table (spreadsheet) where you can enter numbers, text or formulas.

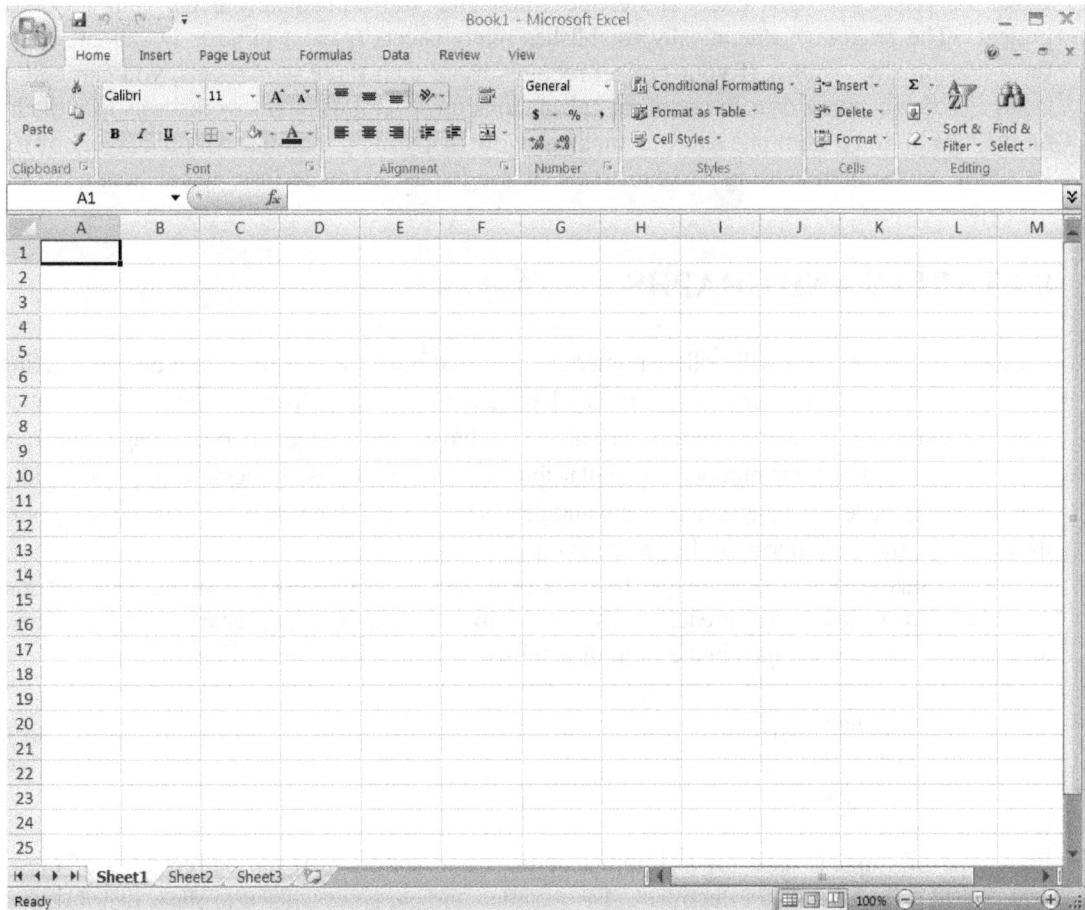

Fig. 8.2 The Excel 2007 spreadsheet showing cell A1 as the active cell

You may move about on the spreadsheet using either the mouse or the keyboard. Because the mouse usually permits the cells to be identified more rapidly, it is the preferred method for placing the cursor over a particular cell. When the pointer is on the spreadsheet, its location is identified with a large plus sign. You may use the mouse to point to any cell on the screen; however, if the cell of interest is not visible on the screen, the scroll bars located below and to the right of the spreadsheet are used to bring this cell into the field of view. When the large plus sign is pointed at the correct cell—click; the cell becomes active and is ready to receive the data you enter. You may also move the active cell by using the arrow keys, the tab key, the shift-tab keys, the enter key, the shift-enter key and the control-arrow keys. Try them and note how the active cell moves about the spreadsheet.

The spreadsheet is much larger than the table shown on the screen. In Fig. 8.2, only columns A through M and rows 1 through 25 are shown. Explore the number of rows on the spreadsheet by holding down the control and the arrow-down key. Note that the number of the last row is 1,048,576. Depressing the control and the arrow-right keys moves the active cell to the far right of the spreadsheet. There are a total of 16,384 columns. You have a total of $(1,048,576)(16,384) = 17,179,869,184$ cells on the spreadsheet. Your spreadsheet is huge—let's hope you never have to use all of the available cells!

At the bottom of the spreadsheet, in Fig. 8.2, you will find three tabs identifying sheet numbers 1, 2 and 3. When you open Excel and create a new file, you immediately begin with a workbook for a project. The program automatically establishes many sheets (pages) in your workbook, although only three are initially visible. If your project requires more than three worksheets, you can add more very easily. You may enter data and perform calculations on several different worksheets in a file by clicking on the tabs to move from one sheet to another. Now that you are familiar with the layout of Excel 2007, let's turn our attention to using Excel to create tables, graphs, and to perform calculations.

8.3 TABLES AND GRAPHS

On many occasions you will collect numerical data that must be presented at a design briefing or in an engineering report. At other times, you may have extensive results from a mathematical relationship that must be expressed in a meaningful format. You have two choices in presenting large quantities of numerical data—show the numbers in tabular form or on a graph. Both methods of presentation, the table and the graph, have advantages and disadvantages. The method that you select depends on your audience, the message, and the purpose of the presentation.

In this section, the technique for arranging data in a table using a spreadsheet program will be described. Data entry into a spreadsheet is also a convenient method for preparing graphs. Three common methods for representing data in the form of graphs will be described, including:

1. Pie charts
2. Bar charts
3. X-Y graphs

Each type of graph is used for a different purpose and selection of the correct type of presentation is essential in communicating effectively. For instance, the pie chart is used to show distributions, whereas the bar chart is used to compare one set of data with another. X-Y graphs, the most frequently used type of graphic in engineering, shows the variation of some dependant variable Y with some independent variable X. Regardless of the type of table or graphic required, this section will describe how Excel 2007 can assist you in creating high-quality, professional graphics.

Tables

Very early in the design of a hovercraft, your team will be faced with the requirement of making a special table called a parts list. The parts list provides an estimate of the overall weight of your vehicle, which is needed in order to size your levitation fan. A complete parts list should include every component required to construct your team's hovercraft. For each item listed, the parts list should include the quantity each component required, the vendor, the part number, a weight estimate and its cost. An example of a parts list created in Excel 2007 is presented in Table 8.1.

Table 8.1
Sample hovercraft parts list (albeit incomplete)

	A	B	C	D	E	F
1	**Item Description**	**Vendor**	**Unit Cost ($)**	**Quantity**	**Total Cost ($)**	**Weight (lbs)**
2	Foamular Styrofoam	Lowes	7.15	3	21.45	0.85
3	Liquid Nails (adhesive)	Lowes	1.57	1	1.57	0.10
4	Hobby Plywood (lower deck)	GPA Hobbies	3.99	2	7.98	0.45
5	Balsa Sheet (centrifugal fan)	GPA Hobbies	2.25	1	2.25	0.15
6	APC 4.2x2 Sport Propeller	GPA Hobbies	1.75	2	3.50	0.05
7	Super Speed 9-18V Hobby Motor	GPA Hobbies	5.29	3	15.87	0.30
8	MOSFET N-CH 60V 8A Transistor	Radio Shack	1.89	3	5.67	0.05
9	Lego Light Sensors	Lego Educational	16.50	3	49.50	0.10
10	TENERGY Li-Ion 14.8V 2200 mAh Battery	GPA Hobbies	32.99	1	32.99	1.00
11	RCX Rental	UMD	25.00	1	25.00	0.86
12	**TOTAL**				**165.78**	**3.91**
13						
14	**Sample formulas used to create table above:**					
15	Cell E2: =C2*D2					
16	Cell E12: =sum(E2:E11)					

A number of operations are required to construct the table shown above. First, each cell must be formatted. For instance, you may have noticed that each cell along the top row of the table (row 1) is shown in bold font type. Cells can be made bold by highlighting the desired cells and clicking the standard Microsoft icon for Bold font found under the Home tab on the main ribbon. Under this tab, the font style, size, alignment, formatting, color, etc. may be changed for each cell and each character within the cell. Under the home tab, there is subsection called Alignment. The top right icon in this subsection is called "Wrap Text". This icon can be used to allow the descriptions written in a cell to wrap and create multiple rows within the cell instead of being cut off at the start of the adjacent cell text (see cells A7, A8 and A10). Another important set of commands can be found under the Home ribbon tab in the Number subsection. These icons are used to set the number of decimal places shown in the highlighted cells and/or the style of number to be used.

Because the costs were given in U. S. Dollars ($) at the top of the parts list, each of the numbers in columns C and E were formatted to display two decimal places, which gives a precision of ± one cent. Similarly, column F, showing weight (in lbs), is specified within ± 0.01 lbs because the scale used to weigh these components was accurate to within ± 0.01 lbs. It makes no sense to list a weight to a higher level of certainty than it can be measured than it does to provide a cost to a higher precision than $\pm$ one cent. You should not list numbers in a table with a higher level of precision (number of decimal places) than your certainty in reporting.

There are a number of other formatting operations with which you must become comfortable to make professional looking tables. These include adding borders and adjusting the widths of each column/row (which is as simple as left clicking and dragging the line separating the column/row headers), It is advised that you use some time early in your academic career to better familiarize yourself with the commands required to make highly professional looking tables and graphs.

Another set of operations is required to create the parts list shown in Table 8.1. The use of mathematical formulae simplifies the creation of this table. For instance, because the unit cost is provided in column C and the quantity in column D, the total cost (column E) for a specified part is found by multiplying the numbers in columns C and D together. Excel has a built in mathematical library to make these types of repeated calculations efficiently. From row 2 of the parts list, the unit cost for the Foamular Styrofoam is \$7.15 and the quantity is 3, so the total cost is \$7.15 $\times$ 3 = \$21.45. In Excel, this calculation is made very simply by typing "=C2*D2" into cell E2. The "=" sign indicates to Excel that a formula will follow and the "*" sign indicates the multiplication operation. The basic algebraic operations use the following self-explanatory symbols: +, −, *,and /. For a much more complete set of formulae available in Excel, explore the Formulas tab on the Excel ribbon.

To enter equations more efficiently, you can select a cell to be used in a formula by simply clicking on it with the mouse instead of typing its location in a cell. Another very useful feature of Excel is that after the first equation has been typed into a cell, it can be used to automatically compute any number of similar calculations. For instance, let's automate the computation of the Total Cost column (column E). To accomplish this, copy the formula in cell E2 by pressing CTRL-C and then highlight the cells in column E from rows 2 through 11 where you are to perform a similar calculation. Next, paste this formula into each of these cells by typing CTRL-V. Excel automatically performs all of these calculations using an indexing system with the cells E# =C#*D#, where the # sign indicates the numbers of the row you have highlighted. To check the accuracy of these calculations, double click on any of the boxes in column E to show the formula hidden within that cell.

Excel has hundreds of different mathematical functions that can be used in equations, ranging from trigonometric functions like "=sin(Cell)" to statistical functions like "=average(Cell Range)". In creating Table 8.1, one additional computational technique was used in Excel. Instead of adding up all of the cells in the column showing the total cost ("=E2+E3+…+E11), the summation tool was used. To employ this tool, type "=sum(" and then highlight the cells of interest—in this case the total cost of each component—and then type ")". The summation will be performed automatically when you hit the enter key. If you double click on the cell showing the total cost, you will observe "=sum(E2:E11)" as the mathematical operation. Here, "E2:E11" provides the Cell Range over which the mathematical operation was performed. Similarly, by copying the formula for the total cost, and pasting it into cell F12, automatically computes the total weight of the entire hovercraft. If you double-click on this cell, it will show "=sum(F2:F11)". Excel indexes to the F column from the E column when the copy/paste function is used across columns in a similar way that it indexed row numbers when calculating the Total Cost as the product of the Unit Cost and the Quantity.

The user has control over the way in which Excel indexes when the copy/paste commands are employed. For instance, an entire column of numbers (let's say measurements in feet) can be converted into inches by multiplying each cell in feet by a conversion factor (12 in / 1 ft). To accomplish this conversion, the factor 12 can be placed in a cell, say B2, and is defined as \$B\$2. The "\$" before B locks in column B and the "\$" before 2 locks in row 2. When mathematical operations are indexed across columns and rows, the quantity in cell B2 will always be used for the calculations if the $ sign precedes the column and row indicators.

Excel is an extremely powerful computational tool that can be used for much more than making simple tables like the parts list shown in Table 8.1. Again, it is recommended that you devote time to further explore the procedure for inserting equations and familiarize yourself with the built-in formulas available in Excel. These can be found under the "Formulas" tab located on the main Excel ribbon. If you cannot find the formula you are seeking, do not be discouraged. Excel 2007 allows the user to define custom formulas. With some practice, you will find Excel that benefit you in many of your engineering classes.

Graphs

Excel is an extremely useful tool for quickly creating high quality graphics. The first graphic that we will create using Excel is a pie chart. A pie chart is usually employed to show distributions in data. To apply this to the hovercraft project, the data in Table 8.1 has been rearranged to create a table showing the cost to construct each of the hovercrafts major subsystems. This cost data is presented in Table 8.2.

Table 8.2
Hovercraft cost and weight breakdown among subsystems

	Cost ($)	Weight (lbs)
Structure & Levitation	38.54	1.65
Power & Propulsion	52.74	1.3
Sensors & Controls	74.50	0.96
TOTAL	165.78	3.91

Based on Table 8.2, Excel will be used to create a pie chart to illustrate which sub-system requires the largest portion of the $165.78 budget. To begin this task, click on the Insert tab located on the main ribbon. The subsection titled Charts is used to insert any one of the many different built-in chart formats available in Excel. At this time, let's focus on the task of creating a simple pie chart by selecting "Pie" as the chart format and choosing "Exploded pie in 3-D" as the specific type. After this selection has been made, a blank chart appears on the screen and a new feature on the ribbon appears that is called "Chart Tools" with the "Design" subsection selected.

The procedure for creating this a pie chart is very similar to the procedure used to create any type of chart. The data, which you will typically have in tabular format, must be selected. To make the selection, click on the "Select Data" icon under the "Data" subsection of Chart Tools tab. Then a pop-up box similar to the one presented in Fig. 8.3 appears.

Creating the chart is as simple as selecting the appropriate set of data to be shown. To do this, click the box to the right of the "Chart data range" entry box of Fig. 8.3. Next, highlight the appropriate set of data. In this case, it is the cell range containing the cost to construct each of the three major subsystems of the hovercraft design. The next step requires providing the appropriate set of labels to explain the data in the pie chart. To do this, simply click the Edit box underneath the "Horizontal (Category) Axis Labels" box of Fig. 8.3. Similarly to before, highlight the three sets of labels from Table 8.2 (Structure & Levitation, Power & Propulsion, and Sensors & Controls). Though not appropriate in the construction of a pie chart, the "Legend Entries (Series)" box shown in Fig. 8.3 allows the user to add multiple sets of data (or series) to a single graph. This is often used in the construction of engineering X-Y graphs.

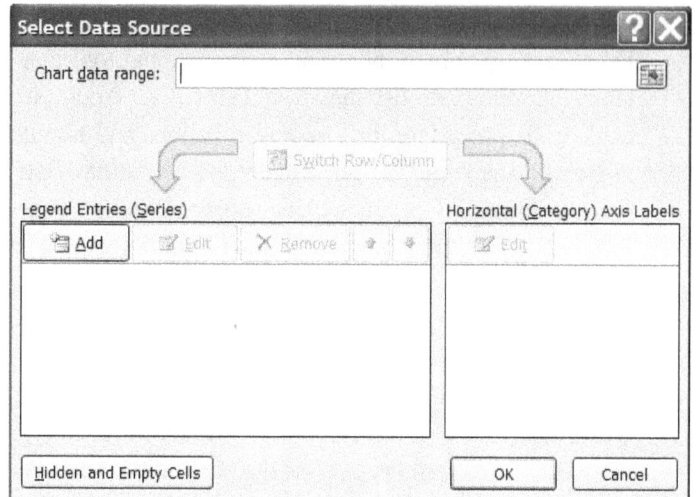

Fig. 8.3 Data source entry pop-up screen

The final information that should be added to every graphic you create includes the title of the graph, axis labels, legend labels, etc. To make these additions, switch to the "Layout" tab under the Chart Tools tab. The Labels subsection allows you to add the final information to the graph. Even for the very simple pie chart constructed, it is necessary to add a title. To add a title, select the "Chart Title" command and choose "Above Chart" as the type. Next, type a descriptive title for the chart such as "Hovercraft Cost Breakdown". Now the only information missing from this graphic is the magnitude of each section of the pie chart. To add this information, select "Data Labels" and choose "Best Fit" as the type. With some minor formatting, this simple procedure allows you to create a very professional looking pie chart similar to the one shown in Fig. 8.4.

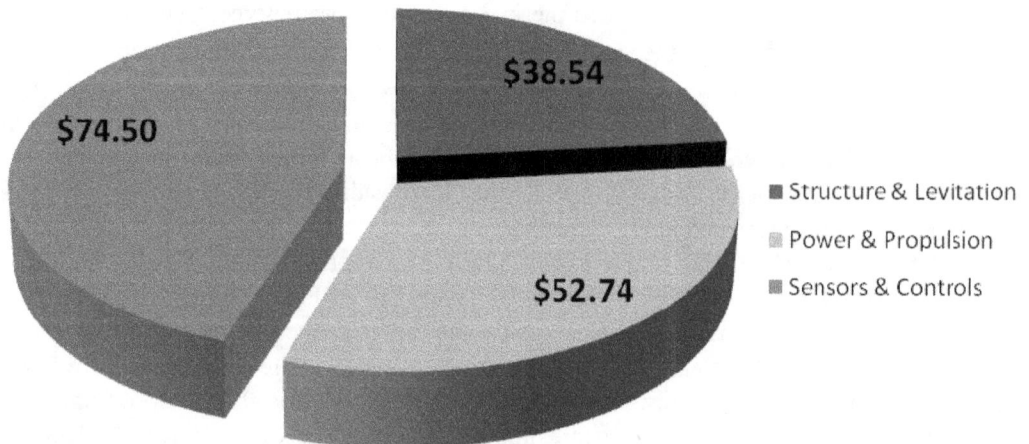

Hovercraft Cost Breakdown

Fig. 8.4 Pie chart made in Excel showing the sub-system cost distribution for a hovercraft

The process described above for inserting a pie chart can be used to insert many different types of charts and graphs in your spreadsheet or in other documents. Recall, this chapter will limit the discussion to pie charts, bar charts, and X-Y graphs, but you should explore the other types of charts that can be created using Excel. Bar charts can also be used to show distributions, but they are better suited for illustrating comparisons. As an example of a bar chart, consider the results of the hovercraft competition during the 2006-07 academic year at the University of Maryland, which is presented in Fig. 8.5. At a glance, the reader can note that the number of hovercraft constructed in the fall was significantly larger than the number of hovercraft built during the spring. At a second glance, the reader can observe that the rate of success was significantly higher in the spring than it was in the fall. Finally, the reader can conclude that less than 50% of the hovercrafts built were able to meet the product specifications during either semester. A lot of very valuable information is contained within the very simple bar chart shown as Fig. 8.5.

2006-07 Hovercraft Competition Results

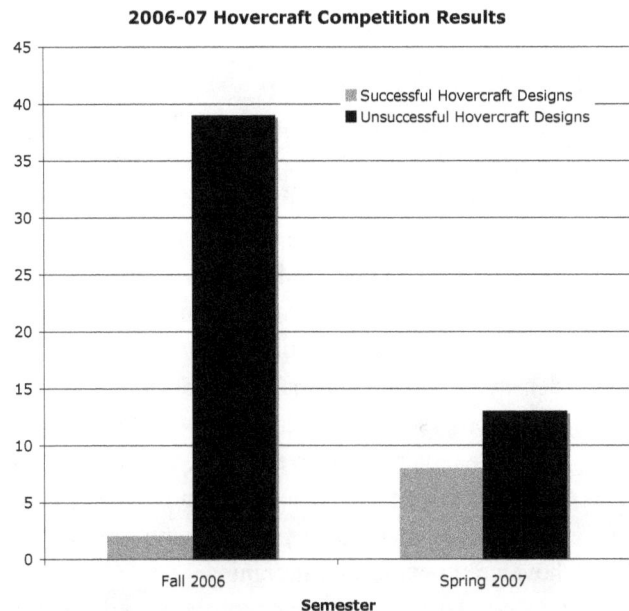

Fig. 8.5 Bar chart illustrating results of hovercraft competition

The final graphic format described here is the X-Y graph. It is inserted in your spreadsheet using techniques that are similar to those used to create pie and bar charts. First, create a table of data. Next, select insert a chart and then select X-Y or X-Y scatter as the type. In engineering, X-Y scatter plots are often used because they show data points without lines drawn through the points. If the data is to convey a story, it is customary to fit a trend line to the points. Finally, label all axes (with units when applicable), provide the graph with a descriptive title, and draft a legend when necessary. The creation of an X-Y graph will be described in the next section where the use of Excel in conducting design trade-off studies for your hovercraft is discussed.

8.4 EXCEL AS A DESIGN TOOL

As you may have realized, design is a highly iterative discipline. This fact is better understood when you consider the number of times your team's hovercraft weight estimate, size and shape has changed over the first few weeks of the semester and how often you were required to solve the static force balance equation $p_{req} = \dfrac{W}{A_D}$ to determine the required pressure (p_{req}) your team's levitation fan must create to ensure that your vehicle can hover. Similarly, how many times has your team solved $Q = h_{gap}\boldsymbol{L}\sqrt{\dfrac{2p_{req}}{\rho}}$ to determine the required air flow rate (Q) your levitation fan must provide to hover with the desired gap height (h_{gap})? Excel can make calculations like these much less tedious, as well as provide guidance on the ideal size and shape for your vehicle. Let's consider the design of a hovercraft with a square planform of dimension L × L, as shown in Fig. 8.6.

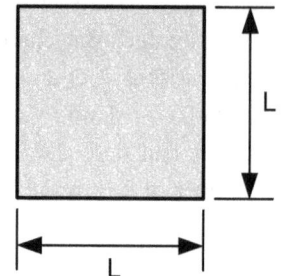

Fig. 8.6 Generic hovercraft configuration with a square planform section

For a simple shape like a square, it is very easy to determine the plenum area ($A_D = L^2$) and leakage perimeter L = 4L. Using Excel, a levitation-system sizing calculator can be created to solve the pressure and flow rate fan requirements for a range of vehicle sizes. To begin assume a vehicle weight of 3.91 lbs (17.4 N), because it is the weight estimate given in Table 8.1. While the assumption that the weight of the vehicle is constant regardless of vehicle size (L) is certainly a flaw in this analysis, it is a reasonable assumption because the hull weight represents a small proportion of the overall vehicle weight (~20%). An improved sizing calculator can be created with size-specific weight estimates by considering both the density of the hull material selected and the volume of material required. Table 8.3 illustrates a hovercraft levitation-system sizing calculator created in Excel. This table calculates the levitation fan pressure and flow rate requirements over a range of vehicle sizes from L = 0.23 m (~9.0 in) to L = 0.65 m (~25.6 in) for a hovercraft with a fixed weight and a constant hover height. It is recommended that your team create a calculator similar to this one for the configuration (size and geometry) and weight estimate for the hovercraft your team is designing.

To truly appreciate the power of the Excel, consider the impact that adding a second battery would have on the size and type of levitation fan required. By hand, a small addition to your team's part list (and weight estimate) would require a timely and tedious recalculation effort. However, using Excel, a modified weight estimate can easily be considered by simply updating the value in cell D3 of Table 8.3. Excel allows the iterative nature of design to be managed in a swift and efficient manner. In addition, it provides the user with the ability to conduct design trade off studies with little delay. How would the flow rate from the levitation fan change if you wanted to hover at a gap height of 3 mm (0.003 m) instead of 2 mm (0.002 m)? Simply change the value of cell D4 in Table 8.3 and note how the remainder of this table automatically updates.

The sizing calculator is also valuable because it allows a designer to quickly perform parametric design studies. For instance, examine column I in Table 8.3. For a constant vehicle weight and hover height, how does the required flow rate change with an increase in vehicle size (L)? Clearly, column H indicates that the required flow rate is independent of the vehicle size. Similarly, how does the required pressure change with an increase in vehicle size? By examining column D, it is clear that the pressure a fan must create decreases as the vehicle size (L) is increased. But how much does it decrease? To establish a better idea of the impact of this change, let's create an X-Y scatter plot of vehicle defining length L (column A) versus required plenum pressure (column D), as shown in Fig. 8.7 To prepare this graph, follow the procedure outlined in Section 8.3 for inserting a chart. The X-data series consists of cells A8:A22 and the Y-data series consists of cells D8:D22. This data can be inserted into the chart by choosing "Add" under the "Legend entries (Series)" section shown in Fig. 8.3.

Table 8.3
Hovercraft sizing calculator

◇	A	B	C	D	E	F	G	H	I
1	**Hovercraft Sizing**								
2									
3	**Weight Estimate (N):**		17.4				**Conversion Factors:**		
4	**h_{gap} Req. (m)**		0.002				0.00401	**inH$_2$0 / Pa**	
5	**Density (kg/m^3):**		1.225				2119	**CFM / m^3/s**	
6									
7	**L (m)**	**l (m)**	**A (m^2)**	**P_{req} (Pa)**	**P_{req} (inH$_2$0)**	**Gap Area (m^2)**	**V$_e$ (m/s)**	**Q$_{req}$ (m^3/s)**	**Q$_{req}$ (CFM)**
8	0.23	0.92	0.053	328.9	1.32	0.00184	23.2	0.0426	90.4
9	0.26	1.04	0.068	257.4	1.03	0.00208	20.5	0.0426	90.4
10	0.29	1.16	0.084	206.9	0.83	0.00232	18.4	0.0426	90.4
11	0.32	1.28	0.102	169.9	0.68	0.00256	16.7	0.0426	90.4
12	0.35	1.40	0.123	142.0	0.57	0.00280	15.2	0.0426	90.4
13	0.38	1.52	0.144	120.5	0.48	0.00304	14.0	0.0426	90.4
14	0.41	1.64	0.168	103.5	0.42	0.00328	13.0	0.0426	90.4
15	0.44	1.76	0.194	89.9	0.36	0.00352	12.1	0.0426	90.4
16	0.47	1.88	0.221	78.8	0.32	0.00376	11.3	0.0426	90.4
17	0.50	2.00	0.250	69.6	0.28	0.00400	10.7	0.0426	90.4
18	0.53	2.12	0.281	61.9	0.25	0.00424	10.1	0.0426	90.4
19	0.56	2.24	0.314	55.5	0.22	0.00448	9.5	0.0426	90.4
20	0.59	2.36	0.348	50.0	0.20	0.00472	9.0	0.0426	90.4
21	0.62	2.48	0.384	45.3	0.18	0.00496	8.6	0.0426	90.4
22	0.65	2.60	0.423	41.2	0.17	0.00520	8.2	0.0426	90.4
23									
24	**Sample formulas used to create table above:**								
25	Cell B8 =4*A8				Cell F8 =$D$4*B8				
26	Cell C8 =A8^2				Cell G8 =SQRT(2*D8/$D$5)				
27	Cell D8 =$D$3/C8				Cell H8 =F8*G8				
28	Cell E8 =D8*$G$4				Cell I8 =H8*$G$5				

Having created an X-Y scatter plot, you may wonder why the graph has a curve drawn through the data points. That is an excellent question. The data points shown in Fig. 8.7 are certainly indicating a trend. To better understand that trend, a line has been fit to the data points. To construct this line, select the "Layout" tab found in the "Chart Tools" section of the ribbon. There is an icon for Trend line in the Analysis subsection. Click on the Trend line icon and select "More trend line options…". For the trend type, select Power. Also, put an X in the boxes next to "Display Equation on chart" and "Display R-Squared value on chart". Select "Close" and note how the chart has been updated with a trend line through each point on the graph.

By choosing to display the equation of the power series and the R^2 value on the plot, valuable information is provided about the nature and accuracy of the curve fit. The R^2 value is a measure of how well a curve fits a set of data points and typically ranges from 0 to 1. An R^2 value of 1 represents a perfect fit. Excellent experimental data may approach an R^2 value of 1, but rarely ever reaches this level of perfection. As a design engineer, you will need to look at charts similar to the one shown in Fig. 8.7 to make informed design decisions. What pressure is required for an extremely small vehicle (L $\Rightarrow$ 0)? What pressure is required for an extremely large vehicle (L $\Rightarrow$ ∞)? What is the benefit in designing a vehicle with a defining length L of 0.35 m instead of 0.25 m? What happens when you define a length of 0.65 m instead of 0.55 m?

Required Plenum Pressure vs. Vehicle Size

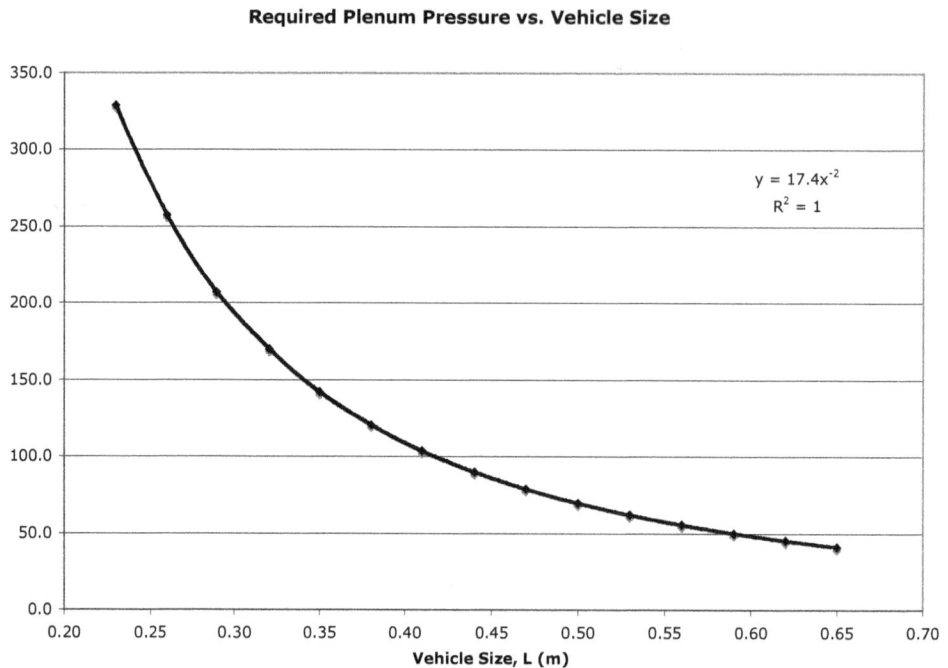

Fig. 8.7 An X-Y scatter graph created in Excel with a power series trendline added.
The Y axis (the ordinate) shows the required pressure in Pa.

Hopefully, you are beginning to understand the power and utility of spreadsheets. You are encouraged to use Excel to aid your team's hovercraft design. With design being an inherently iterative process, you will not regret the initial investment of time required to create a robust hovercraft sizing calculator for your team.

8.5 SUMMARY

Microsoft Excel has been introduced and procedures for creating tables and graphs have been described. Tables represent an excellent method for reporting numerical data. They are well suited for the presentation of precise numerical data, such as the data required in a parts list. If you need to convey numbers in a presentation or a report with six-figure accuracy, it is easy to obtain that accuracy with tabular representation.

Many different types of graphs are used to aid in visualizing data. Graphs are much less precise than tables, but they show trends, comparisons and distributions much more effectively. The pie chart is used to indicate the distribution of some quantity. The procedure for creating a pie chart using Excel was described in detail. The bar chart, used to compare the magnitudes of two or more quantities, was introduced. The bar chart was demonstrated and the methods used in its construction were discussed.

Trends are best-illustrated using X-Y graphs. X-Y graphs effectively show the trend of one variable with respect to another. The procedure for constructing this type of graphic was described, along with the steps necessary to add a trend line.

Finally, the use of Excel in the design process has been introduced. An example of sizing trade-off studies for the levitation system was given. However, similar analysis can be made for many of the other hovercraft sub-systems. The Excel spreadsheet greatly aids the iterative design process by providing a means for conducting automated design trade-off studies with results that are displayed immediately.

EXERCISES

8.1   Using Microsoft Excel, create a parts list for the hovercraft your team is developing. Remember to include the manufacturer, model number, description, cost, and weight for each component on your team's hovercraft on the parts list.

8.2   Using Microsoft Excel, prepare a table similar to the one presented as Table 8.2 showing the breakdown of weight and cost between the main sub-systems of your hovercraft design. This table should be based on the parts list created in Exercise 8.1.

8.3   Use the data from Exercise 8.2 to construct a pie chart showing the distribution of your team's cost budget between each of the main sub-systems.

8.4   Use the data from Exercise 8.2 to construct a pie chart showing the distribution of your hovercraft's weight between each of the main sub-systems.

8.5   Use the data from Exercise 8.2 to construct a bar chart showing the distribution of your team's cost budget between each of the main sub-systems.

8.6   Use the data from Exercise 8.2 to construct a bar chart showing the distribution of your hovercraft's weight between each of the main sub-systems.

8.7   Using Excel, create a table to calibrate an ultrasonic proximity sensor. The data should include distance in inches from an object as well as the associated NXT reading.

8.8   Prepare an X-Y scatter graph showing the data found in Exercise 8.7. What trend can be observed in the graph of the data?

8.9   Fit an appropriate trend line to the X-Y scatter graph created in Exercise 8.8. Include the R^2 value on the graph. Was the trend line you established represent a good fit?

8.10   Create a levitation sizing calculator, similar to the one described in Section 8.4, for the bullet shape configuration shown in Fig. Ex 8.10. Note that the bullet shape consists of an aft $L \times L$ square with a semi-circle nose in the bow.

Fig. Ex 8.10

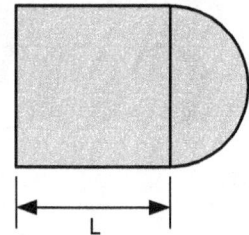

8.11   Based on the results of Exercise 8.10, create an X-Y scatter plot of the area (A_D) versus the required plenum pressure (p_{req}).

8.12   Create a sizing calculator for your team's power sub-system that computes the required battery capacity based on the current flow required for each electrical component on your hovercraft.

PART III

COMMUNICATION

CHAPTER 9

TECHNICAL REPORTS

9.1 INTRODUCTION

It is important that you enjoy writing, because engineers often have to prepare several hundred pages of reports, theoretical analyses, memos, technical briefs and letters during a typical year on the job. It is essential that you learn how to effectively communicate—by writing, speaking, listening and employing superb graphics. Communication, particularly good writing, is an extremely important skill. Advancement in your career will depend on your ability to write well. It is recognized that you will be taking several courses in the English Department and other departments in Social Sciences, Arts and the Humanities that will require many writing assignments. These courses should help you immensely with the structure of your composition and the development of good writing skills. Most of the assignments will be essays, term papers, or studies of selected works of literature. However, there are several differences between writing for an engineering company and writing to satisfy the requirements of courses such as English 101 or History 102.

In college, you write for a single reader—your instructor. He or she must read the paper to grade it, subsequently determining how well you are doing in class. In industry, many people within and outside the company, and with different backgrounds and experience, may read your report. In class, the teacher is the expert, but in industry the writer of the report is supposed to be the individual with the knowledge. Many of those working in industry do not want to read your report because they are busy: their phone is ringing continuously, meetings are scheduled back to back, and many important tasks must be completed before the end of the day. They read—if not skim—the report only because they must be aware of the information that it contains. They want to know the key issues, why these issues are important and who is going to take the actions necessary to resolve them. In writing reports in industry, you cut to the chase. There is no sense in writing a 200-word essay when the facts can be provided in a 40 to 50-word paragraph that is brief and cogent. An elegant writing style, so valued by instructors in the arts and humanities, is usually avoided in engineering documentation. In writing an engineering document, do not be subtle; instead be obvious, direct and factual.

In the following sections, some of the key elements of technical writing, including an overall approach, report organization, audience awareness and objective writing techniques will be described. Then a process for technical writing, which includes four phases: composing, revising, editing and proofreading is discussed.

9.2   APPROACH AND ORGANIZATION

The first step in writing a technical report is to be humble. Realize that only a very few of the many people who may receive a copy of your report will read it in its entirety. Busy managers and even your peers read selectively. To adapt to this attitude, organize your report into short, stand-alone sections that attract the selective reader. Three very important sections—the title page, summary and introduction are located at the front of the report so they are easy to find. At the front of your report, they attract attention and are more likely to be read. In college, you call the page summarizing your essay an abstract, but in industry it is often called an executive summary. If it is prepared for an executive, perhaps a manager will consider it sufficiently important to take time to read it. Follow the executive summary with an introduction and then the body of the report. A common outline to follow in organizing your report is provided below:

- Title page
- Executive summary
- Table of contents, list of figures and tables
- Nomenclature—only if necessary
- Introduction
- Technical issue sections
- Conclusions
- Appendices

The title page, the executive summary and the introduction are the most widely read parts of your report. Allow more time polishing these three parts, because they offer the best opportunity to convey the most important results of your investigation.

Title Page

The title page provides a concise title of the report. Keep the title short—usually less than ten words; you will have an opportunity to provide more detail in the body of the report. The authors list their names, affiliations, addresses and often their telephone and fax numbers and e-mail addresses. The reader, who may be anywhere in the world, should be provided with a means to contact the authors to ask questions. For large firms or government agencies, a report number is formally assigned and is listed with the date of release of the report on the title page. A title page of a report on a computer code for designing combustors is shown in Fig. 9.1. This report was prepared by several authors from different organizations and was sponsored by NASA (National Aeronautics and Space Administration) and ARL (Army Research Laboratory).

Executive Summary

The executive summary should, with rare exceptions, be less than a page in length (about 200 words). As the name implies, the executive summary provides, in a concise and cogent style, all the information that the busy executive needs to know about the content in the report. What does the big boss want to know? Three paragraphs usually satisfy the chief. First, briefly describe the objective of the study and the problems or issues that it addresses. Do not include the history leading to these issues. Although there is always history, the introduction is a much more suitable section for historical development and other background information related to the problem. In the second paragraph, describe your solution and/or the resolution of the issues. Give a very brief statement of the approach that you employed, but reserve the details for the body of the report. If actions are to be taken by others to resolve the issues, list the actions, those responsible and the dates of implementation. The final paragraph indicates the importance of the study to the business. Cost savings, improvements in the quality of the product, gains in the market share, enhanced reliability, etc. can be cited. You want to convince the executive that your work was worth its cost, valuable to the corporation and that your group performed admirably.

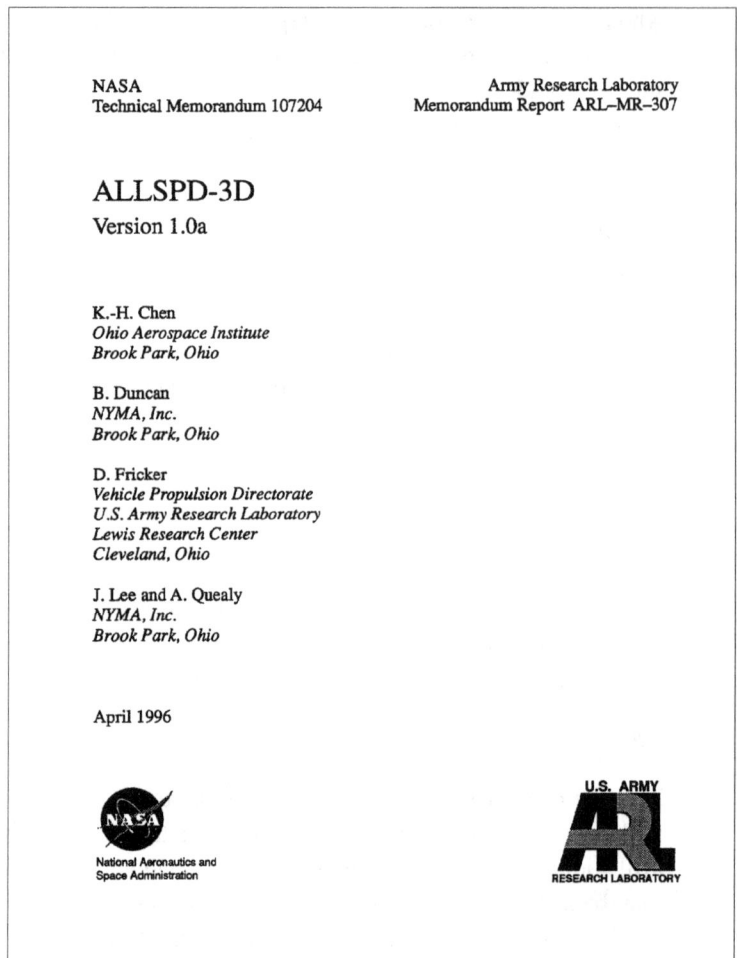

NASA
Technical Memorandum 107204

Army Research Laboratory
Memorandum Report ARL–MR–307

ALLSPD-3D
Version 1.0a

K.-H. Chen
Ohio Aerospace Institute
Brook Park, Ohio

B. Duncan
NYMA, Inc.
Brook Park, Ohio

D. Fricker
Vehicle Propulsion Directorate
U.S. Army Research Laboratory
Lewis Research Center
Cleveland, Ohio

J. Lee and A. Quealy
NYMA, Inc.
Brook Park, Ohio

April 1996

NASA
National Aeronautics and
Space Administration

U.S. ARMY
ARL
RESEARCH LABORATORY

Fig. 9.1 Illustration of a title page of a professional report.

Introduction

In the introduction, you restate the problem or issues. Although the problem was already defined in the executive summary, **redundancy is permitted in technical reports**. Recognize that you have many readers, so important information, such as a clear definition of the problem is placed in different sections of the report. In the introduction, the problem statement can be much more complete by expanding the amount of information describing the problem. A paragraph or two on the history of the problem is in order. People reading the introduction are more interested than those reading only the executive summary; therefore, more detail is appropriate.

Following the problem statement, establish the importance of the problem to both the company and the industry. Again, this is a repeat of what was included in the executive summary, but your arguments are expanded. You can cite statistics, briefly describe the analyses that lead to the alleged cost savings, give a sketch of customer comments indicating improved quality, etc. In the executive summary, you simply stated why the study was important. In the introduction, arguments are presented to convince the reader that the investigation was important and worth the time, effort and cost.

In the next segment of the introduction, you briefly describe the approach followed in addressing the problem. You begin with a literature search, followed by interviews with select customers, a study of the products that failed in service, interviews with manufacturing engineers and the design of a modification to decrease the failure rate and improve the reliability of the product. In other words, tell the reader what actions were required to solve the problem.

The final paragraph in the introduction outlines the remaining sections of the report. True, that information is already covered in the table of contents, but again, you repeat to accommodate the selective reader. Also, placing the report contents in the introduction gives you the opportunity to add a sentence or two describing the content of each section. Perhaps you can attract additional reader interest in one or more sections of the report and convince a few people to study the report more completely.

Organization

An organization for the report was suggested by the bullet list shown at the beginning of this section. However, you should recognize that an organization exists for every section and every paragraph in the report. In writing the opening paragraph of the section, you must convey the reason for including it in the report and the importance of its content. You certainly have a reason for including this section in the report, or you would not have wasted the time required to write it. By sharing this reason with your readers, you convince them that the section is important.

A paragraph is written to convey an idea, a thought or a concept. The first sentence in the paragraph must convey that idea, clearly and concisely. The second sentence should describe the significance or importance of the idea. The following sentences in the paragraph support the idea, expand its scope and give cogent arguments for its importance. State the idea in the first sentence and then add more substance in the remaining sentences. If the report is written properly, a speed-reader should be able to read the first two sentences of every paragraph and glean 80 to 90% of the important information conveyed in the report.

As you add detail to a section or a paragraph, develop a pattern of presentation that the reader will find easy to read and understand. First, tell the reader what he or she will be told, then provide the details of the story and finally summarize with the action to be taken to implement the solution. Do not try this approach in the theme that you have to write for English class, because the instructor will not appreciate your redundant style. Technical writing is not the same as theme writing. In technical writing, you will often reiterate facts to insure that the message is conveyed to the readers. Often one or more of the readers must take corrective actions for the benefit of the company and the industry.

You will frequently include analysis sections in engineering reports. In these sections, the problem is stated and then the solution is described. After the problem statement and the solution, you introduce the details. The reader is better prepared to follow the details (the difficult part of the analysis) after they know the solution to the problem.

9.3   KNOW YOUR READERS AND OBJECTIVE

Another important aspect to technical writing is to know something about the readers of your report. The language that you use in writing the report depends on the knowledge of the audience. For example, I am writing this book for engineering students. I know your math and verbal skills from the SAT scores required for admission. I recognize that you have very good language skills and even better math skills. You are bright and articulate. I believe that the content in this book closely matches your current abilities. However, there are a few sections in the book that deal with topics usually found in the business school. Profits and costs are important in product development, and although they may not be as interesting as a computer programming or circuit analysis, they need to be clearly understood.

When you write a report for a class assignment, you understand that the instructor is the only reader with which you need to communicate. You also know that he or she is knowledgeable. In that sense, the writing assignments in a typical college class are not realistic. For example, suppose your manager asks you to write an assembly routing (a step by step set of instructions for the assembly of a certain product) for the production of your new widget in a typical factory in the U. S. What language would you use as you write this routing? If the plant is in Florida or the Southwest for example, Spanish may be the prevalent language. You should also be aware that a significant fraction of the factory workers in the U. S. are functional illiterates and many others detest reading. In this case, it might be a good approach to use fewer words and many cartoons, photographs and drawings.

Know the audience **before** you begin to write, and adapt your language to their characteristics. Consider four different categories of readers:

> Specialists with language skills comparable to yours.
> Technical readers with mixed disciplines.
> Skilled readers, but not technically oriented.
> Poorly prepared readers who may be functional illiterates.

Category 1 is the audience that is the least difficult to address in your writings; the audience in the last category is the most difficult. Indeed, with poorly prepared readers, it is probably better to address them with visual presentations conveyed with television monitors and videotapes.

A final topic, to be considered in planning, is the classification of the technical document that you intend to write. What is the purpose of the task? Several classifications of technical writing are listed below:

- Reports: trip, progress, design, research, status, etc.
- Instructions: assembly manuals, training manuals and safety procedures.
- Proposals: new equipment, research funding, construction and development budget.
- Documents: engineering specifications, test procedures and laboratory results.

Each classification of writing has a different objective and requires a different writing style. If you are writing a proposal for development funding, you need persuasive arguments to justify the costs of the proposed program. On the other hand, if you are writing an instruction manual to assemble a product, arguments and reasons for funding are not an issue. Instead, you would prepare complete and simple descriptions with many illustrations to precisely explain how to accomplish a sequence of tasks involved in the assembly.

9.4   THE TECHNICAL WRITING PROCESS

Whether you are writing a report as part of your engineering responsibilities in industry or as a student in college, you will face a deadline. In college, the deadlines are imposed well in advance and you have a reasonable amount of time to prepare your report. In industry, you will have less time, and an allowance must be made for the delay in obtaining approvals from management before you can release the report. In both situations, you have some limited period of time to write the report. The idea is to start as soon as possible. Waiting until the day before the deadline is a recipe for disaster.

Many professionals do not like to write; consequently, they suffer from writer's block. They sit at their keyboard hoping for ideas to occur. Clearly, you do not want to join this group, and there is no need for you to do so. A technique is described in this section that should help you to avoid writer's block.

Define Your Task

Start your report by initially following the procedures described in the previous section. Understand the task at hand, and classify the type of technical document that you have to write. Define your audience; before starting to write, establish the level of detail and the language to employ. Prepare a skeletal outline of the report. Start with the outline presented previously and add the section headings for the body of the report. This outline is very brief, but it provides the structure for the report. This structure is important because it has divided the big task (writing the entire report) into several smaller tasks (writing one section of the report at a time).

Gather Information

It is impossible to write a report without information. You must generate the information to be presented in the report. You can use a variety of sources: interviews with peers, instructors or other knowledgeable persons who are willing to help; complete literature searches at the library; read the most suitable references; conduct additional Internet searches for information. Take notes as you read, gleaning the information applicable for your report. Be careful not to plagiarize. While you can use material from published works, you cannot copy the exact wording; the statements must be in your own words. If the report has an analysis, go to work and prepare a statement of the problem, execute its solution and make notes about interpreting the solution.

Organize Data

When you have collected most of the information that is to be discussed in your report, organize your notes into different topics that correspond to the section headings. Then incorporate your topics in an initial outline that will grow from a fraction of a page to several pages as you continue to incorporate notes in an organized format.

Compose Document

You are now ready to write. Writing is a tough task that requires a great deal of discipline and concentration. It is suggested that you schedule several blocks of time and reserve them exclusively for writing. The number of hours that should be scheduled will depend on the rate at which you compose, revise, edit and proofread. Some authors can compose about a page or so per hour, but most students need more time.

In scheduling a block of time for composing, you should be aware of your productive interval. Most writers take about a half an hour to come up to speed, and then they compose well for an hour or two before their attention and/or concentration begins to deteriorate. The quality of the composition begins to suffer at this time, and it is advisable to discontinue writing if you want to maintain the quality of your text. This fact alone should convince you to start your writing assignments well before the deadline date.

While you are writing, avoid distractions. Writing requires deep concentration—you must remember your message, the supporting arguments, the paragraph and sentence structure, grammar, vocabulary and spelling. Find a quiet, comfortable place and focus your entire concentration on the message and the manner in which you will present it in your report.

Naturally, some sections of a report are easier to write than others. The easiest are the appendices, because they carry factual details that are nearly effortless to report. The interior sections of the body of the report carry the technical details that are also easy to prepare, because describing detail is less concise and less cogent. Do not become careless on these sections because they are important; however, each sentence does not have to carry a knockout punch.

The most difficult sections are the introduction and the executive summary. It is advisable to write these two sections after the remainder of the first draft of the report is completed. Usually the introduction is written before the executive summary. Although the introduction contains much of the same information as the executive summary, it is more expansive. The introduction can be used as a guide in preparing the executive summary.

9.5   REVISING, EDITING AND PROOFREADING

Writing is a difficult assignment. Do not expect to be perfect in the beginning. Practice will help, but for most of us, it takes a very long time to improve our skills because writing is such a complex task. Expect to prepare several drafts of a paper or report, before it is ready to be released. In industry, several drafts are essential because you often will seek peer reviews and manager reviews are mandatory. In college, you have fewer formal requirements for multiple drafts, but preparing several drafts is a good idea if you want to improve the report and your grade.

First and Subsequent Drafts

The first draft is focused on composition, and the second is devoted to revising the initial composition. Use the first draft for expressing your ideas in reasonable form and in the correct sections of the report. In the second draft, focus on revising the composition. Defer the editing process, and concentrate on the ideas and their organization. Make sure the message is in the report and that it is clear to all of the readers. Polish the message later in the process.

Several hours should elapse between the first and second drafts of a given section of the report. If you read a section over and over again, you soon lose your ability to judge its quality. You need a fresh, rested brain for a critical review. In preparing this textbook, the author composes on one day and revises on the next day. Revising is always scheduled for an early morning block of time. When you are rested, concentration is usually at its highest level.

Let's make a clear distinction between composing, revising, editing and proofreading. Composition is writing the first draft where you formulate your ideas and organize the report into sections, subsections and paragraphs. Unless you are a super talented writer, your first draft is far from perfect. The second draft is for revisions where you focus on improving the composition. The third draft is for editing where errors in grammar, spelling, style, and usage are corrected. The fourth draft is for proofreading where typographical errors are eliminated.

Revising

When you revise your initial composition, be concerned with the ideas and the organization of the report. Is the report organized so that the reader will quickly ascertain the principle conclusions? Are the section headings descriptive? Sections and their headings are helpful to the reader because they help organize his or her thoughts. Additionally, they aid the writer in subdividing the task and keeping the subject of the section in focus. Are the sections the correct length? Sections that are too long tend to be ignored or the reader becomes tired and loses concentration before completely reading them.

Question the premise of every paragraph. Are the key ideas presented together with their importance before the details are included? Is enough detail given or have you included too many trivial items? Does the paragraph contain a single idea or have you tried to include two or even three ideas in the same paragraph? It is better to use a paragraph for every idea even if the paragraphs are short.

Have you added transition sentences or transitional phrases? The transition sentences, usually placed at the end of a paragraph, are designed to lead the reader from one idea to the next one. Transitional phrases, embedded in the paragraph, are to help the reader place the supporting facts in proper perspective. You contrast one fact with another, using words like **however** and **although**. You indicate additional facts with words such as **also** and **moreover**.

Editing

When the ideas flow smoothly and you are convinced that the reader will follow your concepts and agree with your arguments, begin to edit. Run the spell checker, and eliminate most of the typographical errors and the misspelled words in the report. Search for additional misspelled words, because the spell checker does not detect the difference between certain words like **grate** or **great**, or say **like** and **lime** and between **from** and **form**.

Look for excessively long sentences. When sentences become 30 to 40 words in length, they begin to tax the reader. It is better to use shorter sentences where the subject and the verb are close together. Make sure that the sentences are actually sentences with a subject and a verb of the same tense. Have you used any comma splices (attaching two sentences together with a comma)? Examine each sentence and eliminate unnecessary words or phrases. Find the subject and the verb, and attempt to strengthen them. Look for redundant words in a sentence, and substitute different words with similar meaning to eliminate redundancy. Be certain to employ the grammar check incorporated in the word processing program. It is not perfect, but it does locate many grammatical errors.

Proofreading

The final step is proofreading the paper to eliminate errors. Start by running the spell checker for the final time. Then print out a clean, hard copy to use for proofreading. Check all the numbers and equations in the text, tables and figures for accuracy. Then read the text for correctness. Most of us have trouble reading for accuracy because we read for content. We have been trained since first grade to read for content, but we rarely read for correctness.

To proofread, read each word separately. You are not trying to glean the idea from the sentence, so do not read the sentence as a whole. Instead, read the words as individual entities. If it is possible, arrange for some help from a friend—one person reading aloud to the other with both having a copy of the manuscript. The listener concentrates on the appearance of each word and then checks the text against the spoken word to verify its correctness.

9.6   WORD PROCESSING SOFTWARE

Word processing software is a great tool to use in preparing your technical documents. The significant advantage of using word processing software is the ability to revise, edit and make the necessary changes without excessive retyping. You can mark, delete, cut, copy and move text. You can easily insert words, phrases, sentences and/or paragraphs.

While working with word processing software, it is difficult to revise, edit, and proofread on the screen because only a portion of the page is visible. Better results are obtained if you print a hard copy and view the entire page. With the entire page, you can see several paragraphs at once and can check that they are in the correct sequence. Use double spacing when printing the first few drafts to give adequate space between the lines for your modifications.

After you have completed the revisions on the hard copy, make the required modifications to the text using word processing software and save the results to a memory stick. It is recommended that you keep only the most recent version (draft) of the document. If you save several versions of the document, it is necessary to keep a logbook of the changes to each version. If the writing takes place over several weeks, it is easy to lose track of what changes you made and which version of the hard copy goes with which electronic copy. You will find it easier to keep one electronic file (on your memory stick) of the most recent draft with a hard (paper) copy to serve as a back up.

Word processing software has several features that are helpful in editing. The spellchecker finds most of your typographical errors and provides suggestions to correct many misspelled words. The search or find command permits you to systematically examine the entire document so you may replace a specific word with a better substitute. A thesaurus is available to help you with word selection, but you must be careful when using it. Make sure you understand the meaning of the word that you select; do not try to impress the reader by using long or unusual words. Short words that are easily understood by the reader are preferred in technical writing.

Formatting the report is another significant advantage of word processing software. You can easily produce a document with a professional appearance. The formatting bar permits you to select the type of font, the point size (the height of the characters) and emphasis such as bold, underline or italic.

You can also format the page with four different types of line justification, which are commonly available. I am using word processing software to prepare this textbook, and applying "justify" alignment for both the right and left margins. The word processing software automatically added the spaces required in each line and aligned both margins. In designing your page, use generous margins. One-inch margins all around are standard unless the document is to be bound; then the left margin is usually increased from 1.00 to 1.25 inch.

Tables and graphs can be inserted in the text. Take advantage of the ability to introduce clip art into the text or to transfer spreadsheets and drawings produced in other software programs into your report. Position the tables and figures as close as possible to the location in the text where they are introduced. Try to avoid splitting a table between two pages. If the table is longer than a page or two, consider placing it in an appendix. Identify figures and photographs with a suitable caption. The caption, placed directly beneath the figure, is to reinforce the message conveyed by the illustration.

9.7   SUMMARY

Writing is a difficult skill to master; most engineers experience significant problems when they prepare reports early in their careers. Unfortunately, the writing experiences in college do not correspond well with the writing requirements in industry. In college, you write for a knowledgeable instructor. In industry, you write for a wide range of people with different reading abilities. Moreover, the audience often varies from assignment to assignment. In both college and industry, you write to meet deadlines imposed by others. While writing may not always seem to be fun, there are many techniques that you can employ to make writing much easier and more enjoyable.

The first technique is organization; an outline for a typical report was suggested. Gather information for the report from a wide variety of sources, and generate notes that you can use to refresh your memory as you write. Sort the notes and transpose the information to expand the outline for your report.

When you organize the outline, but before you begin to write, determine as much as possible about your audience. They may be technically knowledgeable regarding the subject or they may be functionally illiterate. The language that you use in the report will depend on the reader's ability to understand. Also understand the objective of your document. Is it a report, an extended memo, a proposal or an instructional manual? Styles differ depending on the objective, and you must be prepared to change accordingly.

There is a process to facilitate the preparation of any document. It begins with starting early and working systematically to produce a very professional document. Divide the report into sections and write the least difficult sections (appendices and technical detail portions) first. Defer the more difficult sections, such as the executive summary and the introduction until the other sections have been completed.

The actual writing is divided into four different tasks—composing, revising, editing and proofreading. Keep these tasks separate:

- Compose before revising

- Revise before editing
- Edit before proofreading
- Proofread with great care.

Multiple drafts are necessary with this approach, but the results are worth the effort. Word processing software does not substitute for clear thinking, but it is extremely helpful in preparing professional documents. Word processing saves enormous amounts of time in a systematic editing process. It enables a mix of art, graphics and text neatly integrated into a single document. Word processing software has a thesaurus and word search features that are helpful in editing. It essentially turns a computer and printer into a print shop so that you have wide latitude in the style and appearance of your professional documents.

REFERENCES

1. Eisenberg, A., Effective Technical Communication, 2nd Edition, McGraw Hill, New York, NY, 1992.
2. Goldberg, D. E., Life Skills and Leadership for Engineers, McGraw Hill, New York, NY, 1995.
3. Elbow, P. Writing with Power, 2nd Edition Oxford University Press, New York, NY, 1998.
4. Alreb, G. T., C. T. Brusaw and W. E. Oliu, The Handbook of Technical Writing, 7th Edition, St. Martins Press, New York, NY, 2003
5. Struck, W. C. Osgood and R. Angell, Elements of Style, 4th Edition, Allyn & Bacon, Needham Heights, MA, 2000.
6. Pickett, N. A. et al, Technical English: Writing, Reading and Speaking, Longman, Reading MA, 2000.

EXERCISES

9.1 Prepare a brief outline of the organization for the final report describing the development of the hovercraft that your team is designing.
9.2 Prepare an extended outline of the final report for the development of the hovercraft that your team is designing.
9.3 Write a section describing one of the major subsystems for your team's hovercraft.
9.4 Write a section covering the testing of a subsystem of the hovercraft that your team is designing.
9.5 Write an Introduction for the final report on the hovercraft that your team is developing.
9.6 Write the Executive Summary for the final report on the hovercraft that your team is developing.
9.7 Revise the Introduction that another team member wrote.
9.8 Edit the Introduction after it has been composed and revised by others.
9.9 Proofread the Introduction after it has been composed, revised, and edited by others.
9.10 Edit the final report after other team members have completed it.
9.11 Proofread the final report after other team members have completed it.

CHAPTER 10

DESIGN BRIEFINGS

10.1   INTRODUCTION

You communicate by writing, speaking and with a number of different visual methods. All these modes of communication play a role as you attempt to convey your thoughts, ideas and concerns to others, ands they all are important. Let's focus your attention on speaking in this chapter. You use speech almost continuously in your daily life. Why do you need to study about design briefings? There are several reasons. You usually speak with your friends and family in an informal style. You know them and feel comfortable talking with them. They know you. They are concerned with your well being, genuinely like you and are usually interested in what you have to say. A professional presentation is different. It is a formal event that is usually scheduled well in advance. The audience may include a few friends, but usually it consists of peers (coworkers) and strangers. The time you have in which to convey your message is limited, and the audience may not be interested in your message. A person or two in the audience may be managers who control your advancement in the company. There are many reasons for tension headaches as you prepare for a professional presentation.

The design briefing is extremely important to both the product development process and to your career. Information must be effectively transmitted to your peers, management and any external parties involved in the project. Clear messages accurately defining problems that the development team is effectively addressing are imperative. On the other hand, ambiguous messages are often misunderstood, hinder the definition of the problem and lead to delays in implementing timely solutions. Clearly, such messages should be avoided. The design briefing provides an opportunity to review the status of a specific product development. It also permits peers to share their ideas with you and affords management an open forum for assessing the quality of your work and the progress made by your development team. Because the professional presentation is critically important, you should become knowledgeable about some effective techniques for accurately delivering your messages to a mixed group of strangers, peers and acquaintances.

10.2   SPEECHES, PRESENTATIONS AND DISCUSSIONS

To begin, let's distinguish between three types of formal methods of oral communication—speeches, presentations and group discussions.

Speeches

The speech is the most formal of the three modes of communication. In the year of a presidential or senatorial election, there are always many examples of speeches. Everyone is besieged with political addresses that clearly illustrate the key features of speeches. These speeches are given to large audiences in huge rooms, stadiums or coliseums. The setting is usually not appropriate for visual aids although the speaker usually uses a teleprompter. The audience is diverse with a wide range of concerns. Those listening to the speech are of widely different ages with different interests and persuasions. The speech is carefully scripted with the speaker usually reading or closely following the script. Ad-libbing is carefully avoided. Communication is one way—from the speaker to the audience. Questions are usually not appropriate and generally not appreciated. Time is strictly enforced unless you are the President of the United States giving the State of the Union address. Fortunately, engineers are rarely called upon to make speeches; hence, this topic will not be developed further in this book.

Presentations

Professional presentations differ from speeches. Presentations are made to smaller audiences in smaller rooms that are usually equipped with electric power, light controls and projection devices. You depend on visual aids, demonstrations, simulations and other props to help convey your message during a typical presentation. The audience is knowledgeable (about your topic) and usually has many common characteristics. The presentation is very carefully prepared because of its importance. It has order and structure, but it is not scripted. The flow of information is largely from the speaker to the audience; however, questions are permitted and often encouraged. The speaker is considered the expert, but discussion, comments and questions give the audience an opportunity to share their knowledge of the topic. Time is carefully controlled and is often insufficient from the speaker's viewpoint. The professional presentation is a vitally important mode of oral communication that you must quickly learn to master.

Group Discussions

Group discussions are also very important to the design engineer. The audience is smaller with much more in common. The group may all be members of a development team. The topic being discussed is narrowly focused. The speaker (central person) serves as a moderator, and he or she is an expert on the topic being considered. However, members of the discussion group (audience) may be as knowledgeable as the speaker (moderator). The moderator works in two modes. He or she may act as a presenter giving brief background information to frame an issue that prompts any member of the group to discuss the topic. The moderator may also direct questions to a group member known to be the most knowledgeable concerning the issue being addressed. The speaker (moderator) controls the flow of the information, but the flow is clearly multi-directional—from speaker to the group and from one group member to another. When a group (team) discusses a design issue, time is difficult to control and the content and range of coverage are strongly dependent on the skill of the moderator. Group discussion is very important in industry because the leader of a development team will often use this method of communication to identify problems and to initiate efforts directed toward their solution. Group discussion methods will not be described in this textbook; however, it is recommended that you watch *Washington Week in Review* on the PBS television network to gain some insights. While the participants are journalists, you can adopt many of moderator's clever techniques to guide engineering discussions.

Design Briefing

A design briefing is a type of professional presentation. It is the method you will use in speaking before the class and will be an educational objective emphasized in this course. Indeed, you probably will be required to make presentations describing your team's product development on at least two occasions— the preliminary and the final design reviews. In these two design reviews, you will employ visual aids to assist in delivering the information required to describe the status of your project with conciseness and clarity.

10.3   PREPARING FOR THE BRIEFING

A design briefing is too important to take casually. You should prepare very carefully to insure that you will accurately report the status of your team's development and to identify problems or unresolved issues. There are three aspects that you should consider in your preparations:

- Identify the audience
- Plan the organization of the presentation and your coverage of each topic
- Prepare interesting and attractive visual aids.

Identifying the Audience

In a typical college classroom, it is much too easy to identify the audience. You have your peers, the instructor and perhaps a visitor or two from industry. In an industrial setting, the audience will be more difficult to identify because it will be much more diverse. The size and diversity of the audience depends on the magnitude of the development project. A small briefing will have 10 to 20 people in attendance with most participants from within the company. A large briefing may include 50 to 100 representatives who are both internal and external to the company. The characteristics of the audience are important because you must adapt the content of the presentation to the audience. Classify your audience with regard to their status, interest and knowledge before you begin to plan the style and content for your presentation.

 The status of the audience refers to their position in the various organizations they represent. Are they peers, managers or executives? If they are a mix, which is likely, who is the most important? If you are preparing a design briefing for high-level management, it must be concise, cogent and void of minor detail. Executives are busy, stressed and always short on time. High-level executives and managers are impatient and rarely interested in the technical details that engineers love to discuss in their presentations. Recognize these characteristics and adjust the content and the timing of your presentation accordingly. The executives are interested in costs, schedule, market factors, performance, risk or any other critical issue that will delay the development or increase projected costs of either the development or the product. Usually you will have only 5 to 10 minutes to convey this information at an executive briefing.

 First and even second level managers have more time and are more interested in you and the designs that you are creating. If you prepare the presentation for this lower level of management, you can safely plan for more time (10 to 20 minutes). These managers are likely to be engineers, and they will share your knowledge of the subject. In fact, they probably will be more experienced, expert and knowledgeable than you. They will want to know about the schedule and costs because they share responsibilities with the higher-level management for meeting these goals. However, they are also interested in the important details, and you can engage them in a discussion of subsystem performance. The first and second level managers usually control the resources for the development team. If you need help, make sure that they receive the message in enough detail to provide the assistance required. Do not hide the problems that your team is encountering. Managers do not like surprises. If you have a problem, make sure they understand it and are able to participate in its solution. If you hide a problem from management and it causes a delay or escalation in costs, the manager will take the heat and you will be in a very cold doghouse.

Peer Reviews

Design briefings for peers are often less tense, and you will have more time (20 to 30 minutes). You will address schedule and cost issues because everyone needs to know if your team is meeting the milestones on the Gantt charts. However, the main thrust of your presentation will pertain to design details. If you are working on a subsystem that interfaces with other subsystems, you will cover your subsystem in sufficient detail for all the team members to understand its geometry, interfaces, performance, etc. It is particularly important that the interface issue be fully addressed in the presentation. If four subsystems are to be integrated in a given product, many details about all four subsystems must be addressed to ensure that the integration goes smoothly. Suppose that your team is developing a power tool with a motor that draws a current of 10 amps, and some team member plans to employ a switch, rated at 6 amps, to turn the power on/off. The switch will probably function satisfactorily in the short-term tests with the prototype, but it will malfunction in the field with extended usage sometime after the product has been released to the market. Clearly, the two persons responsible for the interface between the motor and the switch did not communicate properly. The integration of the two subsystems failed.

The most important purpose of design briefings with peers is to make certain that all of the details have been addressed. You strive to integrate the subsystems without problems arising when the prototype is assembled and tested.

Peer reviews in the absence of management are very beneficial in the development process. The knowledge of most of the team members is about the same although some members are more experienced than others. The topic is usually a detailed review of one subsystem or another. The audience is fresh and capable of providing a critical assessment of the technology employed. Your peers may check the accuracy of your analysis, comment on the choice of materials, discuss methods to use for manufacturing and assembly and relate their experiences with similar designs on previous products. Peer reviews afford the opportunity for the synergism that makes good products better. They also provide a forum for passing on important lessons from the more experienced designers to beginning designers in a friendly, stress-free environment.

The subject of the presentation is self-evident in a design review. You are developing a product, and the content of the design briefing will deal with one or more issues regarding this development. There is a choice of topics and considerable latitude regarding the content. Although advice has been given in previous paragraphs about matching content with the interests of your audience, you must understand the importance of knowing who will be in your audience. Executive reviews serve a different purpose than peer reviews, and the content and the time allotted for the presentation is adjusted accordingly.

For every type of review, there are two **absolute** rules that you must follow. First, **you must know the material absolutely cold**. It takes an audience about two seconds to understand that you are faking it. Also, some presenters fail to rehearse adequately, and the audience often misinterprets this lack of preparation for insufficient knowledge. In either event, when they realize that you are not the expert you allege to be. Your presentation fails. You have **lost** the opportunity to effectively communicate your message. Second, **be enthusiastic**. It is all right to be calm and cool, but do not be dead on arrival. You must command the attention of your audience or they will quickly turn off. You control the attention of the audience only if you are enthusiastic and knowledgeable in presenting your material (story).

10.4   PRESENTATION STRUCTURE

There is a well-accepted structure for professional presentations. If you are familiar with PowerPoint®, a graphics presentation program (GPP) marketed by Microsoft, you are already aware of the templates included in this program for various types of presentations. An excellent structure for a design briefing is shown in the outline for a professional presentation presented below:

* Title
* Overview
* Status
* Introduction
* Technical Topics
  * First Topic
    * Background
    * Status
  * Second Topic
    * Background
    * Status
  * Final Topic
    * Background
    * Status
* Summary
* Action Items

Let's examine each of these topics individually and discuss the content that should be included on the visual aids that accompany your presentation. Recognize that the visual aids (35 mm slides, overhead transparencies or electronic projection) control the flow and the information that you will present. For this reason, the topics listed above in the context of information included on the presentation slides will be addressed.

Title Slide

The title slide, illustrated in Fig. 10.1, obviously carries the title of the presentation. However, it also states your affiliation and the names of the team members that contributed to the work. Because you may be reporting on the work of the entire team, it is necessary for you to acknowledge their contributions on the title slide and in your opening remarks. A brief title—ten words or less—is a good rule to follow in drafting the title for your presentation. It is also a good idea to use descriptive words in the title. For example, you could use **PRELIMINARY DESIGN BRIEFING: HOVERCRAFT SYSTEM: LEVITATION SUBSYSTEM**. Seven words and you have informed the audience of the topic (a design briefing or review), the type of briefing (preliminary), and the product being developed (a levitation subsystem for a hovercraft). A significant amount of information is successfully conveyed in seven words.

Below the title, the team name together with the name of each team member involved in the development is listed. It is essential that you share the ownership of the material covered in your presentation. Often a single member of the team covers the work of several people in a design briefing. It is not ethical to make the presentation without acknowledging the contributions of others. Students typically use first names in identifying team members, but in industry the complete names of the team members are used. If necessary, division and department affiliations are also stated.

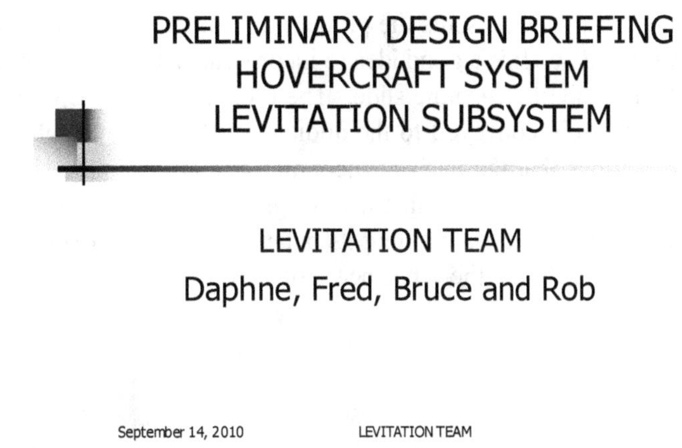

PRELIMINARY DESIGN BRIEFING
HOVERCRAFT SYSTEM
LEVITATION SUBSYSTEM

LEVITATION TEAM
Daphne, Fred, Bruce and Rob

September 14, 2010   LEVITATION TEAM

Fig. 10.1 Illustration of an informative
and descriptive title slide.

It is also a good idea to date the title slide and to cite the occasion for the presentation. If you make many presentations, this information is useful when referring to your files, allowing you to easily reuse some slides after selective editing and/or updating.

Overview Slide

The second slide presents an overview of the design briefing. The overview for a presentation is like a table of contents for a report. You list the main topics that you intend to cover in the presentation. In other words, you tell the audience what you plan to tell them in the next 10 to 20 minutes. The number of topics should be limited. In a focused presentation, you can cover about a half dozen topics. If you press and try to cover ten or more topics, you will encounter difficulty maintaining the attention of the audience throughout the presentation. Constrain the tendency to tell all—focus on the important issues. In a design review, the topics usually are organized to correspond with the subsystems involved in the product under development. For a component redesign, you concentrate on the issues guiding the component modification being considered. An example, of an overview slide for a preliminary design review of a levitation subsystem system is illustrated in Fig. 10.2.

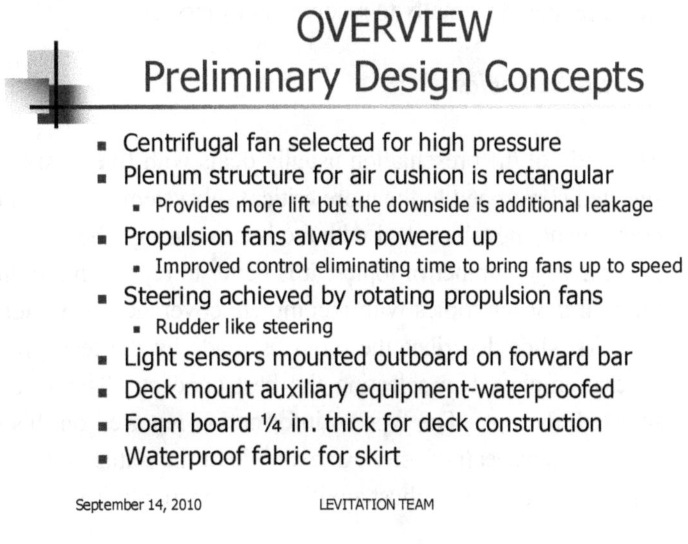

OVERVIEW
Preliminary Design Concepts

- Centrifugal fan selected for high pressure
- Plenum structure for air cushion is rectangular
  - Provides more lift but the downside is additional leakage
- Propulsion fans always powered up
  - Improved control eliminating time to bring fans up to speed
- Steering achieved by rotating propulsion fans
  - Rudder like steering
- Light sensors mounted outboard on forward bar
- Deck mount auxiliary equipment-waterproofed
- Foam board ¼ in. thick for deck construction
- Waterproof fabric for skirt

September 14, 2010   LEVITATION TEAM

Fig. 10.2 The slide with the overview
indicates the content in your
presentation.

Status Slide

The third slide presents the product development status. You should report on the progress made by the team to date. It is a good idea to incorporate a Gantt chart showing the development schedule either on the status slide or a separate slide. The progress of the team on each task is defined on the Gantt chart and is immediately conveyed to the audience. If there are problems or uncertainties, this is the time to air the difficulties that the team is encountering. An example slide, shown in Fig. 10.3, illustrates uncertainties that the team has regarding the design of the system for mounting all of the equipment on the deck of the hovercraft while maintaining its balance. The team expressed concern over the rotation of the propulsion fans affecting the balance of the hovercraft.

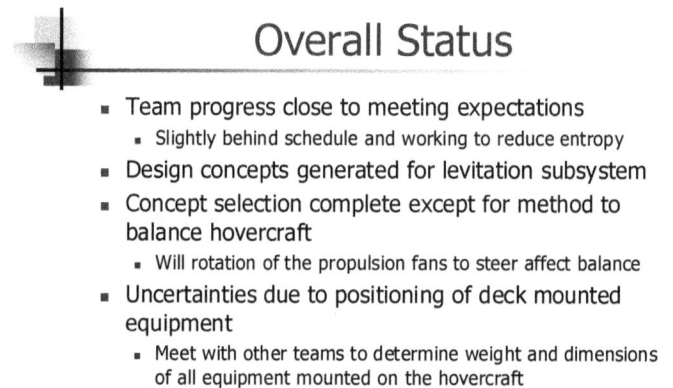

Overall Status

- Team progress close to meeting expectations
  - Slightly behind schedule and working to reduce entropy
- Design concepts generated for levitation subsystem
- Concept selection complete except for method to balance hovercraft
  - Will rotation of the propulsion fans to steer affect balance
- Uncertainties due to positioning of deck mounted equipment
  - Meet with other teams to determine weight and dimensions of all equipment mounted on the hovercraft

September 14, 2010   LEVITATION TEAM

Fig. 10.3 Status slide indicates overall progress and accomplishments while signaling potential problems.

Introductory Slide

The fourth slide covers the introductory material such as background, history or previous issues. It is important that the audience understand the product, the main objectives of the development, the role the team has in that development and the most pressing of the current issues. This introductory slide permits you to set the stage for the body of the presentation that follows. The audience responds better to your presentation when they know in advance the background and the topics that you will cover. They are better prepared mentally to receive your message.

Technical Topics

The body of the presentation usually deals with five or six technical topics. In most instances, the topics selected correspond to the more critical subsystems involved in designing the product. In the redesign of a component, the topics deal with the issues arising when considering new design concepts. Avoid trying to cover a large number of topics; it is better to report on a limited number of topics thoroughly than to rush through a dozen topics with incomplete coverage. It is suggested that you use two slides for each topic. The first slide describes the progress made by the team in generating design concepts and indicates the criteria employed in selecting the best concept. The second slide covers the status of developing the selected concept. The type of information reported on this slide includes accomplishments, outstanding issues, lessons learned, etc. An example of the status of the design of a lift subsystem including a powerful centrifugal fan and a robust auxiliary power supply is presented in Fig. 10.4.

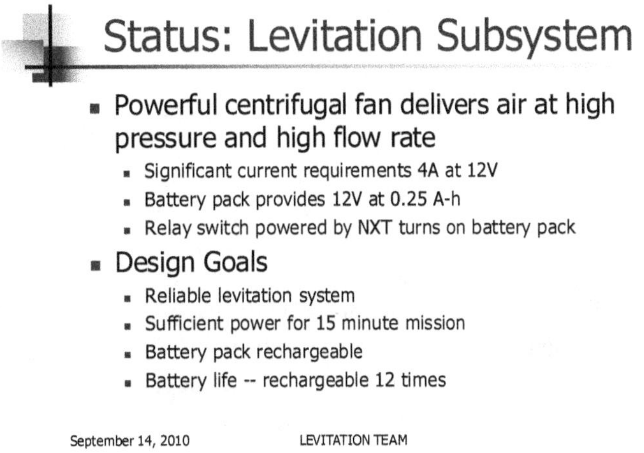

Status: Levitation Subsystem

- Powerful centrifugal fan delivers air at high pressure and high flow rate
  - Significant current requirements 4A at 12V
  - Battery pack provides 12V at 0.25 A-h
  - Relay switch powered by NXT turns on battery pack
- Design Goals
  - Reliable levitation system
  - Sufficient power for 15 minute mission
  - Battery pack rechargeable
  - Battery life -- rechargeable 12 times

September 14, 2010   LEVITATION TEAM

Fig. 10.4 Information presented in describing the early design of a reliable lift subsystem for a hovercraft.

Concluding Slides

After the technical topics have been covered, you must conclude the presentation with a decisive pair of slides. Suggestions for two concluding slides are presented in Figs. 10.5 and 10.6. The next to last slide in the presentation, shown in Fig. 10.5, is titled "Key Issues." This is your opportunity to identify very important issues (questions) that your team has discovered. Do not hesitate. This is the time to introduce the uncertainties and to come forward and seek help. Design reviews are not competitive. In a design review, you seek help regardless of your status or role as a member of the audience. Managers will arrange support for your team if it is required. Peers will make suggestions and introduce fresh approaches to unresolved problems that the team may find useful. The design review is a formal process in a product development in which everyone participates. Take advantage of this review to identify uncertainties and to seek help whenever your team needs assistance.

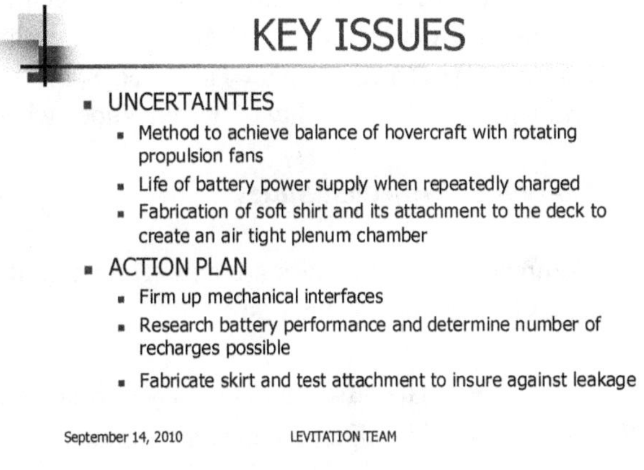

KEY ISSUES

- UNCERTAINTIES
  - Method to achieve balance of hovercraft with rotating propulsion fans
  - Life of battery power supply when repeatedly charged
  - Fabrication of soft shirt and its attachment to the deck to create an air tight plenum chamber
- ACTION PLAN
  - Firm up mechanical interfaces
  - Research battery performance and determine number of recharges possible
  - Fabricate skirt and test attachment to insure against leakage

September 14, 2010   LEVITATION TEAM

Fig. 10.5 The "key issues" focuses the team and management on your most significant problems.

The final slide presented in Fig. 10.6 is to present your plans for the future. Management is interested in how you intend to solve the issues that you have disclosed and the amount of time and money required for the solutions. Your peers will be interested in the technical approaches that you intend to follow. You should also define any requirement that you have of anyone in the audience. If you need help from your peers or more resources from the manager, be certain this assistance is clearly indicated on this final slide. You should distinctly identify those individuals who have agreed to provide the requested support. If action items are listed, clearly identify the responsible individual and estimate the time to completion. The planning incorporated in the final slide is very important. It is the prescription for your team's get-well program.

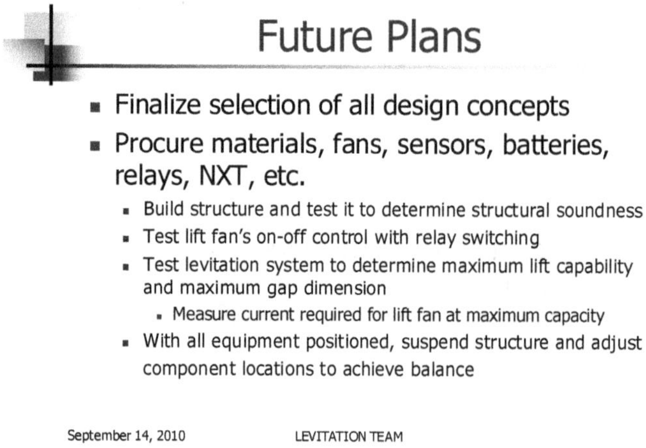

Future Plans

- Finalize selection of all design concepts
- Procure materials, fans, sensors, batteries, relays, NXT, etc.
  - Build structure and test it to determine structural soundness
  - Test lift fan's on-off control with relay switching
  - Test levitation system to determine maximum lift capability and maximum gap dimension
    - Measure current required for lift fan at maximum capacity
  - With all equipment positioned, suspend structure and adjust component locations to achieve balance

September 14, 2010   LEVITATION TEAM

Fig. 10.6 The "future plans" slide gives you the opportunity to describe your approach for solving problems that the team is encountering.

10.5   TYPES OF VISUAL AIDS

Because visual aids control the content and the flow of the information conveyed to the audience, they are essential for a technical presentation. Slides provide visual information that reinforces the verbal information. With visual aids you communicate using two of a person's five senses. Take time to carefully select the type of visual aid that will most effectively convey your message.

In selecting visual aids, you are usually limited—first, in your ability to produce the visual aids within the time and financial constrains imposed on the development team and second with regard to the availability of equipment and the room in which the presentation will be made. Is the room equipped with a 35 mm slide projector, an overhead projector or a computer-controlled projector? Another important consideration is the control of the light intensity in the room. Most rooms have light switches used to control the overhead lights. However, some rooms do not have blinds and bright sunshine will cause problems with the visibility of 35 mm slides when they are projected.

Computer Projected Slides

Computer projected slides are an excellent choice if the room can be darkened. Computer projection of slides has several advantages:

- The presentation is easy to prepare using PowerPoint.
- The slides are in color are more effective than black and white illustrations.
- Advanced visual effects (slide transition and slide building feature) enhance the impact of the presentation.
- It is also easy for you to copy your presentation onto a floppy disk or a memory stick.

A computer is attached to a digital projection system that stores your presentation material. You employ a mouse to click through your slide line by line. Finally, low-cost handouts with four or six slides per page can be prepared that help the audience follow your presentation. The only disadvantage of computer-projected slides is that sometimes a relatively dark room is required for projection. With a darkened room, eye contact with the audience or even reading their body language is not possible. You lose your visual contact with the audience that is so important in reading their response to your presentation style.

Overhead Transparencies

Projecting overhead transparencies is probably the most common method of presenting visual material. Overhead projectors are relatively inexpensive (a couple of hundred dollars compared to about a thousand dollars for the computer controlled projectors). Hence, overhead projectors are available in most rooms. Overhead projectors are also the best choice if the room cannot be darkened. The projectors are sufficiently bright to provide quality images without extinguishing the lights. Bright rooms have an advantage because they aid you keep the audience alert. Overhead transparencies are also easy to make on a copy machine or a laser printer. If you are willing to sacrifice the benefits of color, utilize inexpensive black and white transparencies. If you want to use color in the presentation because it is more effective, the costs increase somewhat when using a color laser printer or a color copier to prepare the overheads. However, color ink-jet printers can be used to produce good quality color transparencies at a reasonable cost.

35-mm Slides

The 35 mm slide presentation is a third option for visual aids; however, the equipment required is not always available. Projectors for 35-mm slides are common, but in an engineering college, they usually are much more difficult to find than the overhead projector. A slide presentation is the preferred option if you have many photographs to project because the quality of colored 35-mm slides is excellent. The slides are inexpensive and easy to prepare if you have a camera and the time necessary for the film processing. It is possible to prepare colored slides from your PowerPoint files, but conversion from the digital format to the 35-mm format requires special equipment. Again, a significant disadvantage is that the room must be darkened to view the image from a 35-mm projector. Also, the relatively long distance required from the typical projector to the screen precludes the use of very small conference rooms for presentations.

Photographs taken with a high resolution digital camera are presented with a computer driven projector. These images are clear, and if the resolution of the camera was sufficient, they closely match the quality of the images projected from 35 mm slides. The ability to present both slides and photographs from the same projector is a significant advantage.

Although ineffective, sometimes speakers will employ two or three different types of projectors. Such a tactic should be avoided as switching from one projector to the other is disruptive. If possible, use only one type of projector. If you must use more than one type of projector to properly present your message, try to minimize the number of times you switch from one device to the other.

Other Types of Visual Aids

There are several other visual aids sometimes used in professional presentations. Videotapes are common; unfortunately, the small size of TV monitors often makes viewing difficult for a large audience. However, if you want to show motion or group dynamics, video clips are clearly the best approach. Video cameras are readily available and after some practice you can become a reasonably good video producer.

Motion picture films, particularly of older material with historical interest, are very effective in developing a long-term prospective. However, finding an 8-mm or 16-mm motion picture projector may be difficult—allow time in your schedule for making the necessary arrangements.

Hardware, or other materials used in the product development, is sometimes passed around the audience during a presentation. The hands-on opportunity for the audience is a nice touch, but you pay for it with the loss of attention by some members of the audience during the inspection period. It is recommended that you defer passing materials to the audience during your presentation. Instead, after the presentation, invite the audience to inspect exhibits that are placed on a strategically located table. This approach maintains the attention of the audience throughout the presentation while giving those interested in the hardware time for a much more thorough examination.

10.6 DELIVERY

Excellent presentations require well-prepared slides and a smooth, well-paced delivery. There are many aspects to the delivery part of the presentation process including: dress, body language, audience control, voice control and timing.

Attire

Let's start with dress. Broadly speaking, there are four levels of dress. The highest level, black tie for men and formal gowns for women, is a rare event and is never appropriate for design briefings. The next level, business attire for men and women, is sometimes appropriate for design briefings. A conservative Eastern company may have a dress code requiring business attire; whereas, a less formal Western company would encourage more casual attire. Casual attire should not be confused with sloppy attire. Casual attire is neat and tasteful, but without suits, ties and white shirts for men and suits for women. Although becoming more common in the workplace, sloppy attire (old jeans and a sweatshirt) is strictly taboo. You should select tasteful business or casual attire when you dress for the presentation. While it is recognized that most students these days prefer a relaxed, collegiate style of dress, you should resist the impulse to dress too casually. If you look like a homeless person, who will believe your message?

Body Language

Posture is another important element in the presentation. In the words of former President Reagan, "stand tall." Your body language signals your attitude to the audience—you are the presenter and in control. Make sure you are calm, cool and collected. Nervous gestures with your hands, rocking on the balls of your feet, scratching your head, pulling on your ear, etc. should be avoided.

Audience Control

Before you begin your presentation, take control of the audience. One approach is to pause a moment before projecting the title slide and immediately make eye contact with the entire audience. How do you look at everyone in the house? You scan the audience from left to right and then back looking slightly over their heads. Occasionally drop your eyes and make eye contact for a second or two with one individual and then another. Pause long enough for the audience to become silent (10 to 20 seconds). If someone is rude and continues to talk, walk toward them and politely ask for their attention. When you have everyone's attention, project the title slide and begin your delivery.

If you have rehearsed, you will not need notes. The slides carry enough information to trigger your memory. If they do not, you have not rehearsed long enough to remember the issues. Scripting the presentation is not recommended. People who script will eventually start to read their comments, which is a deadly practice. Rehearse until you are confident. Make notes to help you rehearse, but do not use them during the presentation.

Voice Control

When you begin to speak, be certain everyone can hear you. If you are not sure of how far your voice carries, ask those in the back of the room if they can hear. If you use a microphone with an amplifier and speakers, be careful; the tendency is to speak too loudly. Try to control the loudness of your voice within the first minute of the presentation when you introduce the topic with the title slide. The audience will read that slide with anticipation, and they will bear with you as you adjust the volume of your voice.

Do not mumble. Speak slowly, clearly and carefully enunciate each word. It is better to present your material too slowly than too rapidly. If you can improve your enunciation, you will be able to increase the rate of your delivery without losing the audience.

Don't run out of breath while speaking. Learn to complete a sentence, pause for breath (without a gasp) and then continue with the next sentence. There is nothing wrong with an occasional pause in the presentation, provided the pause is short.

Timing

If you forget some detail, do not worry. Skip it, and move to the next point that you are trying to make. If the detail is critical, count on one of your team members to raise the issue in the question and answer period. Most speakers forget some of the material that they intended to present and introduce some additional items on an ad-hoc basis. Usually, the audience will never be aware of your omissions or additions.

Studies have clearly shown that people can listen to conversation faster that the presenter can speak. The trouble with listening to someone who speaks very rapidly is not their rate of speech, but their enunciation. Many people who speak rapidly tend to slur their words, and the audience has trouble understanding poorly pronounced words. The trouble you have in listening to a person speaking too slowly is to stay with his or her message; while you are waiting for the next word, your mind goes off on a mental tangent. You anticipate that your mind will return to the speaker in time for the next word, but unfortunately the mind is sometimes tardy. You miss keywords or even complete sentences before your attention returns to the speaker.

Listening is a skill that many people have not developed. As a speaker, it is your responsibility to keep the listener, even the poor ones, on the topic. You may use several techniques to maintain the attention of the audience. Employing the screen is probably the most effective tool for this purpose. As you project the slide, walk to the screen and point to the line on the slide that corresponds to the topic that you are addressing. If a few members in the audience are coming back from their mental tangents, you reset their attention on the current topic.

When referring to the image projected on the screen—and good speakers use this technique—stand to the left of the screen as shown in Fig. 10.7. When you stand to the side of the screen being careful not to block anyone's view. Face the audience and maintain eye contact. When you must look at the slide, turn your head to the side and read over your shoulder. The 90° turn of your head and the glance at the slide should be quick because you want to continue to face the audience. Under no circumstance should you stand with your back to the audience and begin reading from the slide as if it were a script.

As you develop the discussion, point to key phrases to keep the audience focused on the topic. Engage both of their senses. Deliver your message through their eyes and ears simultaneously. When you point to a line on the slide, point to the left side of the line. People read from left to right—so start them from the initial point on the left side.

As you speak, modulate your tone and volume. Avoid both a monotone and a singsong delivery. A continuous tone of voice tends to put the audience asleep. A singsong delivery is annoying to many listeners. If you have a high-pitched voice, try to lower the pitch because a high-pitch tone is bothersome to many people

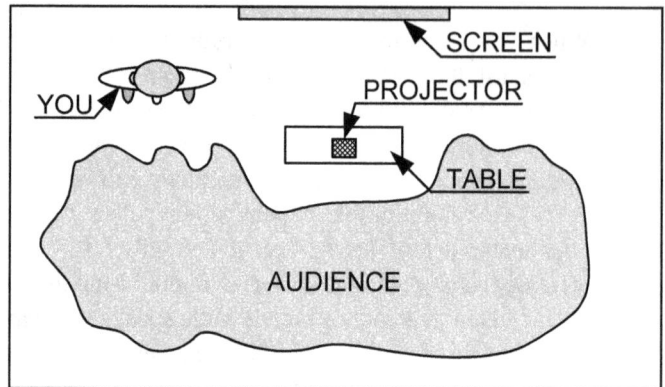

Fig. 10.7 Typical room arrangement for presentations. Position yourself to the left of the screen and face the audience.

If a question is asked, answer it promptly unless you intend to cover material later in the presentation that addresses the question. The answer should be brief, and you should try to avoid an extended dialog with some member of the audience. If a member of the audience is persistent with several related questions, simply indicate that you will speak with him or her off-line immediately after the presentation. The timing of your presentation can be destroyed by too many questions. Some questions are helpful because they permit audience participation, but an excessive number of questions cause the speaker to lose control of the topic, the flow and the timing of the presentation. Too many questions cause the format of the communication to change from a presentation to a group discussion. Group discussions have a purpose; however, they are different than a design briefing.

10.7   SUMMARY

The design briefing is extremely important both to successful product development and to your career. Information must be effectively transmitted to your peers, management and to external parties involved in the project. Clear messages, which accurately define the problems that the entire development team can effectively address, are imperative. The design briefing provides the opportunity to inform management and peers of your progress and problems.

There are three types of formal communication: speeches, professional presentations and group discussions. The characteristics of these three types of communication are described. For the engineer, the professional presentation is the most important and the formal speech is the least important. The main emphasis of this chapter is on the professional presentation with a focus on the design briefing.

Preparation for a presentation is essential. Accomplished speakers often spend the better part of a day preparing for a 20-minute briefing. There are three important elements in the preparation. First, know the characteristics of your audience, and be certain that the presentation's content corresponds to their interests. Second, organize the presentation in a manner acceptable to the audience. Insure that the organization is efficient so that you use the allotted time effectively. Prepare high-quality visual aids (slides) that will help you convey your message and rehearse until you command the subject material.

A structure for the presentation is recommended that is commonly employed in design briefings. This outline includes four initial slides—title, overview, status and introduction. The body of the presentation is devoted to technical details using about eight to twelve slides. Two closure slides are employed for a summary and a listing of action items.

The type of visual aid selected is important because the visual transmission of information will markedly affect the outcome of the presentation. Computer projected slides have some significant advantages, but the availability of equipment often precludes their use. Overhead transparencies are the most common medium. However, if you have a presentation that includes many photographs and a large room in which to deliver the presentation, 35 mm slides are the most suitable medium. However, if the photographs have been taken with a high resolution digital camera or if the images have been digitized, they can be shown using a computer projection system.

The delivery makes or breaks a presentation. You have been provided with two pages of details about how to do this—and not that. However, the best way to learn to deliver a design briefing is by practicing your presentation. On your first two or three attempts, have a friend video tape the event. Then review your delivery style, technique and appearance during a trail presentation. You will identify many problems of which you were not aware. Another suggestion is to use one of your free electives to take a speech course. The experience gained in a typical communications course will not help much in crafting the content to include in a design briefing, but it will help you to develop very important delivery skills.

REFERENCES

1. Wilder, L. Talk Your Way to Success, Simon and Schuster, New York, NY, 1986.
2. Eisenberg, A., Effective Technical Communication, 2nd ed., McGraw Hill, New York, NY, 1992.
3. Goldberg, D. E., Life Skills and Leadership for Engineers, McGraw Hill, New York, NY, 1995.
4. Elbow, P. Writing with Power, 2nd Edition Oxford University Press, New York, NY, 1998.
5. Struck, W. C. Osgood and R. Angell, Elements of Style, 4th Edition, Allyn & Bacon, Needham Heights, MA, 2000.
6. Pickett, N. A. et al, Technical English: Writing, Reading and Speaking, Longman, Reading MA, 2000.

EXERCISES

10.1 Write a brief description of the characteristics describing your teammates. Focus on those characteristics that will influence the language and content in your design briefings.

10.2 Prepare an outline for a preliminary design review. Include in the outline the titles of all of the slides that you intend to use.

10.3 Prepare the title slide for your preliminary design review.

10.4 Prepare an overview slide for your preliminary design review.

10.5 Prepare a status slide for your preliminary design review.

10.6 Prepare the key issues slide for your preliminary design review.

10.7 Beg or borrow a video camera. Video tape a practice presentation by a teammate. Critique his or her presentation with good taste. Then have a teammate video tape your presentation and listen carefully to their critique of your performance.

10.8 Practice the delivery of your presentation employing the screen of your laptop computer to display your slides.

10.9 Measure and record the time required to present your design briefing. As you rehearse, continue to measure the time. When the presentation becomes more polished, the time required for the presentation should decrease. Does it?

10.10 Prepare a group of slides for a design briefing that utilizes the slide transition and slide building features of PowerPoint.

PART IV

ENGINEERING AND SUCCESS SKILLS

CHAPTER 11

THE ENGINEERING PROFESSION

11.1   WHY ENGINEERING IS A PROFESSION

You may consider engineering from two different viewpoints—as a course of study pursued in an accredited college of engineering and as an occupation. Let's discuss these two different viewpoints, because they are both important in establishing engineering as a profession.

As a student in an engineering college, you are presented with a structured curriculum. You are required to take many credit hours of mathematics, chemistry, physics and perhaps biology in your first two years of study. Mixed with mathematics and science courses are several additional courses in engineering science. These are essentially applied science courses taught by engineering faculty members. In the last two years of the curriculum, many discipline oriented engineering courses are required. These courses provide the basic analytical methods necessary for success after graduation when you begin to practice engineering. Also included in the final two years are courses intended to provide realistic design experiences related to product development. This schedule of courses is designed to prepare you to practice engineering upon graduation. Near the conclusion of your studies, you will be presented with an opportunity to take the first of two examinations leading to a professional license to practice engineering—the Engineering Fundamental Examination. It is advisable to take this examination before graduation because it provides you with an opportunity to test your analytical skills. It is also an attribute that you can add to your resume. Passing the examination is a certification of your fundamental analytical skills.

Your ability to practice engineering immediately upon graduation is the primary reason for such a highly structured curriculum containing so many technical courses. The on-the-job training opportunities available to you in your first position depend upon the policies of the company that is paying your salary. Some large firms provide a transition period where you serve as an engineer trainee with close supervision and rotate from one division to another to provide an overview of the company's operations. However, if you take a position with a smaller corporation, usually there is little formal training, and you are expected to earn your salary almost as soon as you begin working.

In addition to at least three years of mathematics, science, engineering science and engineering courses, you will be required to take several courses in the other colleges on campus. The university has general education requirements, which dictate that all students become familiar, if not proficient, with basic premises in the social sciences, fine arts and the humanities. The engineering profession endorses this out-of college exposure because it broadens one's perspective and encourages important assessments of contemporary issues.

Let's now consider the second viewpoint—engineering assignments undertaken after graduation. Whether you will be working as an engineering professional depends to a large degree upon the position you accept. Some engineering graduates are offered positions with investment banking and brokerage houses at very attractive salaries. Managers at brokerage houses like the analytical skills developed in engineering programs and find that engineering graduates perform well in assessing investment opportunities. These graduates pursue an interesting and lucrative career, but they are not practicing engineering.

At the other extreme, a graduate of a civil engineering program takes a position with a construction firm that designs and builds highways, bridges and large public buildings. In this position, public safety is an important consideration and licensing is essential because State laws require it. The graduate is expected to pass the Engineering Fundamental examination, and then— after several years to gain practical experience—he or she will take the Principles and Practice of Engineering examination to become licensed. During the period between the two examinations, the individual works with licensed engineers who supervise his or her work and certifies it to be accurate and correct.

For those graduates working in industry, licensing is usually less of an issue. Most corporations have a small staff of licensed engineers who certify work when certification is required. However, most of the engineers are working on products in which public safety is not an issue. In these cases, professionalism is not a matter of licensing, but it is more a matter of attitude. Professionals are usually salaried and their remuneration is independent of the hours worked. They do not punch a time clock and they are expected to devote the time necessary to finish a task on schedule. Engineers assume responsibility and are committed to the project and to a development team. They cooperate with management, and freely sharing ideas and concepts to advance the welfare of the corporation and their division. They may decide to move from one corporation to another, but loyalty and respect are important considerations even after they leave.

11.2 CHARACTERISTICS OF ENGINEERING STUDENTS

Students entering colleges of engineering today are bright, with average Scholastic Achievement Test (SAT) scores ranging from 1,800 to 2,100 in most colleges in the U. S. In fact, the higher ranked engineering colleges attract students with average scores of 2,000 or above. In addition, engineering students are usually in the top 5 or 10% of their high school graduating class with grade point averages ranging from 3.8 to 4.0+. Student applications to the admissions office of the university often include strong letters of recommendation from their teachers, counselors and principals certifying the student's stellar abilities.

It is well recognized that the abilities of entering engineering students in mathematics is exceptional. SAT scores for the mathematics portion of the examination approach or exceed 700 for most students. What is less well recognized is that engineering students also have excellent verbal skills. With average SAT scores for verbal skills of exceeding 600, engineering students usually rank above their peers in the colleges of arts, humanities and social sciences.

Most engineering students are men; however, 18.1% of the bachelor degrees in engineering were awarded to women in the 2007-2008[1] academic year. The interest of women in engineering depends on the discipline as shown in Table 11.1. Statistics show that women are attracted to chemical, biomedical, environmental and industrial engineering disciplines, but not to petroleum, electrical, computer or mechanical engineering programs.

The geographic representation of students in a class depends strongly on whether the university is state supported or privately funded. State universities often restrict enrollment of out-of-state students to a small proportion of the class. A tuition differential is also imposed making it

---

[1] See Reference [6].

more expensive for out-of-state students to attend. Consequently, undergraduate classes in state universities tend to be much more homogeneous with students representing the backgrounds and cultures of the region. Privately funded universities usually do not have such restrictions and often vigorously seek students with diverse backgrounds from different states and countries.

Table 11.1
Percentage of BS Degrees awarded to women by discipline (2008)

Discipline	Percent	Discipline	Percent
Biomedical	38.6	Nuclear	17.3
Chemical	34.9	Computer Science Outside Engineering	9.4
Agriculture & Biological	31.4	Aerospace	15.1
Industrial or Manufacturing	30.4	General Engineering	18.4
Engineering Management	21.8	Computer Science Within Engineering	11.4
Metallurgical and Materials	24.7	Petroleum	18.3
Mining	19.6	Electrical/Computer	10.7
Architectural	27.6	Mechanical	11.9
Engineering Science Engineering Physics	19.9	Computer Engineering	9.2
Civil	21.1	Environmental	43.2

Clearly, engineering students are well prepared academically to begin their studies. They have proven themselves in high school as scholars and class leaders. They have gained the respect of their teachers, peers and family. However, the retention rate for engineering students is appalling. Probably one half or more of the entering class of well-qualified students either fail-out or withdraw from the program with most leaving in the first two years. While the exact retention rate varies from college to college across the U. S., the problem is widespread. Why then do so many students leave the engineering program without completing the requirements for a B. S. degree in one of the engineering disciplines? Answers to this question will be explored in Chapter 12.

11.3   SKILL DEVELOPMENT

Pursuing a successful career in engineering is strongly dependent on your ability to develop a number of important skills. Of course, you enter the program with many skills, but it is necessary to improve them and to add new ones to your repertoire. The engineering curriculum is designed to develop these additional skills. As you proceed through the program, you will find that your ability to perform challenging engineering tasks expands each year.

The Accrediting Board for Engineering and Technology[2] (ABET) has developed a list of requirements for all of the disciplines offered in the accredited colleges of engineering in the U. S. Each department in the college must demonstrate to a team of ABET visitors that their graduates

---

[2] The Accreditation Board for Engineering and Technology, Inc., 111 Market Place, Suite 1050, Baltimore, MD 21202, certifies all programs (not departments) leading to accredited engineering degrees in the U. S.

have the skills and knowledge listed in Table 11.2 to maintain the accreditation of each engineering program for which a B. S. degree is offered.

Table 11.2
Listing of knowledge and skills that engineering graduates must demonstrate.
ABET Criteria 2000

(a)   Ability to apply knowledge of math, engineering, and science
(b1) Ability to design and conduct experiments
(b2) Ability to analyze and interpret data
(c)   Ability to design a system, component or process to meet needs
(d)   Ability to function on multi-disciplinary teams
(e)   Ability to identify, formulate and solve engineering problems
(f)   Understanding of professional and ethical responsibility
(g)   Ability to communicate effectively
(h)   Broad education necessary to understand the impact of engineering solutions in a global and societal context
(i)   Recognition of the need for, and the ability to engage in life-long learning
(j)   Knowledge of contemporary issues
(k)   Ability to use techniques, skills, and tools necessary for engineering practice

Before examining this list, let's consider a somewhat different set of skills and abilities for engineering graduates prepared by a study committee for the American Society of Mechanical Engineers (ASME). The ASME listing [1], in priority order, is presented in Table 11.3.

Table 11.3
Skills considered important for new Mechanical Engineers
With B. S. degrees, Priority ranking

1. Teams and Teamwork
2. Communication
3. Design for Manufacture
4. CAD Systems
5. Professional Ethics
6. Creative Thinking
7. Design for Performance
8. Design for Reliability
9. Design for Safety
10. Concurrent engineering
11. Sketching and Drawing
12. Design for Cost
13. Application of Statistics
14. Reliability
15. Geometric Tolerancing
16. Value engineering
17. Design Reviews
18. Manufacturing Processes
19. Systems Perspective
20. Design for Assembly

As you explore both of these lists, the emphasis on skill development in colleges of engineering should become evident. The importance of these skills, as determined by two major engineering societies, should guide your efforts in achieving the proficiencies necessary to insure a successful career.

Design

Most engineering graduates join corporations and participate in some way related to either product or process development. If it is a manufacturing company, the product may be a vacuum cleaner, a lawn mower, an automobile or some other item. The point is you will design, manufacture, assemble, ship and service an array of products. Design is essential to the success of product development. If the design is flawed, the product or service offered will fail, and the company and your career may suffer. If the design is optimal, the product will be profitable, the company will prosper and your career will advance. This is an over simplified viewpoint, but one that has merit.

Design is perhaps the most difficult of the engineering skills to acquire because it is more an art than a science. While there are well-defined design procedures that you will learn (some of them introduced earlier in this book), one often intuitively senses the good and bad elements of a design. Also, experience is important because design is a complex activity that must incorporate many different aspects that cross discipline lines. Examine Table 11.3 and note the various descriptors ASME used with the word design. For example, engineers design for reliability, for performance, for safety, for cost, for manufacture, and for assembly and one could add "ease of maintenance" and others to the list.

As you proceed through the engineering program, the opportunities you have to practice design will be severely limited. The emphasis in engineering education is on developing analysis methods used to predict reliability and performance of products and systems. Some engineering programs are limited to a single capstone design experience in your final year of study. Other programs have two or three design courses where you will have the opportunity to design a new product or to modify the design of an existing product or process. Engineering programs with more than three design courses (a total of nine credit hours in a program of about 128 credit hours[3]) are rare in the U. S.

Many colleges of engineering participate in design competitions that are sponsored by the professional societies. These competitions involve the design of robots, large model airplanes, automobiles, steel bridges, and human and engine powered vehicles, etc. You are encouraged to become involved in one or more of these projects, because they provide you with an opportunity to participate on an interdisciplinary design team and to gain valuable experience in the design process. Usually you can obtain credit towards your gradation requirements for this design competition by registering for a technical elective that has been established by the faculty member in charge of these design projects.

Another method for gaining design experience is by working off campus. Many students work to partially fund their expenses while pursuing their education. If you are able to find a position that involves design, manufacturing or assembly, the experience gained is probably worth more than the salary you earn. Try to obtain a position in a company that produces products rather than with a retail organization. While experience in either type of organization is valuable, the production experience is more applicable to engineering.

Teamwork

In this course, you probably will be required to participate on a development team. There are three reasons for this requirement. First, the project is too ambitious for an individual to complete in the time available. You need the collective efforts of the entire team to develop a hovercraft during the semester. The development teams will be pressed time-wise to complete the project on schedule.

---

[3] The number of credit hours required varies from engineering program to program and college to college. The 128 credit hours cited here is the typical requirement for many programs.

Second, you must begin to learn teamwork skills. Experience has shown educators that most students entering the engineering program are seriously deficient in team skills. From elementary through high school, the educational process has focused on teaching you to work as an individual, often in a setting where you competed against others in your class. However, you should begin functioning as a team member where cooperation, following, and listening are as important as individual effort. Leadership is important in a team setting, but cooperation and following the lead of others are also critical elements for successful team performance.

The final reason is to better prepare you for the real world that you will enter upon graduation. You will probably be assigned to a development team very early in your career if you take a position in industry. A recent study by ASME, the results of which are shown in Table 11.3, ranked teamwork as the most important skill to develop in an engineering program. Teamwork was also the first skill, in a list of 20, considered important by managers from industry. Hopefully, this course will be instrumental in exposing you to team working skills so necessary for a successful career. A detailed discussion of teamwork skills was presented previously in Chapter 1.

Communication

Communication skills are vitally important in every aspect of your life. On the personal side you must be able to accurately convey your thoughts to your family, friends and peers. Professionally, it is just as important to communicate effectively. A great idea is of little value if you cannot express it with sufficient clarity for it to be accepted by management or your associates. With family and friends, most of your communication is by informal conversation. Occasionally, you write letters or more frequently e-mail or text messages. You can be casual in communicating with friends and family because they are accommodating and overlook your shortcomings. You must be much more careful in your professional communications. Messages must be clear and unambiguous. The information conveyed must be accurate and timely. The presentation must be a concise and to the point; long rambling prose is not appreciated.

Three different modes of communication are used in engineering—writing, speaking and graphics. All three are important, and engineering versions of all three modes different to some degree from commonly accepted methods.

It is important that you enjoy writing, because engineers often have to prepare several hundred pages of reports, theoretical analyses, memos, technical briefs and letters during a typical year on the job. Advancement in your career will depend on your ability to write well. You will be taking several courses offered by the English Department and by departments in Social Sciences, Arts and the Humanities that require writing assignments. These courses will help you with the structure of your composition and the development of good writing skills. Most of the assignments will be to write essays, term papers, or to study selected works of literature. However, there are several differences between writing for an engineering company and writing to satisfy the requirements of courses such as English 101 or History 102. These differences will be described in a Chapter 9 titled **Technical Reports**.

The design briefing is important to both the product development process and to your career. Information about the product must be effectively transmitted to your peers, management and others involved with the project. Clear messages that accurately define problems, which the development team will address, are imperative. On the other hand, ambiguous messages are often misunderstood, hinder the definition of the problem and lead to delays in implementing solutions. The design briefing provides an opportunity to review the status of a specific product development. It also permits peers to share their ideas with you, and affords management an open forum for assessing the quality of your work and the progress made by your development team. Because the professional presentation is critically important, Chapter 10 titled **Design Briefings** has been included in this textbook. This chapter describes some valuable techniques for properly delivering your message to different audiences—strangers, peers, team members and management representatives.

Engineering graphics is the most important method of communication for presenting design concepts and details. When attempting to communicate design ideas, you will find writing and speaking insufficient to express your thoughts and concepts. A more visual technique for communicate is required. It is much more effective to present your ideas by means of drawings, sketches, pictures, and different types of graphs. Visuals aids, such as drawings and photographs, convey your ideas quickly and with remarkable accuracy.

There are two general approaches used in preparing drawings and graphs. The first is a manual approach where drawings and graphs are prepared by hand using a few simple drawing instruments. The second utilizes a computer and suitable software programs that greatly facilitate the preparation of drawings or graphs. An extended discussion of computer-aided design (CAD) software programs, used in preparing both two and three-dimensional engineering drawings, was included in Chapter 7.

Design Analysis

In engineering, you will attempt to predict the performance of a product prior to its final design and construction. Obviously, you would not want to design and construct a bridge to have it fail after a few months or years in service. You must be able to predict with confidence that the bridge will not fail over its entire design life (perhaps a hundred years or more). Constructing analytical models and performing analysis is essential in making accurate predictions regarding performance. In modeling, you reduce your structure into a number of different components or subsystems that are amenable to analysis. You then apply analytical methods that enable you to accurately determine performance parameters for each model. The analytical results permit verification of the adequacy of the design. As such, analysis and design are coupled. You usually begin with a design concept; then you subject this design to an analysis. The results of the analysis are then used to improve this design enhancing its performance.

A significant portion of the engineering curriculum is devoted to developing your analytical skills. About one year, of the four-year curriculum, is devoted to courses in mathematics and science that provide the foundation for engineering courses. In most colleges of engineering, at least another year is devoted to engineering and engineering science courses where analytical and computational methods are taught. In a typical engineering program, your analytical and computational skills will be honed more than any other skill listed in either Table 11.1 and 11.2.

Your performance (read this as grade point average, GPA) will depend on your ability to solve problems posed on hourly or final examinations. These problems are crafted to test your understanding of a few engineering principles[4]. The mathematics involved is usually not overly complex and the physics entailed is even more straightforward. You will be provided with several useful tips on problem solving methods in Chapter 12.

---

[4] In practicing engineering, you will find that analyses are based on only a dozen or so engineering principles. The methods developed in engineering courses may appear complex, but they are based on a very limited number of fundamental concepts.

Experimental Analysis

In predicting the performance of a structure or machine component, you will often conduct experiments and measure the performance parameters. While it is usually more expensive to perform experimental studies than analytical ones, the costs are warranted when you are not certain that the analytical modeling accurately reflects reality. Sometimes converting the physical reality of a product to an analytical model requires many assumptions. You will often perform carefully designed experiments to verify these assumptions and to insure the successful performance of the product. Extensive tests on almost all products are performed before they are released to the market. The purpose of these tests is to insure safety of the product, and to provide management assurance that the product will meet the guarantees specified in the warranty pertaining to performance, life and safety. In this class, you will be required to test a prototype of a hovercraft that your team has developed. The test is to determine if the product's performance meets its specification.

There are four different phases of an experimental program. The first phase is to design an experiment so that it provides the answers required to verify the modeling or to insure the adequacy of a design. The second phase is to conduct the experiment using appropriate instrumentation that will provide accurate and appropriately timed measurements of the controlling performance parameters. The third phase is to analyze the data obtained using suitable statistical methods that predict and bound experimental errors. The final phase is to correctly interpret the data to verify the adequacy of a design or to provide information that may be used to enhance the design.

The laboratory classes offered in conjunction with chemistry and physics courses are not designed to provide you with skills for designing experiments or for making measurements. These are highly structured laboratory experiences intended to reinforce theories and principles introduced during lecture. The curriculum of many of the engineering disciplines provides a course on circuits and instrumentation. This course introduces linear circuit theory and gives some information about sensors, transducers and both analog and digital instrumentation. A course of this type is strongly recommended because it provides some of the basic information needed to plan and design experiments.

You are encouraged to enroll in other courses that provide additional exposure to experimental methods. This may be possible with a careful selection of technical electives that are available in the third and fourth year of your program. You might also consider an assignment working in a laboratory directed by one of the professors in your department. Many professors lead significant research programs, and in conducting these studies, they often perform new and novel experiments. The experience gained in such a setting is often more valuable than that obtained in a traditional engineering measurements course.

Understanding Professional and Ethical Responsibilities

When you graduate and begin practicing engineering, you will be expected to follow a professional code of ethics. An example of the ABET code of ethics is presented in Chapter 15 titled Ethics, Character and Engineering. While there are many other codes, each endorsed by a professional society, they all are similar. Basically the codes ask you to uphold and advance the integrity, honor and dignity of the engineering profession. The codes then list fundamental principles and canons that should govern professional behavior. The fundamental principles advanced by ABET are:

1. **To be selective in the use of our knowledge and skills so as to ensure that our work is of benefit to society.**
2. **To be honest, impartial and serve our constituents with fidelity.**
3. **To work hard to improve the profession.**
4. **To support the professional organizations in our engineering discipline.**

Personal behavior out-of-class or away from the work place is also important. The results of several polls indicate that Americans are less ethical today than in previous generations. Sixty-four percent of individuals in a sample of 5000 admit to lying if it does not cause real damage. An even larger percentage of those sampled (74%) will steal providing the person or business that is being ripped off does not miss the item pilfered [3]. Many students (75% high school and 50% college) admit to cheating on an important exam [4]. Many students have not developed a moral code to use as a guide for their behavior. Relatively minor transgressions of a generation ago have been replaced with more serious problems involving mass murder, drug and alcohol abuse, pregnancy, suicide, rape, robbery and assault. Some of the reasons for these changes are explored later in Chapter 15.

Ethical behavior is important in both your professional and personal lives. Recall the golden rule—do unto others as you would have them do unto you—to guide your behavior. It is a very simple rule, and it is effective.

Committing to Life-Long Learning

Most engineering programs require successful completion of about 128 credit hours for graduation with a B. S. degree. This number is higher by about eight credit hours than the requirements for a B. S. degree from any other college on campus. Engineering educators have always required an extra effort measured in credit hours for their students prior to graduation. This belief is based on the concept that the student should be capable of practicing engineering immediately after graduation.

During the past twenty years the number of credit hours required for graduation has decreased by about 10%, while the amount of important material necessary to practice has increased significantly. Moreover, the scope of the fields in all of the engineering disciplines continues to evolve and to expand. Engineering educators recognize that they cannot increase the number of credit hours to accommodate the evolution and expansion of the knowledge base for each discipline. Instead, it becomes essential that all engineering graduates recognize the need to continue their studies after graduation.

Upon graduation many of you will take a position in industry and be challenged with a new environment and new assignments. You will be able to measure your strengths and weaknesses against this new setting. Perhaps you will conclude that you need advanced courses in your discipline, or basic courses in some other engineering discipline. With more time serving in a company, you will probably be given an opportunity to lead a small engineering group and become a first level engineering manager. In this case you may decide to enroll in a few courses in the business school or to pursue a masters program in business administration. No one really understands with certainty what knowledge will be important in the future. The idea is to stay flexible and be prepared to devote time each week both to study and professional development.

You are fortunate that many opportunities exist for continuing education. Some companies bring instructors into its facilities to focus on topics of immediate concern to the organization. Professional societies offer short courses on timely topics in their discipline. Well-known universities offer short courses each summer with outstanding professors teaching in their specialty. Local universities offer a wide array of courses on a regular basis that lead to M. S. and Ph. D.

degrees in a number of engineering disciplines. Professional societies offer many well-written books that cover advances within their discipline. Distance learning opportunities abound, and this mode of education is growing rapidly.

It is essential that you recognize graduation with a B. S. degree as the beginning of your studies and not the end.

11.4   THE ENGINEERING DISCIPLINES

11.4.1 Enrollment and Salaries

The U. S. Military Academy at West Point, established by an act of Congress in 1802, was the first college of engineering in the America. Military engineering was the first discipline, but in the 19th century several other programs were introduced, and the field of engineering began its division into many different disciplines each with its special emphasis [5]. The number of different disciplines grew slowly in the 19th century and included:

1. Civil engineering
2. Mechanical engineering
3. Electrical engineering
4. Chemical engineering
5. Industrial engineering

In more recent years, the number of engineering disciplines has continued to grow with more and more technical specialties converted into degree programs. Recently the American Society of Engineering Education (ASEE) [6] published the undergraduate enrollment by engineering discipline during the 2008 academic year. These results are presented in Table 11.4. The total undergraduate enrollment increased from 2007 to 2008 by 4.5% to 403,191.

Table 11.4
Full-time undergraduate enrollment by engineering discipline, (2008)

Discipline	Enrollment	Discipline	Enrollment
Mechanical Engineering	85,249	Architectural Engineering	4,033
Electrical/Computer Engineering	73,343	Metallurgical & Materials Engineering	4,312
Civil Engineering	50,167	Engineering Science Engineering Physics	2,643
Other	38,468	Agricultural Engineering	3,1918
Computer Science within Engineering	27,766	Petroleum Engineering	4,029
Chemical Engineering	28,636	Environmental Engineering	3,180
General Engineering	17,011	Civil/Environmental Engineering	2,546
Aerospace Engineering	17,561	Nuclear Engineering	1,832
Biomedical Engineering	17,798	Engineering Management	1,224
Industrial & Manufacturing Engineering	13,476	Mining Engineering	881

The total number of B. S. degrees awarded in engineering in 2008 was approximately 74,170. A large number of degrees were awarded to graduates from mechanical, electrical and computer engineering programs; however, in recent years many departments of electrical engineering have divided their program and they are now offering two different degrees—an electrical engineering degree and a computer engineering degree. Current trends show students tending to favor the electrical engineering program when they enroll. In addition to the B. S. degrees in engineering, 38,986 M. S. and 9,086 Ph.D. degrees were also awarded in engineering in 2008.

Mechanical and civil engineering programs also have large enrollments. The curriculum in both of these programs is relatively stable in comparison with electrical and computer engineering. However, some civil engineering departments are establishing a separate degree program in environmental engineering, and this separation necessitates significant changes in the curriculum.

A large number of engineering programs are classified as other in Table 11.4. This category contains many programs that are offered by only a few accredited colleges of engineering in the U. S. and include:

- Ocean engineering
- Marine engineering
- Fire protection and safety engineering
- Ceramic engineering
- Systems engineering
- Geological engineering

The enrollment in each discipline reflects perceived demand for the graduates and, to a small degree, the starting salaries for graduates from each program. Average starting salaries for University of Maryland graduates from several of the engineering disciplines with B. S., M. S. and Ph D. degrees are listed in Table 11.5.

Table 11.5
Annual starting salaries for B. S. M. S. and Ph. D. graduates in engineering
Source: Survey by the A. James Clark School of Engineering (2009)

Discipline	B. S.	M. S.	Ph. D.[5]
Aerospace Engineering	$58,400	$75,875	$88,333
Bioengineering	$59,000		
Chemical Engineering	$62,200		
Civil & Environmental Engineering	$51,800	$64,400	$86,000
Computer Engineering	$69,300		
Electrical Engineering	$62, 100	$80,000	$92,071
Fire Protection Engineering	$56,200	$65,750	
Materials Engineering		$96,000	
Mechanical Engineering	$61,300	$72,538	
Reliability Engineering		$50,000	$96,500
Systems Engineering		$97,000	
Telecommunications Engineering		$63,500	
Mean Salary for Majors	$61,000	$73,896	

---

[5] Industry average annual salaries.

Table 11.6
Annual starting salaries for B. S. M. S. and Ph. D. graduates in engineering
Source NACE Salary Survey (Spring/Summer 2009)

Discipline	Ave. B. S.	Ave. M. S.	Ave. Ph. D.[6]
Aerospace Engineering	$56,311	$68,061	$66,667
Bioengineering	$54,158	$72,071	
Chemical Engineering	$64,902	$70,084	$75,659
Civil Engineering	$52,048	$54,163	$63,100
Computer Engineering	$61,738	$72,771	$74,750
Electrical Engineering	$60,125	$71,454	$81,297
Environmental Engineering	$48,535		
Materials Engineering	$57,349		
Mechanical Engineering	$58,766	$66,158	$69,034
Systems Engineering	$57,438		
Mean Salary for Majors	$61,000	$73,896	

The difference in starting salaries for engineers in different disciplines is not large and the data differs to some degree depending on its source. These differences vary somewhat from year to year with the petroleum, chemical and computer engineers usually leading the other disciplines by 5 to 10% and civil and environmental engineers usually trailing the other disciplines by 5 to 10%. Starting salaries for engineering graduates are significantly greater than comparable salaries for other majors such as business administration and liberal arts.

College graduates with a bachelor's degree in an engineering discipline have among the highest annual earnings of all major disciplines. A recent study on employment trends by economists N. Fogg and P. Harrington, and psychologist T. Harrington [7], contains the results of a study of 150,000 college students, data from the U.S. Bureau of Labor Statistics and extensive economic research and predictions.

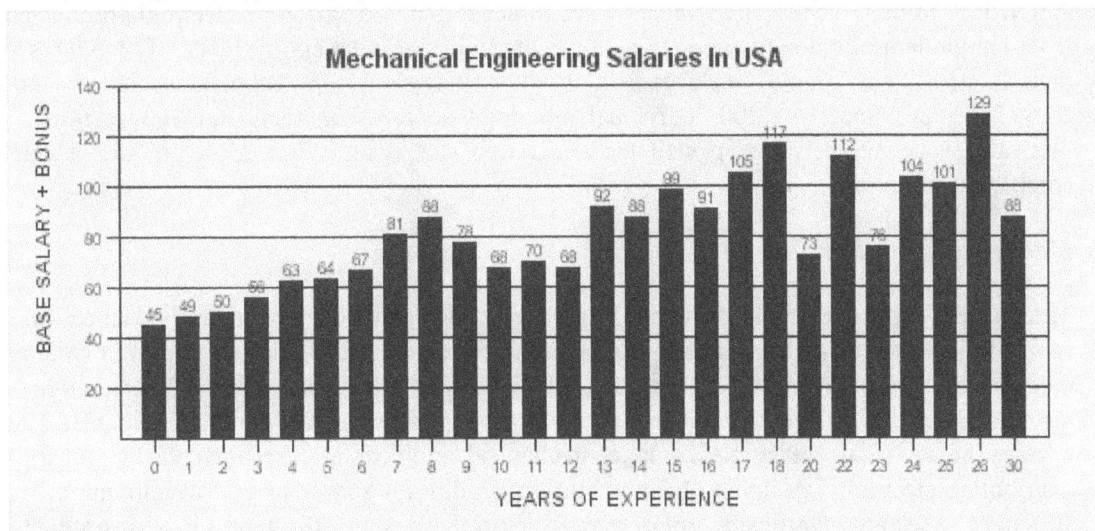

Fig. 11.1 Mechanical engineering salaries in the U. S. with years of experience.

---

[6] Industry average annual salaries.

Median income with discipline of engineering differs considerably from the starting salaries as shown in Fig. 11.1. The median income is higher than the starting salaries because the engineers' compensation increases time on the job and the experience gained. The median salaries depend on the engineering discipline with higher salaries commanded by those working in petroleum, nuclear, computer and aerospace engineering. Those working in civil, environmental and industrial engineering tend to earn lower salaries. However, after about 20 years of service, salaries tend to level out for most engineers with small incremental gains from year to year in the later stage of their careers. The demand for all disciplines of engineering is expected to grow in the coming decade as the baby boomers begin to retire.

11.4.2 Description of Engineering Programs

Electrical Engineering

The **electrical engineering** program provides the basic background needed to work in broad fields of electronics, communications, automatic control, computers, materials processing, electromagnetic, signal processing, and power utilization. The program places emphasis on the fundamentals of mathematics and science leading to careers in research, development, design, or operations in such diversified areas as electronics, control systems, computers, communications systems and equipment, biomedical instrumentation, radar and navigation, power generation and distribution, consumer electronics and industrial devices.

The Institute of Electrical and Electronics Engineers (IEEE), the professional society representing electrical engineering, describes the purpose of electrical engineering as:

> Helping advance global prosperity by promoting the engineering process of creating, developing, integrating, sharing, and applying knowledge about electrical and information technologies and sciences for the benefit of humanity and the profession.

About 11.4 million electrical engineers work to develop a wide array of electronic products ranging from simple household appliances to sophisticated missile intercept systems. The field is so large that the IEEE has divided its organization into 10 regions, 36 technical societies, 4 technical councils, approximately 1,200 individual and joint society chapters, and 300 sections. If this discipline is of interest to you, visit the IEEE web site at http://www.ieee.org/ for a much more complete description[7].

Mechanical Engineering

Mechanical engineers apply the fundamental principles of mechanics, thermo-sciences and design to fulfill the needs of society. These principles are covered in detail in the subjects of solid and fluid mechanics, thermodynamics, heat transfer, design, systems analysis, control theory, vibrations and machine design. Mechanical engineers work on a wide variety of different assignments. One of the primary areas is the energy field where mechanical engineers are responsible for generating and distributing electricity, designing heating and air conditioning systems and developing refrigeration systems. A second significant area of activity is in the design of a wide range of product such as vehicles, appliances, office equipment and machinery. Finally, many mechanical engineers work in production facilities developing manufacturing processes and overseeing production.

---

[7] The Web sites for all of the professional societies are listed in Table 11.7.

The American Society for Mechanical Engineering (ASME International), the professional society representing mechanical engineering, describes its mission as:

> To promote and enhance the technical competency and professional well being of our members, and through quality programs and activities in mechanical engineering, better enable its practitioners to contribute to the well being of humankind.

Mechanical engineering is one of the oldest disciplines with a broad range of industrial activities. Mechanical engineering and electrical engineering are the largest engineering disciplines. While membership in the ASME is about 125,000, employment by mechanical engineers in industry probably approaches one million. The organization of the ASME is divided into 39 technical divisions that publish technical papers and sponsor technical conferences.

Civil Engineering

Civil engineers are concerned with many societal problems including environmental quality, building and maintaining the nation's infrastructure and developing safe and rapid transportation systems. Topics of study include structures, soil mechanics, construction materials, environmental engineering, water resources, pavements and transportation. This program emphasizes fundamentals of mathematics, sciences, engineering principles and engineering methods leading to careers in engineering practice and/or graduate education. An option in **environmental engineering** or a separate degree in this discipline is often available in civil engineering departments.

The American Society for Civil Engineering (ASCE), the professional society representing civil engineering, describes its function as:

> When new developments occur in research and practice, ASCE's technical divisions and councils and their respective technical committees bring information to the membership through conferences and workshops and by publications such as manuals of practice, pre-standards, journal articles and policy statements.

The technical activities of the ASCE include more than 14 annual conferences and workshops and the publication of 10 journals. More than 5,000 ASCE members participate by serving on technical committees. Their work addresses programmatic thrust areas such as sustainable environment, sustainable transportation, extreme environment and information technology and construction and materials.

Computer Engineering

Computer engineering is a relatively new discipline that is still emerging in the many colleges of engineering in the U. S. Many electrical engineering faculties are dividing the traditional electrical engineering curriculum into two separate programs—one dealing with power, electronic systems and communications and the other concerned with computers and information technology. Computer engineers work in the computer industry designing hardware and writing software. The discipline combines technical knowledge from digital electronics, signal transmission and processing with a programming emphasis of computer science.

The IEEE Computer Society, an element within the IEEE organization, has emerged as the professional society representing computer engineers. The mission statement for the IEEE Computer Society is given below:

The IEEE Computer Society is dedicated to advancing the theory, practice, and application of computer and information processing technology. Through its conferences and tutorials, applications and research journals, local and student branch chapters, technical committees and standards working groups, the society promotes an active exchange of information, ideas, and technological innovation among its members. In addition, it accredits collegiate programs of computer science and engineering in the United States.

With nearly 100,000 members, the IEEE Computer Society is the world's leading organization of computer professionals. Founded in 1946, it is the largest of the 36 societies organized under the umbrella of the Institute of Electrical and Electronics Engineers (IEEE).

Chemical Engineering

Chemical engineering is concerned with the processing of materials and the production or utilization of energy through molecular or sub-molecular changes. Chemical or atomic reactions and physical changes are included in this field. Chemical engineers begin their process designs with well-known chemical reactions previously discovered by chemists. They use this information to design large-scale chemical plants to produce sizeable quantities of raw materials, petroleum products, pharmaceuticals, synthetics, etc.

The American Institute of Chemical Engineers (AIChE) is a nonprofit organization providing leadership to the chemical engineering profession. Representing 57,000 members in industry, academia, and government, AIChE provides forums to advance the theory and practice of the profession, upholds high professional standards and ethics and supports excellence in education. Institute members range from undergraduate students, to entry-level engineers, to chief executive officers of major corporations. The AIChE mission statement is given in the bullet listing below:

- Promote excellence in chemical engineering education and global practice;
- Advance the development and exchange of relevant knowledge;
- Uphold and advance the profession's standards, ethics and diversity;
- Enhance the lifelong career development and financial security of chemical engineers through products, services, networking, and advocacy;
- Stimulate collaborative efforts among industry, universities, government and professional societies;
- Encourage other engineering and scientific professionals to participate in AIChE activities;
- Advocate public policy that embraces sound technical and economic information and that represents the interest of chemical engineers;
- Facilitate public understanding of technical issues;
- Achieve excellence in operations of the Institute.

Fire Protection Engineering

Fire protection engineering is a unique profession that builds upon the basic tools of several other engineering disciplines. The fire protection engineer may be responsible for analyzing the level of fire safety in modern or historic buildings, nuclear power plants and aerospace vehicles; designing fire protection systems for high-rise buildings and industrial complexes; conducting post-fire investigations; and researching new technologies for fire detection and suppression.

Fire protection engineers apply engineering principles to protect people and their environment from the unwanted consequences of fire. These principles are applied to understand the nature and characteristics of fire, fire growth and products of combustion, as well as to consider the response of structures, people, processes, materials and systems to fire.

The professional society in the field, the Society of Fire Protection Engineers (SFPE), has 51 chapters in 8 countries. According to the SFPE, fire protection engineers have always been in great demand by corporations, educational institutions consulting firms and government bodies around the world. Graduating from a fire protection engineering program usually guarantees an immediate place in the workforce, most likely at a high starting salary. The median salary for fire protection engineers is significantly above that of other engineering disciplines.

Industrial Engineering

Industrial engineers design the systems that organizations use to produce goods and services. In addition to working in manufacturing industries, industrial engineers serve to insure quality and productivity in places such as medical centers, communication companies, food service, education systems, government, transportation companies, banks, urban planning departments and an array of consulting firms. Industrial engineers educate and direct these groups in the implementation of Total Quality Management (TQM) principles. Today, especially popular functions are manufacturing, health care, occupational safety and environmental management. Industrial engineering is the most people-focused discipline in engineering. Those who pursue careers in this discipline usually have strong leadership skills and a commitment to working with teams of managers, scientists and other personnel to solve important problems. They enjoy helping organizations serve human needs and accommodate many different concerns

The Institute of Industrial Engineers (IIE) is the society that serves the professional needs of industrial engineers and others involved with improving quality and productivity. Its 24,000 members stay current with new developments in their profession through Institute's life-long-learning approach, as reflected in the educational opportunities, publications, and networking opportunities offered. Members also gain valuable leadership experience and enjoy peer recognition through numerous volunteer opportunities.

Aerospace Engineering

Aerospace engineers are concerned with the physical understanding, related analyses, and creative processes required to design aerospace vehicles operating within and beyond planetary atmospheres. Such vehicles range from helicopters and other vertical takeoff aircraft at the low-speed end of the flight spectrum to spacecraft operating at thousands of miles per hour during entry into the atmospheres of the earth and other planets. In between are general aviation and commercial transports flying at speeds below and close to the speed of sound, and supersonic transports, fighters, and missiles which cruise at speeds greater than the speed of sound. Among the subjects studied are aerodynamics, flight dynamics, flight structures, flight propulsion, and the synthesis of all these principles into one system with a specific application such as a complete transport aircraft, a missile, or a space vehicle.

The nonprofit American Institute of Aeronautics and Astronautics (AIAA) is the principal society serving the aerospace profession. Its primary purpose is to advance the arts, sciences and technology of aeronautics and astronautics and to foster and promote the professionalism of those engaged in these pursuits. Although founded and based in the United States, AIAA is a global organization with nearly 30,000 individual professional members, over 50 corporate members, thousands of customers worldwide and an active international membership.

Materials Engineering

Materials engineering involves the study of mechanical, physical, and chemical properties of engineering materials, such as metals, ceramics, polymers and composites. The objective of a materials engineer is to predict and control material properties through an understanding of atomic, molecular, crystalline, and microscopic structures of materials. A materials engineer is an essential member of a team responsible for synthesis and processing of advanced materials for manufacturing. A graduate's work may vary from automobile or aerospace applications to microelectronics manufacturing. Opportunities are available through these industries in areas of research, quality control, product development, design, synthesis and processing operations.

The American Society for Materials (ASM International) serves as one of the technical societies representing materials engineering. For many years it has provided a means for exchanging information and professional interaction. Benefits to members include:

- Access to information for extending and maintaining professional skills.
- Opportunities for professional development through continuing education.
- Network locally and worldwide.
- Achieving professional recognition.

Biomedical Engineering

Biomedical engineering is the newest engineering discipline, integrating the basic principles of biology with the fundamentals of engineering. With the rapid advances in biomedical research and the economic pressures to reduce the cost of health care, biomedical engineering will play an important role in the medical environment of the 21st century. Over the last decade, biomedical engineering has evolved into a separate discipline bringing the quantitative concepts of design and optimization to problems in biomedicine. The opportunities for biomedical engineers are wide ranging. The medical device and drug industries are increasingly investing in biomedical engineers. As gene therapies become more sophisticated, biomedical engineers will play an important role in bringing these ideas into real clinical practice. Finally, as technology plays an ever-increasing role in medicine, there will be a larger need for physicians with a solid engineering background. From biotechnology to tissue engineering, from medical imaging to microelectronic prosthesis, from biopolymers to rehabilitation engineering, biomedical engineers are in demand.

The Biomedical Engineering Society (BMES) serves as the professional society representing the interests of biomedical engineers. They are a relatively new society having incorporated in 1968. Its purpose is to promote the increase of biomedical engineering knowledge and its utilization

Engineering Science and Engineering Mechanics

Engineering mechanics is a field of study that permeates all major engineering disciplines. Solid mechanics deals with analytical formulation of direct and reaction forces on structures and the resulting response of the structures to these forces. The discipline of engineering mechanics is based on applied mathematics and large scale computing. Fail-safe design is another dimension in this discipline. Because of the general applicability of this subject to many aspects of engineering design and manufacturing, opportunities are available in a broad spectrum of industries and government laboratories.   The American Society for Mechanical Engineering (ASME International), the professional society representing mechanical engineering, also represents engineering mechanics.

Agricultural Engineering

Agricultural engineering is in transition with more emphasis currently being placed on biological aspects rather than traditional agriculture topics. As such many programs carry some reference to biology in their titles—for example, **BioResource and Agricultural Engineering**. Agricultural engineers are trained to creatively apply scientific principles in the design and development of new products, systems, and processes for the conversion of raw materials and power sources into food, feed and fiber. They are also concerned with protecting the environment and worker health and safety. The diversity of knowledge and skills that agricultural engineers possess is valuable in the agricultural and agribusiness industries. The agricultural engineer develops skills in design and problem solving that are based on fundamental principles in the engineering sciences including mathematics and physics, computer tools, communication, teamwork, instrumentation and biology. Agricultural engineers are distinguished from other engineering disciplines by their commitment to meeting human and animal needs for food, feed, fiber and ensuring a sustainable, safe living and working environment. Biological and economic constraints will continue to make this a challenging career opportunity.

The American Society of Agricultural Engineers (ASAE) is a professional and technical organization dedicated to the advancement of engineering applicable to agricultural, food, and biological systems. Its 9,000 members, representing more than 90 countries, serve in industry, academia and the public.

Petroleum Engineering

Petroleum engineering is primarily concerned with producing oil, gas, and other natural resources from the earth through the design, drilling, and use of wells and well systems. The petroleum industry develops methods for conveying fluids in to, out of, or through the earth's subsurface for scientific, industrial, and other purposes while remaining mindful of the ecological needs for safety. The curriculum in petroleum engineering provides a proper balance between fundamentals and practice. Graduate engineers are prepared for life-long learning but are capable of being productive contributors immediately. Petroleum engineers are currently in high demand in the industry, and their starting salaries are consistently among the top in the nation. The curriculum includes study of design and analysis of well systems and procedures for drilling and completing wells; characterization and evaluation of subsurface geological formations and their resources; design and analysis of systems for producing, injecting, and handling fluids; application of reservoir engineering principles and practices for optimizing resource development and management; and use of project economics and resource valuation methods for design.

The Society of Petroleum Engineers (SPE), with more than 50,000 professionals from oil and gas-producing regions around the world, represents the technical interests of this discipline. Its mission is to provide the means to collect, disseminate and exchange technical information concerning the development of oil and gas resources, subsurface fluid flow and production of other materials through well bores for the public benefit. The Society also provides opportunities through its programs for interested individuals to maintain and upgrade individual technical competence in these areas. The SPE accomplishes this mission through an international schedule of meetings and exhibitions, periodicals, short courses, books, electronic publications and section programs.

Mining Engineering

Mining engineering involves prospecting for mineral deposits; planning, designing, and operating profitable mines; processing and marketing the extracted minerals; insuring safe and healthy working conditions; and protecting and restoring the land during and after a mining project. Mining engineers use technologically advanced equipment, machines, robotics and computers every day. Those in charge of mine design, plan and test mines using computer simulators before ever breaking ground. Mining engineers in management use mine scheduling software to plan mining activity when operation is underway. Surface mining operations use larger mobile equipment than any other industry in the world. Mining methods and equipment are also applied to the removal of earth and rock outside the mining industry. Each year, an individual requires an equivalent of 40,000 pounds of new minerals and energy equal to that produced by burning 30,000 pounds of coal. With each different mineral comes a different mine site and a different location, giving the mining engineer a very diverse work environment, and limitless horizons to work anywhere in the world.

The Society for Mining, Metallurgy, and Exploration (SME) is an international society of 16,000 professionals with members in nearly 100 countries. Services available to SME members include: publications, professional registration, peer-review of technical papers, future leader and college accreditation programs, meetings and exhibits, public education and SME short courses.

Nuclear Engineering

Nuclear engineering is concerned with the science of nuclear processes and their application to the development of various technologies. Nuclear processes are fundamental in the medical diagnosis and treatment fields, and in basic and applied research concerning accelerator, laser and super conducting magnetic systems. Utilization of nuclear fission energy for the production of electricity is the current major commercial application, and radioactive thermal generators power a number of spacecraft. For the longer term, electricity production based on nuclear fusion is expected to become an increasingly important segment of the field. Nuclear engineers are concerned with maintaining expertise in the design and development of advanced fission reactors, performing basic and applied research in the development and ultimate commercialization of fusion energy, developing both institutional and technical options for radioactive waste and nuclear materials management and in fostering research in nuclear science and applications, with emphasis on bioengineering, detection and instrumentation and environmental science.

The discipline is represented by the American Nuclear Society (ANS). The ANS has a diverse membership composed of approximately 11,000 engineers, scientists, administrators and educators representing corporations, educational institutions, and government agencies.

Manufacturing Engineering

Manufacturing is a prime generator of wealth and is critical in establishing a sound basis for economic growth. Manufacturing education is important in achieving and maintaining a long-term competitive position for U. S. industry in the current global economy. The program covers concepts such as design for manufacture, flexible manufacturing, new communication and information networks and the impact of automation on human experience. Achieving and maintaining a long-term competitive economic position requires students to be able to analyze manufacturing systems for overall technical, economic and environmental performance. Manufacturing engineers should also be able to quantify trade-offs between these performance characteristics in relationship to the technology deployed in the manufacturing system.

The Society of Manufacturing Engineers (SME) is the world's leading professional society serving the manufacturing industries. Through its publications, expositions, professional development resources and member programs, SME influences more than 500,000 manufacturing executives, managers and engineers. The SME has some 60,000 members in 70 countries and supports a network of hundreds of chapters worldwide

Table 11.7
Web site addresses for the professional societies

Professional Society	Web Site Address
Electrical and Electronics Engineers (IEEE),	http://www.ieee.org/
American Society for Mechanical Engineering (ASME International)	http://www.asme.org/
American Society for Civil Engineering (ASCE)	http://www.asce.org/
IEEE Computer Society	http://www.computer.org/
American Institute of Chemical Engineers (AIChE)	http://www.aiche.org/
Institute of Industrial Engineers (IIE)	http://www.iienet.org/
American Institute of Aeronautics and Astronautics (AIAA)	http://www.aiaa.org/
American Society for Materials (ASM International)	http://www.asm-intl.org/
Biomedical Engineering Society (BMES)	http://mecca.org/BME/BMES/
American Society of Agricultural Engineers (ASAE)	http://asae.org
Society of Petroleum Engineers (SPE),	http://spe.org/
Society for Mining, Metallurgy, and Exploration (SME)	http://www.smenet.org/
American Nuclear Society (ANS).	http://www.ans.org/
Society of Manufacturing Engineers (SME)	http://www.sme.org/
Society of Fire Protection Engineers (SFPE)	http://www.sfpe.org

11.5   ENGINEERING FUNCTIONS

Engineers from all disciplines are involved in a wide variety of functions ranging from sales to maintenance. The more common engineering functions will be briefly described in the following subsections.

Design and Development

Companies continuously work on their product lines to maintain or increase their market share. Their ability to release a stream of high-quality, high-performance, reliable products that are competitively priced dictates their success and the profits they generate. **Design and development engineers** work to create new and profitable products. These "new" products are often redesigned models of an older product. Important in these redesigns are improvements in performance and reliability at lower costs. On rare occasions design and development engineers create an entirely new product that establishes a new market. Apple's IPod is a recent example of such a new product.

Design and development engineers interact with marketing personnel to establish the customer's needs. They convert the customer's needs to a product specification and then consider a large number of different design concepts that fulfill this specification. The best design concept is selected from among these options, and then the product is developed and manufactured. During this process, the design and development engineers work with manufacturing engineers, production engineers and test engineers to insure a timely and cost effective approach for producing a safe and reliable product.

Testing

Society today is increasingly litigious. It is a common occurrence for trial lawyers to seek remedies for damages due to what may be perceived as unsafe products. Product liability is a very serious concern. When a company releases a product for the market, it is imperative that it be safe. The only questions are—how safe and under what conditions? **Test engineers**, as the name implies, test products to ascertain if their performance specifications have been met. They conduct tests that enable them to predict the life of the product and the number of failures that will occur with time in service. They provide management with the data needed to determine warranty costs. During the design and development phase, they may test subsystems to provide data helpful in designing the product. Basically test engineers verify analytical methods for predicting performance. When the analytical methods are not sufficient, the test results are essential to prove the adequacy of the product.

Design Analysis

Almost every design and development team includes one or more members with excellent skills in design analysis. **Design analysis engineers** are responsible for modeling—where components from the product are converted to simplified models that may be analyzed. The analysis conducted may be in closed form where well-known mathematical formulas are used to produce results that predict performance or insure safety. In many instances, closed form solutions are not available and numerical techniques, such as finite element models, are used to generate solutions. In recent years, a large amount of specialized software has become available that enables the design analysis engineers to solve increasingly complex problems—rapidly and accurately.

The close interaction of the design analysis engineer with the design team during the development cycle is vitally important. If a problem is identified by analysis early in the design cycle, it can be corrected quickly and the costs of an error are minimal. However, if the problem goes undetected until hardware is produced and tested, the costs to correct an error are dramatically higher and the time required correcting the problem may delay introducing the product to market.

Manufacturing

A **manufacturing engineer** serves on the design and development team to provide expertise on tooling and manufacturing methods. As the components of a product are designed, many decisions are made that affect the cost of manufacturing and assembly of the product. It has been established that a major fraction of the total life-cycle cost of a product is committed in the early stages of design [8]. Quality cannot be manufactured or tested into a product; it must be designed into the product. Manufacturing engineers provide the design and development team with the knowledge of manufacturing equipment and processes available within their facilities and available from qualified suppliers. As the design proceeds, they identify manufacturing requirements and begin the design of any specialized tooling required for production. They design the manufacturing cells for producing component parts. They certify vendors who will supply externally purchased components. They design the lines, cells and the tooling used in assembly.

Manufacturing engineers also work with the quality control department to establish procedures to insure the quality of each component manufactured. They select machine tools and instrumentation to insure timely adjustments for manufacturing processes to remain within the control limits established by quality control personnel.

Production

Production engineers also participate on the design and development team. Their role is to ensure the flow of material and components to the manufacturing facilities and to the assembly lines. Often there is a need to purchase components or materials that require long lead times for delivery. In some cases, the lead-time is comparable to the design time. The production engineer must identify these items, anticipate the number required and place the order before the design is complete. After the design is complete, they work with purchasing personnel to insure delivery of components just-in-time to be used in production. They attempt to minimize the work-in-progress inventory. They work with personnel from the sales department to match production to anticipated demand for the product. They work with personnel from shipping to ensure that the product is removed from the production facilities and delivered to customers in a timely manner. Their goal is to meet the demand of the customers with a minimum of work in progress and finished product in inventory.

Maintenance

Most students own an automobile or have one available to them. Have you ever tried to travel from point A to point B and found the car would not start? This problem improves in your understanding of the need for maintenance. Manufacturing facilities often operate on a 24/7 schedule (this schedule implies the equipment operates 24 hours a day seven days a week). **Maintenance engineers** work to keep the equipment in operation. They seek to avoid breakdowns with scheduled maintenance. They plan for periodic lubrication of bearings, replacement of parts subjected to high rates of wear, replacement of belts and scheduled inspections. They develop a recording system that provides a database for predicting the time between failures and identifies the equipment that will probably fail. They design monitoring instrumentation that enables the facilities to be shut down in a controlled manner prior to failure of any critical parts. They supervise the maintenance crews that perform the repair work. If you enjoy high stresses associated with the inevitable breakdown, consider this engineering function for a career.

Sales

Sales engineers provide an interface between the customer buying technical products and the company selling them. They work with the engineers and purchasing agents representing the customer and provide technical information necessary for them to intelligently purchase a product or a system. Many engineering products and/or systems are complex with the availability of several different options. Sale engineers must know the details of the product and/or the systems, provide an overview of the options, and answer questions from the customer's representatives. People skills are as important as analytical skills in this position because the effectiveness of a sales call is often affected by personalities.

The sales engineers also seek information regarding new products from the customer. In some cases, the product line offered does not satisfy the requirements of the customer. When this occurs, it is important to ascertain the customer's requirements and the price he or she would be willing to pay for the product if it were to become available. This information is passed on the marketing representatives and is used in planning for new product developments.

Management

After demonstrating the ability to perform engineering functions, many relatively young engineers are promoted into **management** positions. If the company you are with is organized along functional lines, the first management position is usually "section head." A section is a small group of engineers (8 to 12) that specializes in a certain area such as motors, controls, transmissions, power, safety, etc. The section manager assigns tasks to the engineers in the group and provides supervision and support to aid them in performing the work. A department manager (2nd level of management) supervises several section managers. The section manager's role is a mix of both technical and administrative skills.

In companies that are organized with product development teams, program managers are responsible for the design and production of a certain product. Often an engineer becomes the program manager. He or she oversees the team, leads team meetings, makes assignments, arbitrates decisions and is responsible for the development schedule and the budget. The program manager usually reports to a vice president for development. For larger development teams, the program manager's administrative skills are usually more important than his or her technical skills.

Education

Many engineering students continue their studies after completing the B. S. degree and pursue advanced engineering degrees. Those graduating with a M. S. degree may go into industry to serve in assignments demanding a higher level of analytical skills. Others take faculty positions in educational institutions (usually a community college or a four-year college). In these educational institutions, they usually teach three or four undergraduate classes each semester. They also perform many service functions for the college and community.

Some students complete the Ph. D. and take a faculty position with a college of engineering in what are known as research universities. They are responsible for teaching, research and service in these positions. The teaching is limited to one or two courses per semester usually divided between graduate and undergraduate courses. They engage in active research programs and seek funding from external sources (federal and state governments, industry and foundations) to finance their research. They guide the studies of graduate students working in their research topic. Publication in technical journals and writing textbooks is an important part of their scholarly activities. Participation in technical societies, by presenting papers, organizing sessions and serving as officers of the society, are ways for performing community service. Many faculty members are also active consultants to both industry and government.

Consulting

Consulting engineers work on a retainer from government agencies or industry. They are independent businessmen and women who basically offer engineering services at a price. In years past, it was unusual for an engineer to select this career path. Most companies and government agencies hired their own specialists. The downsizing of companies and government has changed this practice. Many companies find that outsourcing engineering work is more cost effective. They call on engineering consultants with a given specialty as required and pay fixed prices for well-defined engineering tasks. The advances in communications—fax, cell phones, e-mail and the Internet—have enhanced the flow of information essential to conducting a successful consulting engineering business. Being in an office at a company's location, eight hours a day five days a week, is not required. A consultant can solve a problem or create an engineering document from a distance while maintaining communication with any client in any country for as long as is necessary to complete an assignment. The availability of high-speed, low-cost computers has enabled the consultant to compete effectively with an in-house analyst if the consultant has access to the same software and the data needed to perform the assignment.

11.6 REWARDS

For many years, engineering graduates have received relatively high starting salaries and most have usually found interesting positions available upon graduation. However, salaries later in a person's career depend upon many different factors including:

- Years of experience
- Level of technical and supervisory responsibility
- Type of employer
- Discipline and/or function
- Degree level
- Strength of the economy

Salaries also depend on current salary-wage policies, individual mobility, working conditions, and local and regional salary scales. Many significant factors, such as challenge, productivity, creativity, technical and supervisory responsibility and job commitment are dependent on the individual's initiative. However, other factors affect salaries that the engineer cannot control. These include company size, company policies, fringe benefits and promotion opportunities. Industry and company growth, labor-management relations, and research and development funding also affect earnings. There is considerable variation, even for engineers with similar position, responsibility and experience.

It is not possible to consider all of the parameters that affect salaries in this discussion. However, the three most important factors affecting average engineering salaries—degree level, years of experience and management responsibilities are presented in Fig. 11.2 and Fig. 11.3.

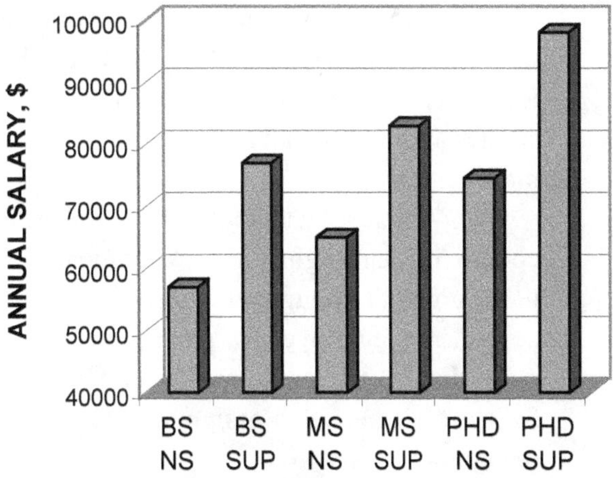

Fig. 11.2 Salaries for all engineers as a function of degree and management responsibilities.
"Engineers Salaries: Special: Industries Report, Engineering Workforce Commission of the AAES, 1997.
NS—non supervisory: SUP—supervisor

The results, presented in Fig. 11.2 and Fig. 11.3, are clear. Engineers with advanced degrees command a higher average salary than those with only a B. S. degree. Also, those with Ph. Ds earn more than engineers with a M. S. degree for the same class of work. Finally, for all degree levels, management responsibilities are rewarded with higher salaries. The differences are significant with manager's pay exceeding non-manager's pay by about 35, 28 and 25% for the B. S., M. S. and Ph. D. degrees, respectively.

The role of experience and degree level is illustrated in Fig. 11.3. All engineers gain in salary with years of experience. The gains are more rapid in the early years, with increases of about twice the increase in the cost of living regardless of the degree level. However, after about 20 to 25 years of experience, the rate of increase in annual salary decreases. In the last decade of a typical engineering career, an individual's compensation may not keep up with the increases in the cost of living. These data are averages, and you may prove to be an exception with either higher or lower rewards for your services. But the message in Fig. 11.3 is clear. Make your mark early in your career. After you reach the age of 45 to 50, significant salary gains are difficult for many engineers to achieve.

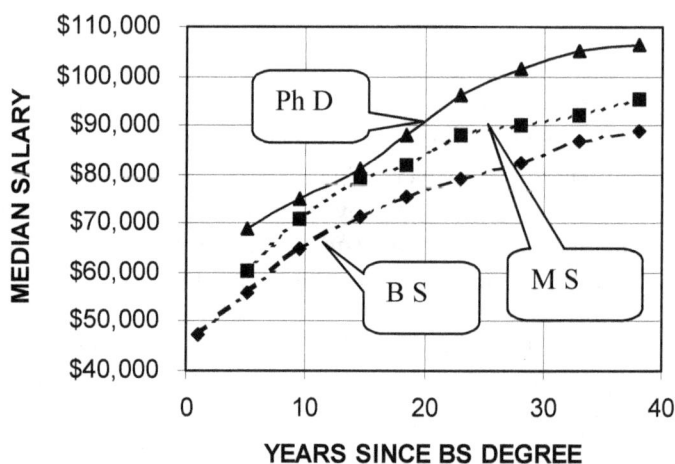

Fig. 11.3 Salaries for all engineers as a function of degree and years of experience.
Engineering Workforce Commission of the AAES, 2000

The data in Fig. 11.2 and 11.3 are from surveys that were conducted 1997 and 2000, respectively. As such, the salaries offered in 2010 and later will be significantly higher than those reported here[8]. Regardless of the exact numbers, the message is clear. If money is important to you—continue your studies and earn an advanced degree. Also, if you have the talents and interests, pursue a position in management.

Another factor not considered in a discussion of salaries is the satisfaction derived from completing an engineering assignment. With sufficient experience, many assignments typically those given to younger engineers lose their challenge. (Been there, done that.) With an advanced degree, an employer is paying you more and has higher expectations of your capabilities. Accordingly, the assignments tend to be more challenging, more visible and more rewarding. Success in one tough assignment leads to another and continued success leads to promotion and recognition.

---

[8] More current data for salaries in select engineering disciplines are presented in Tables 11.5 and 11.6.

11.7 PROFESSIONAL REGISTRATION

Today about 30% of graduate engineers are registered and licensed as "professional engineers." Professional registration is optional for many engineers unlike medicine or law where registration is mandatory before one may practice. When public safety is an issue, registration is required for engineers. Civil engineers often design buildings, highways, bridges and other structures that would endanger the public if they failed in service. Consequently, civil engineers participating in these activities must register and become licensed to practice.

You are encouraged to begin the four-step registration process for the simple reason that you do not know when in the future you will be required to demonstrate your qualifications. Sometimes evidence of your B. S. degree is sufficient for qualification, but in other instances it is not. Also as you approach graduation you are very well prepared to begin the registration process.

State boards of registration control the procedure for registration. There are minor variations from state to state, but the guide shown below is typical of the process required:

- Graduation from a four year engineering program that has been accredited by ABET.
- Achieving a passing grade on the "engineering fundamentals" examination.
- Practicing engineering for a specified number of years to gain engineering experience. The state board of registration specifies the number of years required.
- Achieving a passing grade on the "principles and practice of engineering" examination.

The "engineering fundamentals" examination may be taken before graduation. In fact, it is recommended that you take it in either April or October of your senior year[9]. The examination questions are designed to test your skills in fundamental subjects. Because engineers of all disciplines take the same examination, it is evident that the coverage must be on subjects common to all disciplines—mathematics, physics, chemistry, and engineering science. The examination is scheduled for eight hours. After passing the examination and graduating, you receive a certificate that designates you as an Engineer-in-Training.

The "principle and practice of engineering" examination is discipline specific. The civil, mechanical, electrical, etc. engineers must pass a discipline-oriented examination. While the topics covered are more focused, the questions probe the depth of your knowledge in the discipline. After passing this examination, you receive a license as a professional engineer and a seal that may be used when certifying your work.

11.8 SUMMARY

Two reasons are cited for classifying engineering as a profession. The first is based on a highly structured curriculum that prepares you for a position as an engineer upon graduation. You are expected to be productive almost immediately after assuming a beginning position with an industrial firm. The second reason is based on the type of work performed after graduation. When practicing in industry or for the government, one is engaged in engineering activities. A professional engineer is almost always salaried and is not compensated directly for overtime. Nor are you docked for time off. Engineers assume responsibility and are committed to the project, development team and the company or agency. They devote the time necessary to complete a project on schedule and on budget.

---

[9] The engineering fundamentals examination is offered twice a year in April and October. It is usually offered at several of the colleges of engineering in each State.

Statistics are cited for students entering engineering colleges. The students admitted to colleges of engineering are top notch—student leaders and scholars ranking in the top 5 or 10% of their high school graduating class. Average scores for both mathematics and verbal in the SAT examinations are outstanding. About 19.5% of those graduating from engineering in 2005 were women, but the percent of women in a typical class depends strongly on the discipline (see Table 11.1).

Pursuing a successful career in engineering depends on your ability to develop a number of different skills and understandings. These include:

- The ability to design.
- The ability to perform well on a development team.
- The ability to communicate in many ways and in different settings.
    - o Writing letters, memos, reports, specifications, etc.
    - o Participating in discussions, and presenting design briefings.
    - o Preparing two and three-dimensional drawings.
    - o Preparing slide presentations.
- Performing design analysis and interpreting the effect of the results on the design.
- Designing and conducting experiments and interpreting the data acquired.
- Understanding professional and ethical responsibilities.
- Understanding the need for life-long learning and actively pursuing this goal.

The most popular of the engineering disciplines are described and a listing of the undergraduate enrollment by discipline is shown in Table 11.4. Starting salaries for beginning engineers by discipline shows some differences with chemical and computer engineers leading civil engineers. The technical emphasis of each of the popular disciplines is described. Professional societies representing each discipline are discussed.

Engineers of all disciplines perform many different functions in their professional activities. The tasks commonly conducted in each of these functions include:

- Design and development
- Testing
- Design analysis
- Manufacturing
- Production
- Maintenance
- Sales
- Management
- Education
- Consulting

Rewards (salaries) for engineers depend on many different factors. However, three factors—degree level, years of experience and management responsibilities are the most important. Data showing salaries as a function of these three factors are given in Fig. 11.2 and 11.3. Advanced degrees and management responsibilities are rewarded with significantly higher salaries. Experience is also important, but the salary gains due to experience are only significant for about the first 20 to 25 years. After that time, salary gains are barely equal to inflation.

Finally, the advantages of professional registration are discussed. A guide to the four-step process involved in becoming a registered professional engineer is provided. All engineering students are encouraged to take the eight-hour "engineering fundamentals" examination late in their junior year or early in their senior year.

REFERENCES

1. Valenti, M. "Teaching Tomorrow's Engineers," Special Report, Mechanical Engineering, Vol. 118, No. 7, July 1996.
2. Zhang, G. Engineering Design and Pro/ENGINEER, Wildfire Version 3.0, College House Enterprises, Knoxville, TN, 2006.
3. Patterson, J. and P. Kim, The Day America Told the Truth: What People Really Believe about Everything that Really Matters, Prentice Hall, New York, NY, 1991.
4. Sommers, C. H., "Teaching the Virtues," Public Interest, No. 111, Spring 1993, pp. 3-13.
5. Grayson, L. P., The Making of an Engineer: An Illustrated History of Engineering Education in the United States and Canada, John Wiley & Sons, New York, NY, 1993.
6. Gibbons, M. T. "Engineering by the Numbers," Engineering Statistics, ASEE Web Site at http://www.asee.org.
7. N. Fogg, P. Harrington, and T. Harrington, College Majors Handbook with Real Career Paths and Payoffs: The Actual Jobs, Earnings, and Trends for Graduates of 60 College Majors, 2nd Edition, JIST Publishing, Indianapolis, IN, 2004.
8. Anon, Improving Engineering Design: Design for Competitive Advantage, National Research Council, National Academy Press, Washington, D. C., 1991.
9. Anon, "American Consulting Engineers Council Directory", American Consulting Engineers Council, 10015 15th Street, N. W., Washington, D. C. 20005.
10. Landis, R. B. Studying Engineering: A Road Map to a Rewarding Career, Discovery Press, Burbank, CA, 1995.

EXERCISES

11.1   Write a two-page paper on why engineers should be considered professionals.
11.2   Write a two-page paper comparing the engineering profession to the medical profession.
11.3   Write a two-page paper comparing the engineering profession to the legal profession.
11.4   Write a two-page paper comparing the engineering profession to the accounting profession.
11.5   Write a brief paper giving reasons that you know for why bright well-qualified students for leave engineering after only one semester.
11.6   Why do bright, well-qualified students often fail courses in their first year of study?
11.7   Prepare a list of the skills that you currently have and provide a grade (from 1 to 10 with 10 being exceptional) for your achievement level in each skill.
11.8   Add to the list in Exercise 11.7 your goal for an achievement level in each skill at the conclusion of this course.
11.9   Do you believe that you will learn new skills in this course?
11.10   What is the purpose of the ethical codes endorsed by the professional societies?
11.11   Write a two-page paper describing your current ethical standards. Do you believe these standards will change during your tenure at college?
11.12   Are you committed to life-long learning? If so, why? If not, why? How do you intend to continue your studies after graduation with a B. S. degree?
11.13   Have you selected an engineering discipline? If so, state the discipline and give the reasons why you have selected it. If not, state the reasons for your delay?

11.14 If you are still concerned with selecting an engineering discipline, prepare an action plan to gain the information necessary for your decision?

11.15 If you are concerned that your first choice of an engineering discipline was not correct, prepare an action plan to evaluate your choice and to change disciplines if necessary.

11.16 For the engineering discipline of your choice, describe the mission and the activities of the professional society that represents their interests. Visit their web site. Do they have a student chapter? Do you plan to join?

11.17 Most colleges of engineering offer only a limited number of degree programs. What are the degree programs in your college? Do these programs coincide with your career objectives?

11.18 After reviewing the engineering functions, select the two that best suit your interests. Write a paper discussing why these functions interest you.

11.19 Write a paper discussing why engineering salaries begin to plateau after about 20 to 25 years of experience. Include in your paper how you plan to avoid this problem.

11.20 Salary data is a moving target; consequently, the information provided here is at best an approximation. It is possible to obtain more up to date salary information that takes into account the city where the position is located by visiting the Wall Street Journal Website at http://www.careerjournal.com/salary. Click on the Salary Expert and perform a salary search that shows the differences in salaries for different disciplines and different management responsibilities. Also show the differences in salaries to account for the different cost of living in different locations across the United States.

11.21 Today the retirement age for most engineers is about 67. At what age do you plan to retire? Do you believe the age requirement to receive social security benefits will change prior to your retirement? Why?

11.22 Write a paper arguing either for or against professional registration for engineers. Do you plan to take the "engineering fundamentals" examination when you achieve academic status as a senior?

CHAPTER 12

A STUDENT SURVIVAL GUIDE

12.1 OVERVIEW FOR SURVIVAL

As a new student in the College of Engineering, you are encountering many new challenges and meeting many new friends. You have to arrange for housing, transportation and meals (the basics) in an entirely new setting. You also have to cope with a schedule of classes and find the locations of classrooms, lecture halls and laboratories in a maze of buildings scattered all over campus. Finally, you must learn to study and pass courses in a much more competitive environment than you found in high school.

This chapter will discuss some strategies for coping with the new environment and for procedures that will help ensure your successful completion of your studies for a B. S. degree in engineering.   In writing this survival guide, it is recognized that you are bright and have demonstrated your scholastic and leadership abilities—otherwise you would not have been admitted to the College. Success should be insured, but the fact is that about twenty-five percent of each year's incoming class does not complete the engineering program. Many students leave in the first year because of failing grades or lack of interest in the program. Some procedures to insure your success in completing the program will be discussed in this section, and then several important tools for improving your performance will be described later in this chapter.

Goals and Priorities

Many students fail to establish realistic goals that enable them to control their time and direct their energy in a manner needed to succeed in a competitive environment. Long-term, intermediate-term and short-term goals must be established to focus your activities. Otherwise your efforts are diffused, sufficient progress is not made, and one or more goals are not achieved. The goals should be written in priority order and they should be dated. Your goals should be reviewed and revised periodically with changes made to reflect both progress and realism.

Confidence

Success cannot be achieved without confidence. It is essential that you maintain a positive attitude with a can-do philosophy. If you start to doubt your ability to pass a test, fear will become a significant factor that blocks your ability to think clearly during the examination. It is a well-know fact that fear and/or anger limits a person's ability to think clearly and act promptly. One way to gain confidence is to prepare until you are certain that you understand the course material. You will then become confident in your ability to perform well on competitive examinations under pressure.

Stay Current

The clock is as important in studying engineering as it is in a football or basketball game. The course (game) begins on the first day of class and ends at the last minute allowed for the final examination. Your instructor will try to use all of that time to cover as much material as possible. There is little or no float time in an engineering course. Reading and homework will be assigned for each and every class period. Moreover, the material is cumulative—to understand the second topic it is necessary to master the first topic, etc. If you fall behind in the assignments, it is extremely difficult, if not impossible, for most students to catch up and become current.

Attend Class

As an instructor, I was amazed at the number of students cutting classes. An hour of class time costs the student, or his or her parents, more than a ticket to a rock concert. What student would throw away a ticket to a rock concert? Yet, some students throw away many opportunities to attend class over the semester. Attending class gives you the opportunity to hear the instructor's interpretation of the material and to stay up to date on assignment due dates or exam times. Attending class also gives you the opportunity to ask the instructor questions or hear other student's questions, which can often be extremely helpful. On the same note, attending class is not helpful unless you are paying attention to the lecture or questions. While attending class, you might as well make the most of your time there so that you don't have to rely solely on the textbook or other students to learn the material. You will most likely use your textbook or work with other students to study, but doing these in conjunction with attending class will maximize your opportunities for learning, and allow you to gain a deeper understanding of the course material.

Managing Time

You are accustomed to schedules because you moved from class to class following a prescribed schedule while attending high school. The schedule in college is similar except that a class period for a specific course is only scheduled for two or three hours per week instead of an hour every day. You have the **illusion** of more free time. A typical load of 16 credit hours for a semester entails only 12 or 13 hours of classroom participation and 6 to 8 hours of laboratory involvement. Unfortunately, this is misleading because reading and other assignments require significant amounts of out-of-class time. A rule followed by many instructors is to adjust the material and class assignments to require a **time factor of 3 to 4 on the credit hours for the class**. This adjustment means that if you are taking a three-credit hour class, you should expect to spend 9 to 12 hours each week to attend class, complete the assignments and study for the examinations. Your 16 credit hour schedule really involves a time commitment of **48** to **64** hours per week. Time management becomes important when considering the actual time needed to succeed in engineering courses.

Developing Good Habits

It is self evident that good study habits are essential to successfully completing an engineering degree program. Yet many successful high school students have never developed good study habits. Competition in high school was not intense and you could meet or exceed your teachers' expectations without much out-of-class work. The academic competition is much more intense in a college of engineering and your instructors' expectations are higher than you can imagine. It is essential that you establish and follow a personalized schedule for study periods for each of your classes. The time allocation must be sufficient for you to complete the class assignments. Your schedule must insure that you do not fall behind in each subject[1].

Managing People

While you are not yet an engineering manager, you will still interact with a number of people, and will find it necessary to manage your relations with them. A short list of these people includes your instructors, roommate, classmates, teammates and friends on and off campus. Maintaining relationships requires time, and as you will soon determine, time is a precious commodity. Your instructors should command priority; fortunately they are easy to manage. All you have to do is go to class, and perform well on examinations. Your instructor will respond with good grades. You may also meet many of your friends in classes. This social bonding is an important part of the educational process. Education should be fun and classmates help make it so.

Teammates differ to some degree from classmates, because a team is a more formal group that has been organized to perform a well-defined task such as the design of a prototype of some product, process or system. Teammates also bond socially and often remain friends long after the team has been disbanded. Friends on and off campus are important to your social life. Pursuing an engineering degree will take most of your time and energy, but it is essential that you allow some time for socializing.

Seeking Help

Compared to high school, the university is huge with tens of thousands of students and countless buildings scattered over a very large area. Orientation held in the summer preceding fall admission is recommended. These orientation sessions will provide some of the answers to your questions. Information will be provided about the engineering programs, the locations of the buildings, dormitory life, availability of food and entertainment, etc. However, after you arrive and begin classes, many unforeseen questions will arise. Where do you seek reliable answers?

If your questions are about dormitory life, find the **Resident Assistant (RA)** and he or she will provide guidance. The RA's are carefully selected; they relate well to new students and often anticipate questions. The point is for you to ask questions at your earliest opportunity. The answers will alleviate your concerns.

If your questions are about the engineering program, find the **Office for Freshman Advising**. This office, usually found near the Dean's office, is staffed with professional counselors. They know the details of all the programs in the college, and they understand the needs of the students entering the program. They can provide valuable advice about program requirements that will ease your entry into the program.

If you have questions about a class you are taking, check the instructor's schedule. He or she will have a few hours each week reserved for office hours. You may stop at his or her office during these times and find your instructor. He or she should be willing to take the time to discuss your concerns. Sometimes meeting with the instructor during regularly posted office hours is not possible. In these instances, contact the instructor by phone or e-mail and arrange an appointment. The instructors will help you understand the course material, but you should be prepared to ask

---

[1] In addition to good study habits, you should develop good personal habits. These include eating a well balanced diet and exercising regularly. Arrange your schedule to enable you to sleep six to eight hours each night. It is important to be well rested and to maintain your daily schedule. Finally, remember that drinking excessively may affect your ability to achieve your potential.

clearly phrased questions that define the problems you are encountering. The instructor will be willing to respond to your questions, but he or she will not be willing to tutor you.

12.2 TIME MANAGEMENT

Weekly Schedules

The college schedule is a weekly affair. You may have classes every day of the week, for about 15 weeks, before the final examination period. It makes sense to prepare a weekly schedule showing the time and location of your classes similar to the one shown in Fig. 12.1.

WEEKLY SCHEDULE							
FALL SEMESTER 2008							
NAME			ADDRESS			PHONE NUMBER	
TIME	SUN	MON	TUES	WED	THUR	FRI	SAT
MORNING							
8:00		ENES 100		ENES 100		ENES 100	
9:00		ENES 100		ENES 100			
10:00		MATH 140	MATH 140	MATH 140		MATH 140	
11:00							
12:00		CHEM 135		CHEM 135		CHEM 135	
AFTERNOON							
1:00		ENGL 101		ENGL 101		ENGL 101	
2:00			CHEM 135				
3:00							
4:00							
5:00							

Fig. 12.1 A possible class schedule for first semester electrical engineering students.

The recommended load of 13 credit hours for the fall semester of first year students is less than the typical load of 15 to 18 credit hours for the second and subsequent semesters. The lighter load is to provide you with the opportunity to adjust to a new academic environment. While the times for classes on the schedules will differ from one student to another, the number of credit hours and the number of contact hours will probably remain about the same. The number of contact hours[2] (16) exceeds the number of credit hours because laboratories and recitation periods are not always counted with the same weight as lecture hours.

The next step in time management is to schedule study periods. How much time should you schedule? The general rule most instructors use in designing assignments is that the **average** student spends two hours studying for each hour of lecture and another hour studying for each hour of laboratory or recitation. This rule implies that a student should schedule $3 \times 2 + 1 \times 1 = 7$ hours of out-of-class study time for Chemistry 135 each week. For English 101 an average student would schedule $3 \times 2 = 6$ hours of out-of-class time for studying and writing. Adding the study time for all four courses required this semester gives a total of 27 hours that should be devoted to study for the **average** student. The word **average** is stressed. You may be able to perform well with less than

---

[2] Contact hours are the number of hours you attend class, recitation or laboratory periods in a given week.

the recommended 27 hours. Or you may be having trouble with one or more courses and find 27 hours is not sufficient. The weekly schedule shown in Fig. 12.2 provides for 27 hours of study time.

WEEKLY SCHEDULE FALL SEMESTER 2008 NAME		ADDRESS			PHONE		
TIME MORNING	SUN	MON	TUES	WED	THUR	FRI	SAT
7:00	OPEN	BRKFAST	BRKFAST	BRKFAST	BRKFAST	BRKFAST	BRKFAST
8:00	OPEN	ENES 100	STUDY	ENES 100	OPEN	ENES 100	OPEN
9:00	BRKFAST	ENES 100	STUDY	ENES 100	OPEN	STUDY	OPEN
10:00	OPEN	MATH 140	MATH 140	MATH 140	OPEN	MATH 140	STUDY
11:00	OPEN	LUNCH	STUDY	LUNCH	OPEN	STUDY	STUDY
12:00	LUNCH	CHEM 135	LUNCH	CHEM 135	LUNCH	CHEM 135	LUNCH
AFTERNOON 1:00	OPEN	ENGL 101	STUDY	ENGL 101	OPEN	ENGL 101	SOCIAL
2:00	OPEN	STUDY	CHEM 135	STUDY	OPEN	STUDY	SOCIAL
3:00	OPEN	STUDY	STUDY	STUDY	OPEN	STUDY	SOCIAL
4:00	OPEN	STUDY	STUDY	STUDY	OPEN	STUDY	SOCIAL
5:00	OPEN	OPEN	OPEN	OPEN	OPEN	OPEN	WORKOUT
EVENING 6:00	DINNER	DINNER	DINNER	DINNER	DINNER	DINNER	DINNER
7:00	STUDY	OPEN	OPEN	OPEN	STUDY	SOCIAL	SOCIAL
8:00	STUDY	OPEN	OPEN	OPEN	STUDY	SOCIAL	SOCIAL
9:00	STUDY	OPEN	OPEN	OPEN	STUDY	SOCIAL	SOCIAL
10:00	STUDY	OPEN	OPEN	OPEN	OPEN	SOCIAL	SOCIAL
11:00	OPEN	OPEN	OPEN	OPEN	OPEN	SOCIAL	SOCIAL
NIGHT 12:00-6:00	SLEEP	SLEEP	SLEEP	SLEEP	SLEEP	SLEEP	SLEEP

Fig. 12.2 A complete weekly schedule for a first semester student.

Many assumptions were made in developing the schedule presented in Fig. 12.2. The first assumes you are a morning person because all mornings are packed with classes and study except for Sunday. There are two reasons for carrying a heavy schedule in the morning. First, most people think more clearly in the morning after a good night's rest. Second, if you encounter difficulties in completing an assignment during a study period, time is available to recover later in the day.

Every hour in the weekday morning has been filled[3]. If you have a period or two between classes, you may want to schedule it for studying. Usually it is easy to find a quiet desk in the library and complete an assignment in mathematics or chemistry in an hour. On the other hand, you may decide to work with a small study group in one of the student lounges.

The weekly schedule shows four of the seven afternoons devoted to classes or study periods. Two afternoons are completely open and Saturday afternoon and evening are scheduled for social events. It is important to schedule time for socializing with your friends, just remember to keep everything in moderation. The open periods (unscheduled time) provide much needed flexibility in the schedule. On some weeks you will find the assignments much more demanding than other weeks. When this situation occurs, you will be able to use the open periods to study. Open time before bed time has been scheduled for five evenings of a typical week. A total of 36 hours of open

---

[3] All mornings are filled except for Thursday, which is an open day. Many students find it necessary to work during the academic year. This schedule has been arranged with one open day to enable part time employment.

time is available. There is a significant amount of flexibility to add study periods or even to work part time if necessary.

Attending class and studying is sedentary. You will lose your body tone without interrupting this pattern of class, study and sleep. Try going to the gym and working out at least three times a week. The schedule provides ample open time for working out every day.

You may find that the schedule shown in Fig. 12.2 is not suitable. That's fine. It is not necessary to impose this schedule on you. However, it is imperative that you prepare a weekly schedule that is more suitable, and one that accommodates time for:

- Attending all your classes, laboratories and recitations
- Sufficient study time at suitable times (27 hours is a minimum suggested)
- Adequate uncommitted (open) time to provide needed flexibility in your schedule
- A measure of social time to keep your spirits high
- Enough time for meals, sleep, work outs, and mentally preparing for the day

Daily Schedules

You will prepare a weekly schedule each semester. It is a guide to your daily activity, but it does not contain the detail needed to use your time efficiently. A daily schedule provides the detailed guide for allocating sufficient time to tasks essential to your success in each course. To provide information on preparing a daily schedule, consider a typical Monday as defined in the weekly schedule of Fig. 12.2. A new format for the daily schedule is constructed as illustrated in Fig. 12.3.

Each hour of the day is scheduled and tasks are assigned. Let's review the first portion of this example. At 7:00 AM, you rise, shower, etc. and plan the day. Next you walk[4] to the dining hall and enjoy breakfast. At breakfast you arrange for a study group to meet after the English class in the afternoon. At 8:00 AM you arrive on time for your ENES 100 lab and engage as an active team member in planning the design of the hovercraft. At 10:00 AM you attend your mathematics class before walking to the dining hall for an early lunch. Review the afternoon and evening schedule on your own and determine if this schedule seems reasonable.

DAILY SCHEDULE					
MONDAY, OCTOBER 16, 2006					
NAME					
TIME	**MONDAY**				
MORNING		**AFTERNOON**		**EVENING**	
7:00	PREPARE	1:00	ATTEND ENGL 101 CLASS	6:00	DINNER
	BREAKFAST	2:00	STUDY	7:00	OPEN
8:00	ATTEND ENES 100 LAB		WORK TWO MATH PROBS	8:00	OPEN
9:00	ATTEND ENES 100 LAB	3:00	STUDY ENES ASSIGNMENT	9:00	OPEN
10:00	ATTEND MATH 140 CLASS		COMPLETE READING	10:00	OPEN
11:00	LUNCH	4:00	STUDY CHEM 135 ASSIGNMENT	11:00	OPEN
12:00	ATTEND CHEM 135 CLASS		COMPLETE PROBLEMS	**NIGHT**	
		5:00	WORKOUT & SHOWER	12:00-7:00	SLEEP

Fig. 12.3 An example of a daily schedule for a typical Monday.

---

[4] Remember to allow for time to walk between buildings in preparing your daily schedule. Total walking time probably approaches an hour each day for students on large campuses with scattered buildings.

Your daily schedule indicates that the evening is open. Should you relax or study for a while after dinner? You have efficiently used the three hours of study scheduled during the day to work on assignments in three of your four classes; however, you have not completed the assignment for your mathematics and chemistry classes. To make this decision, examine your class schedule for Tuesday. You have three hours set aside for study in the morning, another three hours later in the day. You can relax Monday evening or take on a part time job if you are able to consistently complete your assignments during the study periods scheduled during the day.

While the class structure is consistent from week to week, the assignments vary from day to day. Sometimes the assignments can be completed quickly with little effort. Other times term papers are due and the library research and writing requires much more time than the usual assignment. A few times each semester an hourly examination will be scheduled in each course. The time to prepare for the examinations is often significant, particularly for those students who fall behind. Because of the changing demands for time from one course to another, it is vital that several open periods are available in the daily schedule to provide you with an opportunity to put in the extra effort needed for success.

12.3   PEER MANAGEMENT

As you move through the program, you will meet hundreds of fellow students. Some of these students will be no more than faces in a common classroom, while others will become close friends. These close friends may mold your life style and behavior. Peer pressure is a recognized factor in affecting behavior. You will have an established set of behavioral and moral patterns before you arrive on campus. As you develop friendships in the classroom, dormitories, fraternities, sororities, gymnasium, etc., your behavioral patterns will change. In fact, you will become associated with several groups of friends each having different lifestyles, interests and personalities. For example, you may have a group of friends from a design team in an engineering class. You may have another group of friends from the dorms and still another from a recreational basketball team.

These groups of friends make up your social life which is important for you to have while pursuing an education. It is healthy to be involved with friends, but there are two dangers of which you should be aware. First, the time involved in pursuing social activities with your friends should not interfere with any of your scheduled study time. You must remember that while pursuing an engineering degree, schoolwork comes first and then social commitments.

Second, the behavior of some social groups may cause concern. Before you become deeply involved with a social group you should compare the group members' moral standards with yours. If members of the group break the law or do illegal drugs it would be in your best interest to find a different social group. It is easy to find trouble if you associate with the wrong people.

12.4   STUDY HABITS

Good study habits are important to successfully completing an engineering degree. A study habit is a routine that you establish to prepare for class. There are several important aspects of study habits which affect your performance. First is timing. Some students prefer to study in the morning when they are rested and their minds are fresh. Others prefer to study very late at night when it is quiet. You should choose the time of day that is the most effective for you. Establishing several blocks of study time each week is an important scheduling habit to develop.

Organization is another important feature for effective study. You will be enrolled in four to six courses each semester with different sets of requirements and different types of assignments. In addition to your time management schedule, it is important to list all assignments. Organizing the assignments, assigning priorities and making time estimates and commitments for each course are critical tasks. Time management, as discussed previously, is an important element to organization; however, priorities must be set so that assignments are completed and handed in on time. If you list all assignments, tasks, and due dates at the conclusion of each class, it is easy to review your notes at the end of each day and establish priorities for your time.

Where you study is just as important as when you study. Single desks in the library or computer labs are a good place to study. It is quiet with little or no chance of interruptions. If you form a small study group, tables in the library will serve your purpose. You may talk quietly and discuss problems and approaches to be used in the solution. You may find your dorm room suitable if your roommate is cooperative. If not—head for the library. Student lounges may be suitable in some instances, but they are often noisy with students talking, eating and moving about.

Developing smart study methods is also important. Most engineering students spend much of their study time working homework problem assignments. A typical assignment for a class may include three or four problems. The instructor probably intends the students to spend 15 to 20 minutes per problem. Students understanding the material will be able to complete the assignment in about an hour. However, if you have difficulty with one or more problems you will need to devote additional time to complete the work. How much time should you spend before asking for help? You probably should not devote more than about 30 minutes trying different approaches for the solution to a difficult problem. If you have not solved the problem by then, seek help from your TA, instructor, or a study group.

Prepare neat homework solutions. If your homework is illegible or has many pencil/pen scratch marks on it, it may not be accepted. If you know what you are doing, most solutions can be completed with about a dozen lines and a diagram or two. Plan your approach and execute the answer showing all work. If you do not show all of your work, you may not get credit either. If you make a mistake, either erase completely or rewrite the solution. Always use a pencil because it is easy to erase after you have make a mistake. Ballpoint pens are not appropriate writing instruments for most students.

12.5 PROBLEM SOLVING METHODS

From your first semester until graduation, you will be required to solve problems that require accurate numerical answers. A large percentage of your grade in science, mathematics, and engineering courses will be your performance in solving these problems. It is important for you to master a proven technique to solve problems. It is recommended that you follow the eight step plan indicated below:

1. Write the problem statement as you begin the solution.
2. Carefully read and understand the problem.
3. Identify the unknown quantity that you are being asked to determine.
4. Write the rule or equation you plan on using in your solution.
5. List the information provided and prepare sketches as required.
6. Execute the solution.
7. Add units as required to the final result.
8. Check the solution for numerical errors.

Let's consider an application problem in a Calculus I course to illustrate this approach.

Step 1: Write the problem statement. ⇒Through how many revolutions does a bicycle wheel with a radius of one-foot turn when the bicycle travels one mile[5].

Step 2: Carefully read the problem and understand the physics involved.

Step 3: It is clear that the unknown quantity is the number of times the wheel turns (without slipping) as the bicycle travels a distance of one mile.

Step 4: The equation to obtain the solution is $S = N (2\pi R)$. (12.1)

Step 5: List the known information pertaining to the problem as:

> $S = 1$ mile $= 5280$ ft is the distance traveled.
> N is the number of turns made by the wheel, which is the unknown.
> $R = 1$ ft. is the radius of the wheel.

Step 6: Execute by solving Eq. (12.1) for N to obtain:

$$N = \frac{S}{2\pi R} = \frac{5280 \frac{ft}{mile}}{2\pi \frac{1 ft}{rev}} - 840.3 \frac{rev}{mile} \tag{a}$$

Step 7: Add the required units to the numerical answer. Perform a unit analysis and add the appropriate units to the numerical result (rev/mile).

Step 8: Check the numerical result. You should mentally perform an arithmetic estimate of the answer and check if it is reasonable. Also, plug the numbers into the calculator a second time to verify the result. Many careless errors are found with this simple procedure. With repeated practice the eight steps will become a routine that you will follow with little thought. You will also find that checking your results for accuracy and for units is important because errors of both types are common.

12.6 COLLABORATIVE STUDY OPPORTUNITIES

Collaborative study is encouraged in college. However, you must recognize the difference between collaborative study and academic dishonesty. Academic dishonesty would be if you blindly copy the work of a fellow student. Collaborative study is when two or more people work together sharing ideas and discussing approaches for solving problems. Collaboration allows you to find errors in solutions and then to discuss these errors. Everyone involved benefits from the discussion. The individual with the correct solution increases the depth of his or her understanding by explaining the proper approach. The individual with the wrong answer discovers the correct approach from a fellow student. The student works the problem again and has another chance to learn and to succeed. By students teaching each other in this manner everyone's depth of knowledge is increased.

---

[5] Problem 40, page 57 of <u>Calculus with Analytic Geometry</u>, by Robert Ellis and Denny Gulick, Sanders College Publishing, New York, 1994.

Regular meetings of small study groups are beneficial and reinforce study habits. The group bonds with time, and learning becomes a social experience. Respect, confidence and esteem grow as group members become comfortable with one another.

Forming small (two to four members) study groups for each of your classes is suggested. Meet at a scheduled time after all members of the group have had an opportunity to work on an assignment independently. Compare results, share ideas and discuss difficulties. Seek knowledge from one another to determine the correct approach for solving each and every type of problem. If all members of the study group are encountering difficulty, seek help from your teaching assistant (TA) or your instructor.

12.7 PREPARING FOR EXAMINATIONS

If you attend class, understand the material presented in class, and successfully complete all of the homework assignments, preparing for an examination is not difficult or time consuming. Exam preparation becomes time consuming if you have fallen behind in your daily preparation for the class and find it necessary to cram. Stress caused by cramming can be avoided if you stay current. Let's assume that you have studied regularly and are reasonably current with the material. What should you do to prepare for the examination?

A confident student with a good set of notes should try to write the examination by anticipating the instructor's questions. Anticipating the type of questions on the examination is not only possible; it is usually easy to do. In the instructor's class, he or she will emphasize a few topics indicating their importance. The instructor will make statements like:

- "This is really very important."
- "If you remember anything remember this."
- "I really like this derivation."
- "This is the most important principle in the course."

The instructor will usually dwell on the important topics and move quickly through the ones considered of less importance. Review your notes and place a star beside the passages where the instructor signaled either importance or his or her interest. In the typical five or six week period for a preliminary examination, you may find about a half a dozen stars. Write a question about the topics you have identified as important and ask one of the members from your study team to solve the problems on your exam. Discuss this process with your study team and arrive at a dozen questions that the team solves. After a few examination experiences, you will gain skill at anticipating the instructor. You will be able to predict the form, if not the substance, of at least half of the examination questions in advance. Sample exams and past examinations for the course should also be reviewed.

12.8 STRATEGIES FOR TAKING EXAMINATIONS

There are strategies that you may exercise in taking an examination that will enhance your performance. Let's discuss some of them:

Confidence: Study and learn until you become confident that you understand the material and that you will perform well on the examination. If you believe you will screw up the examination, you may not be comfortable enough with the material. Confident does not mean cocky, but rather comfortable with the subject and the instructor.

Calm: To be calm, you are at peace with yourself. You know that you have prepared for the examination. You are not nervous because you have demonstrated the ability to perform well on the homework assignments. You look forward with anticipation to the opportunity to demonstrate your problem solving skills.

Plan and Select: Before beginning the examination, read each problem in the entire examination. Be certain that you understand each problem. If not, ask the instructor for clarification. Then arrange an order to complete the problems from the easiest to the most difficult. Many instructors will arrange their examinations in this fashion with the easy problems first to give you confidence and the more difficult problems later to give range to the grades. Other instructors will arrange problem chronologically with the first problem covering material studied early in the examination period and the last problem covering much more recent material.

Most hourly examinations are limited to a small number of problems. The instructor has timed the examination so that prepared students will have enough time to complete it. In many cases, students with experience and a deep knowledge in the subject can complete the examination in 15 to 20 minutes. You have 50 minutes, which is usually more than enough time to complete your solutions, **but not enough time to be studying or learning during the examination**.

Execute: Begin with the easiest problem because you know the approach for its solution. If your initial assessment of the difficulty of the problem was correct, you successfully complete the solution in a few minutes and can begin the solution of the next easiest problem. You are ahead of schedule and full of confidence. You continue in this manner until you encounter a problem and become stuck. After you have spent about five minutes on this problem, stop your efforts and move to the next problem. You should return to this problem only if time remains at the end of the examination.

After you have attempted each problem on the examination, reserve 15-25 minutes to return to the problems causing difficulty. Review the problem statement and your solution to the point of where you got stuck. Perhaps you will find an error, a missing formula, or an omission and be able to complete the solution. If not write the instructor a brief note explaining your difficulty. He or she may reward your efforts with generous application of partial credit.

Checks: Use the final ten minutes of the examination for checking your results and making sure that all of the numerical answers have appropriate units. There are three different mental checks you should go through to detect errors. The first is a simple judgment check. Does the number look reasonable or is it obviously too large or too small? Common sense often is a good guide to finding obvious errors.

The second is a numerical check. Examine the equation and make a mental estimate of the result. With practice you will soon be able to estimate the answer to within $\pm$ 20%. Next, run the numbers in the equation through the calculator again. Rerun the numbers until the answer is confirmed. In many instances, a button may have been pressed by mistake and errors result. Recalculating eliminates these careless errors.

The third check is for dimensional consistency. Sometimes errors occur because you fail to recognize the need for unit conversions or you do not make the conversions correctly. By checking an equation for consistent units, it is possible to locate and correct these errors. Equations used in the solution of engineering problems are independent of the units of measurement. Consider an example from physics where the distance S that a baseball falls after it is dropped from a very tall building is to be determined. You seek the distance traveled by the ball after it has fallen for 6 seconds. If you assume[6] that the drag due to air resistance is negligible, the equation governing the distance the ball falls is given by:

$$S = \tfrac{1}{2}gt^2 \qquad\qquad (12.2)$$

where g is the gravitational constant and t is time of the fall

When applying this equation, you must insure that it is dimensionally homogenous by using units that are consistent on both sides of the equation. To illustrate the various options you have in dealing with Eq. (12.2), consider the three following cases:

Case 1: Use U. S. Customary units of feet for length and seconds for time. The gravitational constant in this system of units is 32.17 ft/s^2. Accordingly, you write Eq. (12.2) as:

$$S = gt^2/2 = (32.17 \text{ ft/s}^2)(6 \text{ s})^2/2 = (32.17 \text{ ft/s}^2)(36 \text{ s}^2)/2 \qquad (a)$$

The time unit "s" cancels out on the right hand side of Eq. (a) and the remaining unit represents a length expressed in feet. Therefore:

$$S = 579.1 \text{ ft} \qquad\qquad (b)$$

Case 2: Use U. S. Customary units of inch for length and seconds for time.

The gravitational constant in this system of units is $g = (32.17 \text{ ft/s}^2)(12 \text{ in./ft}) = 386.0 \text{ in./s}^2$. Accordingly, you write Eq. (12.2) as:

$$S = gt^2/2 = (386.0 \text{ in./s}^2)(6 \text{ s})^2/2 = (386.0 \text{ in./s}^2)(36 \text{ s}^2)/2 \qquad (c)$$

The time unit "s" cancels out on the right hand side of the equation and the remaining unit represents a length expressed in inches. Therefore:

$$S = 6948 \text{ in.} \qquad\qquad (d)$$

Case 3: Use SI units of meter for length and seconds for time.

The gravitational constant in the SI system of units is 9.807 m/s^2. Accordingly, you write Eq. (12.2) as:

$$S = gt^2/2 = (9.807 \text{ m/s}^2)(6 \text{ s})^2/2 = (9.807 \text{ m/s}^2)(36 \text{ s}^2)/2 \qquad (e)$$

The time unit "s" cancels out on the right hand side of Eq. (a) and the remaining unit represents a length expressed in meters. Hence:

$$S = 176.5 \text{ m} \qquad\qquad (f)$$

A careful check of units on each side of the equation will allow you to locate errors caused by using inconsistent units when evaluating equations.

---

[6] The drag is not negligible. It is assumed to be negligible to simplify the mathematical solution leading to Eq. (12.2). As your skills develop, drag forces will be considered and you will have the opportunity to solve the more complex mathematical expressions that result when the drag force is introduced.

12.9 ASSESSMENT FOLLOWING THE EXAMINATION

Errors on an examination can be grouped into three classes.

1. Careless mistakes
2. Execution errors
3. Concept errors

Careless mistakes are when you clearly understood the material but you lacked focus. You may have had the correct numbers substituted into the correct equation but the wrong answer for the final result. You may have accidentally reversed digits in a number or forgot to specify the units. It is recommended that you slow down, and increase your concentration on the problem. The more time you spend checking the results, the more careless errors you will find and correct before the end of the examination.

Execution errors occur when you understand the principles covered on the examination, but lack skill in applying these principles to solve one or more specific problems. Examples include leaving out a force in a physics problem or using inappropriate compound combinations in an equation representing a chemical reaction. Execution errors indicate a need to solve more and different homework problems. You already have the talent necessary to complete your degree since you were admitted to the College of Engineering. However, it is crucial for every student to practice solving homework problems. The more problems you work correctly before the examination, the better prepared you will become for problem solving during the examination.

Concept errors are much more serious. They occur when you do not use the correct principle in your attempt to solve a problem. Concept errors indicate that you do not understand the material. These errors signal that you may have difficulties in earning a passing grade in the course. Hopefully, the instructor will identify this type of error with a comment on the graded examination. If an explicit comment is not made, the instructor may signal his or her concern by giving you zero credit for your unsuccessful efforts toward a solution. If you have concept errors it is important that you seek help from the instructor as soon as possible. It will be necessary to devote more time to the course until you have mastered the material.

It is important to carefully review your performance following an examination. The instructor will usually devote class time to show the correct procedure for solving every problem on the examination. You will have the information necessary to study your mistakes. Classify them and adapt your study habits to improve your performance. If you find an execution error, ask questions to your study group or your instructor until you clearly understand the correct solution to the problem.

Unfortunately, many students do not attempt to learn from their errors. It is essential that you understand why you made the error because it would be senseless to lose points again for the same error. Engineering is a discipline that builds from a basic foundation to more advanced subjects. If you do not have a strong understanding in the foundation, your performance in the more advanced subjects will be severely impaired. You will get more and more behind. Make it your business to identify your errors on an examination, correct your procedures and **never repeat the same error**.

12.10 SUMMARY

An overview for your success as a student in a College of Engineering has been presented. Success as a student of engineering depends on setting goals and priorities, developing confidence in your ability to perform and being dedicated to staying current in each subject. Also important is attending class, managing time, developing good study habits, managing your relationships with many new people and seeking help when the need arises.

Time management is based on schedules with time carefully designated to all activities including attending class, studying, open or flexible time, social meetings, and maintaining your physical well being. Weekly schedules have been described in considerable detail and a sample schedule is given. A daily schedule is also presented that allocates time for specific tasks for the entire 24-hour day. Weekly and daily schedules are very important in college because most students have so many time commitments.

A brief discussion has been given on peer management. You will be interacting with several new groups of friends in the next few years. Your relations with some of these groups will be very helpful. Study groups and design teams provide opportunities for collaborative learning and support. Groups formed to workout in the gym or to play club sports help maintain your physical well-being. Most social groups provide you with enjoyment which is very important to experience during your education. Try to avoid excessive time committed to social activities and avoid groups that cause behavioral concerns. It is important to compare the ethical standards of a group with your own set of standards. Participate in a group in which you are comfortable.

Study habits important to your success have been discussed. Suggestions have been made for when and where you study. Advice has also been given on how to study smart and how to prepare homework assignments that are acceptable and respected when reviewed by your instructor. Study groups have been encouraged because of the many benefits of collaborative study.

An eight-step approach to problem solving was reviewed. While eight steps may seem tedious when you are trying to hurry through an examination, experience shows that the process actually saves time by reducing errors. The eight steps include:

1. Write the problem statement as you begin the solution.
2. Carefully read and understand the problem
3. Identify the unknown quantity that you are being asked to determine.
4. Write the rule or equation you plan on using in your solution.
5. List the information provided and prepare sketches as required.
6. Execute the solution.
7. Add units as required to the final result.
8. Check the solution for numerical errors.

Collaborative study opportunities were described. Study teams or design teams are encouraged for all of your classes. Meet at scheduled times after each team member has had an opportunity to work independently on an assignment, and then compare and discuss approaches and results. Student to student learning is effective. There will be occasions when you cannot determine how to solve a problem on your own but your study partner does know how (or vice versa). Respect, confidence and esteem grow as the group members bond and become comfortable with one another.

There are resources and guidelines in the College of Engineering to help you succeed in your studies. The counselors, instructors and teaching assistants are ready to help you. However, they usually will not initiate the process. It is your responsibility to seek help. Procedures for helping yourself and for seeking help from others in the College have been described. To be successful in a College of Engineering, you will need to:

- Prepare in advance for every class
- Attend class
- Arrive on time
- Stay awake and alert
- Ask questions when you become confused
- Complete homework assignments correctly and on time
- Stay current with all assignments

Next, techniques for preparing for an examination have been discussed. The most important factor in preparing for an examination is to stay current daily and avoid the need to cram. Finally, strategies to follow in taking an examination were described. They include:

- Be confident
- Be calm
- Select problems and plan your approach
- Manage your time on each problem
- Execute the solution
- Check the results

Follow these strategies when you take your next examination and determine if it is helpful.

REFERENCES

1. Combs, P. and J. Canfield, <u>Major in Success: Make College Easier, Fire Up Your Dreams, and Get a Very Cool Job</u>, 3ʳᵈ Ed., Ten Speed Press, Berkeley, CA, 2000.
2. O'Brien, P. S., <u>Making College Count: A Real World Look at How to Succeed In and After College</u>, Graphic Management Corporation, Green Bay, WI, 1996.
3. Hanson, J., <u>The Real Freshman Handbook: An Irreverent and Totally Honest Guide to Living on Campus</u>, Houghton Mifflin, Boston, MA, 1996.
4. Carter, C., <u>Majoring in the Rest of Your Life: Career Secrets for College Students</u>, 3ʳᵈ Ed., Farrar Straus and Giroux, New York, NY, 1999.
5. Davis, L., <u>Study Strategies Made Easy</u>, Specialty Press, New York, NY, 1996.
6. Dodge, J., <u>The Study Skills Handbook: More Than 75 Strategies for Better Learning</u>, Scholastic Trade, New York, NY, 1995.
7. Luckie, W. R., et al, <u>Study Power Workbook: Exercises in Study Skills to Improve Your Learning and Grades</u>, Brookline Books, Cambridge, MA, 1999.
8. Tufariello, A., <u>Up Your Grades: Proven Strategies for Academic Success</u>, vgm Career Horizons, Lincolnwood, IL, 1996.
9. Hjorth, L., <u>Claiming Your Victories: A Concise Guide to College Success</u>, Houghton Mifflin, Boston, MA, 2000.
10. Jensen, E. and T. Kerr, <u>Student Success Secrets</u>, 4ᵗʰ Ed., Barron's Educational Series, Hanppauge, New York, NY, 1996.

EXERCISES

12.1 Prepare three different lists of goals—short term, intermediate and long term. Define the time for accomplishment of each set of goals and provide your motivation for setting each goal.

12.2 Describe your plan for staying current in each and every subject for which you have registered this semester.

12.3 Locate the Office for Freshman Advising and the offices of each of your instructors. The instructors should have their office hours posted. Did you find these postings and did you make a note of them?

12.4 Prepare a weekly schedule for your activities this semester similar to one shown in Fig. 12.2. Also list time reserved for class, study and social activities. How many open hours were you able schedule?

12.5 Describe the provisions in your schedule for taking care of your physical and mental health.

12.6 Prepare a daily schedule for each day of the week. Identify your open periods and discuss a strategy for using them when an examination is scheduled.

12.7 Discuss your balance between working, studying and playing over the weekend.

12.8 How many hours do you believe you could work per week without significantly affecting your performance in class?

12.9 Do you know the procedure to apply for scholarships, grants or financial aid?

12.10 How many blocks (two to three hour periods) of study time have you scheduled each week?

12.11 Where do you study? Which location do you find the most conducive? Why?

12.12 Have you joined a study group (team) for this class? If so, when and where do you study?

12.13 Have you ever visited your instructor's office during office hours? Is he or she ready to receive you? Did you find the help provided useful? Do you plan on visiting again when you encounter difficulties with the course material?

12.14 Which is the most difficult course you are taking this semester? How much time do you allocate for study in this course? Is that sufficient? What other strategies do you employ to improve your performance other than study time?

12.15 Explain the methods you use to check answers during an examination?

12.16 What actions do you plan for managing your time on examinations more effectively?

PART V

ENGINEERING AND SOCIETY

CHAPTER 13

ENGINEERING AND SOCIETY

13.1   INTRODUCTION

It should be made clear from the beginning of this chapter that engineering and society are closely joined. Engineering is an integral part of society, and consequently, engineering must be treated as a social enterprise as we continue to develop sociotechnical systems. When engineers develop a new product or improve an existing system, we change society. The magnitude of the change differs from one product to another, and the effect on society may be trivial or enormous, but be assured that there is an impact. For example, when a few companies introduced a random orbit sander a few years ago, the time, effort, and cost of producing flat, smooth, scratch-free surfaces was reduced. These sanders helped several hundred thousand workers in the finishing business to produce high-quality coatings on furniture, cabinets, counters and floors. The results also benefited many millions of customers who purchase furniture each year or remodel kitchens. This new product had a modest, but beneficial, impact on society. There was, however, a small negative effect because some emissions to the environment occurred during manufacturing the product and virgin resources were consumed. Overall though, society has clearly gained by the introduction of this product.

When transistors were first developed from new ultra-pure semiconductor crystals, the impact was revolutionary. While the first transistor was developed in the mid 1950s, the integrated circuit introduced in 1959, and the microprocessor released in 1971, the technological revolution resulting from the introduction of these three products in the microelectronics industry is still underway. Micro-electronic based products are markedly affecting the way in which society lives today. The media refers to the current period as the "information age". New products, which are introduced almost daily, are driven by technology. Products based on new advances in microelectronics will continue to change the way everyone works and plays for the foreseeable future [1].

You often hear about new products and changing life styles. Television, radio, Internet and even newspaper advertising keep everyone well informed of the availability and prices of both old and new products. However, you rarely hear much about the actual technology involved even when the product is new and revolutionary. The media and the public usually are not interested in learning about the technology employed; this is unfortunate, because it would be easier to introduce new or modified systems if the public were more technologically literate.

Engineers are, in part, responsible for this apparent wall between the public and the understanding of technology. Engineers often speak in such technical terms that those few individuals interested in our work become confused. As engineers, we must clearly recognize the need to effectively communicate with society. We can begin by learning how to explain why and how products work in terms more easily understood by the general public. We must also be more enterprising in understanding society and their changing views relative to technology. Lastly, we need to think on a more global scale. Engineers develop sociotechnical systems that are often large and complex. These systems rely on hundreds of products and many different types of services to function efficiently. Even though, as an engineer, you may be dedicated to developing a single component, subsystem or product that is only a small part in a much larger system, you should understand the entire system. Those development teams that have a thorough understanding of all aspects of the complete sociotechnical system develop the superior designs.

Let's consider an example of a sociotechnical system—the commercial air transport system. Do we have technical products involved in this system? **Yes**, literally hundreds of advanced state-of-the-art products. Do we provide technical services for the airlines? **Yes**. Does an engineer working for General Electric in the development of a new jet engine need to understand what Boeing is doing in the design of a new aircraft or modifications of an existing aircraft? Does that same engineer need to know about the sound levels that the public will accept as the plane lands and takes off? Is the engineer working for Boeing required to understand the causes of each and every airline accident? Again, **yes**.

From these simple questions, it should be evident that sociotechnical systems are extremely complex. It will require significant effort for you to understand the detail of the interaction of the many products incorporated in these systems with the public in general. Nevertheless, engineers that understand and appreciate the improvements, which both the customer and society seek, will develop the most successful and safe products.

Finally, as an engineer, you must understand that the systems are adaptive. They are dynamic and change with time in response to several societal and economic forces. First, society changes. The requirements of society and the customer in the 1950s were much different than either will accept today. The customers, a segment of society (more on this fact later), typically expect that the new model of the product will be much improved over the older model and be lower in price. Others in society, not necessarily the customers, will expect (perhaps demand) that you manufacture the product without degrading any aspect of the environment.

An example is useful to distinguish between the customer and society. Let's consider the automobile as the product. The customer of an automobile demands style, comfort, mobility, convenience, affordability, reliability, performance and value. Society requires (through federal regulation) safety, fuel efficiency and reduced emissions and less pollution. The fact that the requirements from the customer and society often conflict should be recognized. Conflicting requirements challenge engineers to create improved designs that advance the state of the art in all areas of technology.

The governments (Federal, State, County and City) are also involved in the relationship between engineers and society. Various government agencies frequently issue regulations that affect both the sociotechnical systems and the product. Consider the commercial air transport system again. Obviously, safety is a very serious issue. The Federal Aviation Administration (FAA) is responsible for issuing regulations to insure safe operations of commercial aircraft. However, 1996 was not a good year for the FAA or the commercial airlines because several very serious accidents occurred with significant loss of life. As a consequence of a fire in the cargo bay, which caused the crash of a Boeing 737 aircraft in the Florida Everglades and the loss of all aboard, the FAA is now requiring the airlines to retrofit all planes with a fire suppressant system. Regulations from government agencies drive a portion of the changes to any sociotechnical system.

This understanding of the relationship between engineering and society is relatively new. If you examine the five-volume tome, *A History of Technology*, [2] published by Oxford University Press over the period from 1954-58, you will find a chronological, deterministic catalog of technological development. However, the editors almost completely ignore the relation between engineering and society. In his review of this tome, Hughes [3] states that the editors belatedly conclude their five-volume treatise with an afterthought essay on technology and the social consequences. Today, the relationship is more clearly recognized although educators have not yet managed to include adequate treatment of this very complex relationship in a typical engineering curriculum. A few examples of the relationships between engineering and society will be described in this chapter. For a more complete treatment, read reference [4].

13.2   ENGINEERING IN EARLY WESTERN HISTORY

Technology has impacted the way that we live since the beginning of recorded history. Long before the first engineering college was founded, intelligent, hard-working men[1] worked as engineers developing new products and processes. You may study history and cite many developments that changed, and for the most part, benefited society.

Prior to about 6,000 BC, there was little evidence of technology. Men and women were hunters and gatherers. They scrounged for food and lived in caves if they were lucky and shacks or huts if they were not. The significant developments in the period from 6,000 to 3,000 BC were in agriculture with the domestication of animals and the cultivation of grains. With time, it was possible to insure a relatively stable, if limited, food supply and people began to congregate together in villages and small cities.

The first technology involved the building of roads, bridges and boats for local transportation. People clustered in villages and small cities wanted to transport food, water and other materials over short distances. The geographic regions where technology was evident were extremely local. Early Western history involved only Mesopotamia, Egypt, Greece and then Rome. In a few cities, there were aqueducts to transport water and a few primitive sewerage systems. In most regions, only the ruling class lived well. The great monuments of Egypt and Greece were built with slave labor. In fact, slaves (people who were on the losing side in a war or those born into slavery) provided most of the power required for building, mining and agriculture for many thousand years. The rise of religion is often credited with the reduction in the prevalence of slavery. However, the steam engine developed by Thomas Savery, Thomas Newcomen and James Watt in the early 18ᵗʰ century signaled the beginning of the end of the need for power provided by humans (slaves and other unfortunate souls in bondage). As society enters the 21ˢᵗ century, the development of the low-cost, high-performance computers, a wide variety of specialized software programs and effective communication 24×7 by the Internet, portends another beginning of the end for many employees performing repetitive tasks or clerical work.

The Egyptians

The Egyptians built many huge monuments and were masters in moving large blocks of stone from distant quarries to a construction site and then elevating these blocks to their respective positions in gigantic structures. Some estimates indicate that 20,000 to 50,000 men were required to drag 20 to 50 ton blocks of stone used in the construction of these monuments. In spite of the majestic grandeur and durability of these monuments, the new innovations introduced by the Egyptian engineers were not outstanding. They constructed temples in a simple fashion using only uprights (columns) and cross pieces (short, deep beams). While the arch was known during this period, it was constructed with mud bricks and never adapted to stone construction. Classical treatments of the history of technology tell us very little about the impact of these developments on society. However, it is apparent that the construction of a pyramid or a temple required a significant segment of the working population for many decades. This effort probably did little to improve the welfare of even the **free** segment of the population. It is also evident that the effort required of the slaves to move the countless large, heavy stones must have been brutal.

---

[1] Historical accounts of engineering achievements indicate that the profession was male dominated. The author is sensitive to the gender issue in engineering education, but it must be recognized that the presence of women in the profession is very recent when considered relative to a historical span of at least 8,000 years.

The Greeks

Although they had very capable engineers, the Greeks are better known for their contributions to science. Thales, Euclid and Aristotle are recognized for their contributions to the development of the basics of geometry and physics. This new knowledge of geometric relations was applied almost immediately in architectural engineering. However, scientific understanding during this period was so limited that nearly 2,500 years elapsed before this knowledge was expanded sufficiently to be useful to the engineering community.

Greek engineers built cities, roads, water (hydraulic) systems and some machinery. Archimedes and Hero were innovators who used the pulley, screw, lever, and hydraulic pressure to develop cranes, catapults, pumps, and several hydraulic devices. The Hellenistic period brought better buildings and local roads that improved living conditions for the upper class. However, the construction of the city infrastructure was still performed largely by slaves because the use of other forms of power was nonexistent. The Hellenistic world was divided into extremely independent city-states. Politics of the period did not foster cooperation between these governmental entities; as a consequence, very few long roads were built. Transportation was largely by sea in boats equipped with oars powered by a galley of slaves. Some of the boats were fitted with a square sail, but these were so poorly designed that they were effective only on down-wind tacks.

The Romans

The Roman engineers followed the Greeks, and while they were considered by some historians to be less inventive, they were masters of detail. As the Roman Empire spread from the Middle East to Scotland, the Roman engineers built long roads connecting the distant cities; roads so well constructed that they were in service for many centuries. The large cities constructed in the conquered lands had paved streets and heated homes with water supplies and sewer systems. The Roman engineers had little or no theoretical knowledge, but they had developed practical methods of construction that served their purposes well. During this period, the subjects of statics and mechanics of materials were not well understood. The theory used to design beams would not be developed for another 1,700 years. The Roman engineers based their construction on experience, and they used very large safety factors. These empirical methods, while crude by today's standards, sufficed to produce many bridges, roads, aqueducts and cities. A great bridge over the Tagus River in Spain supported a road nearly 200 m long with six spans 60 m above the river. Considering it was completed at about 100 AD, this bridge was a remarkable engineering feat. The durability of the Roman construction is well documented by their structures some of which are still in use today—2,000 years after they were placed in service.

The Roman engineers deviated from the Egyptian and the Greeks in their construction techniques by using much smaller building stones and bricks. They were able to produce massive structures with the smaller blocks by utilizing mortar and cement. The use of smaller stones, bricks and cement clearly impacted society. Buildings could be constructed with much less effort because thousands of slaves were not necessary to move a single block of material. Life for the slaves clearly improved although the construction work was still very demanding. Also, the upper classes benefited from the more rapid construction of buildings and houses.

While the Roman engineers were outstanding in the construction of civil infrastructure, they did very little to advance the use of power in driving the few machines in use during that period. Humans were still used in most instances to power pumps, mills and cranes. In rare cases, water wheels were used at power mills to grind grain, but these were of the undershoot design and much less effective than the overshoot design that was developed later. While references on the history of technology are quiet on the issue, the life of a worker powering a pump or a mill must have been very difficult.

Eventually, the vast Roman Empire crumbled, and the world entered a period that the historians classify as medieval civilization from 325 to 1300 AD. The fact that barbarians stripped and destroyed the cities does not imply that progress stopped, but it slowed progress to a snails pace. When Rome collapsed, the Byzantine Empire was the next civilization to follow. Significant engineering achievements were made in this period in developing dome structures and curved dams. The author has omitted discussion of these developments to keep this historical treatment brief.

Early Agriculture and the Use of Animals

In spite of the slow development of engineering and the lack of the construction of monuments during the Middle Ages, an early agricultural revolution occurred. White [5] has described the changes that occurred as agriculture moved from the dry sandy soil of Italy into the heavy alluvial and wet soil of northern Europe. The plows that were effective in the light sandy soils would not turn the sod in the heavy, moisture-laden soil of the north. A new, heavy-wheel-plow was invented that would cut through the grass sod and turn over the soil. However, this plow required eight yoked oxen to provide the force necessary to pull it through the heavy sodded soil. Very few farmers of the day owned eight oxen. They had to combine their land, share their oxen and cooperate in plowing and planting. The farmers merging their land holdings eventually led to a manorial society.

It may be difficult to imagine that the invention of a new plow made such a major change in the way of life. Today farmers are able to produce all of the food needed in the U.S., while exporting excesses in significant quantities, with less than 2% of our population. However, before the advent of engines and motors (steam, gasoline and electrical), men and women struggled to grow enough food to keep from starving. In the Middle Ages, 90% of the population worked in agriculture, and they had much less to eat than we do today.

Oxen were used as a source of power for plowing and transport on the farms. However, oxen are very slow and they consume large quantities of grass and grain. At that time, horses were of limited use because of inadequate harnesses and saddles and the frequent problem of splitting hoofs. (Early harnesses fitted about the horse's neck tended to strangle the animal.) These problems were alleviated when a newly designed horse collar, which transferred the load to the shoulders of the horse, was introduced and when iron horse shoes were fitted to their feet. Horses gradually replaced the oxen because they were 50% faster, worked longer and required less food.

In early warfare, the horse was not very effective in frontal attacks because stirrups for the rider did not exist. Consequently, horse mounted warriors were of limited value because they could not adequately thrust their lance while staying mounted on the horse. With the advent of the stirrup, the horsemen could stand-up, lean forward and thrust the lance through the defenders' shields without losing their seat. The simple addition of stirrups to saddles had a profound influence on society. Feudalism evolved with an aristocracy based on warriors conquering and defending landholdings. A new battle strategy with armored knights mounted on large strong horses was dominant for several centuries [5].

In early history, small technological improvements produced major changes in the form of government, the type of society and the way people of all classes lived. It is remarkable that for about 5,000 years, humans were the prime source of power for most activity. Progress in agriculture and engineering proceeded together to provide more food, better housing, roads and bridges and eventually horse- and oxen-powered plows and wagons.

13.3   ENGINEERING AND THE INDUSTRIAL REVOLUTION

Civilization has always been power limited; even today, space travel is severely constrained by the inadequacies of rocket power. Until the last three centuries, animals, windmills and waterwheels provided the only power available for agriculture, construction, milling, mining and transportation. Progress was often slow because power was not available when and where it was needed. Life, in the power starved 17th century, was rather bleak for the general population. The people worked from dawn to dark as farmers or craftsmen. Only those endowed with landed estates and/or money lived what you would consider a modestly comfortable life.

The Development of Power

The first of several breakthroughs in developing new sources of power occurred in the 18th century. In Great Britain, Savery, Newcomen and Watt developed stationary steam engines. While these early steam engines were woefully inefficient (only 0.5%), they were produced in relatively large numbers (500 existed in 1800) and were employed to pump water and power textile and flour mills.

Because the steam engines were replacing horses as power sources, James Watt conducted experiments to measure the power a "brewery horse" could provide. His measurements indicated the brewery horse generated 32,400 (foot-pounds/minute) of power. The horse's output was rounded to 33,000 (foot-pounds/minute) and this value is still used today as the conversion factor for one horsepower (HP).

Most of the 18th century was devoted to innovations directed toward improving several different models of steam engines. Applications were largely limited to stationary power sources because engines, of that day, were large, heavy and inefficient. However, with improved metallurgy and machining methods, it was possible in the 19th century to reduce the size and the weight of the engines while maintaining their output. These improvements permitted the steam engine to be adapted to power riverboats, ocean-crossing ships, rail locomotives and later even a few steam-powered automobiles.

The development of the steam engine and its adaptation to several modes of transportation formed the foundations of the industrial revolution. The life style of almost everyone was markedly changed with the emergence of steam-powered factories and much more effective transportation. Fewer people were employed in agriculture, but many more were working under appalling conditions in factories and mines. Twelve-hour days, seven days a week, with only Christmas as a holiday during an entire year was the norm. Manufactured products, such as clothing and housewares, were available to a larger share of the population at reduced prices, but the margin between earnings and the money necessary for a factory worker, miner or farmer to stay alive remained very narrow. Steam power had relieved many thousands of men and horses from brutally hard labor, but they were released from one dull job for another. On a more positive note, the progress in agriculture and industry enabled the population to triple during this period (1660 to 1820).

13.4   ENGINEERING IN THE 19TH AND 20TH CENTURIES

For the eight thousand years of recorded history prior to about 1800, the advance of technology was extremely slow. Agriculture dominated society with most of the population growing food that was consumed locally. Factories were being developed and simple products needed on the farm, in the home or by the military were produced. Long distance transportation of goods was usually limited to sailing ships on the seas or horse drawn barges on rivers or canals.

During the 19ᵗʰ and 20ᵗʰ centuries, technology literally exploded. The many scientific discoveries of the preceding millennium expanded knowledge sufficiently to allow countless engineering applications. To illustrate this explosion, let's define modern technology to consist of the following five elements:

> Abundant available power.
> Available food in sufficient quality and quantity to insure a nutritious diet for everyone.
> Transportation by air, land and water with associated infrastructure to provide safe, rapid, convenient and inexpensive access.
> Communication between everyone, anywhere and anytime at a reasonable cost.
> Information (knowledge) storage, retrieval, and processing in seconds at any location, anytime and anywhere at negligible cost.

While the advances in technology have been astonishing, particularly in the last 50 years, there are still significant deficiencies. Rocket power for launching space vehicles currently constrains space travel. Traffic jams in large cities extend the working day for many millions of people commuting to their jobs. Today more than 42,000 people die in accidents on the highways every year. Travel on airlines is becoming unreliable, difficult and often unpleasant. Communication in many parts of the world is still not available to the public. However, the recent introduction of cell phones has greatly improved communication in underdeveloped countries. The cost of cell phone towers is much less than the expense of installing land lines required for conventional phone service. Many people are just beginning to explore the different ways to use the massive amount of information that has recently become available to those fortunate enough to own a personal computer, a modem and a broadband connection to a server.

You may debate the exact year of some invention. However, the exact date is rarely important because an invention usually evolves into a commercially successful product after several iterations with improvements or modifications by several designers. You should not worry about the exact dates. Instead, recognize the continuous progress and observe the shift in emphasis from one period to another. In power, technology evolved from animals, to steam, to internal combustion (gasoline) and to nuclear energy. Power, from gasoline fueled internal combustion engines became mobile, and electricity was widely distributed. In transportation, technology evolved from the horse and buggy and sailing ships to automobiles, airliners, nuclear powered submarines and space shuttles. In communications, technology progressed from the pony express to the telegraph, telephone, facsimile machines, cellular telephones, pagers and the Internet. In the information business, technology is currently evolving from a hard copy library system to a digital multi-media information source available at any time, to everyone, at nearly zero cost.

Have these advances in technology changed society? It is believed that they have had a profound beneficial effect. Farmers produce an overabundance of food with a small number of personnel working in agriculture. Factory workers manufacture most of our consumable products with less than 20% of our work force. The workweek is shorter, and we have many more affordable products. Most families own at least two cars and home ownership is at an all time high. Concerns are expressed regarding the improvements in the standard of living over the past decade, particularly for those people at the lower end of the income spectrum. Technology has probably not helped this group as much as others. Global trade has permitted many of the semi-skilled tasks to be moved to third world countries, where the cost of labor is a small fraction of the labor costs in developed countries. Consequently, many of the previously well-paid factory jobs in manufacturing in the U.S. have been moved offshore and this process has negatively affected those with limited marketable skills.

13.5   GREATEST ENGINEERING ACHIEVEMENTS—20TH CENTURY

Recently the National Academy of Engineering (NAE) working with 27 professional engineering societies complied a listing of the 20 greatest achievements in engineering during the 20th century. What made these achievements great?   They are notable because of the very significant benefits of each accomplishment to all members of society.   An excellent publication outlining the engineering achievements that transformed our lives is presented in Reference [7].

1. Electrification—in the 20th century, widespread electrification gave us power for cities, factories, farms, and homes and forever changed our lives. Thousands of engineers made it happen, with innovative work in fuel sources, power generating techniques and transmission grids. From streetlights to supercomputers, electric power makes our lives safer, healthier and more convenient.

2. Automobile—perhaps the ultimate symbol of personal freedom, the automobile is a showcase of engineering ingenuity and the world's major transporter of people and goods.

3. Airplane—with its speed and efficiency, air travel makes personal and cultural exchange possible on a global scale.

4. Water Supply and Distribution—water is our most precious resource, and improvements in its treatment, supply and distribution have changed our lives profoundly, virtually eliminating waterborne diseases in developed countries and providing clean and abundant water for communities, farms and industries.

5. Electronics—electronics provide the basis for countless innovations including digital cameras, iPods, TVs, and computers to name only a few. Electronic products improve the quality and convenience of modern life.

6. Radio and Television—radio and television were major agents of social change in the twentieth century, entertaining millions and opening windows to other lives to remote areas of the world and to history in the making.

7. Agricultural Mechanization—tractors, combines and hundreds of other agricultural machines increase the productivity of farms, reduce the need for manual labor and improve our ability to feed our citizens and others in the world.

8. Computers—perhaps the defining symbol of 20th century technology, the computer has transformed businesses and lives around the world, increased our productivity and opened up access to vast amounts of knowledge.

9. Telephone and Cell Phone—a cornerstone of modern life, the telephone brings the human family together and enables the communications that enhance our lives, industries and economies.

10. Air Conditioning and Refrigeration—air conditioning and refrigeration make it possible to transport and store fresh foods and adapt the environment to human needs. Once luxuries, they are now common necessities which greatly enhance our quality of life.

11. Highways—vast networks of roads, bridges and tunnels connect our communities, enable goods and services to reach remote areas and encourage economic growth.

12. Spacecraft—the human expansion into space is perhaps the most awe-inspiring engineering achievement of the 20th century. The development of spacecraft has thrilled the world, broadened our knowledge of the universe and led to thousands of new inventions and products.

13. Internet—in a few short years, the Internet has become a vital instrument of social change, transforming business practices, educational pursuits and personal communications. By providing global access to news, commerce and vast stores of information, the Internet brings people together and adds convenience and efficiency to our lives.

14. Imaging—From tiny atoms to distance galaxies, modern imaging technologies have expanded the reach of human vision with tools such as X-rays, telescopes, radar and ultrasound.

15. Household Appliances—household appliances have greatly reduced the labor involved in everyday tasks, giving us more free time and enabling more people to work outside the home.
16. Health Technologies—medical imaging devices, artificial organs, replacement joints and biomaterials are a few of the advanced health technologies that improve the quality of life for millions.
17. Petroleum and Petrochemical Technologies—petroleum and petrochemicals provide fuel for cars, homes and industries and the basic ingredients for products ranging from aspirin to zippers.
18. Laser and Fiber Optics—lasers are used in supermarket scanners, surgical devices, satellites and other products. Today fiber optics provide the infrastructure to carry huge amounts of information via laser light, spurring a growing revolution in telecommunications.
19. Nuclear Technologies—the harnessing of the atom led to dramatic changes in the 20th century by providing a new source of electrical power, improving techniques for medical diagnosis and treatment and transforming the nature of war forever.
20. High-performance materials—from the building blocks of iron and steel to the latest advances in polymers, ceramics and composites, engineering materials are used in thousands of important applications.

13.6   NEW UNDERSTANDINGS

In this discussion of the historical development of technology, we have been optimistic about the very positive effects of technology on society. In the past the public accepted, without serious questions, all forms of technology and the risks that were involved. However, the situation has changed to a remarkable extent in the past 40 years. Our leadership (engineers included) has, to a large degree, lost the public trust. An excellent example of this loss pertains to the technology employed in the generation of electricity using nuclear power. Since the Three Mile Island incident that occurred in Pennsylvania in 1979, the general public in the U.S. has opposed nuclear power until very recently. There is a valid public perception that the technological risks associated with nuclear power were significantly understated. John O'Leary, a deputy director for licensing at the Atomic Energy Commission (AEC), stated that "the frequency of serious and potentially catastrophic nuclear incidents supports the conclusion that sooner or later a major disaster will occur at a major generating facility" [8]. This statement was made several years prior to the accident at Three Mile Island. In spite of this warning of the likelihood of a serious accident, federal, state or local governments have not planned for orderly emergency evacuation of nearby residents or for containment measures to limit the possible contamination of many of power reactors.

The Chernobyl accident, which occurred in the former Soviet Union in 1986, proved that O'Leary was correct. The No. 4 reactor at the Chernobyl facility exploded sending a radioactive plume in the air so high that an alert nuclear plant operator in Sweden detected it. Many died in this radioactive explosion and many more have died since then from radiation poisoning. High radiation levels affected countries as far away from Chernobyl as Italy.

In the recent years many environmental organizations who once opposed nuclear energy have recognized that generating base load electricity with nuclear energy is preferable to its generation by fossil fuels such as coal or natural gas. Nuclear power plants do not emit CO_2 that is believed to contribute to global warming.

The nuclear industry and the government agencies regulating it, world wide, have not been candid with the public. Nor has the public been very interested in an intelligent debate—resorting instead to large, ugly demonstrations to express their displeasure. The public needs an accurate and honest assessment of the risks of any nuclear power project by the industry, the regulators and experts. The risks must be weighed against the benefits to society.

The purpose here is not here to argue the case for or against nuclear power. Nuclear energy is used to illustrate the fact that understating or poorly assessing risk brings a severe penalty—loss of public trust followed by a ban on technical development in the subject area.

As engineers, we must significantly broaden our perception regarding technological risk. Today, many engineers dismiss failure due to human, operator or pilot error because they are not machine errors. This attitude is **absolutely wrong**. The **human operator and the machine** constitute a system and affect public safety accordingly. The machine and its operator must be included in system design since they both constitute an interacting set of weaknesses and capabilities [9].

Another new understanding pertains to geography. Prior to the Second World War (WW–II), international trade was very limited. As a country we produced what we consumed, mined our own minerals and pumped our own oil. Geography was very important because it constrained trade by law and by distance. After the war, laws were changed and trade agreements (GATT, NAFTA, etc.) that lowered or eliminate tariffs were arranged. International trade was encouraged so much so that the U.S. now consistently imports more goods and materials than it exports. The annual deficit in the trade balance for the U. S. was 738.6 800 billion dollars in 2007. This deficit represents the equivalent to more than 18,000,000 high-income jobs paying $40,000 a year.

Previously, the cost of shipping limited trade among countries separated by large distances. However, following WW-II many technological improvements were implemented in the shipping business: the super tanker was developed—greatly improving the efficiency of shipping huge quantities of oil. Also very large containerized ocean vessels were developed reducing the cost and time of loading and unloading cargo. Large efficient diesel engines were installed on the ships permitting increased speed and significant reductions in shipping time and costs.

With the time and costs of shipping greatly reduced and the legal barriers to trade removed, the country's commerce is conducted in a global marketplace. Engineers compete worldwide to develop world-class products or services. Products designed in the U.S. may be produced anywhere in the world, or products designed in Europe or Asia may be produced in the U.S. Today, design and production activities are located to minimize costs and to maximize benefits to the customers and/or to the corporate entities.

The third new understanding is in the role of the governments (federal, state, county and city). Prior to WW-II, governments were small and taxes were relatively low. The primary role of the federal government was to insure national security. The state governments worried mostly about transportation infrastructure, and the local governments concerned themselves with education. Today, the situation is markedly different. Governmental agencies, at all levels and in response to their interpretation of the law (either old or new), issue regulations which pertain to topics of concern to society including the protection of the environment, safety, health and energy conservation.

You can debate the wisdom of many of the regulations, but that is not the issue. The regulations currently exist and new ones will continue to be issued with an alarmingly frequency. As an engineer, you must be prepared to serve society-at-large as reflected by these governmental regulations. If you question the wisdom of some of the regulations, the best approach is to become involved in the political process so that you may have the opportunity to influence them. Engineers tend to resist constraints because they limit the freedom of their designs and add to the cost of the product. You may think of regulations as government imposed constraints, but they are really barriers imposed by society-at-large. Society controls the regulatory process because it has the voting power to change the decisions of the politicians drafting the laws and of the bureaucrats formulating the regulations.

13.7   BUSINESS, CONSUMERS AND SOCIETY

Engineers serve three constituencies—business, the customer and society-at-large. Business leaders (management) look to engineers to develop products, services and processes that meet global competitive challenges. Consumers seek more convenient, reliable, enhanced and value-laden products at reduced prices. Society-at-large, through elected politicians, trial lawyers and public interest groups demand action leading to solutions of problems involving safety, health, energy conservation and preserving or improving the environment. It is an overlapping set of constituencies, because society encompasses business, governments and consumers. To discuss the issues raised by serving three constituencies, consider the U.S. automotive industry as an example. Marina Whitman [10] has described many aspects regarding the three-way demands.

Business demands are simple although they are often difficult to meet. Any business must be profitable to continue to exist. Losses can occur over the short term, but if the red ink is prolonged the company is forced by its creditors into bankruptcy proceedings. Making a well-defined profit every quarter is the usual goal of management. For many years following WWII, it was relatively easy for U.S. businesses to make money. Bombing and shelling during the war had destroyed most of the manufacturing capability and a significant part of the infrastructure in every developed country in the world except in the U.S. It was easy to be a leader when the competition was without factories and adequate infrastructure. For the automobile industry, the golden years began to erode in the late 1960s, and survival became a life or death battle in the 1980s. Problems for the auto industry came from global competition, which continues to this day.

To show the extent of the market penetration by imported automobiles, recall that the imports accounted for only 4% of the market in 1962, but they controlled more than 50% of the market by July of 2007. This devastating loss of market share by US car makers was not limited to the auto industry; several other industries including optics, consumer electronics, ship building, steel, etc. suffered even more serious losses.

Engineers and management in the U.S. firms had lost the competitive edge after decades of modest competition from the Europeans. The Japanese automobile companies (Toyota, Honda, Nissan, etc.) provided more affordable, reliable, appealing, fuel efficient and value-laden cars than the American or European companies. In recent years, the loss market share has been slowed due to improved management and technical methods (concurrent design and systems engineering[2]), but regaining the market share of the 1960s will be a very long and difficult process due in part to very high legacy costs.

The second constituency, the consumers, has been well recognized in the past decade. The importance of keeping the customer pleased is foremost in the minds of both management and engineering. Systematic methods for establishing the consumers' needs have been developed and placed into practice [11]. Many companies in the U. S. currently follow a product development process that begins with the customer interviews and proceeds through all of the phases of product development. The newer methods [11, 12] tend to insure that the customers' requirements are more accurately defined by the product specification before beginning a development.

The third constituency, society-at-large, generates the most challenging requirements. Society-at-large is represented by government agencies and a variety of public action groups. Officials of the government agencies (bureaucrats) produce regulations that define requirements for automobiles that may affect safety, fuel conservation or the environment. The public action groups create pressure and influence politicians in setting public policy that eventually generates new requirements.

---

[2]Systems engineering means many different things to different groups. It is defined here as a cost and time efficient method to specify, design, develop and integrate all of the individual subsystems so that the total system for the product meets the requirements of customers, business and society-at-large.

Engineers often have some difficulties with regulations and/or proposals from the public action groups. Lawyers write the laws and public servants write the regulations. Because lawyers and public servants are often technically illiterate, they sometimes seek technical support prior to drafting binding regulations. However, government bureaucrats sometimes place regulations into effect prior to a thorough study by knowledgeable representatives of the public and engineering communities. The difficulty with this procedure is that the demands of the customer, including style, comfort, affordability, reliability, performance, convenience, etc., may be in serious conflict with societal imposed regulations on safety, fuel efficiency, emissions, etc.

A very recent example illustrates what many believe to be an ill-advised regulation [13]. The National Highway Transportation Safety Administration (NHTSA) mandated requirement for dual airbags in every vehicle licensed in the U.S. Most people consider airbags a valuable safety feature because many thousand lives have been saved in serious frontal accidents since they have been mandated. However, many children and small adults (mostly women) have been killed by rapidly inflating airbags. NHTSA has reported that from 1990 to 2002 that 280 people were killed by rapidly deploying airbags in low severity crashes since the use of this safety feature was mandated [14]. More than 90 percent of the passenger airbag fatalities have been children and infants. They account for about 60 percent of all airbag related deaths. Small women are also at risk particularly if the sit close to the dashboard where the airbags are stored. The statistics cited above are confirmed deaths due to rapidly deploying airbags. In addition, there were many unconfirmed airbag related fatalities where NHTSA did not send an investigative team to the accident scene to confirm the cause of death. Nor did NHSTA report on the number of serious injuries cause by airbags. Robert Brown[3] estimates that disabling injuries due to airbags currently exceed 500 per year.

Airbags in newer model automobiles trigger less violently; nonetheless, passengers must remain at least 10 in. from the bag to avoid injury from the bag in a crash. Injuries such as abrasion of the skin, hearing damage (from the sound during deployment), head injuries, and breaking the nose, fingers, hands or arms can occur as the airbag deploys.

The killing of women and children by a safety device is a **very serious concern**. Moreover, the regulation prohibits anyone other than the car's owner to disable the passenger side airbag[4]. If you are a concerned parent, you can disable it yourself, but few owners have the required skills. Also, if you ruin the mechanism and/or the bag, the cost of replacement is expensive. Because of their high costs, airbags have replaced stereo systems as the theft of choice if your car is robbed.

The issue is not whether you should have airbags or not. They have proven to be a very effective safety device for most adults either driving or riding as passengers. The issue is with the rigidity of the federal mandates and the failure of the regulators to:

Update the regulations to accommodate different behavior of drivers.
Provide requirements that enable more design flexibility.
Provide a means of permitting the driver to activate or deactivate the passenger side bag.

---

[3] See http://www.airbagonoff.com/rbarticle.htm

[4] A recent ruling from the NHTSA allows auto dealers to install an on-off switch to control the passenger side airbag; however, NHTSA requires an extremely complex process for drivers to obtain a waiver, which is necessary before the dealer will perform the necessary modifications to the deployment system. First, you have to obtain a NHTSA request form. Then you must swear—under threat of criminal penalty—that you are too short to be safe with an airbag deployment, or cannot always put the children in the back seat, or have one of five NHTSA approved medical conditions. If you option for the medical conditions you must have a letter from a medical doctor confirming the condition.

The airbag regulation was written in 1977 when only 14% of the population even bothered to use their seat belts. By 1994, when the seat belt mandate was fully effective, 49 of 50 states were enforcing seat belt laws and usage had risen to 82% by 2007. While seat belts do not eliminate the need for airbags, the belts increase the time allowed for the deployment of the airbag. Responding to the controversy about the risk due to airbags, NHTSA issued a new regulation permitting automakers to install airbags 25% to 30% less powerful than the current bags. However, many cars manufactured before 1999 continue to use airbags that deploy at a speed of 200 MPH to protect an unbelted adult male. This is great for the big guys; however, for kids or smaller women, a 200 MPH kick in the face is extremely dangerous.

Industry representatives recognized these problems and warned the NHTSA of the dangers to children. Rather than change the regulation to provide more flexibility in the design of airbag systems, NHTSA drafted a warning label and industry added instructions in the owner's manuals explicitly cautioning against placing a front facing child safety seat in the car's front seat. The manuals indicate that it is safe to place a child in the passenger seat if the child is buckled-in with a seat belt, **but** the seat must be in its full rear position. The manual also states that children should never lean over with their faces near the airbag cover while the car is in motion. The warning labels and instructions in the owner's manual are a result of poor design that was forced on the automobile industry by an inflexible regulation. Warning labels and instructions in manuals reduce the consequences of litigation (there is another law indicating that the user must be warned of possible dangers), but they are totally inadequate to protect children who are often too young to read. The solution is simple—young children should not ride in the passenger seat.

It would be relatively easy to modify the design of the airbag system to include an on/off switch with a warning light on the control panel of the automobile. With this feature the driver could deactivate the passenger side airbag whenever he or she wished to do so. However, the federal regulation did not permit modification then, and only recently has it allowed modification after the death of 256 men, women and children.

The public is concerned, if not angry. The NHTSA has received many letters from owners requesting the waivers required to have trained mechanics at the dealers to disconnect the dangerous airbags. Recently the NHTSA has provided a listing of auto dealers willing and certified to install switches on their Website.

The interaction of the government agencies representing the public-at-large, business and the consumer is woefully inadequate. The public and business needs solutions often in a relatively short time. Yet, the regulatory system is too cumbersome to provide rapid solutions. The awkwardness comes from inflexibility on the part of the government and inadequate cooperation between the automakers. Competition on the part of business to increase market share is understandable, but safety features should be excluded from the normal competitive secrecy. The government agencies must become much more flexible and recognize that they cannot decree away all danger.

The difficulties between the government and business are not new. An interesting summary of the regulatory relationship between government and business was recently published by Jasanoff [15]. She is quoted below:

> Studies of public health, safety and environmental regulation published in the 1980s reveal striking differences between American and European practices for managing technological risks. These studies show that U.S. regulators on the whole were quicker to respond to new risks, more aggressive in pursuing old ones and more concerned with producing technical justifications for their actions than their European counterparts. Regulatory styles, too, diverged sharply somewhere over the Atlantic Ocean. The U.S. processes for making risk decisions impressed all observers as costly, confrontational, litigious, formal and unusually open to participation. European decision making, despite important differences within and among countries, seemed by

comparison almost uniformly cooperative and consensual, informal, cost conscious and for the most part closed to the public.

The assessment by Sheila Jasanoff gives us clear direction. It is important to reduce the adversarial relationship between business and government. With engineers working for both government and business, we should be an important element in the future to produce a regulatory system with more flexibility, cooperation, cost consciousness and consensus.

13.8   CONCLUSIONS

The standard of living in the U.S. will be determined by the interplay of three powerful influences:

- New and rapid technological advances.
- Business (management) response to global consumer demands.
- Social demands as evidenced by legislative and regulatory requirements.

Engineers have always provided the leadership in technological advances, and they are expected to continue to provide business and the public-at-large with cutting-edge, world-class technology.

Business requires capital, technology and astute management to remain competitive. In addition to providing the technological base for a company, engineers frequently serve in management. If a management career path appeals to you, plan on extending your education in a business school. The combination of a Bachelor of Science degree in engineering and a Master of Science in business provides a very solid foundation for a career path in a technically oriented business.

In the past, engineers usually have not been actively engaged in societal issues. They lack patience in dealing with the public and become irritated by the public's lack of technical literacy. They fume at the inefficiencies of government agencies and their lack of flexibility. They become outraged at the arrogance of bureaucrats and at their lack of concern for time and costs. They withdraw from public forums and focus on new developments and products. In the meantime, the public-at-large grows wary of many important sociotechnical systems and the government resorts to litigious processes that impede real progress.

It is unfortunate that engineers have not been a significant force in dealing with societal issues. When society, business and the customer create conflicting demands, engineering expertise is essential in crafting well-balanced solutions. The conflicting demands provide new opportunities and complex challenges for engineers. To take advantage of these opportunities, engineers of tomorrow will require enhanced communication skill, greater disciplinary flexibility, a better understanding of the mechanisms of regulatory agencies, and a much broader perspective relative to societal demands.

REFERENCES

1. Friedman, T., The World is Flat, Farrar, Straus and Giroux, New York, NY, 2005.
2. Singer, C. E. J. Holmyard, A. Hall, and T. Williams, Eds. A History of Technology, Oxford University Press, New York, NY, five volumes, 1954-1958.
3. Hughes, T. P., "From Deterministic Dynamos to Seamless-Web Systems," Engineering as a Social Enterprise, ed. Sladovich, H. E., National Academy Press, Washington, D. C. 1991, pp. 7-25.
4. Sladovich, H. E., ed. Engineering as a Social Enterprise, National Academy Press, Washington, D. C. 1991, pp. 7-25.
5. White, L., Jr., Medieval Technology and Social Change, Oxford: Clarendon Press, NY, 1962

6. Kirby, R. S., S. Withington, A. B. Darling, and F. G. Kilgour, <u>Engineering in History</u>, Dover Publications, New York, NY, 1990.

7. Constable, G. and B. Somerville, <u>A Century of Innovation: Twenty Engineering Achievements that Transformed Our Lives</u>, Joseph Henry Press, Washington, D. C., 2003.

8. Ford, D. F., <u>Three Mile Island: Thirty Minutes to Meltdown</u>, Viking, New York, NY, 1982

9. Adams, R., "Cultural and Sociotechnical Values," <u>Engineering as a Social Enterprise</u>, ed. Sladovich, H. E., National Academy Press, Washington, D. C. 1991, pp. 26-38.

10. Whitman, M. v. N., "Business, Consumers, and Society-at-Large: New Demands and Expectations," <u>Engineering as a Social Enterprise</u> ed. Sladovich, H. E., National Academy Press, Washington, D. C. 1991, pp. 41-57.

11. Clausing, D. <u>Total Quality Development: World-Class Concurrent Engineering</u>, ASME Press, New York, NY, 1994.

12. Schmidt, L., et al, <u>Product Engineering and Manufacturing, 2nd Edition</u>, College House Enterprises, Knoxville, TN, 2002

13. Payne, H. "Misguided Mandate," Scripps Howard News Service, Knoxville News-Sentinel, January 12, 1997, p. F-1.

14. Anon, Q & A: Airbags, Insurance Institute for Highway Safety, http://www.hwysafety.org/safety_facts.qanda/airbags, 2004.

15. Jasanoff, S. "American Exceptionalism and the Political Acknowledgment of Risk," Daedalus, Vol. 119, No. 4, 1990, pp. 61-81.

EXERCISES

13.1 Consider a product that you or your family has purchased in the past month or so. Write a brief paper describing both the positive and negative impacts of that product on society.

13.2 The Federal Aviation Administration enacted a regulation affecting the mailing of packages weighing more than 16 ounces. The regulation requires you to present the package to a postal clerk for mailing and prohibits the mailing of the package from a postbox. Write a paper covering the following issues:

- Describe the regulation in more detail.
- Why is the FAA writing a regulation affecting the U.S. Post Office?
- Do you believe that the regulation will be effective for its intended purpose? Please give arguments supporting your viewpoint.
- What actions will the post offices (40,000 of them) have to take to make the regulation effective? What actions do the post offices actually take with regard to the regulation?
- Will these actions be costly? Estimate the costs to both the public and the individual. Assume that a postal clerk is paid $15/h and that an individual considers his or her free time worth $15/h.
- Give your assessment of the cost to benefit ratio for this regulation?

13.3 Suppose that you were a slave in the time of Rameses II and were assigned to the task of constructing his statue. For those not up to date on Egyptian statues, this one weighs 1,000 tons and is 56 feet tall. Write a paper describing your daily tasks.

13.4 Horses were not used extensively to relieve man from brutal work or to provide a significant advantage in military actions until nearly 1,000 AD. Write a brief paper describing both the social and technical reasons for the very long time required to effectively employ horses in either military or commercial enterprise.

13.5 Suppose that you were living on a manor in England in about 1,300 AD. Describe your lifestyle and indicate how technology affects your work and play. Select in which of the two classes of society that you existed.

13.6 Write a paper comparing the lifestyles, as you imagine them, for a man or woman living today and living in the year 1,000. Did technology make a difference in the quality of life?

13.7 Find the Website for the National Academy of Engineering and click on Greatest Achievements for the 20th Century. Select one of the 20 achievements and prepare a paper describing the history of that particular achievement.

13.8 Find the Website for the National Academy of Engineering and click on Greatest Achievements for the 20th Century. Select one of the 20 achievements and prepare a paper describing the time line for that particular achievement.

13.9 Explain why the public-at-large was opposed to power generation with nuclear energy prior to 2003? Was the public-at-large correct in their collective assessment? Explain why you formed this opinion.

13.10 In recent years the public's perception of nuclear power is slowly changing because of the publicity associated with global warming. Explain why nuclear power is slowing gaining acceptance because of the fear of global warming?

13.11 Why does a global marketplace exist today? Does global trade improve our standard of living? What does our current trade deficit have to do with wages for factory workers? Does technology help or hinder the balance of trade deficit?

13.12 If you were the Director of the National Highway Transportation Safety Administration (NHTSA), what action would you take regarding the public concern about airbags?
- Explain the reasons for your actions.
- Explain how you would convince the major automakers to agree with you.
- Prepare an outline you would follow in the press conference announcing this action.
- Describe how you would handle a public interest group disagreeing with your ideas.
- Describe how you would handle the public interest groups that agree with you.

13.13 Find the Website for the Insurance Institute for Highway Safety and examine the page on Airbag Statistics. Determine the ratio of deaths caused by airbags to the estimated number of lives saved by airbags. Write a position paper about actions that should be taken in the future regarding airbags to insure public safety.

13.14 Recently side airbags have been installed in some automobiles. Are these side airbags a threat to small children seated near a door? What has the NHTSA done to mitigate this threat?

13.15 What social science courses should you take during your undergraduate program to broaden your prospective and enhance your understanding of sociotechnical issues?

13.16 Is there a technical elective in your program of study that deals with sociotechnical issues?

CHAPTER 14

SAFETY AND PERFORMANCE

With Professor Vincent Brannigan[1]

14.1 LEVELS OF POTENTIAL INJURY

When engineers design a new product or modify an existing one, they can create the possibility of injury. Sometimes the potential injury is minimal, but in some instances the injury may be catastrophic. In the late 1950s Lockheed introduced the L-188 Electra. Three of these aircraft crashed due to structural failure in flight before the Federal Aviation Administration (FAA) grounded the entire fleet. An investigation revealed that a torsional resonance occurred under certain operating conditions causing the engines to vibrate violently and break away from the wing. This was a very serious engineering failure which caused the death of over a hundred innocent passengers and significant losses of property. Clearly, this example illustrates a case where the risk far outweighs the benefits of performance of a new aircraft design. It also illustrates the seriousness of design errors and inadequate testing of prototypes of products that pose danger to the public.

14.1.1 Safety Concepts Important for Safe Engineering Design

Hazard and Risk

The first is the "**hazard** "or the type of injury which might occur (also called the **severity of injury***)*. The second is the **probability** of injury under various combinations of circumstances also known as the **rate of injury**.
      The terminology in this field is often confusing. For the purpose of this chapter **"risk"** is roughly the probability of injury times the specific hazard. A risk is *acceptable* if a fully informed society would consider the combination of rate and hazard to satisfy social expectations.

Intrinsic and Extrinsic Safety

We should also distinguish between safety which is **intrinsic** in the desired performance of an object or system and safety which must be "designed in" but is **extrinsic** to specified performance. In a revolver for example the strength of the barrel is *intrinsic* to the performance but the safety catch is *extrinsic*. **Intrinsic** safety requirements tend to show up in the specification for a product, but extrinsic safety

---

[1] This chapter was revised by Prof. Vincent Brannigan of the Clark School of Engineering, University of Maryland.

requires an engineer to fully understand the entire working environment of the product, not merely its performance specifications. **Ex**trinsic safety features require special expertise and always run the risk of being overlooked. For example, the designers of the TITANIC failed to consider the entire environment of the lifeboats as a system, as a result no plan was created to fill the lifeboats, and over 400 people died for whom there would have been places in the boats.

CASE STUDY — Drill Design

You are designing an electric, pistol-grip drill that is powered from the standard 120 volt, 60 cycle single-phase power supply from a local utility company. The **performance** of the drill describes the torque that can be applied to a drill bit. But electrical safety has to be "designed in". The drill operator is at risk from an electric shock, although the probability for injury is very small if the tool is properly designed. It is your responsibility as an engineer to minimize this risk while maintaining the advantage in performance of a new improved model of an electric-powered drill. In this case the hazard is inherent in the use of 110 volt current and the job of the engineer is to reduce that probability to an acceptable level.

How do you protect an operator from electric shock? The operator is holding an electric motor with 120 volts across the armature coils and the field coils in his or her hands. An early answer was to "ground" the tool using the familiar "three prong plug" But that solution relied on operators using the tool correctly and not subverting the grounding system, or using outlets that were improperly wired. A later concept was to build the case for the drill from a tough, durable plastic that is structurally strong and an excellent electrical insulator. The case keeps the operator from touching any part of the electrical circuit powering the motor. In fact, Black & Decker® a major manufacturer of small power tools, uses double insulation to provide two independent insulation barriers to prevent the operator from making contact with any part of the electrical circuit. Operators may still manage to receive an occasional shock, but it will not be easy. Accidental contact of the operator's hands with a live electrical circuit is a rare event. In fact, the probability of electrical shock is so small that a worker routinely picks up a power tool and employs it to perform some task without even a passing thought about the possibility of receiving an electrical shock. Operating such a power tool does involve risk, but it is acceptable because the rate when using well designed tools is sufficiently low.

14.1.2 System Safety Thinking

One of the key problems in designing for safety is the tendency to focus solely on the object being designed rather than on the entire system of which the object is a component. To use the drill example; a system safety approach might be to use battery operated tools that eliminate the hazard completely or ground fault circuit interrupters to further minimize the shock risk. System safety analysis (SSA) examines the entire design process to assess hazards, probabilities and appropriate responses. A key element is human factors engineering, which examines the human /machine interface.

14.1.3 Analyzing Risk

Consider flying in a commercial airliner on a 1,200 mile trip. How safe is it to make this trip? Most people believe that it is reasonably safe to fly commercial airliners, but they realize that there is a small probability of a fatal crash. Statistical analysis may help to refine the probability of an injury. The statistics show the probability of a fatality p_k in an airline accident is about one in a billion passenger miles flown ($p_k = 1/10^9$). Hence, if you make the round trip from point A to B, your probability P_k of being killed due to an airline accident can be described as:

$$P_k = s\, p_k = (2)(1200)/10^9 = 2.4 \times 10^{-6} = 0.00024\%$$

where s is the number of miles in a round trip.

There are a number of assumptions "hidden" in this type of statistical analysis. In particular are all "passenger miles" the same? The answer is clearly no. The risk of a crash may be much more related to take off and landing rather than the miles flown. All statistical safety analysis has the problem of making sure that the correct data is actually being used. In particular correlation is not causation but sometimes statistics are all we have with which to work.

An accident rate of 2.4 chances in a million is considered by many to be a very low probability for a fatal accident; consequently, most people do not worry much about the possibility of dying in a crash when they board an airliner. However, if you are a sales engineer flying 100,000 miles a year, every year for 20 years, your total mileage accumulates to 2×10^6 miles. Your estimated probability of dying in a crash increases to:

$$P_k = 2 \times 10^6/10^9 = 2 \times 10^{-3} = 0.2\%$$

The probability has increased significantly if you accumulate the miles flown by a traveling professional over a 20-year period. Knowing there is one chance in 500 that you will die in an airline crash, would you still want to be a sales engineer flying weekly to meet with customers? Could you tolerate this level of exposure? Would you be apprehensive?

Everyone has some tolerance for risk. People weigh the speed, convenience, cost and risk of flying against that of traveling by train or by car. In almost all instances, travelers select the plane for long distances and the car for short distances. Travelers select the train when it combines relatively good service (high speed with frequent trains) to convenient (downtown) locations. Most people do not contemplate the risks involved in travel because, except for driving a car, fatal accidents are rare events.

Unfortunately, fatal accidents in automobiles are not rare events. Preliminary estimates by the National Highway Traffic Safety Administration (NHTSA) showed that 42,642 occurred on U.S. highways in 2006[2]. In addition over 2.5 million serious injuries occur in accidents each year. Death on the highway is the leading killer of Americans 1 to 35 years old. It is clearly more dangerous to drive than to fly when you compare the probability of fatalities and/or injuries on a per mile basis. How do you handle the higher risk associated with driving? One answer is rationalization—I am an excellent driver, and it will not happen to me. Not necessarily a true statement, but the rationalization of one's superior driving skills alleviates the worry and concern about a fatality or injury producing accident. The real risk remains.

It does recognize that when driving, you are an active factor in determining the risk. Driving habits (high-speed, reckless steering, tailgating, etc.) and driving skills affect, to a large degree, the level of risk involved. When flying, you are a passive participant in determining the rate of injury.

Involuntary Exposure to Risk

Clearly the exposure to risk ranges from clearly voluntary to involuntary depending on the circumstances. Whether a choice is truly voluntary is a matter of social decision. One of the major changes in safety regulation in the 20th century was to consider the exposure of workers to risk to be involuntary and therefore less acceptable. This is incorporated into OSHA and various regulatory standards. Recreational activities such as skiing, scuba diving, horseback riding, hang-gliding, mountain climbing, dirt biking, etc. carry higher levels of risk than other activities Yet, intelligent people swarm to the ski resorts and pay

---

[2] This data implies that 1.42 fatalities occurred per 100 million vehicle miles.

large amounts of money to potentially break their bones. Why? They have voluntarily decided that the pleasure derived from the activity is worth the risk. Also, they can sometimes control the level of risk. They can choose the simpler slope and minimize the risk or the more challenging slope to maximize the thrill. However even in these cases society often sets the boundaries of acceptable risk.

Knowing the social environment in which a product will be used is critical to setting the level of acceptable risk as part of the design process. For example airlines are expected to take very high levels of safety precautions. Some level of risk is involved in almost all products produced but it is imperative that this risk be within the social expectation of safety while if possible maintaining an acceptable level of performance. An appropriate balance between risk and performance must be achieved, or the product simply cannot be used[3]. It is important that engineers, business and governmental regulatory agencies cooperate to provide a realistic assessment of the risk level and a clear statement of what is acceptable. Finally, the public should be made aware of the residual risks involved in a product, although merely warning of a risk is not an acceptable alternative to producing a safer design, especially when the person being warned is not the person at risk.

14.2 MINIMIZING THE RISK

14.2.1 Safety in Design: Performance Analysis

The first step in safe design is to analyze the required performance and determine suitable safety factors and margins of safety. This approach is normally used with well understood systems with well understood materials and well understood failure modes. It is particularly useful for intrinsic safety analysis. As noted above designing for safe performance is not exactly the same as designing for safety, although it shares many similar techniques. Designing for safety means "avoiding injury". In a "fail safe" design the product stops performing before it can cause injury. However the simplest design case is where the performance is identical with the safety concern, such as with the tension member of a bridge.

Tension Member

Let's begin a performance analysis by introducing a tension member as shown in Fig. 14.1.

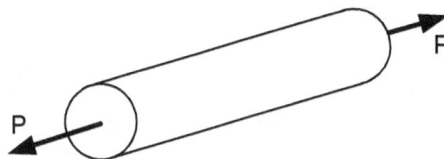

Fig. 14.1 A tension member is a long thin rod subjected to axial load P.

When the tension rod is subjected to an axial load P, an axial stress σ develops that is uniformly distributed over the cross section of the rod. This uniformly distributed stress acting on the rod is illustrated in Fig. 14.2.

---

[3] Cost may sometimes be an issue when balancing safety, risk and performance. As a guide to cost, the Environmental Protection Agency (EPA) uses a worth of $6.1 million for what economist call a "value of statistical life" in their cost benefit analyses for new regulations. However the use of such a figure in the design of product by a private company is more problematic. Indifference to a known and preventable risk is both unethical and may be unlawful.

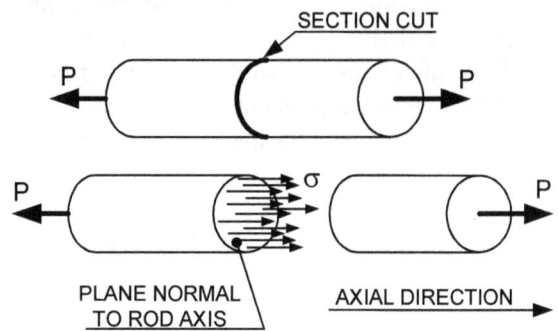

Fig. 14.2 The stress σ is distributed uniformly over the cross-sectional area of the tension rod.

The magnitude of this stress is determined from the formula:

$$\sigma = P/A \qquad\qquad (14.1)$$

where σ is the stress given in units of pound/square inch (psi) or newton/square meter (Pa).

        P is the load expressed in units of pound (lb) or newton (N).

        A is the cross sectional area of the rod specified in square inch (in^2) or square meter (m^2).

Two examples demonstrating the technique for computing stresses in axially loaded tension members are presented below:

EXAMPLE 14.1

If the axial load P applied to a circular tension rod is 10,000 lb, determine the axial stress σ that is developed if the rod is ½ in. in diameter.

    Solution: The cross sectional area is given by:

$$A = \pi r^2 = \pi(0.25)^2 = 0.1963 \ in^2$$

From Eq. (14.1), the axial stress σ is:

$$\sigma = P/A = 10,000/0.1963 = 50,930 \ psi$$

where psi is the abbreviation for lb/in^2.

At this stage of the analysis, it is not possible to interpret the significance of the result for the stress σ = 50,930 psi. More information is needed pertaining to the strength of the material from which the tension member is fabricated. When the strength of the rod's material is known, it is possible to make a comparison of the magnitude of the stress relative to this strength and remark on its significance.

    In this first example, U.S. Customary Units were used to express the axial load in pounds, the cross sectional area in square inches and the stress in psi or pounds per square inch. In the next example, the International System of Units (SI) will be employed where the load is expressed in newton, the cross sectional area in meters, and the stress in Pascal (N/mm^2).

EXAMPLE 14.2

If the axial load P imposed on a circular rod is 50,000 newton (N), and the radius of this rod is 5 mm, determine the axial stress σ.

Solution: First, determine the cross sectional area A as:

$$A = \pi r^2 = \pi (5 \times 10^{-3})^2 = 78.54 \times 10^{-6} \text{ m}^2$$

Then, from Eq. (14.1), the axial stress is given by:

$$\sigma = P/A = (50 \times 10^3 \text{ N})/(78.54 \times 10^{-6} \text{ m}^2) = 636.6 \times 10^6 \text{ Pa} = 636.6 \text{ MPa}$$

where the abbreviation Pa stands for pascal—the unit used for stress in the SI system.

A pascal is equal to a newton per square meter (N/m^2). When calculating stresses in the SI system, it is common to obtain very large numbers for the stress when σ is expressed in pascal. For this reason, MPa (mega pascal) is often employed as the unit for σ where MPa = 10^6 Pa = N/(mm^2).

The stress on a tension member can be accurately determined if you can ascertain its size and the load that it must support. The calculated stress is a number with associated units—either psi or MPa. As a number, the stress is not very informative until it is compared to the strength of the material from which the tension member is fabricated.

Let's suppose that the tension rod used in Example 14.1 is machined from very strong alloy steel exhibiting a yield strength of 100,000 psi. Do you believe the tension rod is safe? Will it retain its shape under load? Will it fail by yielding (stretching under load and not recovering completely when the load is removed)? These questions may be answered by making a simple comparison of the applied stress σ to the strength S of the material. In Example 14.1, the axial tensile stress σ = 50,930 psi and the strength of the material from which the rod was machined is S = 100,000 psi. The comparison shows that the yield **strength** of the rod is **greater** than the applied **stress**; therefore, the rod **will not fail** by yielding or breaking. But how safe is the rod? You know it will not fail; however, some measure of the safety of the rod should be established. Is the safety of the rod marginal, or is the rod too safe? Why would any structural element be too safe?

Safety Factor

To respond to the safety issue for the rod, determine the ratio of the strength to the stress and define this ratio as the safety factor SF:

$$SF = S/\sigma \qquad\qquad (14.2)$$

For the stress imposed on the rod in Example 14.1, the safety factor determined from Eq. (14.2) is:

$$SF = 100,000/50,930 = 1.96$$

The safety factor is unitless. The value of SF = 1.96 indicates that the tension rod is almost twice as strong as it must be to resist yielding under the applied load. You have a reasonable safety factor and can be confident that the rod will perform safely in service. Your only concerns are the accuracy with which the load has been predicted and the quality control for the material used in manufacturing the tension rods. If the factors listed below are satisfied:

- You have extensive experience with the application.
- You have a clear understanding of the failure mode
- You are certain of the magnitude of the load.
- You have confidence in the manufacturing division in tracking their materials.
- You have verified the strength of the materials received from suppliers.

Then you usually can be satisfied that a safety factor of about two is adequate. However, if you are not certain about the load and/or the materials, a safety factor of two is not sufficient. You should increase the safety factor to accommodate your ignorance of either the applied load or your lack of control over the material employed.

If you do not know the failure mode, the safety factor may not be applicable. The TITANIC may have suffered a brittle failure of rivets or plates due to cold water. That failure mode was not well understood at the time and so normal safety factors did not apply.

Safety factors ranging from two to four are commonly employed in design. Safety factors of less than two require considerable care, expense and expertise. The use of relatively low safety factors is justified only for very high-performance applications or for components that are produced in very high volume. In these special situations, the engineering analysis, prototype testing, quality control inspections, maintenance inspections and documentation must be extensive.

Margin of Safety

The margin of safety MS is sometimes used to describe the degree of safety incorporated into the design of a structural component. The margin of safety should not be confused with the safety factor as they are different quantities. The margin of safety is defined as:

$$MS = (S - \sigma)/\sigma = SF - 1 \qquad\qquad (14.3)$$

Using the results from Example 14.1 and Eq. (14.3), the margin of safety is given by:

$$MS = 1.96 - 1 = 0.96 \ \text{ or } \ 96\%$$

The tension rod has a margin of safety of 96%, which implies that the strength of the material exceeds the applied tensile stress by 96%. If the load on a tension member can be determined, it is easy to size (adjust the cross sectional area) of the rod to provide any margin of safety that is deemed necessary to satisfy management's concerns and to meet professional obligations to society-at-large.

System Safety Analysis

Safety factors and margins of safety only apply to well defined materials and failure modes. Numerous engineering disasters have been due to a lack of understanding of the underlying failure paths. This is especially true when the safety failure is extrinsic to performance.

CASE STUDY: Bridge Failure Due to Fire

Consider for example a tank truck fire on the bridge designed in the above section. Steel under tension weakens dramatically when heated.[i] On April 29 2007 a large steel interstate highway bridge collapsed in a tank truck fire when the steel girders supporting the bridge softened when the flames increased their temperature in excess of 1,000 °F.

http://www.sfgate.com/cgi-bin/object/article?f=/c/a/2007/04/29/BAGVOPHQU46.DTL&o=0

The steel columns and beams in buildings are routinely encased in "fire proofing" to provide protection and avoid collapse. The steel in bridges is not. Only a proper system safety analysis can tell if the risk we are running by not protecting the steel is reasonable.

http://www.pbs.org/newshour/bb/science/jan-june07/overpass_05-10.html

The information to develop this case study is available on the Internet at the URLs listed above.

14.3 FAILURE RATE

Component Analysis

In some systems an engineer can minimizes risk by preventing failure of each and every component in the system. This task is not easy as there are several different ways in which components can fail. Parts fail by breaking and aging, by corrosion or fatigue, by overload and burning, etc. In some instances, you may anticipate these failures and replace the parts before they malfunction. This controlled replacement of finite life parts is known as preventative maintenance. In most cases, engineers try to design each component so that failure will not occur during the anticipated life of the product.

Sometimes components fail by burning out, aging or wearing out—not necessarily by breaking or yielding (excessive permanent deformation). When conducting a failure analysis for wear or aging, the safety factor is normally not a relevant parameter. Instead you use some form of time to failure analysis or mean time between failures MTBF. You must cope with wear, aging or burn out by using other methods. To begin, let's recognize that all components do not wear out at the same time. Some people drive their car 25,000 miles before the brakes wear out and others may drive 60,000 miles or more. Some people can use their personal computers for several years before a component fails; yet others have problems after only a few months in service. To analyze these differences in service before failure, a mortality curve for both mechanical and electronic components is introduced in Fig. 14.3a and Fig. 14.3b, respectively.

14.3.1 Computing a Failure Rate for Components

To determine the failure rate FR, records of the performance with time-in-service of a large number of components are maintained. You begin your record of performance with N_0 components that are placed in service at time t = 0. After some arbitrary time t, some number N_f will have failed and the remainder N_s will have survived. At any arbitrary time, you may write:

$$N_f + N_s = N_0 \qquad\qquad (14.4)$$

With time, N_f increases and N_s decreases, but the sum remains constant and equal to N_0. The failure rate FR is defined as:

$$FR = N_f / (N_0 \times t) \qquad \text{when } t > 0 \qquad (14.5)$$

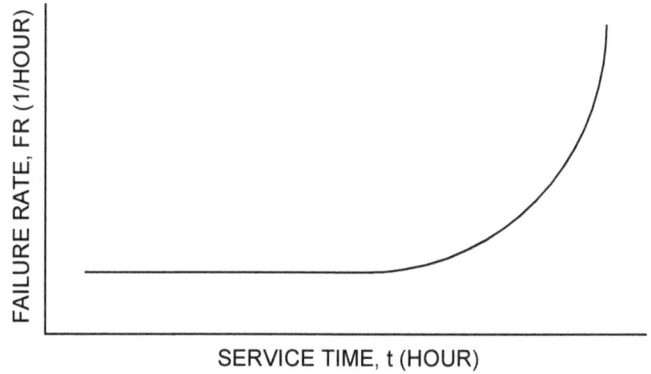

Fig. 14.3a Mortality curve for mechanical components showing failure rates with time in service.

When the results of Eq. (14.5) are plotted with respect to time, the component mortality curve presented in Fig. 14.3a is obtained. For mechanical components, where testing is performed on the assembly line, the failure rate is small when the components are relatively new. However, the failure rate increases non-linearly after some time in service because parts begin to wear, corrode or they are abused.

The failure rate for electronic components is usually characterized by the well-known bathtub mortality curve presented in Fig. 14.3b. This curve exhibits three different regions of interest—each due to a different cause. The high failure rates in the first region, $t < t_1$, are due to components manufactured with minute flaws that were not discovered during inspection and testing at the factory. In the center region of the curve, for $t_1 < t < t_2$, the failure rate FR_0 is small and nearly constant with time. Clearly, this is the best region to operate if you are to maximize the reliability of the product. Near the end of the service life, when $t > t_2$, the failure rate increases sharply with time as the components begin to fail due to the effects of aging.

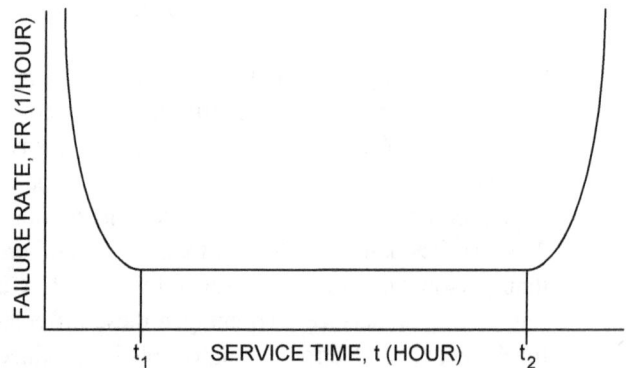

Fig. 14.3b Mortality curve showing the failure rate for electronic components with time in service.

For high-reliability systems, it is important to eliminate the inherently defective components associated with early failures. For mechanical systems, inspecting and testing the system's performance to eliminate flawed components produced in manufacturing accomplishes this objective. This task is not easy for the electronic component manufacturers because the feature sizes on the components are so small that inspection requires very high magnification and detailed examination of relatively large areas. Even with elaborate inspection schemes, not all of the flaws can be detected. A better procedure to eliminate the few inherently flawed components is to conduct what is known as burn-in testing prior to incorporating the components into an electronic product. With burn-in testing, all of the components that are produced are operated for a time t_1 (as defined in Fig. 14.3b). The flawed components burn out (fail) and are eliminated.

The surviving components that will function with a much lower failure rate FR_0 are used in fabricating high reliability products.

Clearly, it is important to design components with low failure rates. If the failure rates are low, then the time between failures will be high. In fact, manufacturers sometimes cite what is known as the mean time between failures (MTBF) in their product specifications. The MTBF and the failure rate are related by:

$$MTBF = 1/FR \qquad (14.6)$$

You should understand the concept of component failure rates and how they vary over the life of the component. How is this information about the failure rate used to establish the reliability of a specific component to perform without failing over the anticipated life of a product? This topic is discussed in the next section.

14.3.2 Component Reliability

To answer the question of component and system reliability, you must introduce the concept of probability or chance of occurrence. In considering probability, let's return to the test conducted to measure the failure rate and use the data collected for N_f, N_s, and N_0. The probability of survival P_s, which varies with time t, may be determined from:

$$P_s(t) = N_s(t)/N_0 \qquad (14.7)$$

And the probability of failure P_f is given by:

$$P_f(t) = N_f(t)/N_0 \qquad (14.8)$$

Both the probability of survival and the probability of failure are functions of time. With increased time in service, the probability of survival decreases and the probability of failure increases.

From Eqs. (14.7) and (14.8), it is evident that:

$$P_s(t) + P_f(t) = 1 \qquad (14.9)$$

Substituting Eq. (14.8) into Eq. (14.9) yields:

$$P_s(t) = 1 - \frac{N_f(t)}{N_0} \qquad (14.10)$$

If Eq. (14.10) is differentiated with respect to time t and the result is rearranged, it is possible to write:

$$\left[\frac{N_0}{N_s(t)}\right]\left[\frac{dP_s(t)}{dt}\right] = \left[\frac{-1}{N_s(t)}\right]\left[\frac{dN_f(t)}{dt}\right] = -FR(t)dt \qquad (14.11)$$

Consider the term $[1/N_s(t)][dN_f(t)/dt]$ and note it is the instantaneous failure rate $FR(t)$ associated with a sample size N_s at time t. Next, integrate Eq. (14.11) and assume that $FR(t) = FR_0$, is a constant. The relation obtained gives the probability of survival as a function of the failure rate.

$$P_s(t) = e^{-(FR_0 t)} \qquad (14.12)$$

where e = 2.71828 is the exponential number. Let's consider an example to demonstrate the method for determining the reliability of a mechanical component.

EXAMPLE 14.3

Suppose a company maintenance records show the number of failures of a certain component over an extended period of time. Analysis of this data indicates that the failure rate is essentially constant with 1 failure per 10,000 hours of service. Does that sound good to you? Will the reliability be adequate? Let's determine the probability of survival (the reliability) of this mechanical component with time.

> **Solution:** Substituting the failure rate FR_0 = 1/10,000 h into Eq. (14.12), yields the results for $P_s(t)$ shown in Table 14.1.

Table 14.1
Reliability $P_s(t)$ of a mechanical component
with increasing service life (FR_0 =1/10,000 h)

Time (1000 h)	Time* (years)	$FR_0 \times t$ (unitless)	Reliability $P_s(t)$
1	0.5	0.1	0.905
2	1	0.2	0.819
5	2.5	0.5	0.606
10	5	1.0	0.367
20	10	2.0	0.135
50	25	5.0	6.738×10^{-3}
100	50	10.0	4.540×10^{-5}
200	100	20.0	2.061×10^{-9}

*The conversion from hours to years of life is based on 2,000 hours/year.

Reliability varies markedly with service life. For a short service life, say a year, the reliability is reasonable with 81.9% chance of surviving. However, for long life, say 10 years, the reliability drops to only 13.5%. For very long life 50 to 100 years, failure is almost certain. While the failure rate of 1 in 10,000 hours appears satisfactory in an initial assessment, the reliability resulting from this failure rate is disappointing. You cannot be certain of a service life of 10,000 hours prior to failure. In fact, there is only a probability of 36.7% to survive for 10,000 hours. If a reliability of 90% is required for a service life of 10,000 hours, the failure rate FR_0 must decrease to about 1 failure in 100,000 hours.

14.4 SYSTEM SAFETY and RELIABILITY

The failure of a component may or may not cause a "safety failure" of a system. When a system is comprised of several components, its SAFETY will depend

1. On the probability of failure of the individual components and
2. Their arrangement.
3. Safety systems put in place to avert negative consequences

14.4.1 Series and Parallel Systems

Conceptually there are two basic system arrangements that are possible—series or parallel.

Series: Failure of a single component causes system failure.
Parallel: Multiple independent failures must occur for system failure.

An automobile can illustrate both the series and parallel arrangement of components. For lighting the highway during the night, an auto is equipped with two headlights. This is a parallel arrangement because the design incorporates two identical components (the headlights) that perform nearly the same function. If one headlight burns-out, you can still drive. Your visibility is impaired to some degree and the system has been compromised; however, it has not failed. You also have notice of the first failure before the second component fails. Similarly automobiles have two or even three independent braking systems, because you may not notice the failure of one system. On the other hand, suppose you intend to start your engine. Consider the components involved in this action:

- Ignition switch
- Battery
- Wiring
- Solenoid relay
- Starter motor
- Starter motor clutch and gear
- Engine ignition components (spark plugs and points)
- Fuel
- Fuel pump
- Fuel lines
- Fuel injectors

If any of these components should fail, the engine will not start when you turn the key or push the start button. A complete series of successful components is required for the system to function. To start your engine, every component must properly function because the failure of a single component will result in a failure of the entire system.

In general for cars inability to stop is considered a greater risk than inability to start. Shutting the engine down rarely causes an accident. However, for an aircraft shutting down the engine is much more hazardous, so they have more redundant ignition systems.

14.4.2 Reliability of Series Connected Systems

Let's consider the reliability of a system involving three components that are in a series arrangement, as shown in Fig. 14.4.

Fig. 14.4 A system comprised of a series arrangement of three components.

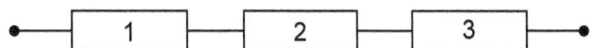

Because all three of the components in this series arrangement must operate successfully for the system to perform satisfactorily, the probability of survival of the system (P_s^s) is a product function given by:

$$P_s^s = P_{s1}\, P_{s2}\, P_{s3} \qquad\qquad (14.13)$$

where the superscript s refers to the entire system of components.

Combining Eqs. (14.12) and (14.13) permits the system reliability to be expressed in terms of the failure rate of the individual components as:

$$P_s^s = e^{-(FR_1 + FR_2 + FR_3)t} \tag{14.14}$$

When the number of components in a system is increased with a series arrangement, the system reliability is decreased markedly. Also, the system reliability continues to decrease with time in service.

EXAMPLE 14.4

Consider the case where n components are arranged in a series where n is a variable that increases from 1 to 1000. Let's also assume that the reliability of each of the n components is the same ($P_{s1} = P_{s2} = \ldots\ldots = P_{sn}$) at some time during the operating life of the system. Evaluate the system reliability for three different values of the component reliability—$P_s = 0.999$, 0.99, and 0.90.

Solution: The results obtained from Eq. (14.13) are shown in Table 14.2.

Table 14.2
Reliability of a system with n series arranged components
as a function of component reliability P_s

Number of Components, n	$P_s = 0.999$	$P_s = 0.990$	$P_s = 0.900$
1	0.999	0.990	0.900
2	0.998	0.980	0.810
5	0.995	0.950	0.590
10	0.990	0.904	0.349
20	0.980	0.818	0.122
50	0.950	0.605	*
100	0.905	0.366	*
200	0.819	0.134	*
500	0.606	*	*
1000	0.368	*	*

* System reliability of less than 1%.

Examination of the results presented in Table 14.2 clearly shows the detrimental effect of placing many components in a series arrangement. For components with a high reliability ($P_s = 99.9\%$), the system reliability P_s^s drops to about 90% with 100 series arranged components. However, with components with a lower reliability ($P_s = 90\%$), the system degrades to a reliability of less than 1% when the number of components approaches 50. The lesson here is clear—if you connect a large number of components together in a series arrangement to develop a system, then the individual component reliability must remain extremely high over the entire service life of the system.

14.4.3 Reliability of Systems with Parallel Connected Components

Recall the description of a highway lighting system on an automobile with a pair of headlights. This system is an example of a parallel arrangement because both headlights must fail before the system fails. It is possible to drive with one light, although the State Troopers might warn the driver of the dangers of doing so. The human body has several parallel systems; you have two hands, two eyes, two legs, two feet and two ears, etc. These parallel systems increase a person's ability to function in the event of a mishap.

A system with a parallel arrangement of three identical components is represented with the model shown in Fig. 14.5.

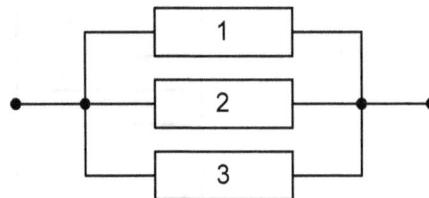

Fig. 14.5 System with a parallel arrangement of three identical components.

Parallel systems are considered to be redundant because all of the components must fail before the system fails. Accordingly, the relation for the probability of system failure is written as:

$$P_f^s = P_{f1}\, P_{f2}\, P_{f3} \qquad (14.15)$$

Substituting Eq. (14.9) into Eq. (14.15) gives:

$$1 - P_s^s = (1 - P_{s1})(1 - P_{s2})(1 - P_{s3}) \qquad (14.16)$$

Expanding Eq. (14.16) and reducing the resulting expression gives:

$$P_s^s = P_{s1} + P_{s2} + P_{s3} - P_{s1}\, P_{s2} - P_{s2}\, P_{s3} - P_{s1}\, P_{s3} + P_{s1}\, P_{s2}\, P_{s3} \qquad (14.17)$$

Let's examine Eq. (14.17) to determine if increasing the number of parallel components improves reliability.

EXAMPLE 14.5

Suppose the probability of success P_s for all the components in a parallel system is the same. However, P_s is considered as a variable increasing from 0.1 to 1.0. Show the effect of component redundancy by considering a system comprised of one, two and three components in parallel.

> **Solution:** The results obtained from Eq. (14.17) are shown in Table 14.3. An examination of the results presented in Table 14.3 shows that the degree of redundancy improves the reliability of a system with a parallel arrangement of components. If you have relatively poor component reliability, say 60%, and employ two of these components in a parallel arrangement, the system reliability improves to 84%. Add a third component to the parallel arrangement and the reliability increases to 93.6%. Adding more components in parallel will continue to improve the reliability, although it is a case of diminishing returns.

Table 14.3
Effect of the degree of redundancy on reliability
$$P_{s1} = P_{s2} = P_{s3}$$

Component Reliability	Single Component	Two Components	Three Components
0.10	0.10	0.19	0.271
0.20	0.20	0.36	0.488
0.30	0.30	0.51	0.657
0.40	0.40	0.64	0.784
0.50	0.50	0.75	0.875
0.60	0.60	0.84	0.936
0.70	0.70	0.90	0.973
0.80	0.80	0.96	0.992
0.90	0.90	0.99	0.999
1.00	1.00	1.00	1.000

The improvement in reliability by employing parallel (redundant) components in designing a system is a distinct advantage. However, one rarely if ever, enjoys a free lunch. The corresponding disadvantages are the increased costs and the added power, weight and size of the system. These are significant disadvantages, and redundant design is used only when component reliability is too low for satisfactory system performance or when a failure produces very serious consequences. For a more complete discussion of component and system reliability, see reference [1].

14.4.4 Redundancy and Safety Systems

Parallel design is only one example of redundancy. A wide variety of types of redundancy can be identified in biological systems that are relevant analogies to technical systems. The biological categories include:

Duplicated organs or systems: Duplication exists in the human body in many ways. Humans have multiple kidneys, genes and other components that can fully function independent of the duplicate unit. A building analogy might be duplicate water supplies, where one or another can carry the full load. Automobiles have multiple braking systems.

Autonomous functioning of coordinated items or systems: Human hearing and sight have duplicate components that function together to provide enhanced capability but each system can operate separately to provide substantial capacity. Headlights are a coordinated system.

Excess capacity: The human liver has such an excess of capacity that large portions can be removed without substantial effect on ordinary living. Similarly human ovaries contain far more eggs than would be needed in a normal lifetime. Higher levels of strength in the frame of an automobile can be described as excess capacity

High reliability systems: The human heart has evolved as a high reliability system with numerous internal checks and carefully supervised functioning. Similarly the electronic controls of cars can be designed for very high reliability.

Automatic reprogramming: The human brain has some capability of "rewiring" itself in response to injury or illness. Smart systems that can "reprogram" in response might fall in this category.

Spare units: Some species reproduce at a high rate so that spare "units" are always available. Like spare tires they ensure that the species as a kind of overall system will keep functioning

Functional redundancy: In addition to structural redundancy humans also have a certain amount of "functional" redundancy, where persons missing arms, sight or other organs develop "functional" substitutes to accomplish the same task. More importantly humans evolved the ability to make and use complex tools that add redundancy to their functional environment. Human factors engineering can identify functional substitutions

14.5 EVALUATING THE RISK ENVIRONMENT

When the space shuttle Challenger exploded during launch on January 28, 1986, a Presidential Commission [2] was established to:

1. Review the circumstances surrounding the accident and determine the probable cause or causes for the explosion.
2. Develop recommendations for corrective or other actions, based on the Commission's findings and determinations.

Richard Feynman, a Nobel Prize winning physicist from the California Institute of Technology, was appointed to the Commission. Dr. Feynman, near the end of his career and his life, took the appointment very seriously and devoted his entire time and energies to the investigation. One of his many contributions during the review was to determine the probability of failure of the space shuttle system.

During the investigation, Dr. Feynman asked three NASA engineers and their manager to assess the probability of failure P_f of a mission due to the failure of one component—the shuttle's main rocket engine. Secret ballots by the engineers indicated two estimates with $P_f = 1/200$ and one with $P_f = 1/300$. Under pressure, the program manager finally estimated the risk at $P_f = 1/100,000$. There was such a large difference between the manager's and the engineer's assessment that Dr. Feynman concluded that "NASA exaggerates the reliability of its products to the point of fantasy" [3].

The safety officer for the firing range at Kennedy Space Center, who had been under considerable pressure to remove the remotely controlled destruction charges from the shuttle, did not believe the reliability figures cited by NASA. He had collected data for all of the 2,900 previous launches using solid rocket boosters. Of this total, 121 had failed. This data provide a very crude estimate of $P_f = (121)/(2900) = 0.042 = 4.2\%$ or about one chance for failure in every 24 launches. The safety officer considered this high risk of failure to be an upper bound, because improvements made since the early launches had improved the reliability of the solid booster motors. Also, the pre-launch inspections on the shuttle were more thorough. His estimate of risk accounting for taking these improvements was $P_f = 1/100$. Dr. Feynman [3], after extensive interviews with engineers, reliability experts, managers from NASA and many subcontractors, concluded that the shuttle "flies in a relatively unsafe condition with a chance of failure of the order of 1%."

After the Challenger accident NASA began a series of comprehensive assessments of risk of flying the shuttle fleet in space. In 1995 the probability of a catastrophic failure was assessed at 1 in every 145 flights. After a series of safety related upgrades to the shuttle system, a new assessment in 1998 placed the probability at one in 245 flights. In October of 2002 NASA most recent assessment placed the probability at one in 265 flights or about 0.39%. How accurate are these assessments? The disintegration of the shuttle Columbia over Texas as it reentered space on February 1, 2003 was the second shuttle failure in 112 flights. This relatively high rate of failure (2/112) = 0.01786 = 1.786% raises questions about the validity of NASA's safety assessments [4, 5].

Clearly, risk assessment is a difficult task. Each component in a complete system must be evaluated to ascertain its failure rate. Then the components are placed in either a series or parallel arrangement, or some combination of the two, to model the system prior to determining its reliability. The system reliability may be lower or higher than the component reliability depending on the number and the arrangement of the components. Testing to determine component reliability is possible in some instances when the components are relatively inexpensive. However, establishing reliability estimates by testing requires a very large number of trials that often destroy the component. If you want to show a probability of failure less than P_f = 1/1,000, you must test considerably more than 1000 components to failure. Obviously, it would not be possible to test more than 1,000 booster rockets when the cost of a single test is several million dollars.

Often the probability of failure of a component must be estimated based on previous experience with similar applications. In some instances, it is possible to compute the probability of failure; however, these calculations require considerable knowledge of the spectrum of loading and the ability of the material to resist fracture. When analytical methods are inadequate and engineering judgment is required to assess the probability of failure, the estimate should be made by a senior engineer with considerable experience and expertise. Even then, the estimate should be pessimistic rather than optimistic. A frank and honest estimate of P_f, based on all of the data and knowledge available, is much better than unrealistic appraisals that give an unwarranted feeling of safety.

14.6 RECOGNIZING RISKY ENVIRONMENTS

When designing a product, there are usually risks involved in either the production of the product or in its use in the marketplace. What can you do to minimize the risk? In previous sections, the concept of safety factor, component and system reliability and evaluating the risk were briefly discussed. There is another component to minimizing risk. Namely, developing an ability to recognize the range of hazards involved. You must clearly recognize potential hazards before taking the necessary precautions in your designs to minimize the risks associated with them.

Human factors

One of the most important and often forgotten areas of safety engineering is human factors. As an example, in 2005 an ATR 72 turboprop aircraft crashed near Palermo Italy killing 16 people. The plane ran out of fuel because the wrong fuel gauges had been installed. The manufacturer had created totally interchangeable fuel gauges for two different aircraft, the ATR 72 and the smaller ATR 42. In addition, the system had no redundancy; the low fuel indicator ran through the same gauge. No technical precautions were taken to make sure that a mechanic did not install the wrong gauge. It is critical that all engineers design products that can be safely used by the real people who will encounter them, some artificially defined "expert user" who never makes mistakes, reads all instructions and has the eye of an eagle and the touch of an artist.

Human factors engineering is a discipline in itself. Especially in consumer products, reasonably foreseeable misuse of the product is a design criterion. In industrial products care must be taken to avoid confusion, mistakes and information overloads. A mass of warning labels and fine print added at the last minute often indicates poor design and inadequate system safety analysis.

14.7 A LISTING OF HAZARDS

Mowrer describes a complete list of hazards and provides an extended discussion of each in reference [6]. You are encouraged to read this chapter and to use the extensive checklists incorporated in his coverage. A very brief excerpt from Mower's reference is provided in this section to assist you in recognizing the many hazards that should be considered when designing a product. This list of hazards includes:

Dangerous chemicals and chemical reactions
Exposure to electrical circuits
Exposure to high forces or accelerations
Explosives and explosive mixtures
Fires and excessive temperature
Pressure
Mechanical hazards
Radiation
Noise

Dangerous Chemicals

Chemicals can be nasty and you must appreciate the extreme dangers of exposures to certain toxic substances. Have you read about the leak that developed in a storage tank in Bhopal, India, in 1984? The storage tank contained the toxic chemical methyl isocyanate used in the manufacture of pesticides. It was a significant leak with about 80,000 pounds of the chemical released to the environment. Three thousand nearby inhabitants were killed, 10,000 permanently disabled and another 100,000 injured. The cause of the failure was not in the design of the tank, but in the training of the individuals responsible for the plant maintenance and operation. Nevertheless, a catastrophic accident occurred because several workers and managers, in positions of responsibility, did not adequately understand the dangers of this very toxic chemical.

Exposure to Voltage and Current

What about exposure to electrical circuits? Almost everyone has been shocked by the standard 120 volt, 60 cycle electrical power supplied by the local utility company. Why worry? You should worry because electrical shocks are **dangerous**. **Yes!** Even the 120-volt supply can cause major problems. Your body acts like a resistor and limits the current flowing from the voltage source through your hands, arms, legs, etc. The problem is that the resistance of your body is variable. It depends on the moisture on your hands, the type of soles on your shoes and even the moisture on the ground. If your hands are dry, your shoes have rubber soles and you are standing on a dry floor when you touch one wire of the circuit with only one hand, then you will probably not be harmed because you have arranged a very high resistance path to ground. Consequently, the current flow through your body will be very small. However, if your hands are wet and you touch both of the wires (the white and the black) from the electrical supply, one with the left hand and the other with the right hand, you have placed 120 volts across your heart. You can be electrocuted with only 120 volts under these conditions. The moisture on your hands greatly reduces the effective resistance of the body.

Do not take chances with electricity. Insulate the operator from the circuits preferably with two independent layers of insulation. High voltages are even more serious than low voltages because the currents flowing through one's body increase dramatically. When the current flow through the human body increases to about 10 to 100 mA (milli-amperes), there are very serious consequences to the respiratory muscles. Higher currents of 75-300 mA produce problems with a person's heart function.

The electrical current I flowing through the body can increase in two ways: first, by increasing the applied voltage V, and second, by decreasing the resistance R offered by the body. Ohm's law gives the relation among the voltage V, current I, and the resistance R as:

$$I = V/R. \tag{14.18}$$

Using adequate electrical insulation in the design of products with electrical power, markedly increases the resistance R and decreases the current I to negligible amounts.

Some people think that low voltages (5 or 10 volts) are safe because they barely feel a tingle when touching a low voltage circuit. However, some low voltage circuits particularly on high performance computers carry substantial (100 or more ampere) currents. If you short a circuit with high current flow, an arc occurs which generates significant amounts of heat. Also, the flash of the arc can damage a person's eyes and the heat may produce serious burns. In your designs, insulate and/or shield even low voltage circuits if they carry significant currents.

The Effects of High Forces and Accelerations

High forces and high accelerations (or decelerations) go hand in hand. Newton's second law requires the connection ($\mathbf{F} = \mathbf{m}\mathbf{a}$). If the brakes are suddenly applied in an automobile, the car decelerates quickly and the passenger (without seat belt or air bags) is thrown into the windshield. You must always be concerned with acceleration or deceleration because of their effects on the human body. Military pilots are trained to withstand high accelerations (high Gs). A good pilot with a well-designed flight suit can pull 7 or perhaps 8 Gs before losing consciousness. However, a civilian will become irritated at less than 2 Gs. If you want to feel an acceleration thrill, go to an amusement park and ride the roller coaster. There is no need to incorporate high accelerations into the design of most new products. A reasonably hot sports car that accelerates to 60 MPH in six seconds requires an acceleration of only 0.46 Gs. You should be careful to keep both acceleration and decelerations low when you design products that move and accelerate.

Explosives and Explosive Mixtures

With the bombing of the U. S. S. Cole, the disaster at the Federal building in Okalahoma City and terrorists bombings at many locations, most people have become aware of the disastrous effects of large explosions. The gas pressures generated by the blast destroy very substantial buildings, break glass over a very large region and kill or injure many people. The population is generally aware of the characteristics of common *detonating* explosives like dynamite and ANFO (ammonia nitrate and fuel oil). In most designs, you do not encounter a need to accommodate these traditional explosives. What you must recognize are other less apparent agents that act like explosives under special circumstances. Fuels such as natural gas, propane, butane, etc. can leak and combine with air to produce a *deflagrating* explosion when ignited. Boating accidents are common when gasoline leaks in an engine compartment. The resulting mixture of gasoline fumes and air can explode when subjected to a small spark, destroying the boat and killing or injuring the passengers. Still another unusual source of fuel for an explosion is dust. When handling large quantities of a combustible solid, dust (fine particles) is generated. If these particles are suspended in air, the resulting mixture will certainly explode when ignited. A grain elevator exploded in Westwego, Louisiana, in 1977 killing 35 people when an explosive mixture of combustible particles (dust) from the grain and air was ignited. More recently on June 8, 1998, the De Bruce grain elevator in Haysville, Kansas exploded killing several workers.

Fires

Over a million fires occur in the U.S. every year. Some are vehicle fires (400,000) and others are structural fires (650,000). Many people die in the fires (4,700) and many more are injured (28,700). Clearly, fire is a serious problem. People are killed, injured or traumatized and property is lost (eight or nine billion dollars per year).

Engineers have to distinguish between the causes of fire *ignition*, which may be natural, negligent or even intentional and the cause of fire *disasters* which is often the result of the failure to take appropriate precautions in the design of products, structures and systems. One major problems that has been identified is the over reliance on inappropriate regulatory tests. Small scale *ignition* tests for example, are totally unsuited for evaluating the contribution of a material to an existing fire. Many serious tunnel fires have resulted from the improper classification of materials using such ignition tests.

Engineers have a responsibility to decrease the number of fires resulting from the products that they design. They need to understand the source of ignition e.g. did these fires start with an appliance or a motor that overheated for some reason? Determine the reason for the overheating and redesign the product so that excessive heat will not be generated. They also need to understand why and how a fire spreads. Some feel that as a nation, we are far too casual about fires. Carelessness in personal practices, poor design in products intended for the home and business and fraud (arson) to collect fire insurance reimbursement for lost property are routine.

Pressure and Pressure Vessels

Pressurized fluids are used for many good reasons, and in most cases, pressure vessels (the containers that hold the fluids) perform very well. It was not always the case. In Boston during the winter of 1919, a huge tank about 90 feet in diameter and 50 feet high fractured. It contained two million gallons of molasses that flooded the local area. Twelve people died and another 40 were injured in this accident.

It is relatively easy to design a pressure vessel today. In fact, the American Society for Mechanical Engineers (ASME) has developed a code that engineers follow in their design to produce pressure vessels that are certified as safe for service. However, on occasion, tanks fail in service. The difficulty is usually with the steel plates that are welded together to manufacture the tank. The steel employed in both the plates and the welds have to exhibit high fracture toughness at low temperatures, and the welds must be free of large flaws. If you have the responsibility of designing a pressure vessel, follow the ASME code, make certain the welding procedures followed in manufacturing produce crack-free welds, and specify steel for the plate and welding rods that is sufficiently strong and fracture resistant. If you intend to become a mechanical engineer, you will have the opportunity to learn how to design pressure vessels, select suitable materials for their construction and specify welding procedures in courses presented later in the curriculum.

Mechanical Hazards

Mechanical hazards are features, which exist on products that may cause injury to someone nearby. Consider as an example an electric fan used to cool a room. Is the fan blade adequately guarded, or is it possible for someone to stick his or her finger into the rotating blades? Suppose the cabinet you recently designed to hold a special tool has a sharp corner at hip level. Can someone walk into the cabinet and break skin or bruise a hip on that corner? You have designed a center post crane to lift material and move it over a 25-foot diameter area. Will the center post be stable under all conditions of loading or will it collapse? You have designed a new pizza machine that rolls the dough into sheets exactly 2 mm thick. Have you provided protection for the entrance to the rolls that prevent the operator from inserting his or her fingers into the rollers? You have designed a wonderful guard that prevents a person from exposing their hands and arms in operating a punch press. However, the guard is attached to the machine with two small screws. Have you used locknuts and/or safety wires to insure that the screws will not loosen during the operation of the press? Are the screws large enough to resist failure by shear? Is there any incentive to remove the guard?

There are many mechanical hazards encountered in designing equipment and products. Always examine each component, and look for sharp points, cutting edges, pinch points, rotating parts, etc. Do not count on peoples' good judgment. If it is possible to insert a finger into the machinery, even if it is a foolish act, you can be certain that sooner or later someone will do so.

Radiation Hazards

Radiation hazards are due to exposure to electromagnetic waves. The damage produced depends on many different factors such as:

- The strength of the source.
- The degree to which the emitting radiation is focused.
- The distance from the source.
- The shielding between the source and the object being radiated.
- The time of exposure to the radiation.

The radiation spectrum is divided into several different regions—very short wave length, visible light, infrared and microwave radiation. The short wave length radiation is the most dangerous (x-rays, gamma rays, neutrons, etc.) with serious health risks (cancer) for overexposures. The hazards due to UV, visible light and IR are usually due to excessive exposure where serious burns to the skin occur. For very intense radiation, even short exposures are detrimental to the eyes. Microwave radiation is absorbed into the body and may result in the heating of one's internal organs. Because the means used by organs to dissipate this heat is not known nor is the effect of the localized increase in temperature known, it is prudent to avoid exposure to microwaves. In America, the Offices of Safety and Health Administration (OSHA) has issued a regulation limiting the power density of microwave energy to 10 mW/cm^2 for an exposure of six minutes or more. In designing products, where operators may be exposed to microwave radiation, shielding should be employed to reduce the power density well below the regulatory levels.

Noise

Noise levels that occur in the environment may be damaging to your hearing, may interfere with your work or play and may degrade the quality of your life-style. Most noise is man-made, although occasionally a storm and Mother Nature provides the sounds of thunder and wind. If you listen occasionally to a band playing rock and roll music, there is a temporary shift in the threshold of audibility to a higher level of pressure. However, if you play in the rock and roll band almost every night for an extended period, the shift in the threshold of audibility becomes permanent and your hearing becomes impaired.

Noise is a pressure disturbance that propagates through some medium such as air. The velocity of propagation through air is 344 m/s at room temperature. The pressure disturbance is oscillatory usually with many different frequencies present. The frequency content of the pressure waves depends on the source of the sound. A note from a violin will have much higher frequencies that a note from a tuba. The intensity of the noise is measured with a sound level meter that consists of a microphone, amplifier and a display meter that provides a reading in decibels (dB).

As a general guideline, the threshold for audibility is less than 25 dB[4] before a person is considered handicapped. In addition, there is a threshold for feeling noise-generated pressure at about 120 dB and another threshold for pain between 135 and 140 dB. When designing a product, the noise level is a serious consideration. Clearly, the feeling and pain levels of noise intensity must be avoided, but what levels are satisfactory? The U.S. Environmental Protection Agency (EPA) has established standards, which provide guidance to engineers in the design of products. For example, the results presented in Fig. 14.6 show the relation among the sound pressure level, the communicating distance and the degree of speech intelligibility.

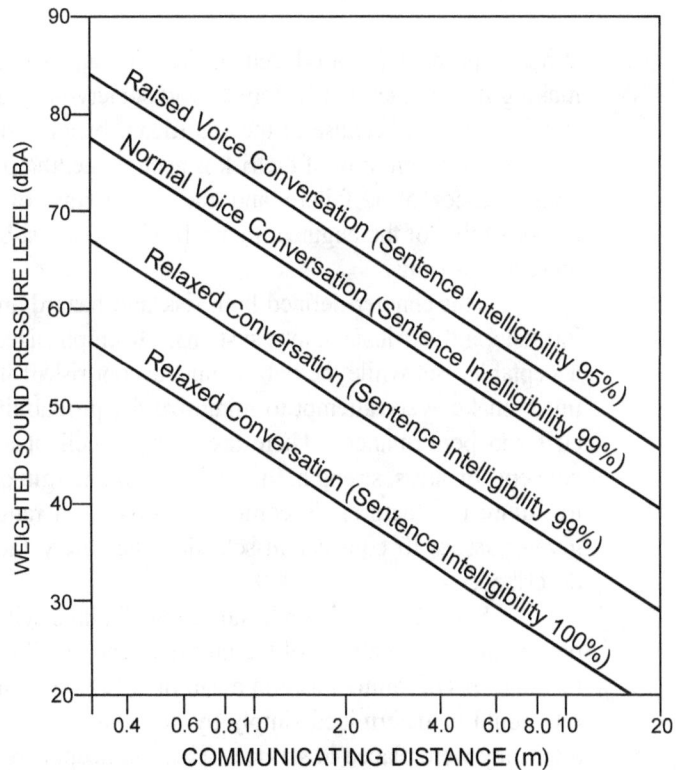

Fig. 14.6 Outdoor distances for intelligible conversation with a background of steady noise [8].

---

[4] What constitutes normal hearing differs from one authority to another. A reference by the American Academy of Ophthalmology and Otolaryngology was used in writing this section.

14.8 PRODUCT LIABILITY

Products liability is the legal doctrine that imposes responsibility on designers, manufacturers and sellers of products. The legal standard for product design in the USA is very high. The most typical formulation is provided by Section 402A of the Second Restatement of Torts:

(1) One who sells any product in a defective condition unreasonably dangerous to the user or consumer or to his property is subject to liability for physical harm thereby caused to the ultimate user or consumer, or to his property if:
   (a) the seller is engaged in the business of selling such a product, and
   (b) it is expected to and does reach the user or consumer without a substantial change in the condition in which it is sold.
(2) The rule stated in Subsection (1) applies although:
   (a) the seller has exercised all possible care in the preparation and sale of his product, and
   (b) the user or consumer had not bought the product from or entered into any contractual relation with the seller.

Suffice it to say that the failure to design for safety can be both financially and professionally disastrous.

14.9 SUMMARY

When a product is introduced in the marketplace, there is usually some risk involved to the workers making the product, the customers and society-at-large. Hopefully, the risks will be small and acceptable to all concerned because of the significant benefits produced for the customers and/or the sociotechnical system. In determining if the risk is acceptable, the public needs an accurate and honest assessment of the consequences of a failure and the probability of the occurrence of a serious accident.   It is the responsibility of the engineering profession to assist business administrators and governmental agencies in these assessments.

This chapter defined both risk and hazard and introduced the distinct problem of designing both intrinsic and extrinsic safety systems. It emphasized that ultimately society decides what risks will be acceptable. But while the public must accept risk if it is to have the benefit of certain products, engineers must make every attempt to minimize the probability of harm within socially acceptable constraints on cost and performance. There are many excellent engineering methods for ensuring safe design. The concepts of stress, strength, safety factor and margin of safety illustrate an approach to minimizing risk due to failure by fracture. Recommendations for a range of commonly accepted safety factors have been given. Issues to consider in selecting the safety factor to employ when sizing components have been described.

Sometimes it is understood that failures will occur in service. In these cases, it is necessary to determine the probability of the failure event. To illustrate methods for determining probability of failure, the concepts of failure rate and mean time between failures were introduced. It was shown that the failure rate could be determined simply by keeping records of the failures of components as a function of time after they were placed in service. These concepts are important, but they should not be confused with the probability of failure or survival.

Component reliability was introduced, and a method for computing reliability from the failure rate was shown and demonstrated. In developing this reliability equation, Eq. (14.12), the failure rate was assumed to be constant over the service life. It is important to observe that the probability of survival decreases as the service life is increased and the probability of failure increases with time in service. For reliable service for a long period of time, the failure rate must be extremely low.

System reliability is different than component reliability. A system is usually composed of many components. If the components are arranged in series, they must all function for the system to operate correctly. The reliability of a series connected system is determined from Eq. 14.13. It is evident from the data presented in Table 14.2 that the number of components arranged in a series markedly lowers the reliability of a system. Sometimes a system must function all of the time. If this is the case, a system failure cannot be allowed, because of its negative consequences. In these instances, engineers design with a number of redundant components arranged in a parallel system. Redundancy improves system reliability, as shown in Table 14.3, but at increased cost and the requirement for more power, weight and size.

The explosion of the space shuttle Challenger and more recently the disintegration of Columbia were described to illustrate the importance of evaluating the risk. It is often a very difficult problem and an analytical solution for the risk is often not possible. Nevertheless, it is a professional responsibility to prepare an intelligent, accurate, honest and frank estimate of the risk, and to insure that all of the principals involved are aware of the consequences of a failure.

Finally, a number of hazards that cause injury or death have been introduced. Unfortunately the listing is relatively long. It is important that you recognize the hazards and be vigilant in your designs to avoid them. Avoidance often does not require extensive calculation from elaborate formulae, but rather a detailed assessment of each component in the system and a good measure of common sense.

REFERENCES

1. Dally, J. W, P. Lall and J. Suhling, Mechanical Design of Electronic Systems, College House Enterprises, Knoxville, TN, 2007.
2. Lewis, R. S. Challenger: The Final Voyage, Columbia University Press, New York, NY, 1988.
3. Feynman, R. "Personal Observations on the Reliability of the Shuttle," Appendix to the Presidential Commission's Report, Ayer Co., Salem, 1986.
4. Mowrer, F. W., Introduction to Engineering Design: ENES 100, McGraw Hill, New York, NY, 1996, pp. 149-172.
5. Whoriskey, P., "Shuttle Failures Raise a Big Question," Washington Post, February 10, 2003, p. A-9.
6. Gugliotta, G and R. Weiss, "Dangers of Gauging Space Safety," The Washington Post, February 17, 2003, p. A-14.
7. Cunniff, P. F., Environmental Noise Pollution, Wiley, New York, NY, 1977, pp. 101-115.
8. "Information on Levels of Environmental Noise Requisite to Protect health and Public Welfare with an Adequate Margin of Safety," EPA Report, March 1974.
9. Powers D. G. and J. Proctor, Editors, Lockheed L-188 Electra, World Transport Press, April 1999.
10. Brannigan, G. P. Corbett, Brannigan's Building Construction for the Fire Service, 4th Edition, National Fire Protection Association, 2007.
11. Tuninter, "The ATR 72 Accident: Safety Recommendations," Addressed to European Aviation Safety Agency, EASA.

EXERCISES

14.1   Have you been involved in one or more automobile accidents since you began to drive? Were you or anyone else injured or worse yet killed? Estimate the total mileage you have driven over this period and calculate your accident rate (number/mile). What were the reasons for the accident or accidents? Comment on your driving behavior and its influence on the accident rate.

14.2   Are the benefits of driving worth the risks of injury or death? Determine your probability of being killed in a fatal accident while driving this year. Hint: Statistics on fatal accidents are always listed in the World Almanac and on the web site for the NHTSA. Make any assumptions necessary for your analysis, but justify each of them.

14.3   Your co-worker designs a tie rod (a tension member) from a 20 mm diameter steel bar with a strength of 400 MPa. If the team leader has indicated that he or she wants to maintain a safety factor of 2.4, determine the maximum load that can be applied to the tie rod.

14.4   If a component is designed with a safety factor of 2.3, what is its margin of safety?

14.5   If engineers are so smart and have all of these great computer programs to determine stresses, why is it necessary to specify a margin of safety or a safety factor when designing a structural member?

14.6   NASA's space shuttle utilizes two booster rockets fueled with solid propellant to provide the thrust required for launch. Is this a redundant system? State your reason for this conclusion.

14.7   If the probability of failure of one of the solid propellant, booster rockets on the space shuttle is 1/1,000, determine the reliability of the solid booster rocket system.

14.8   You are designing a very large computer system to contain the database for a world-wide reservation system. It is estimated that at any instant 500 operators will be accessing the database, and another 4,000 operators will soon be ready with their requests for computer availability. The mainframe computers that you plan to employ each have a MTBF of 8,000 hours. Present a design of the computer system that will insure that all 500 operators have a 99.9 % probability of being served. In your design show all calculations and carefully list your assumptions. Justify the costs involved if each mainframe computer employed in the system is valued at $200,000.

14.9   Janet is an engineer in the transmission department of the Fink Motor Corp. Her job is to record data on the mileage prior to failure of the new lightweight transmission that has been placed in 250,000 of the new 2008 model of the Clunker III. She records the following data:

Mileage (1000 miles)	No. Failures, N_f
0 - 5	65
5 - 10	38
10 - 20	20
20 - 30	18
30 - 40	22
40 - 50	34
50 - 60	90
60 - 70	521
70 - 80	1,364
80 - 90	6,677
90 - 100	15,592

Determine the failure rate as a function of service life expressed in terms of mileage, and prepare a graph showing your results.

14.10 Using the data from the table in Exercise 14.9, determine the probability of survival and the probability of failure of the transmissions as a function of service life measured in terms of mileage. Prepare a graph of your results.

14.11 If the transmissions in the Clunker III were under warranty for 100,000 miles, what would be the consequences for the Fink Motor Corporation?

14.12 Prepare a listing showing the ratio of fatal events per flights for the airlines in the U. S. and Canada since 1980. The data needed to prepare this listing is at http://airsafe.com/airline.htm.

14.13 Based on the results of Exercise 14.12 cite the three airlines with the best safety records. Cite the three airlines with the worst safety records.

14.14 Will the results of Exercise 14.13 affect your planning for air travel in the future?

14.15 Prepare a paper describing the top ten fatal air transport accidents. The Website listed in Exercise 14.12 contains the data necessary for you to write this paper.

14.16 Prepare a paper discussing your reaction to the risk assessment of the space shuttle system by NASA. Give arguments for and against NASA's position and practices regarding safety of the shuttle and its crew both before and after the Challenger and Columbia accidents.

14.17 Describe an incident when you received an electrical shock. Something obviously went wrong to cause this incident. Please indicate the problem. Was there a design deficiency? What could be done to prevent the incident from reoccurring?

14.18 You are designing a cabinet-mounted, self-cleaning oven that requires very high power levels to heat its interior surfaces until they are free of all the splattered, burnt-on grease. As the lead engineer on this design team, what precautions should you take to insure that the oven would not be the source of a fire during its anticipated 15-year life?

14.19 Automobiles are the most commonly used mode of transportation in the U. S. and many other developed countries. Describe a least one-design flaw pertaining to safety of many sports utility vehicles. What government agency is charged with the responsibility of insuring safety of automobiles in America?

14.20 There are many fatal accidents each year involving crashes with large trucks. The Federal Motor Carrier Safety Administration (FMCSA) is the government agency responsible for regulations involving trucks. Write a paper describing possible rules the agency could impose on the trucking industry to reduce the number of fatal truck crashes. Cite reasons for the new rules and predict the benefits that would result if they were implemented. Discuss the economic consequences for your suggestions.

CHAPTER 15

ENGINEERING ETHICS AND DESIGN

With Vincent M Brannigan

15.1   INTRODUCTION TO ETHICS

15.1.1   Professional, Organizational and Personal Ethics

Technology represents the potential for both benefit and harm to society. Engineered products can kill or injure large numbers of people, either "all at once" in catastrophic injury or one at a time in "routine" injuries. Technological development is often "asymmetrical" in that those who benefit from the technology may not be those at risk of injury. Because the developers and users of technology may not internalize the full costs of injuries, they may have little incentive to avoid potentially harmful innovations. The "public" safety problem of engineered products has therefore always been a matter of social, not merely private concern. Before society fully accepts a technology, the public typically demands guarantees that public safety will be protected. Society often creates safety regulations to help protect the public.

But while regulation may be one method of protecting public safety, there are many reasons why it is inadequate. Technology develops with time and there is a need for flexibility in the regulations. If engineers can be trusted to design for safety, society can delegate responsibility for product design. This delegation requires an assurance that public safety will be protected by the engineers without governmental supervision, acting as "professionals" serving the public interest. This concept is the foundation for professional ethics in engineering design. **The Code of Ethics approach does not fully substitute for the legal system, but it provides an important complement to an often overworked, rigid or otherwise inadequate regulatory system.**

Public safety is of course only one of the ethical issues in Engineering, but it can be used to explore many of the most difficult problems of ethics and the practice of engineering. Ethics is a very complex and culturally driven subject. On a **personal** level ethics deals with issues of character, loyalty, honesty, decency and all the other virtues that make interpersonal dealings more comfortable and efficient. On an **organizational** level it describes the informal and formal rules that govern the internal functioning of an organization, which in turn create the "public face" for the organization. Finally on a **professional** level it describes the mutual expectations of those who practice a given profession and the acceptable public expectations which control the profession.

This chapter will primarily focus on **engineering professional ethics** in the design process. This in no way detracts from the importance of other ethical issues for engineers. Many aspects of personal ethics are inherent in the professional ethics of engineers, so the importance of those traits is recognized. The engineer's responsibility within an organization is also part of professional ethics

15.1.2 Professions and Ethics

Professions can be defined as groups of trained individuals who mutually agree to enforce standards designed to serve a "moral" ideal

> *Moral—relating to the practice, manners, or conduct of men as social beings in relation to each other, as respects right and wrong, so far as they are properly subject to rules[1].*

The origin of the concept of professional ethics is generally attributed to physicians. The "oath of Hippocrates" in its many forms is thought to establish the beginnings of a concept of professional Ethics. For example the oath states

> "I will prescribe regimens for the good of my patients according to my ability and my judgment and never do harm to anyone."

This obligation, to keep the **patient's** well being at the center of medical practice, establishes a far different relationship between physician and patient than between seller and customer. Attorneys also developed canons of ethics that governed their actions. The clear goal of such canons was to ensure that their public responsibility outweighed their private interests. As early as the 13th century attorneys in Britain took the Lawyer's oath.

> *You shall doe noe Falshood nor consent to anie to be done in the Office of Pleas of this Courte wherein you are admitted an Attorney. And if you shall knowe of anie to be done you shall give Knowledge thereof to the Lord Chiefe Baron or other his Brethren that it may be reformed. You shall Delay noe Man for Lucre Gaine or Malice*
> > *You shall increase no Fee but you shall be contented with the Old Fee accustomed. And further you shall use your selfe in the Office of Attorney in the said office of Pleas in this Courte according to your best Learninge and Discrecion.*

This oath, which is the ancestor to all those used by attorneys in the USA today, establishes the key components of professional ethics:

1) A professional will be personally honest, and insist that all they deal with be honest.
2) When unethical conduct is found, the professional will "blow the whistle" on the offender.
3) The professional will give those matters, entrusted to his or her care, his or her best professional effort.
4) The professional will deal fairly with the public in carrying out professional duties.

It is very important to understand that a professional often has defined duties owed to persons other than those who are paying the person's salary or fee. Professionals can owe duties to other specific persons or even the public as a whole. These may be described as the *beneficiary* of the professional's actions. In law and medicine for example ethics fundamentally defines the *beneficiary* who is owed the highest duty.

---

[1] Webster's dictionary.

The military lawyers who are defending the "unlawful combatants" in Guantanamo are upholding the highest ethical responsibility a lawyer can have—the obligation to protect your client no matter who is the employer. Similarly corporate physicians who protect the safety of research subjects against the financial interests of their employers are acting in accordance with an ethical code that is over 2,000 years old. Lawyers and physicians take enormous pride in these professional standards and feel betrayed when a member of their profession "lets down the side", as President Clinton did when he misled a court.

A professional cannot do something which violates ethical obligation to the beneficiary even if the employer wants it. A professional is given special privileges in return for a commitment to the public interest, not merely the person's private interest. When a profession is accepted by society, the public is invited to depend on the commitment of professionals to their ethical norms, usually expressed in codes of ethics. Under this "Professional Codes of Ethics" approach, the commitment to the public interest is enforced by professionals themselves.

15.2 ENGINEERING AS A PROFESSION

Doctors and lawyers are clearly professionals, but what about engineers? Do engineers actually accept this same understanding of their role? Do they see any obligation to protect the public from the decisions of their employers or paying clients? Or are they merely consider themselves highly trained technical employees, who owe the public no special duty? Do they think that if they comply with all rules and regulations are they acting ethically even if the product is dangerous?

The answer is not easy. Engineering is clearly an emerging profession, and the concept of serving the public as opposed to private interests is by no means accepted around the world. However, at least in the U. S., the claim of engineers to professional status and commitment to public safety has been made for many years. As one court stated it:

> **An obligation to follow good engineering practice (as distinct from the FDA's Good Manufacturing Practices) may arise for the unlicensed engineer, as a professional obligation. The Court views such professional obligations—as the FDA undoubtedly does—as inherent in all medical manufacturing. [1]**

15.2.1 The History of Engineering Ethics in the USA

Engineering codes of ethics in the USA date to just before the First World War. In the beginning, many such professional codes were anti-competitive and primarily designed to enhance the dignity and income of the profession. However, by the 1970s engineers were clearly making the claim to true professional status directly connected with service and obligations to the public. The bedrock public interest principles of professional engineering ethics were found in statements such as the Canons of Ethics of the Accreditation Board for Engineering and Technology (ABET). Among the most important are the Fundamental Canons:

Engineers, in the fulfillment of their professional duties, shall:

1. Hold paramount the safety, health and welfare of the public.
2. Perform services only in areas of their competence.

These Canons acting together require engineers to both fully understand the products and systems that they are designing and insure those products meet a high standard of public safety. They have become widely accepted around the world. As one author describes Engineer's obligations:

> **Protect the health and safety of citizens in the host country**. The requirement to hold paramount the health and safety of the public is explicitly recognized in most engineering codes, and so it must have the status of a universal norm for engineers.[2]

The profoundly difficult task engineers set for themselves in these Canons should be recognized. Arguably neither attorneys nor physicians make such a sweeping commitment to "hold paramount" the general public interest. Attorneys and physicians are taught that they are normally serving the public interest by devoting their honest and ethical skills to their client or patient. The assumption is that the benefit to the public is derivative of their professional treatment of the patient or client.

Engineers have a much more complex problems, because the beneficiary of their actions is supposed to be the public but the "paying client" is often be the party pressing them to design some product that puts the public at risk. This requirement unquestionably puts the engineer in a difficult position, because unlike the doctor or lawyer they do not have a beneficiary sitting across the room watching their actions. They also may lack strong peer support in protecting the public.

Engineered products routinely function correctly within their intended design envelopes. This is a tribute to the skill and ethics of the designers. Catastrophic failures due to failures in the process of engineering decision making are fairly rare. But when such failures do occur, they force a reexamination of the assumptions and processes that produce the decisions and especially the ethical issues involved.

One peculiarity of Engineering in the United States is that some Engineering disciplines are primarily represented by state licensed professional engineers who are permitted to put the terms PE after their names. These tend to be fields in which engineers work as "consultants "for "clients" or there are regulatory requirements for PE approval. In other fields Engineers tend to work as employees for large firms and PE status is rare or unknown. The question might be raised whether this changes the ethical responsibility of the engineer. This is an ongoing debate. Many but not all non licensed engineering societies have also adopted codes of ethics.

Engineers have a vested interest in preserving and enhancing the confidence of the public. To gain public confidence, it is essential that they all behave in an ethical manner. All three types of ethics are important. **Personal** ethics set a foundation and because engineering developments are sponsored, financed and controlled by businesses, **organizational** ethics are critical. Finally, compliance with **professional** norms is the last and most important step.

15.3 RIGHT, WRONG OR GRAY

Christina Sommers [2] develops an interesting viewpoint regarding what is considered right, wrong and controversial by society-at-large. She argues that there are some ethics related issues that are not clearly defined, and that society has not reached a consensus regarding the correctness of these issues. You can develop cogent arguments for and against a specified viewpoint for controversial issues such as abortion, affirmative action, handguns, gambling, capital punishment, nuclear weapons etc. Sommers refers to these controversial issues as "dilemma" ethics that should be distinguished from "basic" ethics. With "dilemma" ethics, society-at-large is not certain about what is right, wrong or gray. There are several

---

[2]Charles E. Harris^ Jr., Internationalizing Professional Codes in Engineering Science and Engineering Ethics (2004) 10, 503-521.

attitudes prevalent in society—sometimes the issue is so important to segments of our population that demonstrations are organized to elevate one viewpoint or another. Occasionally a group can get its ethical viewpoint enacted into law, but without the full backing of the majority of the society such a law will often fail.

These dilemma issues reflect the public ethical face of a society. The term "moral ecology" can describe the overall environment in which a profession functions. In some instances, society changes its moral ecology and its law. An example of a failed moral change in this country was Prohibition, which was enacted by a constitutional amendment ratified in 1919. The government wanted to make it illegal to consume alcoholic beverages, but society-at-large wanted to drink. Bootlegging, racketeers and speakeasies followed bringing crime and violence to many neighborhoods. When it became evident that the law was causing more problems than it solved, it was repealed in 1933.

From an engineering standpoint, dilemmas include topics such as nuclear weapons development or response to global warming. These are essentially political decisions which engineers have the right and duty to influence as citizens and where they may provide specialized input, but the profession neither has or claims any unique insight. No attempt will be made in this textbook to resolve any of the "dilemma" issues. These issues have been unresolved for decades. These issues will require much more time and understanding before our society is willing to reach the consensus necessary for resolution. It is much more productive to consider "basic" ethics where right and wrong or white and black are much more clearly recognized by the majority of society.

15.4 PERSONAL ETHICS

Many people are confused about the value of personal ethical behavior. The results of several polls indicate that Americans are less ethical today than in previous generations. Sixty-four percent of a sample of 5,000 individuals admits to lying if it does not cause real damage. An even larger percentage of those sampled (74%) will steal providing the person or business that is being ripped off does not miss it [3]. Many students (75% high school and 50% college) admit to cheating on an important exam [4]. Most students have not developed a moral code to use as a guide for their behavior.

Virtues

To understand personal ethics we can start with the four traditional cardinal virtues[3]:

- Prudence
- Justice
- Fortitude
- Temperance

Combined together these virtues have been considered the foundations of "right living" in Western societies for the past 2,300 years.

Prudence refers to careful forethought, good judgment and discretion. In other words, think before you act and exercise care and wisdom. If you act, will your action cause problems, now or later, for you or anyone else? Are you careful about your remarks concerning others? One of my rules is not to

---

[3] These virtues are of course described by their western European traditional names. However, all are represented in both Islamic and Eastern philosophies

speak ill of others under any circumstances. Is the action that you are about to undertake in your best interest? Will that action be appreciated by your family, friends, coworkers, managers and peers?

Justice involves fairness, honor, giving everyone their due. It is clearly wrong to lie, cheat or steal. Justice is not the same thing as legality. Actions may be lawful but unjust, such as slavery.

Fortitude is about courage and persistence. Will you stay with an idea, even if it is not popular, if you know that it is the "right" approach? Showing determination and resolution is sometimes difficult when the risks of failure are high or when peers encourage you to abandon your idea. Fortitude also discourages rashness

Temperance refers to moderation in pleasure seeking. Temperance can also mean control of human passions such as anger, lust, hostility, revenge and spite.

Most individual behavior can be judged by very clear and well-understood virtues that are part of "basic" ethics. Many institutions teach and counsel compliance with these norms as the basis for human interactions.

15.5 ETHICS AND LAW

Governments (federal, state and local) also play a role in defining ethical behavior, because they generate laws and set public policy. Laws (mostly at the state level) are written on the basis of the "police power", the general power to protect the health welfare and safety of the people. Moral or even religious values routinely influence the creation of law. While notions of just and unjust laws are beyond this chapter, the role of ethics in the creation and enforcement of law is very important to engineering design.

The purpose of most laws is to prohibit a well-defined set of vices. For example, it is illegal to sell drugs, rob banks, commit burglary, kill or abuse another human, or engage in prostitution. It is also illegal to park overtime at a parking meter or open a business without a license. However these violations are not the same type of unlawful act. Legal scholars describe two different types of unlawful acts based on essentially ethical considerations.

> **Malum in se** is Latin for "evil in itself" and would normally apply to acts which are universally described as morally and ethically wrong. Such laws would derive from principles of right and wrong and violations are considered as involving "moral turpitude."

> **Malum prohibitum** is Latin for "wrong because it is prohibited." These are acts which are unlawful but not otherwise immoral. Violation of a copyright for example is usually considered malum prohibitum.

Personal ethical violations tend to resemble offenses under **malum in se** concepts.   Rape murder and robbery are all malum in se. Professional ethical violations, such as contempt of court by an attorney, can often be considered **Malum prohibitum**. Former President Clinton, for example, was disbarred for a professional ethical violation that did not meet the more strict standard of the criminal law.

The difference is important because an ethical or legal violation can occur without the individual feeling they did anything "wrong". "I complied with all regulations"—"I was only following orders"—"I was told it was out of my job responsibility" are routine justifications heard after a disaster. But only acts which are **malum in se** require a guilty mind. "Knowing violations" are ethically easy, because they involve the "guilty mind" or **mens rea** that is routine in the law. But other acts without such a guilty mind may be just as unethical.

On the other hand, some acts may even involve legal consequences are not considered by the law to have any moral element at all. For example, in the law there is no "moral element" in breech of contract. You may have to pay damages, but you have done nothing morally wrong in the eyes of the law. Whether such conduct is ethical or not depends on the field.

The relationship between law and ethics is therefore complex. In a few cases, the legal system has overruled the profession, such as on limitations on professional advertising or restraints on competition. In other cases, such as insider trading it has added the force of law to what was already considered an ethical violation. But in the case of President Clinton professional ethical standards are normally much higher than merely "complying with the law". The physician who is legally allowed to perform an operation for which she is unqualified is acting in an unethical manner. An engineering ethical problem however can arise without bad intent due to the decentralized nature of engineering decision making.

15.6 ETHICAL VIOLATIONS WITHOUT "BAD INTENT"

A professional ethical violation can occur even if the professional does not believe they were doing "wrong". A common case is where an engineer "doesn't know what he doesn't know". Recall that the code requires:

> Engineers, in the fulfillment of their professional duties, shall:
> 1. Hold paramount the safety, health and welfare of the public.
> 2. Perform services only in areas of their competence.

For example consider the stairway at the University of Maryland's COMCAST center, shown in Fig. 15.1. The professionals who designed the northeast exit of the Comcast Center created a "funnel" that could cause a lethal crowd crush. A pinch point is obvious halfway down the left side. The engineers who approved the design simply did not recognize the hazard, possibly because their training did not include "crowd crush".

Brannigan[4] has described this ethical issue as "design process failure", and it can occur any time critical design decisions are disaggregated. In such cases, different design concepts or knowledge or assumptions in different groups combine to create unsafe products, even if no individual set out to create the safety hazard. Engineers have an ethical duty to make sure that the entire system of which they a redesigning a part holds paramount the public safety.

Fig. 15.1 Comcast Center as constructed on the University of Maryland campus. Note the two funnels created by the stair rails.

---

[4] Teaching Ethics in the Engineering Design Process: A Legal Scholars Perspective, Vincent M. Brannigan, November 5-8, 2003, Boulder, CO, 33rd ASEE/IEEE Frontiers in Education Conference, S2A-19

15.7 ORGANIZATION ETHICS

Corporations are entities that under law are treated as "juridical persons". A corporation, acting like a person, has the right to conduct business and the obligation to pay taxes. Just as individuals are judged by their ethical behavior, corporations are judged by their ethical conduct [5]. The chief executive officer (CEO) for a corporation sets the moral and ethical tone for the organization. Any large business is organized with several layers of management. If the CEO operates the business with the highest ethical standards, the middle and lower level managers will normally follow the example set at the highest level in the corporation. However, if lax ethical standards are permitted, a pattern is set and questionable ethics will become the corporate style and standard.

In an excellent paper, Baker [5] has developed a list of issues that occur daily in a typical corporation. The manner in which a corporation deals with these issues establishes its character. Baker's list is shown below:

1. How people, employees, applicants, customers, shareholders and the families who live near our facilities are treated?
2. Is everyone free of systemic or individual practices of discrimination?
3. How do we spend our shareholders' money on our expense accounts as an institution and as individuals?
4. What is the level of quality that goes into our products? Do we meet our customers' expectations for quality? Do we meet our own standards for quality?
5. What is our concern for safety, not only for our employees, but also for our customers and our neighbors in communities in which we operate?
6. How "ethically" do we compete?
7. How well do we adhere to the laws of the locales, regions and nations in which we do business?
8. What is acceptable gift giving and gift taking?
9. How honest are our communications to our employees and our public advertising?
10. What are our corporate and personal positions on public policy issues, and how do we promote those positions?

This is a long list, but we certainly could add more items to Baker's catalog of concerns—sustainability and the environment for instance. Adding more items is not as important as recognizing that corporate decisions are made on a daily basis by its managers and employees. These day-to-day decisions determine the corporate character. The ethical judgments that serve as the basis for most of these decisions will depend on individual personal values (virtues) and experience. The quality and consistency of the ethics applied in these decisions will markedly affect the success of the corporation, its management and its employees.

15.8 PROFESIONAL ETHICS FOR ENGINEERS

A personal code of ethics is the cornerstone to achieving a meaningful sense of moral values. Professional ethics builds on a well-understood personal code of ethics. Hopefully, the preceding sections of this chapter will be helpful in any attempt you make to establish a sense of right and wrong.

There are several codes of ethics for engineers. Most of the founding societies of engineering have developed and distributed codes and guidelines for professional conduct and faith statements. The codes of ethics for the different engineering societies are all similar. The fact that each society has their own code is more to insure a complete distribution of the code than to pursue unique ethical issues.

Let's consider the Code of Ethics of Engineers, presented in Fig. 15.2, which was sponsored by the Accreditation Board for Engineering and Technology (ABET). The code is divided into two parts. The first part deals with four fundamental principles, and the second part contains seven fundamental canons. The Engineers' Council for Professional Development first advanced this code in 1977. The fact that the code remains without modification for more than 30 years indicates that professional ethical values are as constant as personal ethical values.

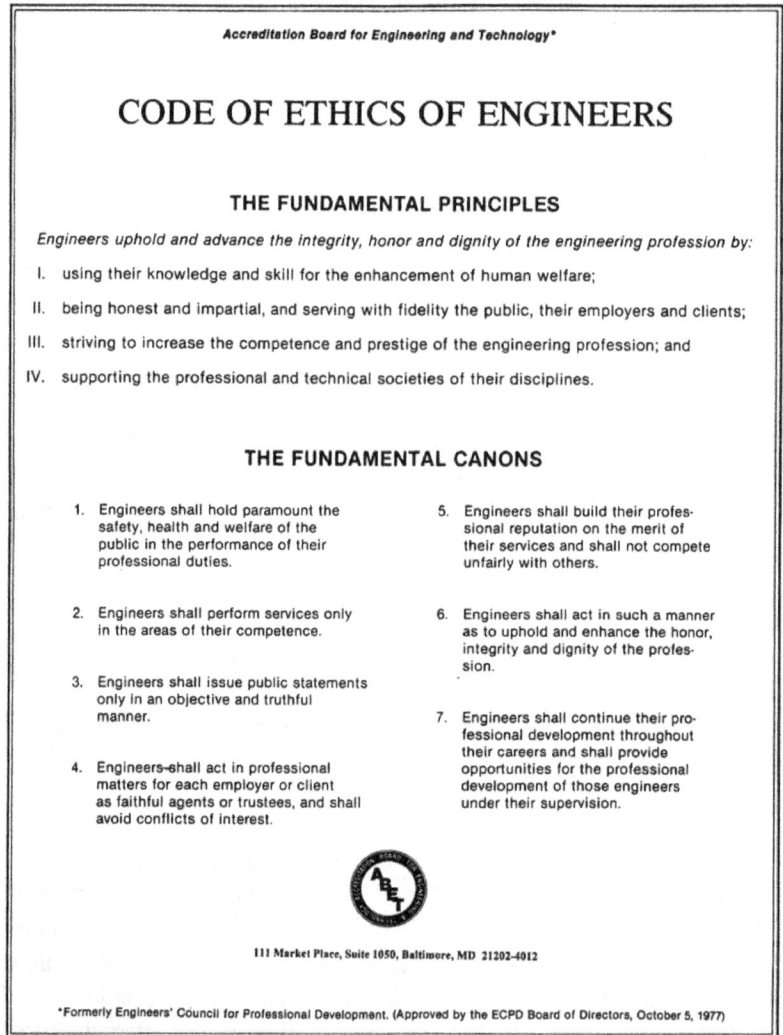

Fig. 15.2 ABET's code of ethics.

The fundamental principles seek to insure that the engineer will uphold and advance the integrity, honor and dignity of the profession. These goals are common to your personal goals that were discussed previously; however, the approach to achieve the professional goals is different as indicated below:

1. We are selective in the use of our knowledge and skills so as to ensure that our work is of benefit to society.
2. We are honest, impartial and serve our constituents with fidelity.
3. We work hard to improve the profession of our discipline.
4. We support the professional organizations in our engineering discipline.

The seven fundamental canons in the ABET Code of Ethics of Engineers, presented in Fig. 15.2, are explained in considerable detail in guidelines that are used to expand and clarify the relatively brief statements that represent the "regulations" that shape our professional behavior.

15.9   ETHICS IN LARGE ENGINEERING SYSTEMS

Engineers design and build many different products each year that are included in large and complex sociotechnical systems.   Usually these products and the systems are conservatively designed with adequate safety factors, carefully tested, and perform well in service for extended periods of time.   They provide a much needed service without endangering either individual or public safety.   However, from time to time mistakes are made in the initial design.   These mistakes are usually detected in prototype testing and eliminated prior to releasing the product to the market place.   In very rare circumstances, a mistake or several mistakes are overlooked, for a variety of reasons, and the system is released and placed in service with an unacceptably high probability for failure.   An undetected mistake often leads to a catastrophic accident.   The space shuttle system clearly falls into this category.   If you are a space supporter, you may argue that the risks are worth the benefits.   But if this is the case, an unprepared and untrained high school teacher should not be invited to ride along to enhance the image of the space program.   It is one thing to order a career astronaut into a peril, which they understand and accept, but totally a different proposition to invite an uninformed civilian to participate in a very dangerous project.

In Chapter 14, the probability of failure (or an accident) has been discussed in considerable detail. You must understand that there is always some risk of failure when designing high performance systems. It is important to learn to accept a trade-off between risk, safety and performance—it is an inherent part of the process.   In most complex sociotechnical systems (air transportation for example), there is a small but finite risk for failure with a subsequent loss of life and property.   The public must know this risk.   It is the responsibility of the engineering community, industry and the government to alert the potential customers to the dangers involved.   Knowing the risk, the customers can decide whether or not they want to use the system.

Some people worry more than others and place a very high value on safety.   They require a probability of failure of nearly zero.   Do you know someone who will not travel by flying?   (John Madden, the popular professional football announcer, travels from game to game each week on a special bus).   Other folks love the thrill of a risk, and they are willing to accept a much higher probability of an accident.   They think hang-gliding or skiing on black diamond slopes is great sport.

15.10 THE CHALLENGER ACCIDENT—A CASE STUDY

With this background on risk, safety and performance, let's begin a discussion of the Challenger accident [6].   The Challenger was one of the original four orbiters built by the National Air and Space Administration (NASA) to serve the space shuttle system.   The rocket fuel (liquid hydrogen) on the Challenger 51-L mission exploded 73 seconds after launch on Tuesday, January 28, 1986.   The crew of six and a schoolteacher passenger were killed, and the space shuttle was lost.   Among the dead was Dr Judith A. Resnik, a graduate of the Clark School of Engineering at the University of Maryland

The Challenger accident has been selected as a case study because it illustrates several ethical issues in the engineering and management of large, complex and inherently dangerous systems: Some issues that will be raised are:

1. The design of the shuttle and the selection of the contractors involved many political considerations [6].
2. The lack of communications between key people and organizations was a significant factor in the accident.
3. The interface between the upper-level administrators (business managers) and the engineers was an important element in the decision to launch on that disastrous morning.
4. Public attention and opinion, not safety, markedly affected the decision-making process.
5. The risk potential was not known by the public and not appreciated by top administrators and politicians directly involved in the launch decision.
6. Christa McAuliffe, a high school teacher and mother of two children, was killed in the accident. Why was she invited to fly on such a dangerous mission?

Background Information

To set the stage for the accident, you have to go back to the early 1970s. NASA had been very successful with the Apollo missions (even with the problems of Apollo 13), and looked forward to larger and more aggressive space endeavors. They proposed an integrated space system that would include a space station, space shuttle, space tug and manned bases on both Mars and the moon. This agenda sounded wonderful until the public examined the price tag. The public did a quick look and wanted no part of it. A poll indicated that the public believed that Apollo had been too costly. The politicians, always driven by the polls, took note and reduced NASA's budget. NASA recognized the need for a new, cost-effective project to follow Apollo that the public (and politicians) would buy. Responding to these political pressures, NASA proposed the space shuttle that would serve the military, the scientific community and the rapidly growing commercial business of placing satellites in orbit. The space shuttle system, illustrated in Fig. 15.3, was marketed as a relatively routine space transport system. Even the name "shuttle" connected with the airline shuttle services that routinely fly every hour from one large city to another.

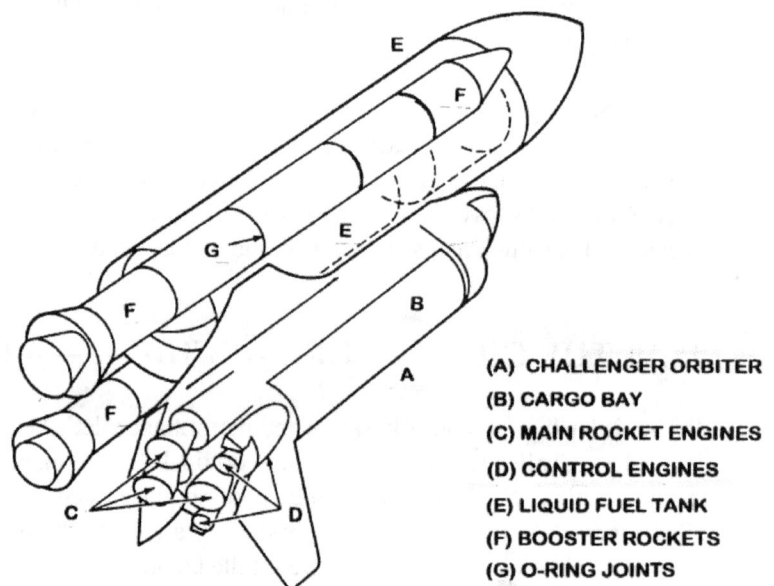

Fig. 15.3 Illustration showing the main components on the NASA space shuttle.

(A) CHALLENGER ORBITER
(B) CARGO BAY
(C) MAIN ROCKET ENGINES
(D) CONTROL ENGINES
(E) LIQUID FUEL TANK
(F) BOOSTER ROCKETS
(G) O-RING JOINTS

To make the space shuttle system a commercial success, NASA proposed a fleet of four orbiters [7], which would eventually fly on a weekly basis (NASA initially set a goal of 160 hours for the turnaround time for an orbiter). They planned on nearly 600 flights in the period from 1980 to 1991. The early estimate of the cost of a launch was $28 million with a payload delivery cost of $100 to $270 per pound. After some operational experience, the cost estimates proved to be completely unrealistic. Launches actually cost on the order of $280 million, and the cost to place a pound of payload in orbit on the shuttle was in excess of $5,200 [8]. Early experience with the space shuttle system indicated that NASA could not hold their schedules and their costs were running more than a factor of 10 higher than the original estimates. Because of its poor performance, NASA was struggling to improve its image as 1985 ended.

Prior to the launch of the Challenger on January 28, 1986, twenty-four shuttle flights had been made. These flights were all successful in that they returned to earth with all crewmembers safe. They also showed the operational capabilities of the shuttle system in placing commercial satellites into orbit, repairing satellites in space and salvaging malfunctioning satellites. However, these flights also indicated that the shuttle system had many serious problems. The three main liquid rocket engines were too fragile with many critical components. Some of the tiles in the heat shield needed repair and/or replacement after every flight. The computers and the inertial navigation systems experienced occasional failures. The brakes and landing gear were stressed to the limit when the orbiter (an 80 ton dead stick glider) landed at speeds ranging from 195 to 240 MPH. Finally, the seals in the solid fuel booster rockets showed distress (sometimes extensive) in 12 of the 24 previous launches.

The record of the shuttle from 1981-1985 showed a consecutive series of successful launches, but with many prolonged delays to repair failing components in a very large, highly stressed system. The maintenance records showed so many problems that NASA estimated it took three man-years of work preparing for a launch for every minute of mission flight time. Clearly, the word shuttle to describe such a transportation system is a misnomer.

The Solid Propellant Booster Rockets

The explosion of the main fuel (liquid hydrogen) tank on the Challenger was due to the failure of the O-ring seals on the solid fuel boosters that were adjacent to the hydrogen fuel tank. To understand the seals and their purpose, it is essential that you appreciate the size and function of the two booster rockets used to provide much of the thrust necessary for the launch. They are enormous cylinders—12 feet in diameter and 149 feet tall. Each cylinder is filled with 500 tons of a solid propellant consisting of a rubber mixture filled with aluminum powder and an oxidizer—ammonium perchlorate. When ignited, the propellant burns to produce an internal pressure in the motor case of about 450 psi (lbs/in^2) at a temperature of about 6,000 °F. The expanding gasses exit the rocket motor case through a nozzle, to produce about 2.6 million pounds of thrust from each booster. These booster rockets are essentially very **big** bottle rockets!

Big was the nub of the problem. The solid rocket boosters were made by Morton Thiokol in Utah, but launched from the Kennedy Space Center in Florida. They were much too long to ship across the country as a single cylinder. To circumvent this problem, the motor casing was fabricated in segments—each 27 feet long. The segments were filled with propellant and shipped by train to the launch site in Florida. They were then assembled in a special facility near the launch pad. This procedure solved the shipping problem, but it created a different problem. The rocket casing is a pressure vessel that must contain very hot gasses at a pressure of about 450 psi. The joints, where the cylindrical segments of the motor case were fitted together, must be sealed so that these hot gasses will not leak and cause damage to adjacent components of the launch vehicle.

The seal was made with a pair of rubber O-rings fitted over the internal finger of a clevis type of joint, as shown in Fig. 15.4. The O-rings, 0.280 inch in diameter, were compressed between the two surfaces affecting a seal that prevents the pressurized fluid from leaking past the joint. Putty like compound was used to prevent the hot gases from eroding the O-rings. The pins locked the segments together (axial constraint only), but did not clamp the clevis fingers about the center finger.

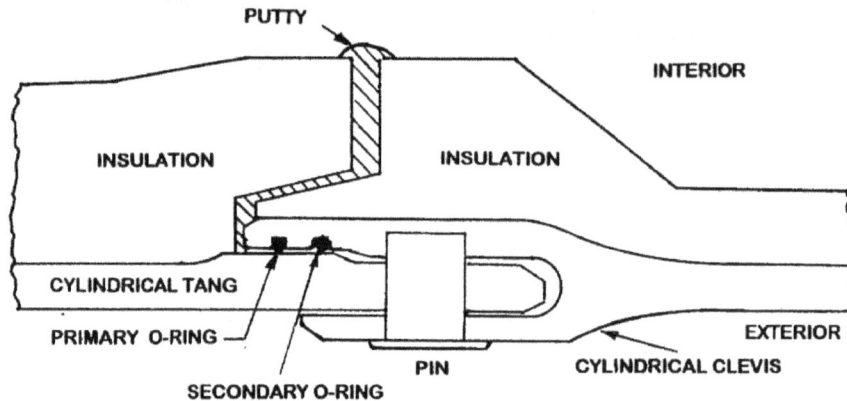

Fig. 15.4 Design of the O-ring seal used at the circumferential joints of the solid rocket motor cases.

The sealing of the segmented cylinders with the O-rings was not a new concept. A single O-ring had been employed previously on the Titan III rocket effectively sealing the circumferential joints on a smaller diameter motor case. Morton Thiokol, the contractor building the boosters, introduced the second O-ring to provide a margin of safety in the event the first O-ring failed. It all sounded feasible, and the initial test firings of the booster in Utah were apparently successful.

Unfortunately, one test does not validate the safety of even a simple system. To insure high reliability of a component, many tests are required. Moreover, this test firing had little or no bearing on problems associated with the reuse of the solid rocket boosters (up to 20 times). After splash down following a launch and recovery from the Atlantic Ocean, the cylinders were shipped back to Utah to be filled and used again.

Failure of the O-ring Seal

Motion photographs taken at the final launch of the Challenger indicated that the O-ring seals failed almost immediately after ignition of the booster. The hot gasses cut through the O-rings and the joint of the booster. A hole was formed in the wall of the booster at the joint, and a flaming jet of white-hot gasses escaping from the booster rocket cut through the wall of the adjacent tank containing the liquid hydrogen. The launch was effectively over. While many things happened in the last five seconds before the devastating explosion—all bad—the penetration of the adjacent tank, which contained liquid oxygen and liquid hydrogen, spelled the disastrous end of the mission.

It was a terrible day, and we as a nation were in shock while we mourned the loss of the crew and the schoolteacher/mother. It was sometime later, when the facts were brought to the public about the space shuttle program that we learned the shuttle should not have been permitted to fly that day. The on-site engineers understood the very high probability of failure of the O-ring seals. Moreover, knowledgeable engineers at Morton Thiokol tried without success to prevent the launch.

The story containing all of the facts about the failure of the O-ring seal is too long to be covered here, but you are encouraged to read references [6 – 11], where very complete and well-written accounts are given. The essential elements leading to the catastrophic failure are listed below:

1. The circumferential joint changed shape when the motor case was pressurized, and the gap, which the O-rings filled, increased markedly in size. A schematic illustration of the new gap geometry is presented in Fig. 15.5.

2. The new gap opening was so large that the back-up O-ring probably could not seal the joint. When the booster case was pressurized, the seal depended on a single O-ring—not two.

3. The hot exhaust gases had eroded the O-rings on 12 of the previous 24 launches indicating some leakage about half of the time, and on a few occasions, significant erosion of the rings occurred. There was also clear evidence that the putty failed to keep the hot gasses from attacking the rubber O-rings.

4. The joints moved during the early launch sequence as the orbiter engines and then the booster engines were ignited. This motion requires the O-ring seals to be flexible and to reseat and reseal continuously during these movements.

5. The temperature the night before the launch dropped to 22°F and had only increased to about 28°F by the time of the launch.

6. The O-ring seals were not certified to operate below a temperature of 53°F by the contractor responsible for the solid rocket boosters.

7. Tests conducted by the contractor Morton Thiokol in 1985, six months before the accident, indicated that the O-ring seal was not effective at low temperatures. (At 50°F the O-rings would not expand and follow the movement in the joint during the period of operation of the booster.) In fact, at room temperature 75°F, it required 2.4 seconds for the O-rings to reseat and seal pressurized gasses after a gap in the joint was opened.

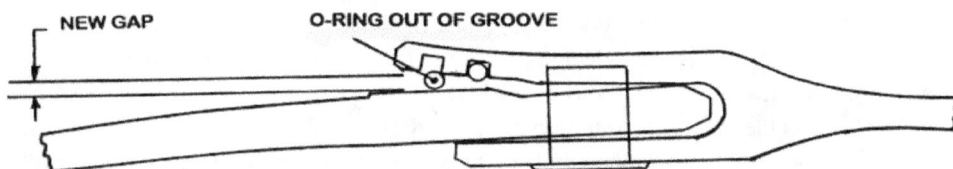

Fig. 15.5 Rotation of the joint due to cylinder pressurization opens the gap in the seal region.

The seven facts listed above show very clear evidence of two problems. First, the design of the seal in the circumferential joint was marginal and should have been fixed much earlier in the program. You do not inspect a piece of burnt and eroded O-ring and walk away from the problem.

Second, the launch should have been postponed due to cold weather for several different reasons. The very low temperatures were well below the limits to which the system had been certified. The rubber in the O-rings becomes very stiff at these low temperatures and cannot respond quickly enough to accommodate joint movement. The O-rings were almost guaranteed to leak. Also, there was considerable ice on the rocket motors and the fuel tank. Pieces of this ice might damaged the rocket motors, the tiles on the heat shield or the wing's leading edges of the orbiter when it separates and falls during the violent vibrations that occur just after ignition but before lift-off.

Ignoring the Problem

Problems usually do not go away when they are ignored. They persist and sooner or later failure will occur. Although the seals had failed on 12 of the previous 24 flights, the failures were not catastrophic because the leaks were small and at locations which did not endanger the adjacent components. The boosters on the shuttle operated for only a few minutes before they are cut loose to fall into the Atlantic Ocean for recovery at a later time. So those concerned with the launch schedule tolerated the small leaks for short periods of time. The trouble with the leak on the Challenger was that it was large, and the jet of hot gas issuing from the leak was pointed directly at the liquid hydrogen fuel tank. The jet of hot gas acted exactly like a cutting torch, and the huge thin-walled fuel tank was penetrated within seconds.

The O-ring seals were inadequate and the joint needed to be redesigned and modified. In fact, during the development of the boosters in 1977–79, several memos were written by engineers at the Marshall Space Flight Center indicating their concern over the design of the joint. They recommended that the joint be redesigned because "the adequacy of the clevis joint was completely unacceptable." The suppliers of the O-rings stated: "The O-ring was being required to perform beyond its intended design and that a different type of seal should be considered."

Apparently there was a major breakdown in communication; none of these memos and reports from the Marshall Space Flight Center was forwarded to Morton Thiokol, the designers and builders of the boosters. A problem, properly detected by engineering at the NASA Center in charge of monitoring the technical aspects of the development of the boosters, was not pursued to its logical conclusion. Instead of insisting on a redesign of the joint, Marshall Space Flight Center approved the boosters in September of 1980. The pressures of cost overruns and repeated schedule slippage sometimes make managers (and engineers) accept flawed designs. It is very shortsighted and not in the best interest of safe design to follow this practice.

NASA accepted a booster rocket with inadequate seals, but the bird flew. On the first launch, with Columbia, there was no reported damage to the seals. The O-ring seals failed on the second flight. One of the O-rings was burnt with about 20% of the thickness of the ring vaporized. The ring failure did not affect the mission, but it was clear that the putty was not protecting the rings from the hot gas. Marshall Space Flight Center reacted to the field experience by reclassifying the joint to a "Criticality 1 category." This was the official recognition on the part of NASA that a seal failure could result in "loss of mission, vehicle and crew due to metal erosion, burn through and probable case burst resulting in fire and deflagration." The failure of the O-rings by erosion continued with high frequency (50% of the missions), but NASA decided to accept the situation and did not recommend remedial action.

It appears that NASA decided to live with the seal problem at least until a new lighter booster motor could be designed with improved joints. These new boosters were to be ready about six months after the Challenger exploded. A very real lesson—in too-little-too-late.

Recognizing the Influence of Temperature

From the initial development phase of the boosters, the O-ring seals had been recognized as a problem. A problem that NASA classified as very serious, but one that they must have erroneously believed did not elevate the risk to unacceptable limits. In January of 1985, a year before the accident, the Challenger was launched on a cold day at a temperature of 51°F. An examination of the joints after the recovery of the boosters showed that a number of O-rings were severely damaged and evidence of extensive gas leakage was observed. The contractor, Morton Thiokol, and Marshall Space Flight Center reviewed the O-ring seal problems. During this review, Morton Thiokol noted that low temperatures were a contributing factor to the inability of the O-rings to seal properly. The condition was considered undesirable, but was acceptable. However, the O-ring problems persisted and NASA eventually placed a launch constraint on the space shuttle.

The launch constraint would prohibit launches until the problem had been resolved or reviewed in detail prior to each launch. Unfortunately, NASA routinely provided launch waivers for each subsequent flight to avoid delays in an unrealistic launching schedule. The problem of O-ring erosion was common; one or more joints on 50% of the launches exhibited significant erosion. Fortunately, the erosion and attendant leaks had not caused any serious difficulty until the Challenger accident. The problem was expected and accepted as routine.

A Management Decision

The very cold weather on the night before the fatal launch was no surprise. The weatherman (or woman) was on the mark in predicting the temperature and the ice that formed on structures that night. The engineers at Morton Thiokol, in Utah, were very concerned about the effects of the very low temperatures on the ability of the O-rings to function properly. Teleconferences took place between the Morton Thiokol engineers and the NASA program managers. The engineers argued to postpone the launch until the temperature increased to at least 53°F.

The NASA program managers were very unhappy about the engineer's concerns. They did not want the extended delay required to wait for warmer weather. Senior vice presidents from Morton Thiokol sensed the displeasure from the customer and took over the decision process. Senior managers decided to keep the customer happy and reversed the decision of the Vice President of Engineering not to launch at these very low temperatures. The engineers at the end of the day, sat silent.

Senior management at Morton Thiokol signed off on the launch ignoring the fact that the temperature at launch time was expected to be 25°F lower than the lowest temperature which engineering believed the seals would be effective (53°F).

The senior management at Morton Thiokol yielded to client pressure. The engineers were unanimous in their opposition to launch. Senior management asked the engineers to prove that the O-ring seals would fail. Of course, the engineers could not state with 100% certainty that the rings would fail. Failure analysis is performed in terms of probabilities. The risk had elevated as the temperature decreased, but the engineers could not prove conclusively that the seals would fail in a manner that would detrimentally affect the mission. Senior managers at Morton Thiokol and program managers at NASA concluded erroneously that the risks were low enough to proceed with the launch.

Approvals at the Top

NASA has a four-level approval procedure to control the launch of the shuttle. This sounds good, but for a multilevel approval process to be of any value, information must flow freely from the bottom of the organization to the top of the chain of command. The technical discussions concerning the ability of the O-rings to function at the very low temperatures took place between level 4 NASA administrators and engineers and program staff at Morton Thiokol. Program managers at Marshall Space Flight Center, level 3 administrators, were also included in these discussions. While the discussion was extended, with clear polarization between engineering and management, not a word of these concerns was conveyed to the top two levels of management at NASA.

Approvals for the launch were given by level 2 administrators for the National STS Program in Houston, TX and level 1 administrators at NASA's Headquarters in Washington D.C. These approvals were essentially automatic because the administrators in charge had been isolated. They were not privy to the management decision to fly with very cold O-ring seals. They were not informed of the engineering recommendation to postpone the launch and to wait until the temperature increased to at least 53°F.

The lesson here is clear—approvals by the very high level managers are worthwhile only if they are informed decisions. If complete information is not presented to the executives for their evaluation, then their approval is meaningless. These administrators are not knowledgeable, and they add no value in an informed decision making process.

15.11 THE COLUMBIA FAILURE

On February 1, 2003 the Columbia shuttle disintegrated as it reentered the atmosphere at the end of its 16-day mission. In Mission Control, Columbia's re-entry appeared normal until 8:54:24 AM[5], when engineers monitoring the sensors informed the flight director that four hydraulic sensors in the left wing were indicating—off-scale low—a reading that falls below the minimum capability of the sensor. At 8:55:00 a.m. nearly 11 minutes after Columbia had re-entered the atmosphere, its wing leading edge temperatures reached their normal level of nearly 3,000 °F. At 8:55:32 a.m. Columbia crossed from Nevada into Utah while traveling at Mach 21.8 at an altitude of 223,400 ft. At 8:58:20 a.m. as Columbia crossed from New Mexico into Texas, a thermal protection tile was lost. At 8:59:15 a.m. the engineers monitoring the sensors informed the flight director that the sensors monitoring the pressure on both left main landing gear tires were lost. The flight director then informed the Columbia's crew that Mission Control was evaluating these sensor readings, and added that the crew's last transmission was not clear. At 8:59:32 a.m. a broken response from Columbia was recorded but it was cut off in mid-word. Videos made by observers on the ground less than a minute later, presented in Fig. 15.6, revealed that Columbia was disintegrating.

Fig. 15.6 A video photograph of Columbia disintegrating as it is passing over Texas while traveling at about Mach 19.5 at an altitude of about 210,000 ft.

Columbia's History

Columbia was the first space-rated Orbiter, and it made the Space Shuttle Program's first four orbital test flights. Because it was the first of its kind, Columbia differed slightly from the other four Orbiters—Challenger, Discovery, Atlantis, and Endeavour. Built to earlier engineering specifications, Columbia was slightly heavier, and was not able to carry sufficient cargo to conduct cost effective missions to the International Space Station. For this reason, Columbia was not equipped with a Space Station docking system; hence, it had more space in its cargo bay for longer pieces of equipment. Columbia usually flew science missions and serviced the Hubble Space Telescope.

The final flight of Columbia was designated STS-107. This was the Space Shuttle program's 113th flight and Columbia's 28th. It appeared that the flight was nearly trouble-free. Unfortunately, there were no clear indications to either the crew onboard Columbia or to personnel in Mission Control that the flight was in trouble as a result an impact by at least one large piece of

---

[5] Times are given as Eastern Standard Time (EST.).

foam insulation that occurred shortly after launch. Mission management failed to note a few clues revealed during launch that the Orbiter was in trouble and take corrective action.

STS-107 was an intense science mission that required the seven-member crew to form two teams, enabling round-the-clock shifts. Because the extensive science cargo and its extra power sources required additional checkout time, the launch sequence and countdown were about 24 hours longer than normal. Nevertheless, the countdown proceeded as planned, and Columbia was launched on January 16, 2003. At 81.7 seconds after launch, when the Shuttle was at an altitude of about 65,600 feet and traveling at Mach 2.46 (1,650 mph), a large piece of insulating foam came off an area where the Orbiter attaches to the external tank. This piece of foam impacted the leading edge of Columbia's left wing at a velocity of about 500 MPH. The foam impact was not detected by the flight crew or observed by ground personnel. However, the next day, detailed reviews of launch camera photography revealed a large piece of foam striking the Orbiter. This foam impact and resulting damage to the left wing of Columbia had no effect on the daily activities performed by the astronauts during the 16-day mission. This mission had met all its objectives prior to reentry.

The Accident Investigation

Immediately following the failure of the Orbiter, NASA established a Columbia Accident Investigation Board to identify the chain of events that caused the Columbia accident [12]. Evidence the Board considered included:

- Film and video during launch.
- Radar images of Columbia in orbit.
- Amateur video of debris shedding during reentry.
- Onboard sensor data from the onboard recorder recovered after the accident.
- Analysis of the debris recovered.
- Computer modeling.
- Impact and wind tunnel tests.

Fig. 15.7 The upper circle shows the foam covered bipod on the external tank and the lower circle shows the impact point on the leading edge of the Columbia's left wing.

The reason for the loss of Columbia and its crew was a breach in the thermal protection system on the leading edge of the left wing. The breach was initiated by a piece of insulating foam that separated from the left bipod ramp of the external tank and struck the wing in the vicinity of the lower half of a reinforced carbon-carbon panel 81.9 seconds after launch. During re-entry, this breach in the thermal protection system allowed superheated air to penetrate the leading-edge insulation. This hot air (plasma) heated and softened the aluminum structure of the left wing and weakened the structure until aerodynamic forces caused loss of control, collapse of the wing and subsequent disintegration of the Orbiter. A photograph of Columbia prior to launch, showing the area of the foam released from the external tank and the location of its impact on the left wing of the Orbiter is presented in Fig 15.7.

The external tank is the largest component of the Space Shuttle. It serves as the main structural component during assembly, launch and supports both the solid rocket boosters and the Orbiter. It also serves as the cryogenic, propellant tank for the space shuttle main engines. It contains 143,351 gallons of liquid oxygen at – 297° F in its upper tank and 385,265 gallons of liquid hydrogen at – 423 ° F in its lower tank.

The Orbiter is attached to the external tank by two umbilical fittings at the bottom and by a "bipod" at the top. The bipod is attached to the external tank by fittings to the right and left of the external tank centerline. The bipod fittings, which are bolted to the external tank, are located near the flange joint at the inter tank as shown in Fig. 15.8.

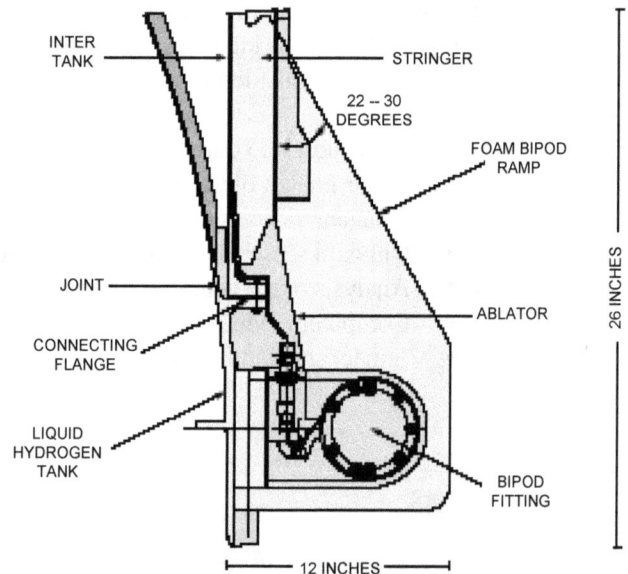

Fig. 15.8 Drawing of the bipod showing the foam insulation that covered its fitting.

The external tank is coated with two materials that serve as its thermal protection system. The inner coating is a dense composite ablator for dissipating heat, and the outer coating is low-density closed-cell foam for high insulation efficiency. The external tank thermal protection system is designed to:

- Reduce heat flow into the cryogenic tanks to minimize the vaporization of the oxygen and hydrogen prior to launch.
- Maintain the temperature of the external surfaces sufficiently high to prevent the formation of frost or ice.

A combination of factors led to the loss of the left bipod foam ramp (see Fig. 15.8) during the ascent of Columbia shortly after launch. NASA personnel believe that pre-existing defects in the foam were a major factor and that these defects were necessary to induce the foam to fail. However, analysis indicated that pre-existing defects were not the only factor responsible for foam loss. Other factors were involved such as the technique for producing and applying the foam, particularly in the bipod region. The foam used to insulate the external tank consists of two chemical components that must be mixed in an exact ratio and is then sprayed according to strict specifications. The foam is applied to the bipod fitting in a manual operation to form the ramp illustrated in Fig. 15.8. This process is probably the primary source of defects in the foam. Dissection of these foam ramps, conducted after the accident, revealed defects such as voids, pockets and debris. These defects are due to a lack of control of several parameters in the manual spraying used to apply the foam. The quality control of the foam is exacerbated by the complexity of the underlying hardware configuration. The dissection studies showed that the defects usually occurred at the boundaries between each layer as the ramp is formed by the repeated application of thin layers.

It is important to recognize that the failure of the foam insulation on previous shuttle missions was not a rare event. Foam loss has occurred on more than 80 percent of the 79 previous Shuttle missions for which imagery was available. Foam was lost from the left bipod ramp on nearly 10 percent of missions analyzed where the left bipod ramp was visible following external tank separation. For many of the missions, it was not possible to determine if foam was lost because the launches were at night or the external tank bipod ramp areas were not in view at the time when the photographs of the external tank were taken.

It is believed that the impact of a large piece of foam damaged the leading edge of Columbia's left wing. The leading edge of an Orbiter's wing is comprised of 22 panels fabricated from a reinforced carbon-carbon (RCC) composite. To prevent oxidation, the outer layers of the carbon substrate are converted into a 0.02 to 0.04 inch thick layer of silicon carbide. The components of the Orbiter's wing leading edge provide the aerodynamic load bearing, structural support and thermal protection with temperatures upon reentry up to 2,300 °F. The design requirements for the leading edge of the wing included 100 missions with minimal rework, limiting the temperature of the aluminum wing structure to less than 350 °F, and withstanding a kinetic energy impact of only 0.006 foot-pounds. The engineering specification clearly indicate that the wing leading edge was not required to withstand impact from debris or ice, because these objects would not pose a threat during the launch sequence.

The risk of micrometeoroid or debris damage to the RCC panels has been evaluated several times. Hypervelocity impact testing, using a variety of projectiles, as well as low-velocity impact testing with projectiles of ice and other materials, resulted in a design change that improved the resistance of the leading edges to impact damage. Analysis of this design change predicted that an Orbiter could survive re-entry with a ¼ to 1-inch diameter hole in the lower surfaces of the RCC panels depending on panel location.

Analysis of maintenance reports indicated that the RCC leading edge panels from all of the Orbiters had been struck by objects throughout their operational life. However, none of the panels had been completely penetrated. A sampling of 21 post-flight reports noted 43 hypervelocity impacts. The largest damage zone was 0.2 inch in diameter. The most significant low-velocity impact was to one of Atlantis' panels. The damaged area was 1.9 inches by 1.6 inches on the exterior surface and 0.5 inches by 0.1 inches in the interior surface. The substrate of the panel was exposed and oxidized. After inspection the severely damaged panel was replaced. A study concluded that the damage was caused by a strike by a man-made object, possibly during ascent. This panel damage to Atlantis was a clear warning of an impending disaster.

Upon completion of its exhaustive study [12], the Columbia Accident Investigation Board concluded that the cause of the accident was a breach in the thermal protection system on the leading edge of the left wing. The breach was initiated by a piece of foam that separated from the left bipod ramp of the external tank and impacted the left wing in the vicinity of the lower half of RCC panel 8. The conclusion that foam separated from the external tank bipod ramp and struck the wing in the vicinity of panel 8 is clearly documented by photographic evidence. Sensor data and the aerodynamic and thermodynamic analyses confirmed that the breach was in the vicinity of panel 8. This data and subsequent analysis also indicated subsequent melting of the supporting structure, the spar, and the wiring behind the spar. The detailed examination of the debris also pointed to panel 8 as the breach site. Impact tests established that a large piece of foam could breach an RCC leading edge panel. These tests and the physical evidence collected over two states convinced those individuals who were discounting of the analytical evidence of the cause of the failure.

Hard Lessons to Learn

Dr. Diane Vaughan, an authority on risk assessment, testified before the Columbia Accident Investigation Board and stated: "What we find out from a comparison between Columbia and Challenger is that NASA is an organization that did not learn from its previous mistakes and it did not properly address all of the factors that the presidential commission identified."

Organizational failures usually occur with higher frequency in larger organizations with several layers of management. Failures happen regardless of the safeguards and systems that an organization employs. Failures are even more common in high-risk organizations like NASA. NASA uses several risk-avoidance systems that are designed to insure that the instruments and astronauts sent into space complete their missions and return safely. However, NASA has failed in three cases to achieve this objective when sending astronauts in space or when preparing space systems to do so. The Space Shuttles Challenger and Columbia tragedies as well, as the Apollo launch pad fire in 1967, are examples of NASA's failures that resulted in the deaths of 17 astronauts. Organizational failures played an important role in all three cases.

As was the case with O-ring damage in the years before the Challenger accident, the failure of Columbia did not represent the first time that foam had detached from the external tank and caused damage to an Orbiter. There had been many reports of impacts on previous flights by pieces of foam that were much smaller than that one which severely damaged Columbia. Orbiters would often return with many damage sites due to both hypervelocity and low velocity impact. Some sites were larger than 1 inch in diameter. There are numerous reports in the Problem Reporting and Corrective Action (PRACA) system that showed some of the thermal barrier damage was probably due to foam failure during launch and assent. Not only was foam impact not considered in the design of the thermal protection system, but the engineering specification for the Orbiter clearly states that nothing should impact the shuttle during launch.

However, foam had been failing and striking the Orbiter with high frequency and a team of chemists and engineers were working on improving the foam and the techniques used for applying it. The studies of impacts with foam concluded that they did not pose a "flight safety" risk because experience showed that the small pieces of foam that had been coming off the external tank only caused small pits and craters on the Orbiters. Unfortunately, the engineers did not consider a large piece of foam traveling at high velocity[6] when they declared foam failure not to be a "flight safety" issue. The impact of such a large piece of foam on a vulnerable area of the Orbiter was unprecedented. (So was the three days of abnormal cold before the Challenger launch.)

---

[6] Photographic analysis revealed that one large piece and at least two smaller pieces of foam had separated from the bipod ramp area of the external tank. The large piece of foam was about 24 in. long and 15 inches wide. It was rotating at 18 RPS and moving with a velocity of about 500 MPH when it impacted the Columbia's wing.

Foam failure that regularly damages the thermal protection system on a large fraction of the flights was not considered to be dangerous because the orbiters were returning safely with damage that could be repaired or ignored. What the engineers or management did not considered was the possibility of a larger or heavier piece of foam hitting the Shuttle in a particularly vulnerable area like the leading wing edge. This is a classic example of Vaughan's "normalization of deviance" [13] where an unpredicted anomaly becomes routine.

Normalization of Deviance or Accepting Problems Instead of Pursuing Solutions

Vaughan [13] has developed the concept of **normalization of deviance** to explain the way technical flaws escape the scrutiny of the various safety boards within a large organization such as NASA. Frequently these boards observe unanticipated problems that continue to occur on a regular basis without leading to an accident. People begin to believe the problems are benign and are lead to the pragmatic notion of "acceptable" deviance. The leadership at NASA found it very expensive and time consuming to establish the cause of some problems and to incorporate added inspections in the regular maintenance cycle of the Shuttle system without strong evidence of "flight safety "risk. Under the pressures of the Shuttle's flight schedule and limited budget, NASA did not commit significant resources on problems that its management did not classify as "flight safety" risks that could cause the loss of an Orbiter. This practice resulted in disincentives for the engineers to determine the source of problems, even though scenarios could be envisioned that would cause significant "flight safety" risks. NASA frequently cleared flights as operational based on previously successful flights that had completed their missions while exhibiting clear evidence of a design problem. This reasoning prompted Richard Feynman [10] in his analysis of the Challenger accident to remark: "When playing Russian roulette the fact that the first shot got off safely is little comfort for the next."

The Role of NASA's Management Structure in Columbia's Accident

NASA's organizational and physical infrastructure was developed in the days of Apollo program when it had a large budget and a sharply focused mission—to land a man on the moon before the Russians. A layered bureaucratic group of middle and upper level managers functioned well and the Apollo program accomplished its goals with only one serious mishap in 1967 and a near catastrophic mission on Apollo 13. After more than 30 years, NASA's organizational structure is nearly the same although its missions have changed markedly. There is reason to question if their organizational structure is sufficiently flexible considering its plans for the future and its ongoing projects.

For the Shuttle program, there should be a mechanism for engineers to bypass the bureaucracy and hierarchy, especially in the pre-launch and launch processes. Suppose the engineers had succeeded in calling off the launch of the Challenger due to the effect of cold weather on the O-ring seals. The flight would have been cancelled, Challenger would have been taken off of the launch pad and the solid rocket boosters disassembled to replace the damaged O-rings. This action would have been expensive but not nearly as costly as the loss of crew and the Orbiter. Similarly, if an engineer needs a certain type of data, there should be a way to bypass the formal bureaucratic procedures and to obtain the data with dispatch. Engineers have many intuitions and hunches that take time and resources to translate into analysis and data. These intuitions need to be respected, given credence, explored and welcomed by upper management [14].

Probability of Failure

Richard Feynman served on the Presidential commission that investigated the Challenger accident that occurred on January 28, 1986. It was near the end of his life, and the famous physicist devoted significant amounts of time to the investigation. After the investigation, Dr. Feynman prepared a paper describing his personal observations on the reliability of the Shuttle [15]. We suggest you read this interesting paper, because we are only reviewing a small portion of this reference. The following paragraphs are direct quotes from Feynman's personal observations.

1. It appears that there are enormous differences of opinion as to the probability of a failure with loss of vehicle and of human life. The estimates range from roughly 1 in 100 to 1 in 100,000. The higher figures come from the working engineers, and the very low figures from management. What are the causes and consequences of this lack of agreement? Since 1 part in 100,000 would imply that one could put a Shuttle up each day for 300 years expecting to lose only one, we could properly ask—"What is the cause of management's fantastic faith in the machinery?"

2. If a reasonable launch schedule is to be maintained, engineering often cannot be done fast enough to keep up with the expectations of originally conservative certification criteria designed to guarantee a very safe vehicle. In these situations, subtly, and often with apparently logical arguments, the criteria are altered so that flights may still be certified in time. They therefore fly in relatively unsafe conditions with a chance of failure of the order of one percent (it is difficult to be more accurate).

3. Official management, on the other hand, claims to believe the probability of failure is a thousand times less. One reason for this may be an attempt to assure the government of NASA perfection and success in order to ensure its annual funding from the government. The other may be that they sincerely believed it to be true, demonstrating an almost incredible lack of communication between themselves and their working engineers.

4. In any event this has had very unfortunate consequences, the most serious of which is to encourage ordinary citizens to fly in such a dangerous machine, as if it had attained the safety of an ordinary airliner. The astronauts, like test pilots, should know their risks, and we honor them for their courage. Who can doubt that McAuliffe was equally a person of great courage, who was closer to an awareness of the true risk than NASA management would have us believe?

5. Let us make recommendations to ensure that NASA officials deal in a world of reality in understanding technological weaknesses and imperfections well enough to be actively trying to eliminate them. They must live in reality in comparing the costs and utility of the Shuttle to other methods of entering space. They must be realistic in making contracts, in estimating costs and the difficulty of the projects. Only realistic flight schedules should be proposed, schedules that have a reasonable chance of being met. If in this way the government would not support them, then so be it. NASA owes it to the citizens from whom it asks support to be frank, honest and informative, so that these citizens can make the wisest decisions for the use of their limited resources.

6. For a successful technology, reality must take precedence over public relations, for nature cannot be fooled.

As one studies the Columbia's fate, we appreciate the significant value of the personal observations made by Richard Feynman nearly 30 years ago. It appears that his message was lost on NASA as they continue to ignore significant operational problems in their efforts to met schedule with a constrained budget. Dr. Feynman estimated that the Shuttle flies in a relatively unsafe condition, with a chance of failure of the order of one percent. With the additional Shuttle flights since the Challenger accident it is possible to test his estimate with the additional flight experience. The Columbia accident was the second in 113 flights; thus the estimate of the probability of failure P_f of future Shuttle flights is:

$$P_f = N_f/N = 2/121 = 1.65\%$$

This estimate is the same order as Dr. Feynman predicted after the Challenger accident and 1,700 times higher than the probability of failure of 0.001% (1/100,000) predicted by NASA managers.

15.12 SUMMARY

The degradation of personal ethical behavior during the past two or three generations has been described. While there are many controversial issues that can be classified as "dilemma" ethics, there are several "basic" characteristics where the difference between right and wrong is clearly defined. The list of basic virtues was described that include:

- Prudence
- Justice
- Fortitude
- Temperance

Laws are written to prohibit, by punishment, a well-defined set of actions. These laws control behavior, but only affect the specific actions on the part of individuals or corporations.

Ethics allows society to trust engineers to design products and systems with a high regard for safety. Engineers have both an individual and a collective interest in keeping that trust. Individual behavior is important, but as we learned in the Challenger explosion, it is not sufficient. Corporations (business) must operate with the highest ethical standards. The chief executive officer (CEO) establishes the standards and the lower level managers act to establish the corporate character. Design systems must also avoid ethical failures due to individuals not understanding what they don't know or information not flowing to those who have the needed knowledge.

Professional ethics were described as following patterns established first in medicine and law. Professionals get special status based on a commitment to the public. Engineering professional ethics was described by introducing the Code of Ethics of Engineers sponsored by the Accreditation Board for Engineering and Technology (ABET). The code is short with four fundamental principles and seven basic canons.

The fatal accidents of the Challenger and Columbia Space Shuttles, which failed during their missions in 1986 and 2003, were discussed. Many ethical issues dealing with individual and corporate behavior were apparent when the events leading to this fatal accident were reviewed.

The bottom line is that society will only allow engineers to design and manufactures products and systems if they act at all times with the public safety as their highest duty.

REFERENCES

1. Goodman, L. S., The Historic Role of the Oath of Admission, American Journal of Legal History, Vol. 11, No. 4, (October 1967), pp. 404-411..
2. Sommers, C. H., "Teaching the Virtues," *Public Interest,* No. 111, Spring 1993, pp. 3-13.
3. Woodward, K. L. "What is Virtue," *Newsweek*, June 13, 1994.
4. George, R. P., Making Men Moral: Civil Liberties and Public Morality, Oxford University Press, New York, NY, 1994.
5. Baker, D. F. "Ethical Issues and Decision Making in Business," Vital Speeches of the Day, 1993
6. Jensen, C., No Down Link: A Dramatic Narrative about the Challenger Accident, Farrar, Straus and Gitoux, New York, NY, 1996.
7. *Time,* February 10, 1986
8. Lewis, R. S., The Voyages of Columbia: The First True Space Ship, Columbia University Press, New York, NY, 1984.
9. *The New York Times* April 23, 1986.
10. Feynman, R. P. What Do You Care What Other People Think? Bantam, New York, NY, 1988.
11. Report to the President by the Presidential Commission on the Space Shuttle Challenger Accident, Ayer Co., Salem, MA, 1986.
12. Anon, Report of Columbia Accident Investigation Board, Volume I, August 26, 2003, See http://www.nasa.gov/columbia/home/CAIB_Vol 1.html.
13. Vaughan, D., The Challenger Launch Decision: Risky Technology, Culture and Deviance at NASA, University of Chicago Press, Chicago, 1996.
14. Hall, J. L., "Columbia and Challenger: Organizational Failure at NASA," Space Policy Vol. 19, 2003, pp. 239-247.
15. Feynman, R. P., Personal Observations on the Reliability of the Shuttle, see Website http://www.fotuva.org/feynman/challenger-appendix.html.

EXERCISES

15.1 Prepare a two page brief that distinguishes the differences among professional, organizational and personal ethics.

15.2 Why was the engineering profession much later than either the medical or legal professions in establishing a code of ethics for professional behavior?

15.3 If you have selected an engineering discipline for your studies, determine the professional society that represents this discipline. From the Internet locate the URL for its Website and then examine its code of ethics. Write a one page brief describing the code and your reaction to it.

15.4 Political leaders are often attacked by the media for lax ethical behavior. Please write a short paper describing three recent lapses of ethical behavior on the part of the leadership in either a State or the Federal government. Did these officials break the law? Is it important that they did or did not break the law?

15.5 Many corporate executives have been involved in serious crimes involving hundreds of millions of dollars. Identify one of these executives and write a paper about his misfeasance and the consequences.

15.6 Have you ever cheated on an important exam in high school? What about a college exam? Do you know of someone who cheated? What was your attitude when you observed this cheating?

15.7 Is there an honor code at the University where you are pursuing your studies? Have you read it? Do you abide by the rules?

15.8   Write a paragraph or two explaining your position regarding:

- Prudence
- Justice
- Fortitude
- Temperance

15.9   If you could write a law that would go on the books tomorrow and be strictly enforced, what behavior would it require or prohibit?

15.10 Why does the CEO of a corporation set the standard for ethical behavior? What are some of the issues that arise on a daily basis that develop corporate character? Is it possible for a corporation with tens of thousands of employees to develop a character like an individual?

15.11 Examine the code of ethics for engineers given in Fig. 15.2, and describe your opinion of the fundamental principles or canons. Can you live with these expectations if you become a professional engineer? Do you believe ABET's code is adequate, or should it be revised to reflect a more modern viewpoint of what is right and wrong?

15.12 Suppose that you were a lead engineer working for Morton Thiokol on the evening of January 27, 1986 involved in the discussion of the O-rings. What would you have done?

- Early in the teleconference.
- At the critical stage of the teleconference.
- After management had taken over the decision process.

There is no right or wrong answer to these questions. The purpose of the exercise is to place you in a professional dilemma. Someday you may be placed in a similar situation where there is a trade-off between safety and corporate business interests. Think about your response in advance and be prepared to deal with such a dilemma and its consequences.

15.13 Suppose you were the engineer responsible for maintaining the foam on the external tank of the Space Shuttle. What would you do to prevent the failure of foam during the launch and ascent of a Shuttle?

15.14 Suppose you were the engineer responsible maintaining the RCC composite panels that form the leading edge of the three remaining Orbiters. What changes in your maintenance plan would you make following the Columbia accident?

15.15 Write a review of Chapter 2 "Columbia's Final Flight" from volume I of the Report of the Columbia Accident Investigation Board. Reference 12 provides the URL for the Website.

15.16 Write a review of Chapter 3 "Accident Analysis" from volume I of the Report of the Columbia Accident Investigation Board. Reference 12 provides the URL for the Website.

15.17 Write a review of Chapter 5 "From Challenger to Columbia" from volume I of the Report of the Columbia Accident Investigation Board. Reference 12 provides the URL for the Website.

15.18 Are there government agencies that are concerned with product safety? Name them and describe the way they function.

APPENDIX A

A.1 HOVERCRAFT PRODUCT SPECIFICATIONS — 2010

- **HC_009_01 Structure and levitation requirements**
  (a) There are no minimum or maximum dimensional restrictions.
  (b) There are no minimum or maximum weight restrictions.
  (c) Only the skirt (rigid or soft) may contact the ground/water.
  (d) Skirts must be fabricated in-house.

- **HC_009_02 Power and propulsion requirements**
  (a) The use of internal combustion engines (gas and glow fuel engines) is prohibited.
  (b) The hovercraft must be able to levitate for at least 10 minutes without replenishing or recharging its energy source and without modulating the lift fans' power.

- **HC_009_03 Sensors and control requirements**
  (a) The hovercraft must be controlled by the Lego NXT microcontroller.[1]
  (b) The hovercraft operation during testing must be autonomous.[2]
  (c) The Bluetooth feature must be disabled on the NXT.

- **HC_009_04 Safety requirements**
  (a) All fans must have protective fan guards to reduce the risk of bodily injury. The fan guards must have a protective "mesh" spacing of 7 x 7 mm or smaller.
  (b) The use of pins and/or needles on your hovercraft is strictly prohibited.

- **HC_009_05 Cost requirements**
  (a) Total as-built replacement cost of the hovercraft must be less than $400.[3]
  (b) The cost must be broken down in a Bill of Materials (BOM), in which the Fair Market Value (FMV) of each component must be listed along with the part number, vendor and quantity.
  (c) Donated and/or used components may be incorporated, but the FMV of a NEW equivalent component must be given in the BOM.

- **HC_009_06 Testing requirements**
  (a) The hovercraft must be capable of autonomously navigating the course layout shown in Fig. 1 in less than 10 minutes without contacting any side walls.

- **HC_009_07 Product deliverables**
  (a) The Hovercraft
  (b) The final Bill of Materials
  (c) Preliminary and final Gantt charts
  (d) The final written design report with all design drawings and schematics
  (e) The final oral presentation (PowerPoint slide file)

---

[1] The NXT can be rented for a $15 rental fee + $35 refundable deposit. You may use your own NXT, but the BOM must reflect a $50 charge for the use of the NXT.

[2] For testing purposes, the NXT program will be manually started, the power to the propulsion/levitation system may be manually started, and no further direct contact or remote control will be permitted.

[3] Cost is calculated by adding the fair market value of all components used in the Hovercraft during the testing phase. It does not include shipping costs or costs for parts that were bought but later discarded/returned. Costs will be shared equally among group members.

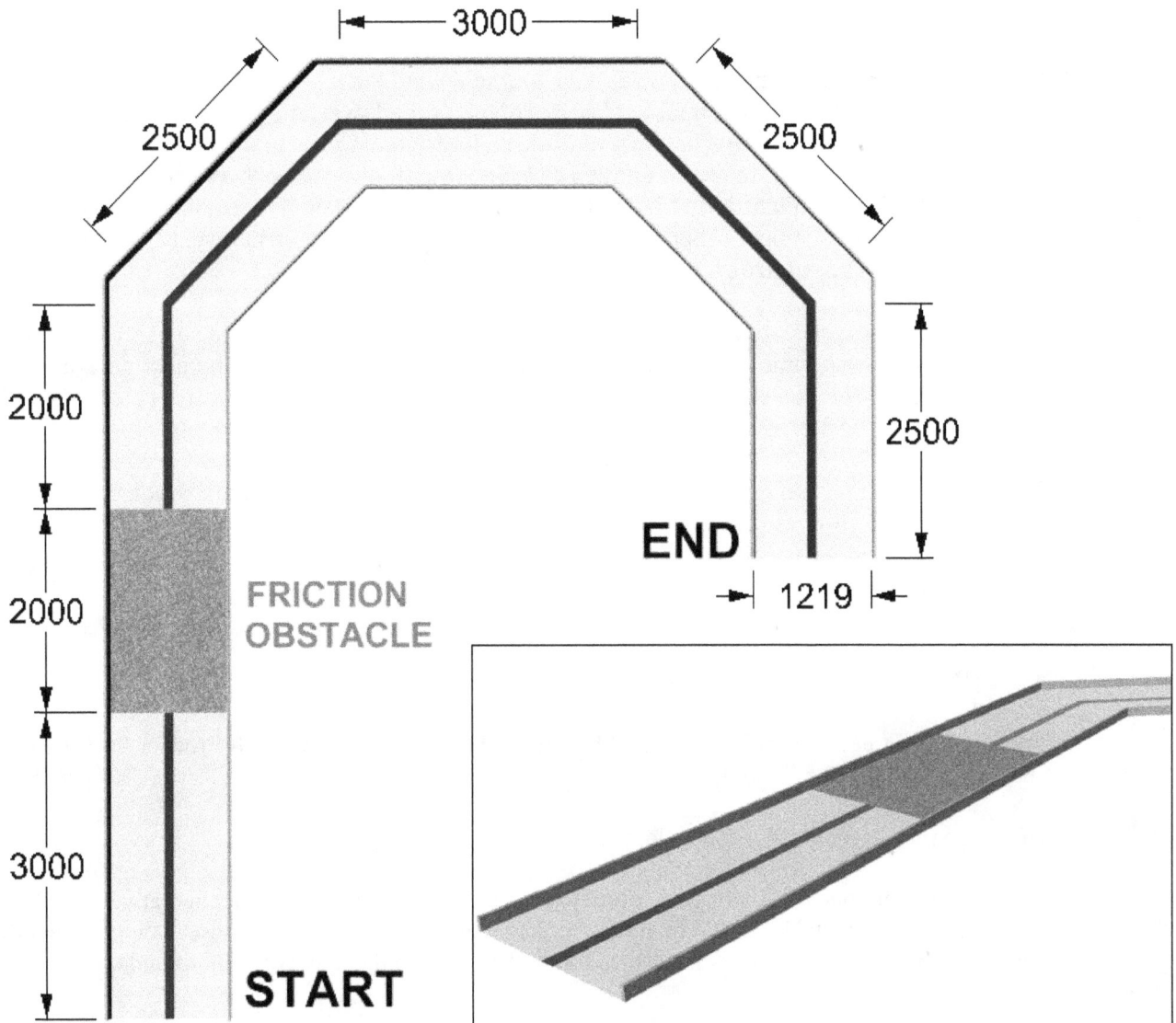

Fig. 1 Test course layout [all dimensions in mm]

The test course will be indoors on a smooth white surface. To aid navigation, there is a single 100 mm wide solid black tape line located along the centerline of the track and two sidewalls that are spaced 1,219 mm apart and constructed with 25 (w) x 89 (h) mm pieces of wood.

The course includes a 2,000 mm rough patch ("the desert"), as shown in Figure 1. This section of track consists of fine grain sand with variable diameters ranging up to ¼ mm. No centerline tape line exists over this section of the track.

A.2 HOVERCRAFT DEVELOPMENT CONTRACT —2010

A major portion of this semester's ENES 100 course will be the design, construction and testing of an autonomous hovercraft capable of performing a timed overland navigation mission. This endeavor will require a semester-long, multidisciplinary team effort in order to be successful. The instructor will create teams of about 10 students with diverse backgrounds and skill sets. Each team will subsequently divide into sub-teams consisting of 2-4 students. Possible sub-teams required to complete this project could be: (1) levitation & structure, (2) propulsion & power, (3) sensors & control, and (4) crossing the friction obstacle.

Each team will use class-time and time outside of the class throughout the semester to design and assemble a hovercraft capable of completing the specified mission. The best designs from each section will then be entered into a competition against the top hovercraft prototypes from the other course sections.

In addition to successfully navigating the specified route, the mission requires that the hovercraft do so in minimal time. Details of the course layout and the design constraints are given in the product specifications. Additional information about scoring and schedule are given below.

A.2.1 Development Constraints

The group must develop, build and test a hovercraft based on the Hovercraft Product Development Specification — 2010.

The group must meet the global development schedule provided in Table A.2.1.

While electric motors are an obvious possibility for propulsion and levitation, alternative propulsion sources may be acceptable. It will be up to the instructor's discretion if the chosen source meets the low noise/pollution requirements needed to compete indoors in the Kim Building rotunda.

Table A.2.1 Global Milestone Schedule

Milestone No.	Task	Date
1	Development Plan Presentation	TBD
2	Preliminary Design Presentation	
3	Preliminary Design Reports	
4	Begin Prototype Fabrication	
5	Preliminary Testing	
6	Final Testing	
7	Deign Team Competition	
8	Final Design Report	
9	Final Design presentation	

Scoring

Scoring for the Hovercraft competition will be based on distance traveled, time and number of faults. The competition winner will travel the course without faults from beginning to end in the least amount of time. Teams may have up to three scoring attempts. The scoring run with the least number of faults, greatest distance and least time will be used for final placement in the overall competition.

Faults

A "legal fault" occurs when any part of the hovercraft comes in contact with a wall/obstacle and immediately "bounces-off". A hovercraft cannot use any portion of any wall as a guide; this will be called an "illegal fault". The judge alone will dictate if a hovercraft has used a wall as a guide. Should this happen, the scoring run is ended at that distance marker and time stamp. The illegal fault that resulted in the end of the scoring run is added to any other faults accumulated during the run. Recall, faults can only occur in the navigation segment of the competition, as your vehicle is allowed to bump any number of walls and the gate during the gate actuation segment of the course.

Distance

The distance a hovercraft travels beyond the gate will be measured. Distance will be measured in number of turns completed only. The forward most portion of the hovercraft will be used to determine distance traveled.

Time

There is a 10-minute time limit per run that a hovercraft may utilize to cover the track. Should a hovercraft exceed 10 minutes, the judge will note the distance and faults while using 10 minutes as the maximum time. Otherwise, during a scoring run, time will be stopped upon completion of the course, illegal fault, low/dead battery and/or team request. The stoppage of time signals the end of the scoring run. The judge will keep time and only the time from the judge's timer will be used for scoring.

Scoring Criteria

To judge the effectiveness of the hovercraft navigation system, the scoring system is tiered based on number of faults. The highest tier is reserved for those hovercrafts that do not fault at all. For example: a hovercraft that travels 16 feet without fault will place higher than a hovercraft that traveled the entire track with 1 or more faults. Distance traveled beyond the gate will be used to sort hovercraft within a tier; lastly, time will be used as a tiebreaker between hovercrafts that have the same traveled distance.

A hovercraft MUST travel through the friction obstacle to be considered for placement. Hovercrafts that do not meet this requirement during a scoring run are not eligible for placement and will lose one of the three allowed scoring attempts.

For example, consider 6 hovercrafts with the following scores:

- No faults – 4 turns, 5minutes 35 seconds
- Six faults - completed course, 4minutes 15 seconds
- One fault – 4 turns, 3minutes 01 seconds
- One fault – 4 turns, 2 minutes 59 seconds
- No faults – 3 turns, 2 minutes 23 seconds
- No faults – unable to cross friction obstacle, 1 minutes 05seconds

Applying the scoring system above, these contestants would place in the following order:

1. No faults – 4 turns, 5 minutes 35 seconds
2. No faults – 3 turns, 2 minutes 23 seconds
3. One fault – 4 turns, 2 minutes 59 seconds
4. One fault – 4 turns, 3 minutes 01 seconds
5. Six faults – complete course, 4 minutes 15 seconds
6. No faults – unable to cross friction obstacle, 1 minutes 05 seconds   NOT ELIGIBLE FOR PLACEMENT

The undersigned students hereby agree to the following rules of conduct for the group work to be completed as a part of the Class project. **We have read and understand this design contract as well as the hovercraft design specifications and we accept the terms of the class project.**

1) When subgroup meetings are created, the remaining members will be informed of the meeting's time, location, and agenda either in person, through e-mail, or by phone. Any members that could not attend the meeting will be informed of the nature of the discussions and the resulting action items that were assigned.

2) All e-mails concerning the meetings and discussions will be checked and acknowledged at least once a day.

3) If a mandatory meeting is missed, it will be the responsibility of the absent member to find out what happened and what his/her assigned tasks were. The team must be notified in advance of any anticipated non-emergency absences.

4) Everyone will take ownership for an equal portion of the work to be completed. Each member will provide deliverables to the group completed and on time. The aspect of the work that each member will be responsible for will be decided by consensus, and documented in the final report.

A.2.2 Signature Page

Name (printed) Signature Date

_____ _____ _____

Name (printed) Signature Date

_____ _____ _____

Name (printed) Signature Date

_____ _____ _____

Name (printed) Signature Date

_____ _____ _____

Name (printed) Signature Date

_____ _____ _____

Name (printed) Signature Date

_____ _____ _____

Name (printed) Signature Date

_____ _____ _____

Name (printed) Signature Date

_____ _____ _____

Name (printed) Signature Date

_____ _____ _____

Name (printed) Signature Date

_____ _____ _____

Name (printed) Signature Date

_____ _____ _____

Name (printed) Signature Date

_____

Approved: _____

Agenda
Weekly Meeting
Date _____ Time _____

Team Member Responsible

1. Weekly status report   _____

2. Review of outstanding action items

    - Action item #_____   _____
    - Action item #_____   _____
    - Action item #_____   _____
    - Action item #_____   _____
    - Action item #_____   _____

3. Report on progress

    - Subsystem _____   _____
    - Subsystem _____   _____
    - Subsystem _____   _____
    - Subsystem _____   _____
    - Subsystem _____   _____

4. Identify new problems

    - Problem #_____   _____
    - Problem #_____   _____
    - Problem #_____   _____
    - Problem #_____   _____
    - Problem #_____   _____

5. Assignment of action items

    - Action item #_____   _____
    - Action item #_____   _____
    - Action item #_____   _____
    - Action item #_____   _____
    - Action item #_____   _____

6. New business
7. Summary
8. Adjourn

Agenda
Weekly Meeting
Date _____ **Time** _____

Team Member Responsible

1. Weekly status report   _____

2. Review of outstanding action items

    - Action item #_____   _____
    - Action item #_____   _____
    - Action item #_____   _____
    - Action item #_____   _____
    - Action item #_____   _____

3. Report on progress

    - Subsystem _____   _____
    - Subsystem _____   _____
    - Subsystem _____   _____
    - Subsystem _____   _____
    - Subsystem _____   _____

4. Identify new problems

    - Problem #_____   _____
    - Problem #_____   _____
    - Problem #_____   _____
    - Problem #_____   _____
    - Problem #_____   _____

5. Assignment of action items

    - Action item #_____   _____
    - Action item #_____   _____
    - Action item #_____   _____
    - Action item #_____   _____
    - Action item #_____   _____

6. New business
7. Summary
8. Adjourn

Action Item Record

Team Meeting of _____

Decisions made	Follow-up required	Responsible member	Date complete
1.	1.	1.	1.
2.	2.	2.	2.
3.	3.	3.	3.
4.	4.	4.	4.
Action at next meeting	**Preparation required**	**Responsible member**	**Date complete**
1.	1.	1.	1.
2.	2.	2.	2.
3.	3.	3.	3.
4.	4.	4.	4.

Team Meeting of _____

Decisions made	Follow-up required	Responsible member	Date complete
1.	1.	1.	1.
2.	2.	2.	2.
3.	3.	3.	3.
4.	4.	4.	4.
Action at next meeting	**Preparation required**	**Responsible member**	**Date complete**
1.	1.	1.	1.
2.	2.	2.	2.
3.	3.	3.	3.
4.	4.	4.	4.

Team Meeting of _____

Decisions made	Follow-up required	Responsible member	Date complete
1.	1.	1.	1.
2.	2.	2.	2.
3.	3.	3.	3.
4.	4.	4.	4.
Action at next meeting	**Preparation required**	**Responsible member**	**Date complete**
1.	1.	1.	1.
2.	2.	2.	2.
3.	3.	3.	3.
4.	4.	4.	4.

CONCEPT SELECTION
PUGH CHART

CRITERIA	CONCEPTS			
	B	C	D	A
				DATUM

S - CONCEPT EQUAL TO DATUM

+ CONCEPT SUPERIOR TO DATUM

− CONCEPT INFERIOR TO DATUM

The Seven Metric Base Units

Symbol	Name	Quantity
m	meter	length
kg	kilogram	mass
s	second	time
K	kelvin	temperature
A	ampere	electric current
mol	mole	amount of substance
cd	candela	luminous intensity

Derived Units with Special Names

Symbol	Name	Quantity
Bq	becquerel	radioactivity
C	coulomb	electric charge
F	farad	capacitance
Gy	gray	absorbed dose
H	henry	inductance
Hz	hertz	cycle frequency
J	joule	energy
kat	katal	molar flow
lm	lumen	luminous flux
lx	lux	illuminance
N	Newton	force
Ω	ohm	resistance
Pa	pascal	pressure
rad	radian	plane angle
S	siemens	conductance
sr	steradian	solid angle
Sv	sievert	dose equivalent
T	tesla	magnetic flux density
V	volt	voltage
W	watt	power
Wb	weber	magnetic flux

SI Prefixes

Multiplication Factor	Prefix Name	Prefix Symbol
10^{18}	exa	E
10^{15}	peta	P
10^{12}	tera	T
10^{9}	giga	G
10^{6}	mega	M
10^{3}	kilo	k
10^{2}	hecto*	h
10^{1}	deka*	da
10^{-1}	deci*	d
10^{-2}	centi*	c
10^{-3}	milli	m
10^{-6}	micro	μ
10^{-9}	nano	n
10^{-12}	pico	p
10^{-15}	femto	f
10^{-18}	atto	a

*To be avoided when possible.

U. S. Customary Units and Their SI Equivalents

Quantity	U. S. Customary Unit	SI Equivalent
Acceleration	ft/s^2	0.3048 m/s^2
	in./s^2	0.0254 m/s^2
Area	ft^2	0.0929 m^2
	in.^{2}	645.2 mm^2
Energy	ft-lb$_f$	1.365 J
Force	kip	4.448 kN
	lb$_f$	4.448 N
Impulse	lb$_f$ -s	4.448 N-s
Length	ft	0.3048 m
	in.	25.40 mm
	mi	1.609 km
Mass	lb$_m$	0.4536 kg
	slug	14.59 kg
Moment of a Force	lb$_f$ –ft	1.356 N-m
	lb$_f$ –in.	0.1130 N-m
Moment of Inertia of an Area	in.^{4}	0.4162×10^6 mm^4
Power	ft-lb$_f$ /s	1.356 W
	hp	745.7 W
Pressure and Stress	lb$_f$ /ft^2	47.88 Pa
	lb$_f$ /in.^{2} (psi)	6.895 kPa
	1000 lb$_f$ /in.^{2} (ksi)	6.895 MPa
Velocity	ft/s	0.3048 m/s
	in./s	0.0254 m/s
	mi/h (mph)	0.4470 m/s
	mi/h (mph)	1.609 km/h
Volume, Solids	ft^3	0.02832 m^3
	in.^{3}	16.39 cm^3
Volume, Liquids	gal	3.785 L
	qt	0.9464 L
Work	ft-lb$_f$	1.356 J

Both L and l are accepted symbols for the liter. Since "1" is easily confused with the numeral "1", the symbol "L" is recommended by NIST (The National Institute of Standards and Technology) in NIST Special Publication 881, 1991.

www.ingramcontent.com/pod-product-compliance
Lightning Source LLC
Chambersburg PA
CBHW080648220326
41598CB00033B/5144